International Conference on Materials Physics and Mechanics 2017

Journal of Physics: Conference Series Volume 914

Langkawi, Malaysia
23 July 2017

ISBN: 978-1-5108-5067-5
ISSN: 1742-6588

Printed from e-media with permission by:

Curran Associates, Inc.
57 Morehouse Lane
Red Hook, NY 12571

Some format issues inherent in the e-media version may also appear in this print version.

Printed by Curran Associates, Inc. (2017)

For permission requests, please contact the Institute of Physics
at the address below.

Institute of Physics
Dirac House, Temple Back
Bristol BS1 6BE UK

Phone: 44 1 17 929 7481
Fax: 44 1 17 920 0979

techtracking@iop.org

Additional copies of this publication are available from:

Curran Associates, Inc.
57 Morehouse Lane
Red Hook, NY 12571 USA
Phone: 845-758-0400
Fax: 845-758-2633
Email: curran@proceedings.com
Web: www.proceedings.com

Table of contents

Volume 914

International Conference on Materials Physics and Mechanics 2017
23 July 2017, Langkawi, Malaysia

Accepted papers received: 14 October 2017
Published online: 30 October 2017

Materials Science And Engineering

Mechanics

Preface

Nik Hisyamudin Muhd Nor, Al Emran Ismail, Amir Khalid, Mohd Azlis Md Jalil, Mohd Faizal Mohideen Batcha, Izzuddin Zaman, Sofian Mohd, Mohd Rashidi Ibrahim, Ahmad Mubarak Tajul Arifin, Aida Mustapha.

The 1st International Conference on Materials Physics and Mechanics 2017 (MPM2017) was held in conjunction with the 7th International Conference on Mechanical and Manufacturing Engineering (ICME2017) in Langkawi, Malaysia. It was aimed to provide a better platform for communication and discussion among academics, researchers, industrial players as well as post-graduate students from various research fields. It was also aimed to provide a place for sharing knowledge and ideas, networking, presenting research findings and exploring other potential benefits that might be contributed for a better future.

Faculty of Mechanical and Manufacturing Engineering (FKMP), Universiti Tun Hussein Onn Malaysia (UTHM) would like to express fullest gratitude and appreciation all participants, respective authors and technical reviewers for sharing their knowledge and experience for this conference.

IOP Conf. Series: Journal of Physics: Conf. Series **914** (2017) 011002 doi:10.1088/1742-6596/914/1/011002

List of Reviewer

The organizing committee would like to express our fullest gratitude for the support and great advice from the following reviewers:

1. Ahmad Mubarak Tajul Arifin
2. Ainun Rahmahwati Ainuddin
3. Akmal Nizam Bin Mohammed
4. Al Emran Bin Ismail
5. Amir Bin Khalid
6. Anika Zafiah Mohd Rus
7. Aslam Bin Abdullah
8. Azmahani Sadikin
9. Azriszul Mohd Amin
10. Azwan Sapit
11. Azzura Ismail
12. Badrul Aisham Md Zin
13. Sia Chee Kiong
14. Koh Ching Teng
15. Erween Abd Rahim
16. Hamidon Salleh
17. Hamimah Abd.Rahman
18. Ishkrizat Taib
19. Izzuddin Zaman
20. Kamarul Ariffin Zakaria
21. Kamarul Azhar Kamarudin
22. Khir Harun
23. Mohamed Nasrul Bin Mohamed Hatta
24. Mohammad Kamil Abdullah
25. Mohammad Sukri Mustapa
26. Mohd Azahari Razali
27. Mohd Azlis Sani Md Jalil
28. Mohd Fadhli Zulkafli
29. Mohd Fahrul Bin Hassan
30. Mohd Faizal Bin Mohideen Batcha
31. Mohd Nasrull Bin Abdol Rahman
32. Mohd Rasidi Bin Ibrahim
33. Mohd Shahir Bin Yahya
34. Mohd Shukor Salleh
35. Mohd Yussni Hashim
36. Muhamad Zaini Bin Yunos
37. Muhammad Akmal Bin Johar
38. Muhammad Mujtaba Asad
39. Muhammed Nafis Osman Zahid
40. Musli Mohammad

IOP Conf. Series: Journal of Physics: Conf. Series **914** (2017) 011002 doi:10.1088/1742-6596/914/1/011002

41. Ng Chuan Huat
42. Nik Hisyamudin Muhd Nor
43. Nor Azali Bin Azmir
44. Noradila Abdul Latif
45. Nur Azam Badarulzaman
46. Omar Mohd Faizan Marwah
47. Mohd Hilmi Othman
48. Raja Zuraidah Raja Mohd Rasi
49. Reazul Haq Bin Abdul Haq
50. Rosniza Hussin
51. Ruzila Ismail
52. Saifulnizan Bin Jamian
53. Salihatun Bt Md Salleh
54. Sh Salleh Sh Ahmad
55. Shahiron Bin Shahidan
56. Sharifah Adzila Syed Abu Bakar
57. Shazarel Shamsudin
58. Siti Aida Ibrahim
59. Sofian Bin Mohd
60. Sulaiman Hassan
61. Sulastri Sabudin
62. Suzairin Bin Md Seri
63. Suzi Salwah Jikan
64. Syariful Syafiq Shamsudin
65. Wan Nur Azrina Binti Wan Muhammad
66. Yusri Yusof
67. Zakiah Kamdi
68. Zaleha Binti Mohamad
69. Zamani Ngali
70. Zamri Noranai

Peer review statement

All papers published in this volume of *Journal of Physics: Conference Series* have been peer reviewed through processes administered by the proceedings Editors. Reviews were conducted by expert referees to the professional and scientific standards expected of a proceedings journal published by IOP Publishing.

IOP Conf. Series: Journal of Physics: Conf. Series **914** (2017) 012001 doi:10.1088/1742-6596/914/1/012001

UAV telemetry communications using ZigBee protocol

T H Nasution[1], I Siregar[2], M Yasir[1]

[1]Department of Electrical Engineering, Universitas Sumatera Utara, Medan, Indonesia
[2]Department of Industrial Engineering, Universitas Sumatera Utara, Medan, Indonesia

Corresponding author: tigor.nasution@usu.ac.id

Abstract. Wireless communication has been widely used in various fields or disciplines such as agriculture, health, engineering, military, and aerospace so as to support the work in that field. The communication technology is typically used for controlling devices and data monitoring. One development of wireless communication is the widely used telemetry system used to reach areas that cannot be reached by humans using UAV (Unmanned Aerial Vehicle) or unmanned aircraft. In this paper we discuss the design of telemetry system in UAV using ZigBee protocol. From the test obtained the system can work well with visualization displays without pause is 20 data per second with a maximum data length of 120 characters.

1. Introduction

Along with the development of the times and technology needs of rapid information is needed in various fields, agriculture, industry, medical, meteorological stations, and remote area [1] so that it can support the performance of the field. The technology is typically used for controlling devices and data monitoring [2]. However, in monitoring and measurement, not all conditions are possible to be carried out directly due to geographical and distance factors, it can inhibit the data. Constraints on these measurements can be overcome by remote measurement method (telemetry) using UAV (Unmanned Aerial Vehicle) [3,4]. The telemetry process is the measurement of the parameters of an object (objects, space, natural conditions) whose measurement results will be sent elsewhere either via cable or wireless [5,6]. The telemetry communications UAV is carried out wirelessly.

Previously there were several studies on UAV telemetry. Some researchers have discussed the model of communication or transmission media [7,8]. Research on the telemetry model in the UAV for monitoring has also been conducted [9]. Some have discussed the use of telemetry in UAVs to monitor the disaster area [10]. These researches continue to grow along with the development of wireless communication technology.

In contrast to previous studies, in this paper we designed a telemetry communication system in UAV using ZigBee protocol. Zigbee is widely used for wireless technology, because it has many advantages including ultra-low power consumption, low cost, and eases of design [11]. The ZigBee protocol is adopted for wireless communication to achieve high integration, applicability, and portability [12]. The system consists of a GCS (Ground Control Station) that will communicate with UAVs. The GCS interface is built using Microsoft Visual Studio with C # language running on the computer and connected to the UAV wirelessly using XBee Pro RF Transceiver 2.4 GHz. GCS can display visualization of UAV flight parameters through instrument panel that displays airspeed, heading (yaw), cornering and pitch and roll, barometric height, turn and bank rate, as well as climb speed according to data Obtained from IMU10ODF sensors as well as data from GPS.

2. Materials and Method

In this study, the functionality of the system created focused on the functionality of the data telemetry system. To support maximum telemetry system UAV data function required RF devices, long range antenna and GCS. The RF device used is the XBee module (ZigBee) and the antenna used is the 2.4 GHz yagi antenna. While the required GCS device has the following specifications:

- Visual instrument panel featuring basic flight instruments such as altitude indicator, airspeed indicator, artificial horizon, vertical speed indicator (climb speed), attitude indicator / artificial horizon, indicator Direction (compass / heading indicator), and GPS lock status.
- Control engine on UAV (ON, FIRE, STOP / PARACHUTE).

The GCS system is made up of two major parts of hardware and software. GCS hardware part serves to capture data sent by UAV via RF transmitter (ZigBee) and convert to form data that can be processed by software. In addition to processing data obtained telemetry to visualize, the software also serves to store and provide feedback in the form of commands such as turning on, off, and activate the data transmission. The design scheme of this system can be seen in figure 1.

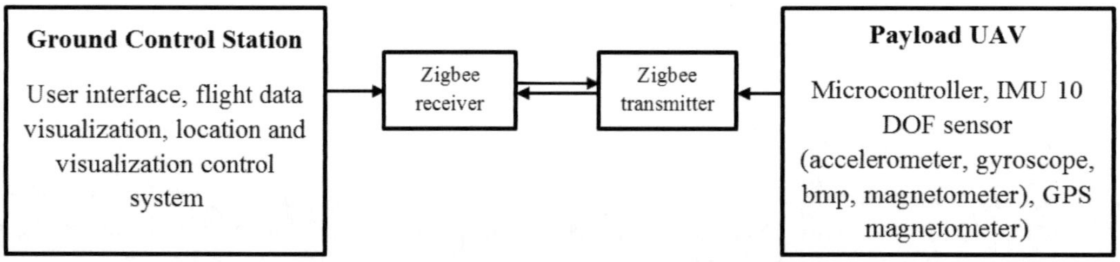

Figure 1. The design scheme of this system

2.1. *XBee Pro*

XBee Pro is a module that allows Arduino Uno to communicate wirelessly using ZigBee protocol. ZigBee operates using on IEEE 802.15.4 specifications operating at 2.4 GHz, 900 and 868 MHz frequencies. XBee Pro can be used instead of serial cable. XBee Pro is expected to reduce costs and become low-power connectivity for equipment that requires batteries to live for several months to several years, but does not require high data transfer rates. XBee Pro enables wireless communications within reach of up to 100 meters indoor and 1500 meters outdoor. In this we use the XBee S2Pro radio module as shown in figure 2.

Figure 2. XBee S2Pro

IOP Conf. Series: Journal of Physics: Conf. Series **914** (2017) 012001 doi:10.1088/1742-6596/914/1/012001

Remote monitoring and measurement systems consist of two of the same XBee Pro modules previously programmed as a receiver-transmitter and transmitters [13]. There are several topology forms commonly used, among others, mesh topology, peer, star, and cluster trees. Topology on XBee Pro network Topology pair is a simple network using only two XBee or node. One node must be the coordinator so that the network can be established. And the other is configured as a router or end device. We use the XCTU application to setup both modules.

2.2. *Sensor IMU 10 DOF*

The IMF sensor module (inertial measurement unit) 10 DOF gy-80 sensor consists of 4 sensors namely, 3 Gyro axis, 3 Axis Accelerometer, 3 Axis Magnetometer, and barometer as shown in figure 3. All sensors communicate using I2C bus protocol so it takes 4 Paths is:

1. GND – Ground
2. 2Supply voltage or source voltage +3.5 V to +5 V.
3. 3SCL for I2C clock
4. SDA for I2C data.

Figure 3. Sensor IMU 10 DOF

2.3. Designing Software

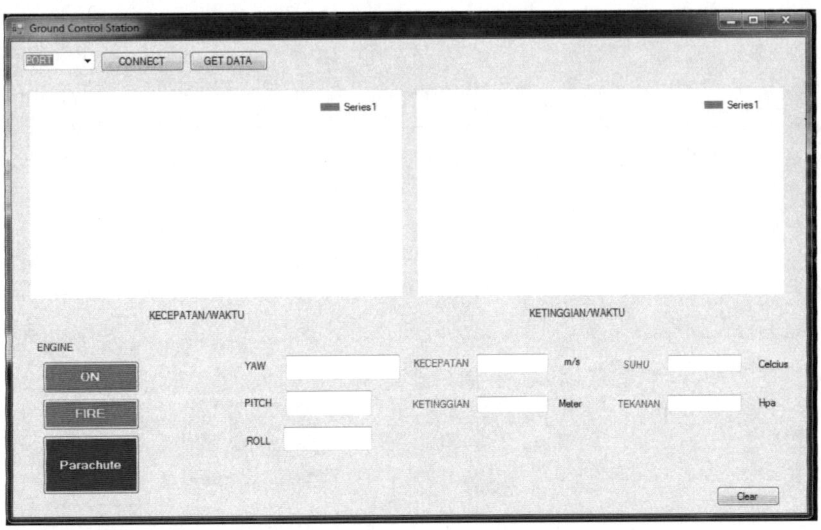

Figure 4. User Interface

IOP Conf. Series: Journal of Physics: Conf. Series **914** (2017) 012001 doi:10.1088/1742-6596/914/1/012001

To meet the needs in accordance with the needs, the software interface designed to be made into several parts such as figure 4. Speed graph of time will display acceleration UAV to monitor the stability of flying. The time height graph is designed to show the vertical speed of the object. Engine is a feature that functions as ON, Fire, Stop / Parachute. While data sent sensors and GPS will be displayed as well.

As planned, software that is also a graphical user interface with UAV is developed using C # language in Microsoft Visual Studio 2010 Ultimate software licensed for free from MSDNAA

In addition to being displayed visually through the visual instrument panel, flight parameter data obtained from IMU10ODF is also displayed numerically. As shown in the figure 4 is a display of IMU10DOF data located under the visual instrument panel

3. Results and Discussions

The functional test results of all instruments and visualizations on the GCS interface can respond well to the entire range of data provided through the serial port. Maximum data rate that can be processed so that displaying visualization without pause is 20 data per second with maximum data length of 120 characters. The results of the data responses obtained can be seen in Table 1.

Table 1. The results of the data responses

YAW	PITCH	ROLL	temperature	PRESSURE	ALTITUDE	LATITUTE	LONGITUDE	ALTITUDE(gps)
20.18	-39.84	-18.31	30.16	788.49	2227.34	3.568529	98.655136	54.75
-6.42	-24.83	-25.71	30.16	820.03	1863.53	3.568530	98.655136	55.17
-13.80	-9.14	-27.96	30.12	851.57	1515.84	3.568531	98.655136	54.67
-18.02	-13.33	-33.03	30.12	883.12	1183.30	3.568530	98.655136	55.17
-43.84	-19.23	-0.21	30.08	914.66	864.52	3.568530	98.655143	54.75
-50.79	-26.35	-19.03	30.08	946.21	558.68	3.568530	98.655143	55.09
-76.76	-43.69	-10.95	30.06	977.75	264.75	3.568529	98.655143	54.17
-88.44	-75.42	3.41	30.06	1009.30	-18.12	3.568527	98.655136	54.75
-90.55	-39.52	19.75	29.99	1009.31	-18.24	3.568526	98.655136	54.41
-80.03	11.20	16.11	29.99	1009.32	-18.30	3.568525	98.655136	54.75
-109.13	-37.27	2.70	29.98	1009.33	-18.41	3.568525	98.655143	55.17
-130.43	24.25	-48.27	29.98	1009.33	-18.43	3.568525	98.655143	55.09
-149.17	27.33	-62.23	29.99	1009.34	-18.50	3.568525	98.655143	55.09
-129.09	48.95	-25.40	29.99	1009.34	-18.47	3.568523	98.655143	55.25
-145.83	21.81	-13.79	29.97	1009.34	-18.48	3.568526	98.655136	55.33
-178.39	60.05	-0.64	29.97	1009.34	-18.49	3.568531	98.655143	56.09
-115.09	65.98	-22.11	29.94	1009.34	-18.53	3.568534	98.655143	56.09
-140.48	24.70	-5.07	29.94	1009.33	-18.37	3.568537	98.655151	56.17
-127.56	21.16	-4.19	29.94	1009.31	-18.24	3.568536	98.655151	56.17
-125.72	34.37	-10.97	29.94	1009.29	-18.09	3.568538	98.655166	56.17
-168.86	14.50	49.57	29.90	1009.28	-17.93	3.568539	98.655174	56.34
-105.48	27.38	39.16	29.90	1009.26	-17.78	3.568538	98.655181	56.34
-98.14	29.89	20.79	29.90	1009.24	-17.64	3.568536	98.655189	55.75
-60.56	6.40	13.84	29.90	1009.23	-17.54	3.568535	98.655197	55.67
-65.98	6.14	-23.27	29.87	1009.22	-17.45	3.568534	98.655204	55.67
-63.83	-0.51	-34.10	29.87	1009.21	-17.38	3.568533	98.655212	55.67
-23.97	7.72	-4.92	29.87	1009.21	-17.34	3.568531	98.655220	55.25

20.41	3.03	-16.62	29.87	1009.20	-17.31	3.568528	98.655227	55.75
34.16	17.63	-35.00	29.81	1009.20	-17.28	3.568530	98.655235	55.50
2.30	-15.22	-38.66	29.81	1009.19	-17.22	3.568531	98.655242	55.75
-18.55	-25.34	-59.63	29.82	1009.19	-17.18	3.568527	98.655250	56.17
-22.07	-14.32	-44.55	29.82	1009.18	-17.13	3.568523	98.655265	56.09

When testing is obtained the calculation of the average number of incoming data every second is 8 data with a data width of 90-120 characters. For data variation is given treatment on sensors in IMU10DOF module to see the response. Variations in attitude data variables (yaw pitch and roll) are obtained by changing the position of the IMU Razor 9 DOF sensor connected to IMU10DOF.

4. Conclusion

From the tests performed, the system can run as expected. GCS can communicate with UAV by receiving data obtained from IMU10DOF sensors. The system can work by displaying data on the GCS interface with 20 data per second with a maximum data length of 120 characters. From these results it can be concluded that the ZigBee protocol can be used for telemetry in the UAV.

Acknowledgments
The authors gratefully acknowledge that the present research is supported by Ministry of Research and Technology and Higher Education Republic of Indonesia. We thank the Vice Rector for Research, Service, and Cooperation University of Sumatera Utara that have supported this research activity. All the faculty, staff members and lab technicians of Electrical Engineering Department, whose services turned my research a success. My Wife, Erna Kurnia and children, Zafran Ahmad Nasution and Zaura Aisya Nasution, family members and friends, without whom I was nothing; they not only assisted me financially but also extended their support morally and emotionally. The support is under the research grant TALENTA USU of Year 2017.

References

[1] R. Rahmat, M. Syahputra, and M. Lydia, "Real Time Monitoring System for Water Pollution in Lake Toba," in *2016 International Conference on Informatics and Computing (ICIC)*, Mataram, 2016, pp. 383-388.

[2] T. H. Nasution et al., "Electrical appliances control prototype by using GSM module and Arduino," in *2017 4th International Conference on Industrial Engineering and Applications (ICIEA)*, Nagoya, 2017, pp. 355-358.

[3] T. Raj and F. H. Hashim, "Development of an autonomous flight controller circuit with real time data transmission," in *2016 International Conference on Advances in Electrical, Electronic and Systems Engineering (ICAEES)*, Putrajaya, 2016, pp. 115-119.

[4] J. A. Alfredsen, T. A. Johansen and K. R. SkØien A. Zolich, "A communication bridge between underwater sensors and unmanned vehicles using a surface wireless sensor network - design and validation," in *OCEANS 2016*, Shanghai, 2016, pp. 1-9.

[5] S. Ponomarev and T. Atkison, "Industrial Control System Network Intrusion Detection by Telemetry Analysis," *IEEE Transactions on Dependable and Secure Computing*, vol. 13, no. 2, pp. 252-260, March 2016.

[6] Komkrit Chooruang and Pongpat Mangkalakeeree, "Wireless Heart Rate Monitoring System Using MQTT," *Procedia Computer Science*, vol. 86, pp. 160-163, March 2016.

[7] Li Bing, "Study on Modeling of Communication Channel of UAV," *Procedia Computer Science*, vol. 107, pp. 550-557, January 2017.

[8] M. Rice, M. S. Afran, and M. Saquib, "Equalization in aeronautical telemetry using multiple transmit antennas," *IEEE Transactions on Aerospace and Electronic Systems*, vol. 51, no.

3, pp. 2148-2165, July 2015.

[9] M. Satrio, A. A. Rafsanjani and R. R. T. Sadewo G. Nugroho, "Avionic system design Unmanned Aerial Vehicle for disaster area monitoring," in *015 International Conference on Advanced Mechatronics, Intelligent Manufacture, and Industrial Automation (ICAMIMIA)*, Surabaya, 2015, pp. 198-201.

[10] J. Tristancho, C. Barrado, S. P. Mansilla, and E. Pastor, "A telemetry modeling for intelligent UAV monitoring," in *2009 IEEE/AIAA 28th Digital Avionics Systems Conference*, Orlando, 2009, pp. 7.C.1-1-7.C.1-4.

[11] K. L. Chen, Y. R. Chen, Y. P. Tsai, and N. Chen, "A Novel Wireless Multifunctional Electronic Current Transformer Based on ZigBee-Based Communication," *IEEE Transactions on Smart Grid*, vol. 8, no. 4, pp. 1888-1897, July 2017.

[12] L. H. Wang, T. Y. Chen, K. H. Lin, Q. Fang, and S. Y. Lee, "Implementation of a Wireless ECG Acquisition SoC for IEEE 802.15.4 (ZigBee) Applications," *IEEE Journal of Biomedical and Health Informatics*, vol. 19, no. 1, pp. 247-255, January 2015.

[13] J. C. Chou et al., "Wireless Sensing System for Flexible Arrayed Potentiometric Sensor Based on XBee Module," *IEEE Sensors Journal*, vol. 16, no. 14, pp. 5588-5595, July 2016.

Small Horizontal Axis Wind Turbine under High Speed Operation: Study of Power Evaluation

Magedi Moh.M. Saad[1], Sofian Bin Mohd[2], Mohd Fadhli Bin Zulkafli[3], Aslam Bin Abdullah[4], Mohammad Zulafif Bin Rahim[5], Zulkhairi Bin Subari[6], Nurhayati Binti Rosly[7]

[1,2,5,6,7]Aircraft System and Design Research (ASDR), Faculty of Mechanical and Manufacturing Engineering, Universiti Tun Hussein Onn Malaysia
[3,4]Aerodynamic and Propulsion Research (APR), Faculty of Mechanical and Manufacturing Engineering, Universiti Tun Hussein Onn Malaysia

Abstract. Mechanical energy is produced through the rotation of wind turbine blades by air that convert the mechanical energy into electrical energy. Wind turbines are usually designed to be use for particular applications and design characteristics may vary depending on the area of use. The variety of applications is reflected on the size of turbines and their infrastructures, however, performance enhancement of wind turbine may start by analyzing the small horizontal axis wind turbine (SHAWT) under high wind speed operation. This paper analyzes the implementations of SHAWT turbines and investigates their performance in both simulation and real life. Depending on the real structure of the rotor geometry and aerodynamic test, the power performance of the SHAWT was simulated using ANSYS- FLUENT software at different wind speed up to 33.33 m/s (120km/h) in order to numerically investigate the actual turbine operation. Dynamic mesh and user define function (UDF) was used for revolving the rotor turbine via wind. Simulation results were further validated by experimental data and hence good matching was yielded. And for reducing the energy producing cost, car alternator was formed to be used as a small horizontal wind turbine. As a result, alternator-based turbine system was found to be a low-cost solution for exploitation of wind energy.

1. Introduction

Nowadays, electrical energy from wind turbines is classified as most efficient and worthier than other green energy resources. This remarkable development can be primarily attributed to the development in the large wind turbines technology. However, small wind turbines are still in their developmental stage. In the past decades, the installed power capacity of large wind turbines displayed an annual growth rate of about 30% whereas the small wind turbines revealed only 9% [1]. Hence, there is immediate need to explore the huge potential of small wind turbines to exploit towards generations of electricity with lesser expense as there has been no great concern in power generation sector as small wind turbines are costly.

Generator is the main component of small wind turbines, which converts the mechanical energy into electrical one. Thus, by reducing the cost of generators in such turbines would certainly lower their overall cost, making them cost competitive. A major reason for the lessened utilization of big wind turbines is due to the fact that the current systems make used of high quality generators that are expensive. The earlier study of the energy has successfully yield commercially available small wind turbines which shows that the cost of generated electricity the from small wind turbines are the most cost effective [2].

The objective of this paper is to evaluate the power produced from small wind turbine by using automotive alternator, and determines the feasibility of installing a small horizontal axis wind turbine on top of an automotive car and subsequently validating the results by using computational fluid dynamics (CFD) software. The validation between numerical simulation and experimental test is based on wind speed, in order to achieve good matching between numerical simulation and experimental test which are the main contributions of this paper.

Utilizations of a vehicle alternator as wind turbine generator offers multifarious applications including automobiles, tractors, and other industrial transportation systems with wide operating speed range (400 to 8000 rpm) [5] that are available in many standard output voltages (12, 24, 32 or 48 V). According to Ani et al. [6], the efficiency of the alternator is generally low, where the maximum efficiency (~ 54%) occurs at lower speeds (~1500 rpm). This is advantageous for various low speed applications since the efficiency at higher speed (~2500 rpm) reduces significantly (~45%). Typically, the output power of a vehicle alternator operating at 2500 rpm with 14 V DC is ~ 600 W. The power output achieved at 1000 rpm was found to be 150 W with maximum efficiency (~54%) at 1300 rpm [1, 8].

In this study, a 14 V DC alternator was used as a cheaper alternative. The output voltage is usually maintained at 14 V DC by an internal regulator that samples the voltage and adjusts the field current accordingly. The regulator maintains the desired voltage by varying the duty-cycle of the pulse width modulated (PWM) voltage applied to the field winding. By increasing the electrical load in the vehicle, the output voltage drops as more currents are drawn from the alternator. Whenever the electrical load drops, the regulator detects it and raises the voltage by enhancing the duty cycle to increase the field current and vice versa [9]. The rotor speed measurement was taken using of Tachometer, while the measuring of current output is done after connecting load resistance to the alternator for evaluating power output performance for the system.

Most of the aerodynamics problem involving wind turbine are being tackled by CFD using the Reynolds-averaged Navier–Stokes (RANS) method with different turbulence models [3]. To solve such problems, various numerical schemes are developed including SIMPLE, SIMPLEC, PISO and COUPLED [4]. In this study, the SIMPLE scheme of the ANSYS FLUENT software is used to solve the flow around the SHAWT.

The simulation is evaluated using the following expression for the power:

- Experimental Power formulas:

$$Power\ Output = (I) * (V) \tag{1}$$

where V is the voltage (Volts), I is the current (Ampere)

- Calculation Power formulas:

$$theoretical\ torque = 0.5 * \rho * A * V^2 \tag{2}$$

$$actual\ torque = 0.5 * \rho * A * (\omega * r)^2 \tag{3}$$

$$angular\ velocity\ (\omega) = (\tfrac{2*\pi}{60}) * rpm \tag{4}$$

$$torque\ coefficient\ C_q = \frac{actual\ torque}{theoretical\ torque} \tag{5}$$

$$Tip\ Speed\ Ratio\ (\lambda) = (angular\ velocity\ (\omega) * r)/V_{in} \tag{6}$$

$$power\ coefficient\ C_p = \lambda * C_q \tag{7}$$

$$power\ output = 0.5 * \rho * A * C_p * V^3 \tag{8}$$

where A is the surface area (m^2), v is the wind speed ($m\ s^{-1}$), ρ is the air density ($kg\ m^{-3}$) [10, 11].

IOP Conf. Series: Journal of Physics: Conf. Series **914** (2017) 012002 doi:10.1088/1742-6596/914/1/012002

2. Experimental procedure

In the experiment, the fabricated SHAWT (rotor diameter of 0.36 m, rotor width of 0.04 m, blade span of 0.117 m, and blade chord of 0.035 m) is fixed on a car and positioned to face the incoming wind. Figure 1 illustrates the final position of the entire mode on the car. The experimental data obtained from this experiment is then compared with the simulation results obtained from the CFD simulation.

Figure 1. Installation whole wind turbine system on car body

3. CFD analysis

The aerodynamic behavior of a seven-blade SHAWT is observed by the simulation conducted using ANSYS FLUENT. The wind speed at computational domain entrance was kept at a constant rate. The rotational speed of the turbine (rotating zone) was measured to evaluate the turbine performance.

Three dimensional Reynolds averaged Navier-Stokes equations for viscous, incompressible, and continuous flow was employed. In this flow pattern, the temperature variations over the flow field were kept nearly constant. Thus, the energy equation was entirely eliminated [12].

The wind turbine model was developed by using SolidWorks, which comprised of stationary domain (cylinder) for the surrounding air, the distance from the rotor to the inlet, top and outlet was 3 times, 3 times and 7 times the diameter of blade respectively as shown in Figure 2. (b) and a rotor turbine as shown in Figure 2. (a) [13-14].

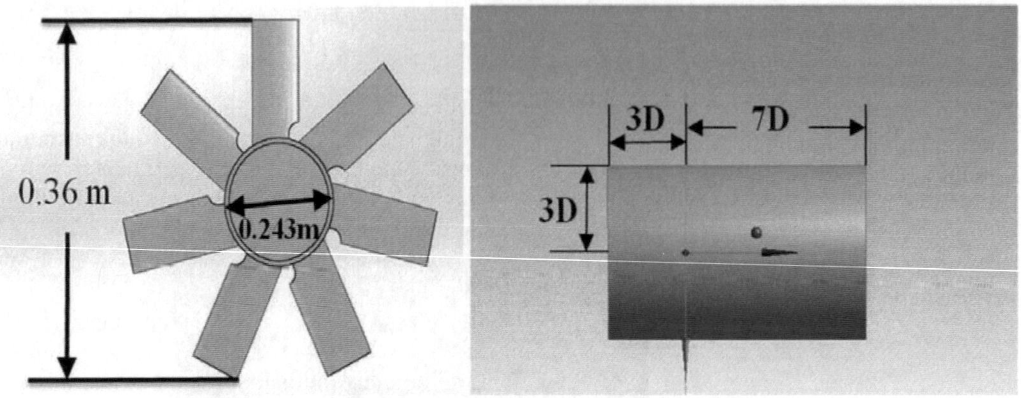

(a) (b)

Figure 2. (a) Rotor dimension (b) Stationary domain dimension

In this study, a non-uniform, unstructured mesh was used with a considerably medium mesh implemented in regions of high gradients. A tetrahedron element was selected as the mesh for both the stationary and rotating domain.

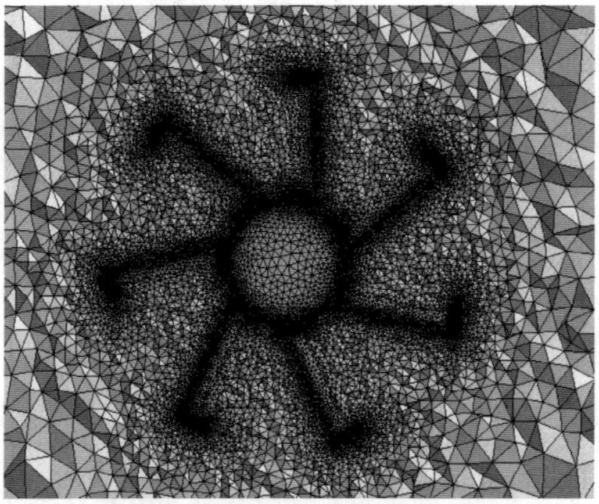

Figure 3. Tetrahedral meshes in the model

Figure 3 depicts the planer cross-sectional view of the mesh layout for the turbine and the cylindrical domain. The total number of mesh elements were 354622 (tetrahedron) and 3979649 (triangular interior faces) respectively. Figure 4 displays the boundary conditions applied to the computational domain. Three types of boundary conditions were used; namely walls, velocity inlet, and pressure outlet [15]. The flow properties at sea level was applied to all regions. Table 1 enlists the simulated feature of flow field.

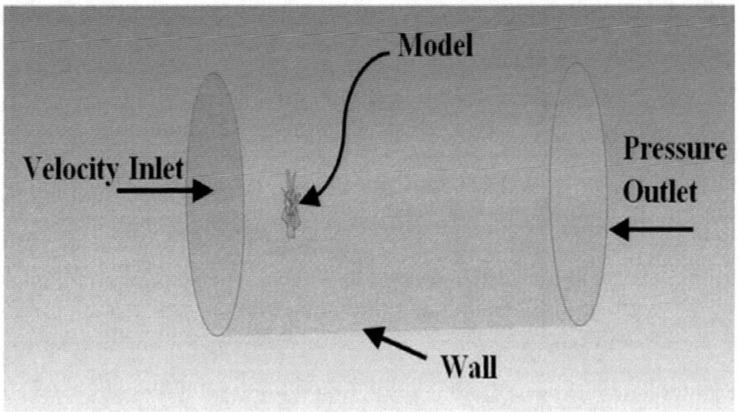

Figure 4. Boundary conditions setup for ANSYS simulation

Table 1. Simulated flow field characteristics

Pressure [Pa]	101,325
Density (ρ) [k gm^{-3}]	1.225
Velocities (m s^{-1})	22.22 m/s 25.3 m/s 27.778 m/s 30.556 m/s 33.33 m/s

The rotation of the SHAWT model was made possible by using dynamic mesh and User Defined Function (UDF). The physical properties such as mass properties and rotation axis of the SHAWT are defined in the UDF written in C language.

4. Results and discussions

4.1. Simulation Results

Figure 5 illustrates the relationship between the wind speed and rotor rpm at different time step for two cases; incoming wind speed of 33.33 m/s and 22.22 m/s. For both cases in the simulation, at 0.002 sec the rotor started to rotate and the rpm kept on increasing as the time step increases.

The blades are specially curved-shape, similar to the airfoil wings on a plane. When the wind blows the blades, it moves them upward by the force produced due to pressure difference, leading the turbine blades to start spinning. Figure 6 illustrates the different pressure between wind speed 22.22 m/s and 33.33 m/s when the rotor started to rotate, in Figure 6 (a) the maximum pressure at the front rotor hub in black circle around 1100 Pa and around -4000 Pa at tip blade as clearly in black circle while in Figure 6 (b) in black circle the maximum pressure at the front rotor hub around 590 pa and around -2000 pa at tip blade in circle.

Concentration effect can be clearly seen by wind speed increasing due to flow separation at the sharp corners of the blades when flow passes through the rotor. Axial direction wind speed increases due to curvature blade then decreases slowly in the downstream due to the blockage effect by blade walls as clearly illustrated in first rotation in black circle in Figure 7 (a). After the rotor started to rotate, the vortices started to develop in the rotor downstream at its second and third rotation as shown in Figure 7 (b) and (c). These vortices affected the aerodynamic performance, and hence turbine power output.

IOP Conf. Series: Journal of Physics: Conf. Series **914** (2017) 012002 doi:10.1088/1742-6596/914/1/012002

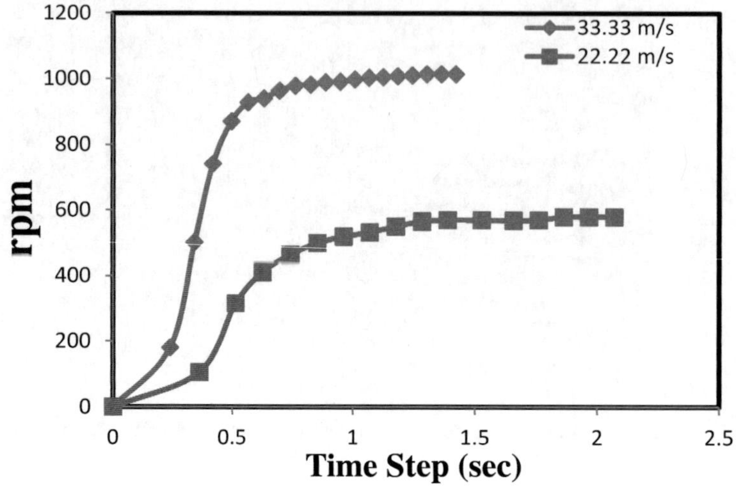

Figure 5. RPM at different time step

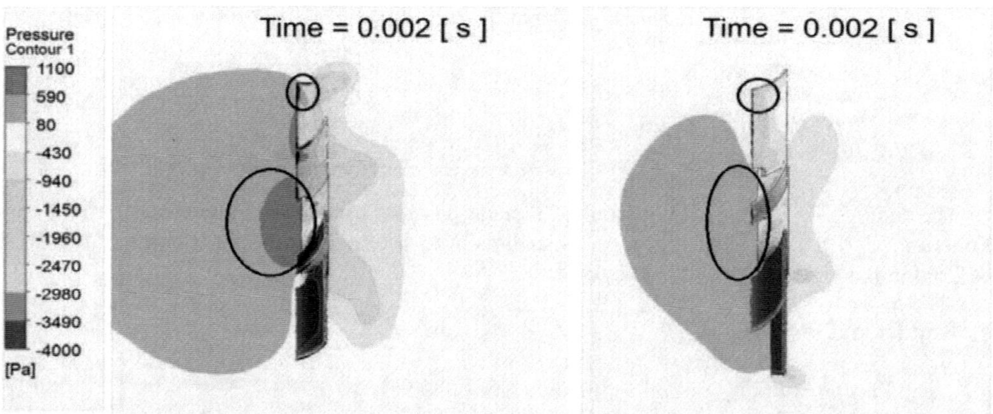

(a) (b)

Figure 6. Pressure contour in the Y-Z plane at (a) Wind speed 33.33m/s (b) Wind speed 22.22 m/s

first rotation second rotation third rotation

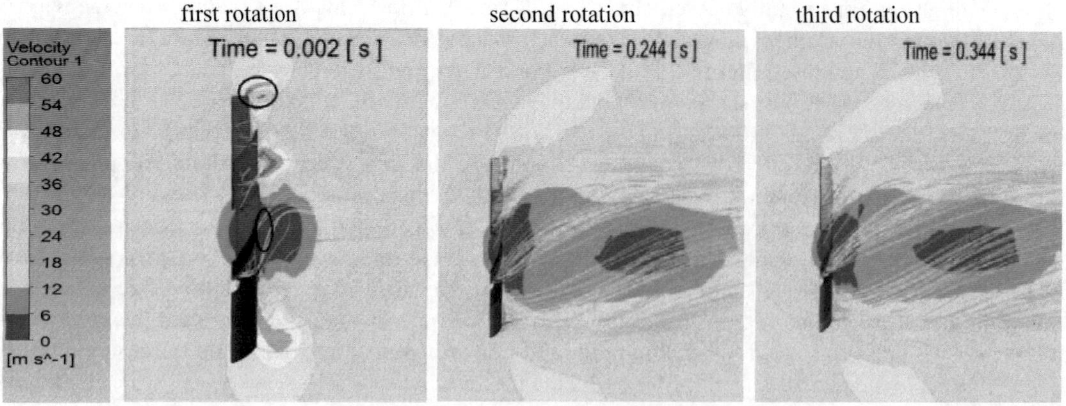

(a) (b) (c)

Figure 7. Wind speed contour and streamline in the Y-Z plane at different time step

4. 2. Experimental Results

Experimental data from small axis wind turbine was used to validate the numerical results. This experiment was conducted using a car driven on Johor Bharu (Malaysia) highway, about 12 km from a start point to an end, the car speed was consistently set similar to numerical simulation speeds. Figure 8 exhibited the whole turbine system installed on the car's body for the test

Figure 8. Installed turbine during test

In wind turbine, the rotor efficiency according to the number of blade rotation was achieved at this speed, and in addition to that, drag, torque and the power output heightens with the increasing wind speed. The maximum drag was produced numerically at maximum speed (33.33 m/s) is 59 N, whereas this value of drag did not greatly affecting the performance of the car. Figure 9 demonstrates the wind speed dependent variation in the effective rpm and torque that obtained in experiment test.

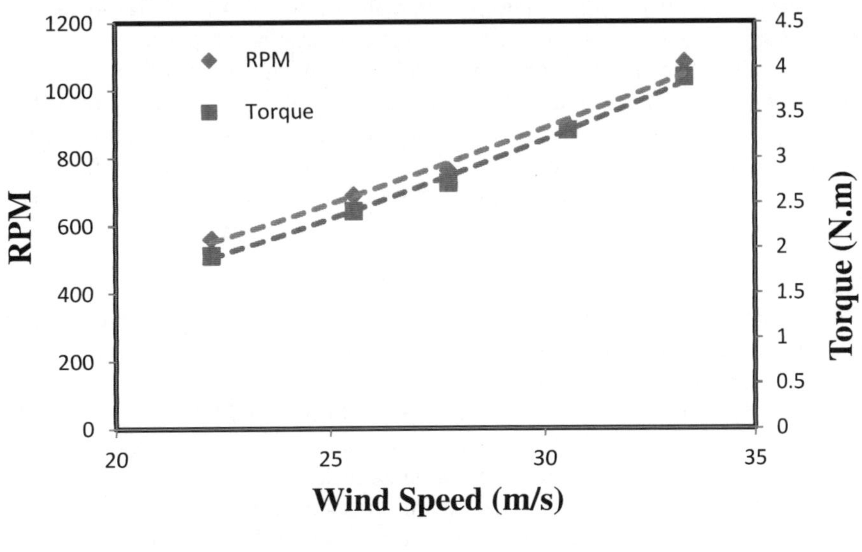

Figure 9. RPM and Torque of the turbine rotor

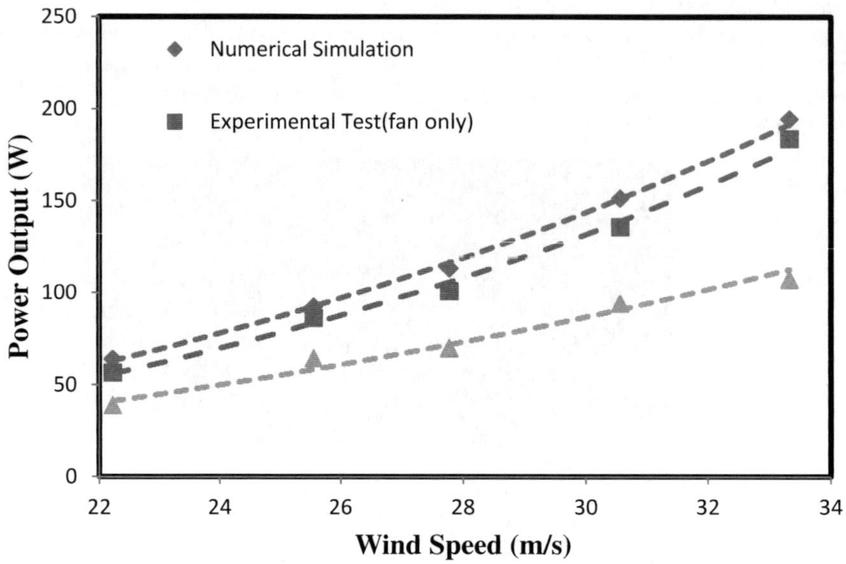

Figure 10. Comparison of power output from experimental and numerical results

Figure 10 compares the values of power output obtained from both the numerical simulation and experiments as a function of wind speed. As the wind speed increase, the power output was also increase as illustrated. Both numerical simulation and experimental test show a similar trend in this Figure. The maximum power produced by CFD was discerned to be 194.84W at the maximum speed of 33.33 m/s, while the maximum power produced by the system without the alternator was 183.674W at same speed depending on the measurement of rpm, whereas, the maximum power produced by the entire system (fan connected with automotive alternator) was 107.46 W. The observed difference of the experimental results from the numerical simulation was mainly attributed to the following: (i) design of automotive alternator for high rpm application; (ii) wire resistance and internal resistance for automotive alternator; (iii) In automotive alternator, a belt system is used to increase the performance, whereas this model was made without any belt system due to lack of equipment; (iv) regulator and rectifier inside the automotive alternator prevent the increase in the voltage [1, 6]. From the experimental results, the 24 V DC alternator is more efficient to be utilize as generator for small wind turbine applications to fit the high-power output produced.

5. Conclusion

This paper evaluated the power output of a SHAWT under high wind speed operation. An automotive alternator was used as generator for SHAWT application. Such turbine was constructed and installed on a top of a car for further experimentation. The maximum power produced experimental was 183.67 W at wind speed 33.33m/s, For the every characteristic of the used rotor geometry and wind speed in the validation, simulations was conducted numerically and the maximum power output was found to be 194.48 W, hence noting that the numerical simulation results were found satisfactory and matching the practical outcomes. Therefore, it is noteworthy to deduce that the simulation environments are sufficiently accurate to fit our work. The idea of evaluating a SHAWT at high speed is demonstrated to be beneficial in terms of low cost wind energy generation, devoid of skilled manpower and easy maintenance. It is established that despite its inherent low efficiency, the vehicle alternator system can be a viable alternative for high wind speed application.

Acknowledgement
The present work has been supported by the Ministry of Education Malaysia under Exploratory Research Grant Scheme (Vote No. E034).

References

[1] Ani, Samuel Ofordile. Low cost small wind turbine generators for developing countries. TU Delft, Delft University of Technology, 2013

[2] Goh, Seach Chyr, Sethu Raman Boopathy, Chidambaresan Krishnaswami, and Jörg Uwe Schlüter. "Tow testing of Savonius wind turbine above a bluff body complemented by CFD simulation." *Renewable Energy* 87 (2016): 332-345.

[3] Gebreslassie, Mulualem G., Stephanie O. Sanchez, Gavin R. Tabor, Michael R. Belmont, Tom Bruce, Grégory S. Payne, and Ian Moon. "Experimental and CFD analysis of the wake characteristics of tidal turbines." *International Journal of Marine Energy* 16 (2016): 209- 219..

[4] Cao, Han. Aerodynamics analysis of small horizontal axis wind turbine blades by using 2D and 3D CFD modelling. Diss. University of Central Lancashire, Preston, 2011

[5] Fernández, H., A. Martínez, V. Guzmán, and M. Giménez. "Wind turbine generation system implemented with a car alternator for use in isolated locations." In *International conference on renewable energies & power quality, ICREPQ, Barcelona*, p. 6. 2004.

[6] Ani, Samuel Ofordile, Henk Polinder, and Jan Abraham Ferreira. "Small wind power generation using automotive alternator." Renewable Energy 66 (2014): 185-195.

[7] Arafat, Md Yeasin, Mahtab Murshed, Md Mahamudul Hasan, and M. Abdur Razzak. "Design aspects and performance analysis of inner and outer rotor permanent magnet alternator for direct driven low-speed wind turbine." In *Advances in Electrical, Electronics, Information, Communication and Bio-Informatics (AEEICB), 2016 2nd International Conference on*, pp. 604-609. IEEE, 2016.

[8] Boldea I. Variable speed generators. Boca Raton, FL: CRC Press; 2006.

[9] Kassakian IG, Wolf H-C, Miller JM, Hurton CJ. Automotive electrical systems circa 2005. IEEE Spectrum August, 1996:22e7.

[10] Manwell, James F., Jon G. McGowan, and Anthony L. Rogers. Wind energy explained: theory, design and application. John Wiley & Sons, 2010.

[11] Magedi. Saad,, Sofian Bin Mohd, Mohd Fadhli. "A Survey on the Use of Ram Air Turbine in Aircraft". Presented on 7th international conference on mechanical manufacturing engineering, Yogyakarta 01-3 August 2016.

[12] Sanderse, Benjamin, S. P. Pijl, and B. Koren. "Review of computational fluid dynamics for wind turbine wake aerodynamics." Wind Energy 14.7 (2011): 799-819.

[13] Lee, Nak Joong, et al. "Performance study on a counter-rotating tidal current turbine by CFD and model experimentation." Renewable Energy 79 (2015): 122-126.

[14] Park, Sewan, Sunho Park, and Shin Hyung Rhee. "Influence of blade deformation and yawed inflow on performance of a horizontal axis tidal stream turbine." Renewable Energy 92 (2016): 321-332.

[15] Hartwanger, David, and Andrej Horvat. "3D modelling of a wind turbine using CFD." NAFEMS Conference, United Kingdom. 2008.

Preliminary Design of Aerial Spraying System for Microlight Aircraft

Zamri Omar[1], Nurfazliawati Idris[2], M.Zulafif Rahim[3]

[1,2,3] Faculty of Mechanical and Manufacturing Engineering, Universiti Tun Hussein Onn Malaysia, 86400 Parit Raja, Batu Pahat, Johor, Malaysia

Corresponding author: nurfazliawati.idris@gmail.com

Abstract. Undoubtedly agricultural is an important sector because it provides essential nutrients for human, and consequently is among the biggest sector for economic growth worldwide. It is crucial to ensure crops production is protected from any plant diseases and pests. Thus aerial spraying system on crops is developed to facilitate farmers to for crops pests control and it is very effective spraying method especially for large and hilly crop areas. However, the use of large aircraft for aerial spaying has a relatively high operational cost. Therefore, microlight aircraft is proposed to be used for crops aerial spraying works for several good reasons. In this paper, a preliminary design of aerial spraying system for microlight aircraft is proposed. Engineering design methodology is adopted in the development of the aerial sprayer and steps involved design are discussed thoroughly. A preliminary design for the microlight to be attached with an aerial spraying system is proposed.

1. Introduction

Aerial spraying is the most efficient and practical technique for pesticide application on large crop area. Because of the importance of agricultural sectors in providing nutrients for human, it is crucial to ensure crops production is protected from any plant diseases and pests. Depending on which part on the globe, there are diverse range of crops production in very large area such as soybean, corn and wheat in North America, wheat, wool and sugarcane in Australia, and oil palm, paddy and banana in Southeast Asia. Aerial application of spraying is very effective way to deliver the treatments to large crops or plantation areas in very short time. Typical aircraft carrying the spraying chemical is able to spray the crop of several hectares in just one hour, whereas it may takes around two weeks if using ground-based spraying system.

Most of the aerial application providers are using high horse power turbine engine aircraft with high cruise speed to carry out the task. Hence, the fuel consumption, maintenance and other operational cost of the aircraft are relatively high. The labour cost would be higher because greater horse power and higher speed aircraft requires more experience pilot to operate. Furthermore, a longer, wider and costly runway is required for the aforesaid aircraft to take off and landing.

Besides, spray drift is another major problem of today's agricultural aerial spraying works. It is a very undesirable situation where the chemicals are deposited on the wrong areas that can cause serious lost such as wasted spray material, damage of sensitive adjoining crops, damage to susceptible off-target areas, environmental contamination, illegal chemical residues and health risks to human and animals. The spraying variables that affecting the spray drift are spray release height, airspeed and propeller wash effect.

Therefore, microlight aircraft has a good potential to be used for aerial spraying in order to minimize all these challenges. The use of microlight aircraft for aerial spraying is because it can move at low speed and ease of spraying evenly on the crop. Microlight aircraft can usually land on a relatively small site to refill the chemical tank rather than flying back to an airport. Furthermore the use of microlight aircraft is expected to increase spraying effectiveness because the ability of low flying speed and low altitude allows the spraying can be done more evenly on the crops and minimize the spray drift. Additionally, because of microlight aircraft is using normal gasoline fuel, this will further reduce the operational cost. The feasibility of using Microlight for aerial spraying is discussed in [7,8].

The paper is to present a preliminary design of aerial spraying system that is to be used on a microlight aircraft. Suitable components of aerial spraying system are reviewed and selected, and the CAD model of aerial spraying system is developed into Solidworks®. The platform that is suitable to be attached the aerial spraying system is basically any two tandem seats microlight aircraft such as the *Quicksilver GT500* and *Challanger* microlight. These microlights are the most suitable platform for aerial spraying system because tandem arrangement of the seats allows the spraying tank to be placed on thc passenger seat behind the pilot. In the effort to achieve the aforementioned objectives, the scope of this study is to develop the CAD model for aerial spraying system until the embodiment design stage.

2. Design Methodology

Function structure is compulsory for the design of aerial sprayer system as it will illustrate all functions of the designed product. The function of the system is designed based on the problem statement. The strengths and weaknesses of the functional synthesis can be obtained as strength shown by creating function structures forces representation into a language that is useful for the manipulation for mechanical design problems. The function structure is a flow diagram where it is the connection of different functions performed by the system. Figure 1 is the black box of aerial spraying system structure where it shows the input-output relationship. Here the system is having the wind energy which drives the pump to generate pressurised spray chemical to be sprayed on the target area. The detail process of the aerial spraying system is describes in the function structure diagram as shown in Figure 2.

Figure 1. Black box of crop sprayer system on microlight aircraft.

2.1. Concept Generation by Morphological Chart

A morphological chart is a table based on the function analysis. It is recommended to generate several feasible designs by using different mechanisms for each function of the proposed concepts. The morphological chart enables these solutions to be expressed and provides the structure for considering alternative combinations. This can enable the early consideration of the product architecture through the generation and consideration of different combinations of sub-solutions that have not previously been identified [3]. Table 1 shows the morphological chart of aerial spraying system where it provides options for types and material for the six components of the aerial spraying system which are tank, pump, boom, control valve, spray nozzle and power source.

2.2. *Concept Generation by Brainstorming*

Brainstorming is a group creativity technique by which efforts are made to find a conclusion for a specific problem by gathering a list of ideas spontaneously contributed by its members. Analysis, discussion, or criticism of the proposed ideas is allowed when the brainstorming session is over and evaluation session begins. Figure 3 shows the map concept of design product. Here we are focusing on five elements in the design process which are functionality, safety, innovative design, low cost and ergonomic.

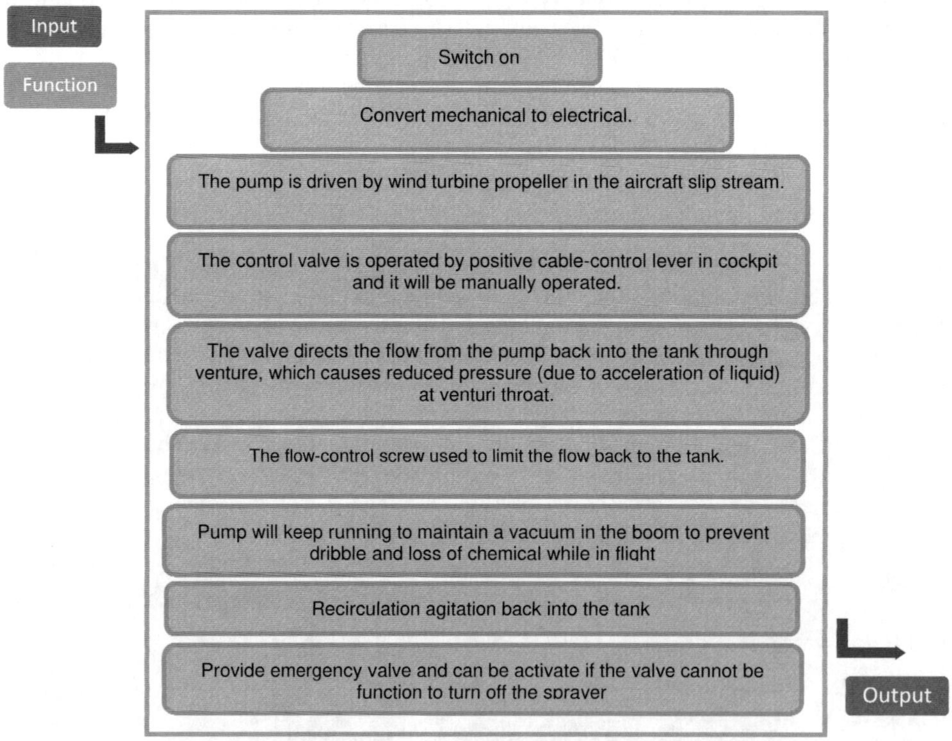

Figure 2. The function structure of aerial spraying system.

2.3. *Concept Evaluation*

The concept evaluation stage represents the convergence stage of design development so we will start by evaluating the concepts developed for the lowest level of function decomposition. As we progress, we will begin to combine the best concepts into sub-systems and then to evaluate the sub-systems using the same procedures. Weighted rating method is technique used to rank the options from highest priority to lowest priority as indicated by the assigned values of Importance Weight. Table 2 shows the weighted rating method of the aerial spraying system design. The rating criteria included in Table 2 are for each map concept elements presented in Figure 3. By using weighted rating method, it is found that the maximum score is 4.61, therefore the combination 2 of the concept alternative is selected.

IOP Conf. Series: Journal of Physics: Conf. Series **914** (2017) 012003 doi:10.1088/1742-6596/914/1/012003

2.4. Concept Selection

Concept selection is the result of the best selection that is obtained from the weighted rating method which in our case it has the score of 4.61 as depicted in Table 2. The selected concept of the design of aerial spraying system is given in Table 2. The selected material for tank is polyethylene, wind driven centrifugal pump is used and controlled by 3-ways valve, stainless steel boom and single nozzle sprayer are selected.

2.5. Spraying Performance Calculation

The required capacity of the spray pump is the total rate of nozzle discharge required at a given swath width and rate of travel as shown in Eq. (1).

$$Q_T = \frac{SQ_AV}{600} \quad (1)$$

where,

Q_T = Total output in litres per minute

Q_A = Applied rate in litres per hectare

V = Aircraft velocity in kilometres per hour

S = Swath width in metres

Table 1. Morphological chart of aerial spraying system.

NO	FUNCTION	POTENTIAL MECHANISM		
		Alternative 1	Alternative 2	Alternative 3
1	Tank	Polyethylene 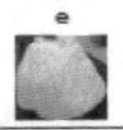	Stainless Steel 	Acrylic Glass 
2	Pump	Centrifugal Pump 	Plunger Pump 	Diaphragm Pump 
3	Boom	Carbon Fiber 	Polyvinyl chloride(PVC) 	Stainless Steel 
4	Control Valve	One-way Control Valve 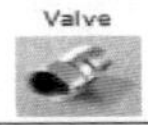	Automatic Control Valve 	Three-ways Control valve 
5	Sprayer Nozzle	Twin Nozzle 	Adjusted Nozzle 	Single Nozzle 
6	Power Source	Solar 	Wind 	Fuel powered/engine driven 

2.6. Design Sketching

As a tool or skill, sketching has its role in the design process. That role will vary depending on the end-product being created, the size and scope of the project, the individual designer's style, experience, and workflow, and the client's expectations [4]. Figure 4 shows the sketching of various views of the aerial spraying system that has been selected by using weighted rating method. In Figure 4(a), it shows the perspective view of the aerial spraying when attached on the microlight aircraft. Figure 4(b) shows the proposed wind-driven centrifugal pump that attached under the belly of the microlight. Figure 4(c) and 4(d) show the front and side views of the spraying system where the assembly of all components are clearly seen.

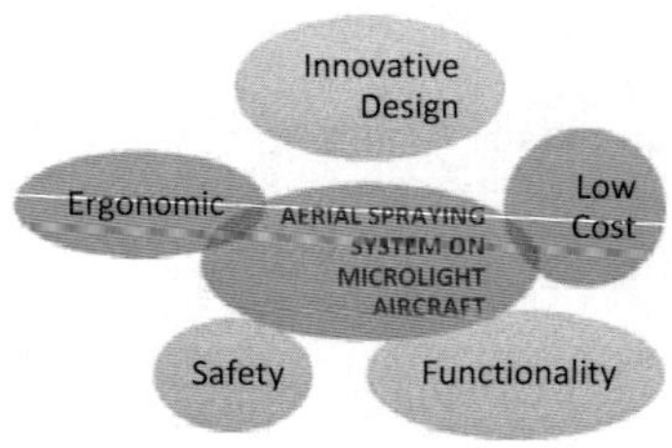

Figure 3. Map concept of aerial spraying system design.

Table 2. Weighted rating method of the design of aerial spraying system.

No	Criteria	Important Weight	Concept Alternative								
			Combination 1			Combination 2			Combination 3		
			Magnitude	Score	Rating	Magnitude	Score	Rating	Magnitude	Score	Rating
A	Design										
	Anti-rust Material	0.06	Yes	4	0.24	Yes	5	0.30	Yes	4	0.24
	Durable Material	0.06	Yes	3	0.18	Yes	5	0.30	Yes	4	0.24
	Sustainability	0.09	Yes	5	0.45	Yes	5	0.45	Yes	3	0.27
	Medium Weight	0.09	Yes	3	0.27	Yes	4	0.36	Yes	5	0.45
B	Ergonomic										
	Easy to operate	0.05	Yes	5	0.25	Yes	4	0.20	Yes	5	0.25
	Area	0.05	Yes	3	0.15	Yes	4	0.20	Yes	4	0.20
C	Low Cost										
	Low labor cost	0.06	Yes	4	0.24	Yes	5	0.30	Yes	4	0.24
	Low material cost	0.04	Yes	5	0.20	Yes	4	0.16	Yes	3	0.12
	Low maintenance cost	0.04	Yes	3	0.12	Yes	5	0.20	Yes	4	0.16
	Low fuel cost	0.06	Yes	3	0.18	Yes	5	0.30	Yes	5	0.30
D	Functionality										
	High efficiency	0.06	Yes	4	0.24	Yes	5	0.30	Yes	4	0.24
	Can spray crops evenly	0.06	Yes	5	0.30	Yes	4	0.24	Yes	4	0.24
	Can move with lower speed	0.08	Yes	3	0.24	Yes	5	0.40	Yes	3	0.24
E	High Stability										
	Low injury	0.10	Yes	4	0.40	Yes	4	0.40	Yes	4	0.40
	Low fire risk	0.10	Yes	3	0.30	Yes	5	0.50	Yes	3	0.30
	TOTAL	1.0			3.76			4.61			3.89

Table 3. Selected concept of the design of aerial spraying system.

NO	FUNCTION	SPECIFICATION
1	Tank	Polyethylene
2	Pump	Centrifugal Pump
3	Power Source	Wind
4	Control Valve	Three-ways control valve

5	Boom	Stainless Steel
6	Sprayer Nozzle	Single Nozzle

3. Results and Discussion

3.1. Bill of Material (BOM)

Bill of materials or most often called as BOM is a list of parts, components, sub-assemblies and assemblies required in developing the product. The BOM is always the core for any manufacturing process as it provides all information required to build the product. Developed to provide manufacturing information, the Manufacturing Bill of Materials (MBOM) differs from the Engineering Bill of Materials, which is prepared to describe the design intent of the product [5]. Table 4 shows the list of components.

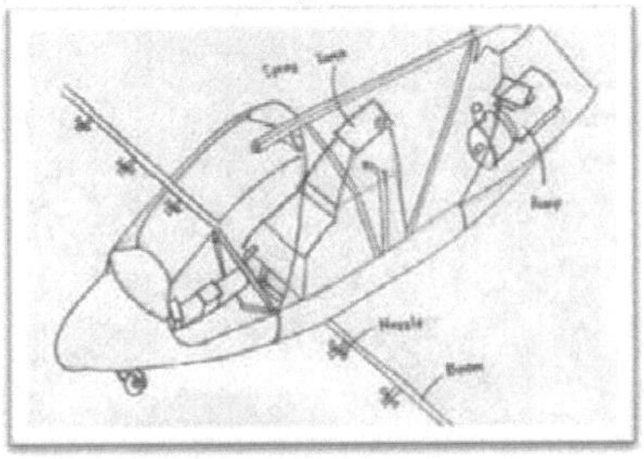

(a) Perspective view.

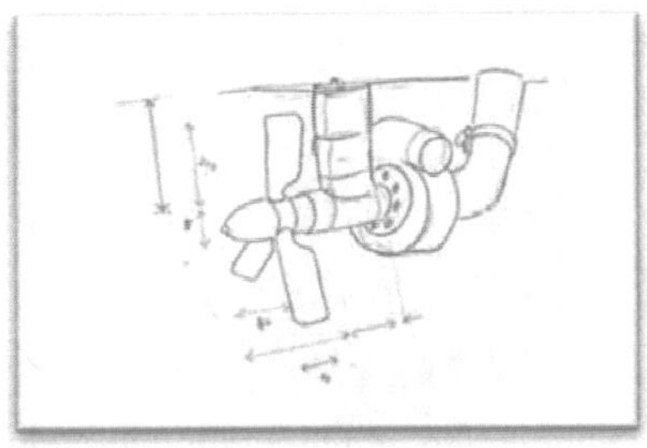

(b) Wind driven pump.

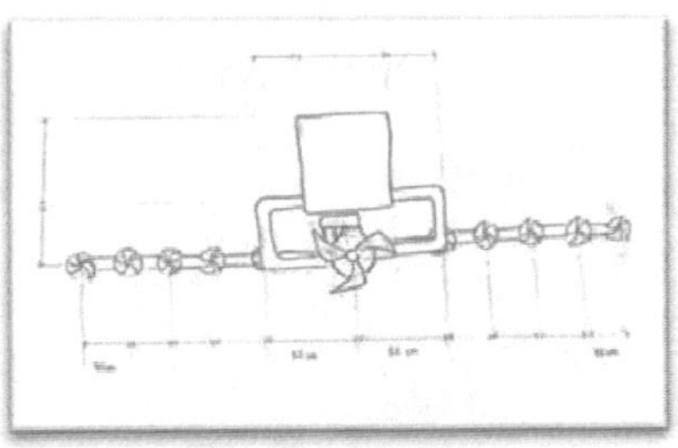

(c) Front view.

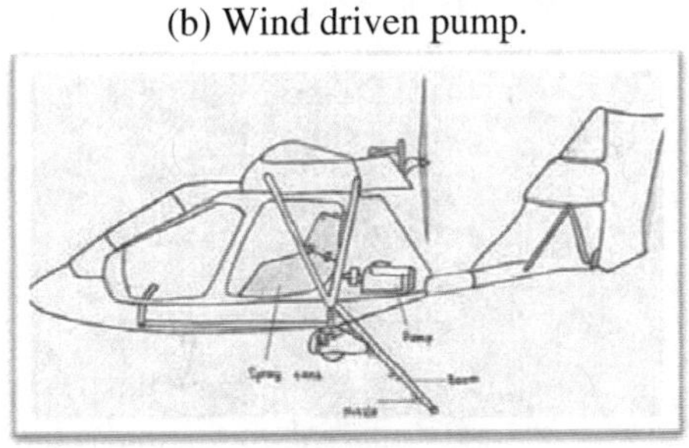

(d) Side view of assembly sketch.

Figure 4 The sketch of aerial sprayer system.

3.2. CAD Model and Design Review of Aerial Spraying System

A design review is a milestone within a product development process whereby a design is evaluated against its requirements in order to verify the outcomes of previous methods and identify issues before committing to and if need to be reprioritize for further work. The ultimate of design review, if successful would be triggers the product launch or product release. The conduct of design reviews is compulsory as part of design controls [6].

Table 4. List of components

No.	Components	Quantity
1	Tank	1
2	Hose	4
3	Boom	1
4	Pump	1
5	Wind-driven Pump	1

IOP Conf. Series: Journal of Physics: Conf. Series **914** (2017) 012003 doi:10.1088/1742-6596/914/1/012003

6	Nozzle	10
7	Three-ways Valve	1
8	Emergency Valve	1
9	Clip	4
10	Atomizer	10

Figure 5 shows the computer-aided-design (CAD) model of the components for aerial spraying system developed by using Solidwork®. The tank shown in Figure 5(a) is designed to be able to carry 140 liters spraying liquid. The selection of tank capacity is based on the allowable microlight payload. It is designed in that shape in order to accommodate the rear seat of the microlight aircraft. The tank is secured in place by using existing seat harness on the rear seat. As stated previously, the material for the spraying tank should be polyethylene.

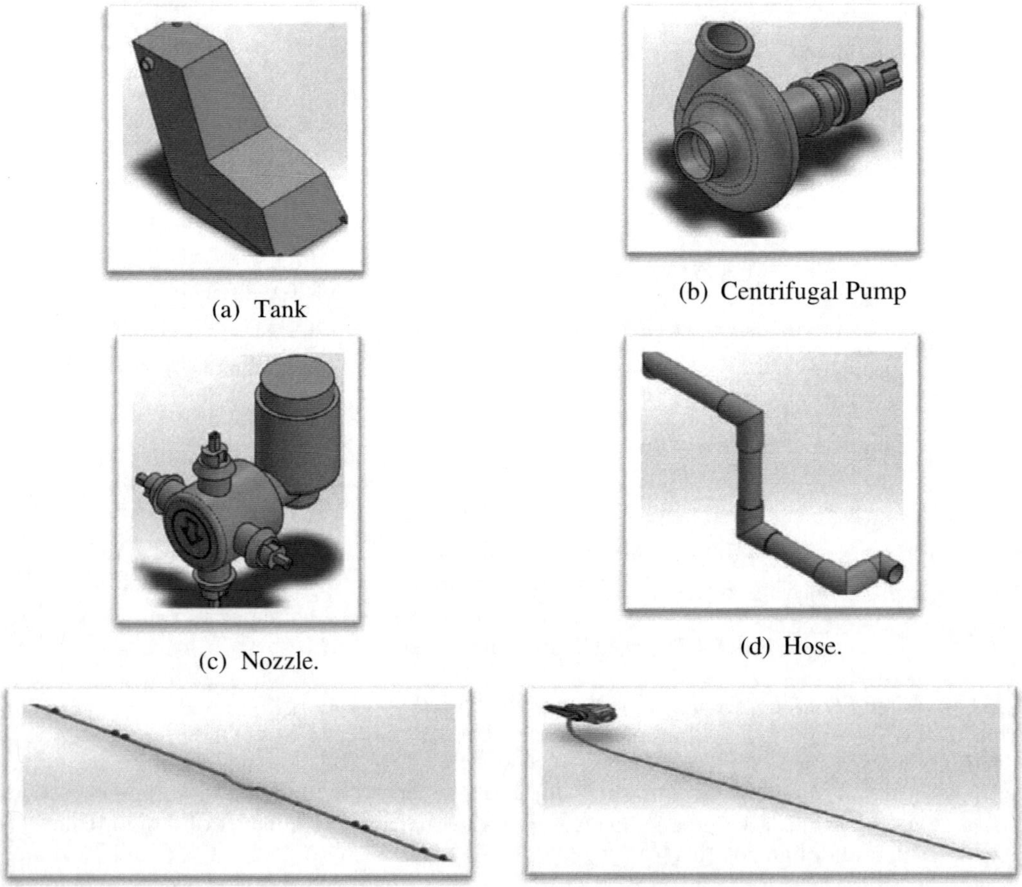

(a) Tank (b) Centrifugal Pump

(c) Nozzle. (d) Hose.

(e) Boom. (f) Three-way control valve.

Figure 5. CAD Model of Aerial Spraying System.

3.3. Assembly Model of Aerial Spraying System

Assembly model is completed drawings that are put together after each of the aerial sprayer system parts is produced. It is the components combination drawings that shows an overview on how the system would be attached on the microlight aircraft as shown in Figure 6. The aerial sprayer system that is installed on this aircraft is user-friendly to setup and easily handled. The sprayer tank that replaced the rear seat eases the farmer to do the spraying without disassemble or reconfigure the existing aircraft system.

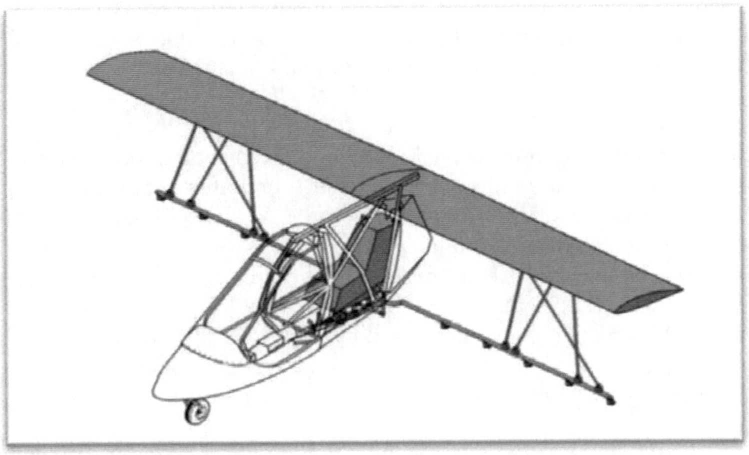

Figure 6. Assembly and position of aerial sprayer system on microlight aircraft.

3.4. Final Design Specification

The crops aerial sprayer that is to be attached on microlight aircraft has been designed to make the farmer getting easy to do their works on spraying. The spraying equipment are selected based on suitable criteria that have been analyzed. Table 5 shows the final design specification of the aerial sprayer.

Table 5. Final Design Specification

Criteria	**Product Specifications**
Performance	Produce a more uniform spray and has four types of spraying that can be changed without removing the nozzle.
Operation Temperature	17°C to 40°C
Material	Fiber Glass, Carbon fiber, Nylon
Power Source	Air (Wind Turbine)
Maintenance	Less (E.g. : Filter conversion)
Safety	There is an open-close emergency valve that are made of materials that are not harmful
Ergonomic	Easy to operate, disassembled and assembled. Tank that can be separated.
User targets	Farmers who have acreage that is not too broad

4. Conclusion and Recommendation

Design is a subject of vital importance. The preliminary design of aerial sprayer system for microlight aircraft project is intended to achieve the main objectives which are to increase the spray coverage area and improvement of existing spray equipment that available in the market.

This study set out to determine and solving the problems that has been faced by the farmers. The development of aerial sprayer system on microlight aircraft need to be followed by the right design methods. The idea of this design would be compared to the existing system used on typical light aircraft to identify the advantages and disadvantages of that system. The project can be summarized as follows;

i. The spraying from the existing sprayer system is uneven and not fully sprays to the crops.
ii. Most of the aerial application providers are using a minimum, high horsepower turbine engine aircraft with high cruise speed of to carry out the task. Hence, fuel consumption, maintenance and cost of aircraft are relatively high.
iii. The hilly and large acreage of the farms cause a difficult process of spraying to the crops.
iv. The existing sprayer is difficult to assemble and dismantle.
v. A conventional spraying process requires more labor to do the spraying.
vi. The knapsack sprayer is ineffective and does not have ergonomic features that can cause pain in the waist and back bone. Also, it could not accommodate the works in spraying on large acreage.

To identify the problems encountered, the proposed system is developed based on the methods of engineering design. The results obtained in the process of designing the aerial sprayer applied on microlight aircraft can be shown by;

i. The use of the aerial spraying system on microlight aircraft can move at low speed and ease of spraying evenly on the crop.
ii. Microlight aircraft can usually land on a relatively small site to refill the chemical tank rather than flying back to an airport.
iii. The spraying works can be done in a short time so it will be more economic such as it will increase the crops in the country.
iv. The sprayer system development by wind turbine to run the pump will reduce the usage of fuel.
v. Make a detailed drawings and matrix calculation to determine the types of material that is needed to design this aerial sprayer system.
vi. Less of workers are needed to handle the crop management when the aerial sprayer system is attached on microlight aircraft.

Finally, this study can conclude that the design of aerial spraying system for microlight aircraft has achieved the objectives. It also fulfills the farmer needs. This sprayer system is ergonomic compared than conventional spraying methods. Besides that, the aerial sprayer system on microlight aircraft will easily get a place in the market due to the needs of higher crop yield nowadays in addition to further facilitate the spraying process. There are some suggestions that can be taken into consideration for improvements to enhance the operability and effectiveness of the attached spraying system. The suggestions include designing a liquid storage tank that can accommodate a large volume of liquid to avoid refilling work done in many times. In addition, the existing tank fastener can be replaced by the passenger belt with a sturdier and neat design. We can also create a design of aerial sprayer system that can be applied to several aircraft without changing the shape of the sprayer. Besides that, the idea of innovate the nozzles can be done by design it to have a multiple choices on the boom so it can be switch to any mode when to make spraying in a different conditions.

References

[1] Silviculture Section, Ministry of Natural Resources and Forestry "*Aerial Spraying for Forest Management*" , Ontario, June 1991
[2] Spectrum Electrostatic Sprayers, Inc "Spectrum – Aerial Sprayer System" , Texas USA, 1987.
[3] James Moultrie, Design Management Group "Morphological Charts" , Cambridge
[4] Sean Hodge, "The Role of Sketching in Design Process" , Orlando. 2008.
[5] Alpesh Patel, Engineering Exchange, "Importance of Manufacturing Bill of Materials (MBOM)", 2015.
[6] George E.Dieter, "Engineering Design-A Material and Processing Approach", McGraw Hill, New York, 1999.
[7] Omar Z, Qiang KY, Mohd S and Rosly, "CFD Simulation of Aerial Crop Spraying". In IOP Conference Series: Materials Science and Engineering Vol. 160, No. 1, p. 012028, 2016.
[8] A.A Wahab, "The Design of UTMAS-1 Spraying Mechanism For Microlight", Universiti Teknologi Malaysia, Jurnal Teknologi, Bil. 28, p. 73-81, 1998.

Acknowledgments

The authors would like to acknowledge assistance or encouragement from colleagues, special work by technical staff or financial support from UTHM and Ministry of Higher Education. Specifically, this project is funded under Fundamental Research Grant Scheme Vot number 1548.

IOP Conf. Series: Journal of Physics: Conf. Series **914** (2017) 012004 doi:10.1088/1742-6596/914/1/012004

Numerical Analysis of Kraft Paper Honeycomb subjected to uniform compression loading

Nurdina Abd Kadir[1], Yulfian Aminanda[1*], Mohd. Sultan Ibrahim Shaik Dawood[1], Hanan Mohktar[1]

[1]Mechanical Engineering Department of IIUM, Jln. Gombak, P.O. Box 10, 50728 Kuala Lumpur, Malaysia

Corresponding author: yulfian@iium.edu.my

Abstract. The objective of this study is to investigate the crushing strength of Kraft paper honeycomb with different thickness of cell wall and different cell size subjected to compression loading. At first, experimental tests were carried out on Kraft paper honeycomb with 10mm cell size and 0.6mm thickness of paper with different mass density (RO). Then, a finite element model was developed in order to study the effect of cell size and thickness of cell wall on crushing properties of Kraft paper honeycomb. Based on the simulation results, the cell size and thickness of cell wall of Kraft paper honeycomb influence the crushing properties of the honeycomb. Crushing strength increases with decreasing the cell size which for the range of the study, cell size 10mm shows the highest strength. Meanwhile, 0.4mm thickness of cell wall results the higher crushing strength of honeycomb. Also, higher density gives higher crushing strength where in this study is found for 2.236x 10^6 kg/mm^3.

1. Introduction

Over the years, low technology industry such as furniture and load bearing has been increasingly adopting more and more lightweight sandwich panel type of structure in order to reduce the materials cost and light weight. Recently, some manufacturers are aiming to use paper honeycomb panel or structure for load bearing application such as floor, decks, transportation pallet, load bearing wall and partition due to lower cost and low material usage of paper honeycomb compared to solid wood based panel [1–3].

The honeycomb core can have various configurations. The core type and density, thickness of the core and skin and their respective materials properties will determine the structure's strength and stiffness. Honeycomb, a typical two-dimensional cellular material, its cell can be characterized by four main geometrical parameters including foil thickness, width, length and elevation angle [4]. Petrone et al [5] reported that energy absorption in honeycomb cores depend on the geometrical parameters of the honeycomb, essentially the cell wall thickness to length ratio (t/l) or the relative density of the core. In case of compressive loads, the critical buckling load is dependent on the bending of the cell wall.

Numerical simulation has always been an effective means for crashworthiness analysis of honeycomb structure. FE method is one of most effective approaches for analyzing stress in structure construction. In this way, some important results and conclusion have been reported [6–8]. Aktay et al. [9] modeled a detailed honeycomb structure which showed that fine micromechanical models can provide better correlation with compression tests in both aluminum and aramid paper honeycomb. However, only limited number of FE model dealt with Kraft paper materials.

FE models developed in this study intend to simulate crushing properties of Kraft paper honeycomb with different cell size and thickness of cell wall. 8 cell of hexagonal cell were considered. The model is isotropic model.

2. Experimental test

In order to investigate the effect of mass density (RO), cell size of Kraft paper honeycomb (s) and thickness of cell wall (t), experimental test was performed which then followed by finite element simulation. Three different mass density (2.236x 10^3 kg/m^3, 1.564 x 10^3 kg/m^3, 8.58x10^2 kg/m^3) were used in the experimental test. The cell size, thickness of cell wall and height for all specimens was 10mm, 0.5mm and 45mm, respectively. The geometric configuration of a hexagonal cell structure is shown in figure 1. The Kraft paper honeycombs were subjected to quasi-static compression loading by using SHIMADZU Autograph AG-X 250 compression machine. As the bottom plate was fixed and the upper plate was subjected to vertical downward displacement of 22.5mm. The force acting on the specimens was measured by load cell, and the force-displacement curve was plotted. The rate of displacement was 3 mm/min.

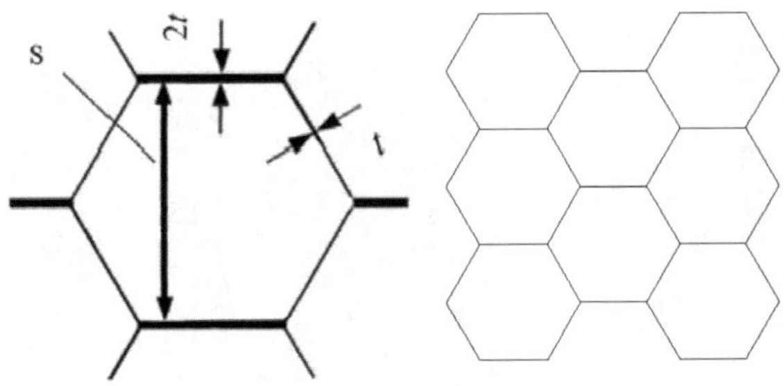

Figure 1. Geometric configuration of a hexagonal cell of honeycomb

3. Finite element modeling

Three-dimensional finite element models of the Kraft paper honeycomb were developed using Ls-Dyna software. The honeycomb model was generated using Belytschko-Tsai shell elements. Due to the complex mechanical behavior of honeycomb, five through integration points were defined for each of the shell elements. The thickness of the shell elements was chosen equal to the actual thickness of the Kraft paper honeycomb used in the experimental test. The binding material between cell walls was ignored in this study since the adhesive has full capability to attach every cell wall. Automatic single surface to surface contact was applied to avoid penetration between cell wall. Bottom nodes of the honeycomb were fixed in all of its degree of freedom. While, Movable plate on the top of honeycomb was modelled as rigid parts via analytical rigid feature and constrained to move only in the compression direction as shown in figure 2. Explicit/solver was implemented for compression analysis.

Through converge analysis completed by employed FE models having different mesh sizes, an element size of 1mm, 1.5 mm and 2mm are found to produce sufficiently accurate stress result for honeycomb with 10mm, 15mm and 20mm cell sizes, respectively (Figure 3).

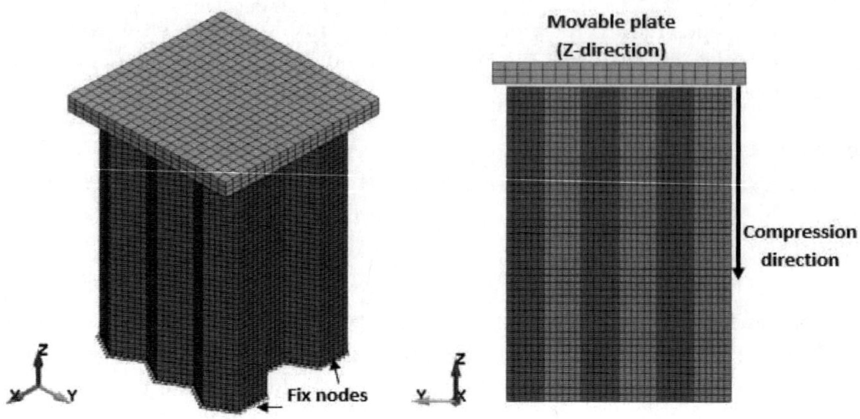

Figure 2. The finite element model of compression test for Kraft paper honeycomb

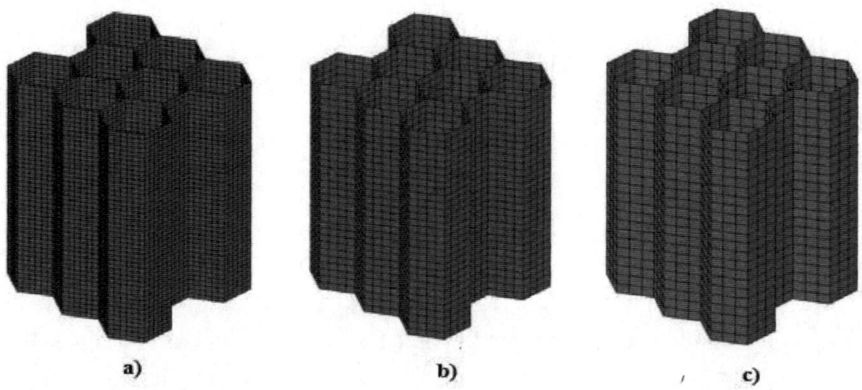

Figure 3: The three-element size of honeycomb for 10mm, 15 mm and 20mm cell size a)1 mm (*s*=10mm) b) 1.5mm (*s*=15mm) c) 2 mm (*s*=20 mm)

4. Results and Discussion of Compression Test

4.1. Finite element model validation

The crushing response of FE model was compared with experimental results in order to validate the numerical model. Load-displacement curves of honeycomb which obtain from quasi-static analysis were implemented to determine the force (N), young's modulus (E) and yield stress (σ_y) of the Kraft paper honeycomb. The elastic modulus was calculated by the ratio stress strain curve in the elastic deformation region. Figure 4 shows the force-displacement curve of the Finite element analysis for different mass density (RO) where RO1, RO2, RO3 refer to 8.58x10^7 kg/mm^3, 1.564 x 10^6 kg/mm^3 and 2.236x 10^6 kg/mm^3, respectively. Based on the graph, it is observed that initially the structure behaves elastically and load increases in a steady manner, then an initial peak force occurs followed by decreases in force value. After that, the post buckling phase develops with secondary peaks which directly related to the formation of subsequent local buckling occurrences in honeycomb structure as crushing progresses. The subsequent peaks in stress history have significantly lower magnitudes in comparison with initial peak. Meanwhile, the peak force values are listed in table 1 and compared with experimental results. It is clear in that the obtained values by FE model are in good agreement with the experimental results and the maximum different are about 12%.

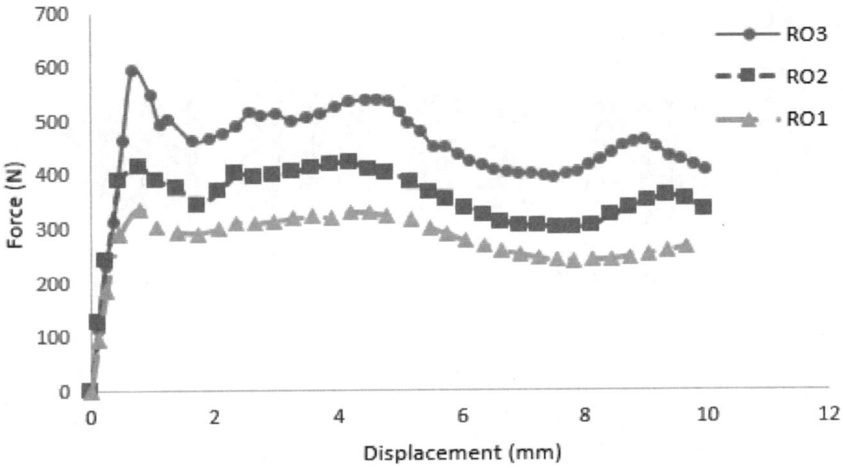

Figure 4: The three-element size of honeycomb for 10mm, 15 mm and 20mm cell size a)1 mm (*s*=10mm) b) 1.5mm (*s*=15mm) c) 2 mm (*s*=20 mm)

Table 1. The peak force (N) of Kraft paper honeycomb obtained by FE model and Experimental tests.

ρ (kg/mm^3)		Force (N)		
		Experimental	FEA	Difference (%)
8.584e^{-7}	RO1	343.07	332	3
1.564e^{-6}	RO2	478.37	423.3	12
2.236e^{-6}	RO3	650.13	597	8

Figure 5 shows the deformation of Kraft paper honeycomb. It observed that the localized fold firstly occurs at the top surface of the honeycomb (free face). The Kraft paper honeycomb initially buckles, and triggers the honeycomb wall to deform. During the first folding process, the second fold start to form simultaneously. With the increase of compression displacement, the cell walls of Kraft paper honeycomb folded in layer by layer up to the compaction of compressed to 22.5mm. Since local buckling occurs early, the honeycomb structure works in a post buckling mode. Castenie et al. [6] found that the stiffener in honeycomb is the vertical edge where it formed by intersection of three cell walls. They reported that, for honeycomb cell, the compression is mainly taken by vertical edge since the buckling of the cell wall occurs earlier. Then the collapse of the stiffened structure corresponds to the buckling of the stiffeners.

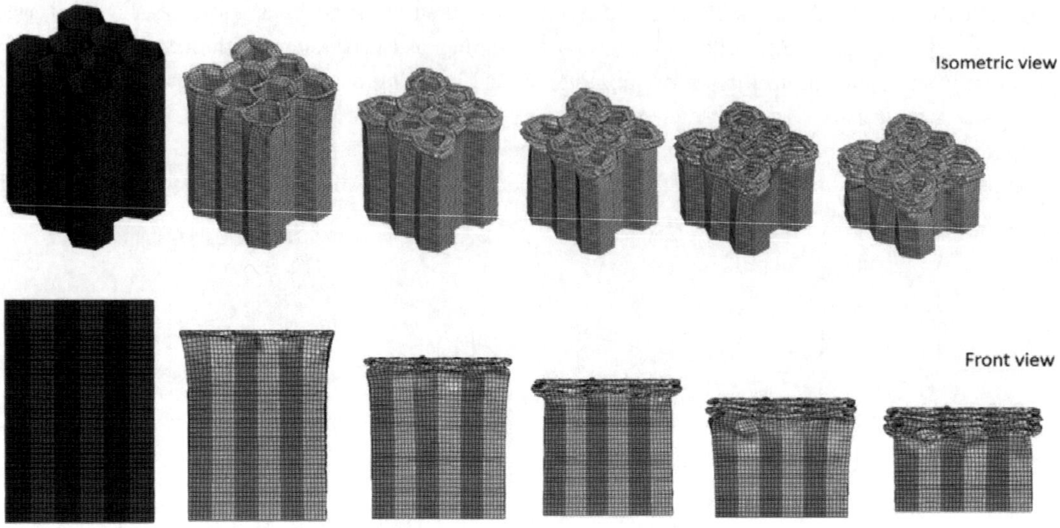

Figure 5. Failure mechanism of Kraft paper honeycomb.

4.2 The effect of cell size of honeycomb

In order to study the cell size effect of Kraft paper honeycomb, crushing of honeycomb structure is simulated by changing the cell size of honeycomb to 10mm, 15mm and 20mm (Figure 6). Figure 7 and 8 show the effect of different mass density (RO) and cell size on crushing strength. It is observed in the graphs that the honeycomb cores with small cell size had the larger crushing strength. Meanwhile increasing the density also increase the crushing strength. According to simple compression failure mechanism by Zhang and Ashby [10], the buckling load is determined by the second moment of inertia of the wall and cell side size as follow:

$$P_{crit} = \frac{KE_s t^3}{(1-v_s^2)l} \tag{1}$$

where, K is an end constant factor, E_s is young's modulus of solid cell wall materials, V_s is Poisson's ratio of the solid cell wall materials, t is the thickness of honeycomb materials and l is the length of cell side. Based on the eq. 1, buckling load decrease with increasing the length of cell wall. When local buckling of the structure is in lower forces values, crushing strength of the honeycomb decline.

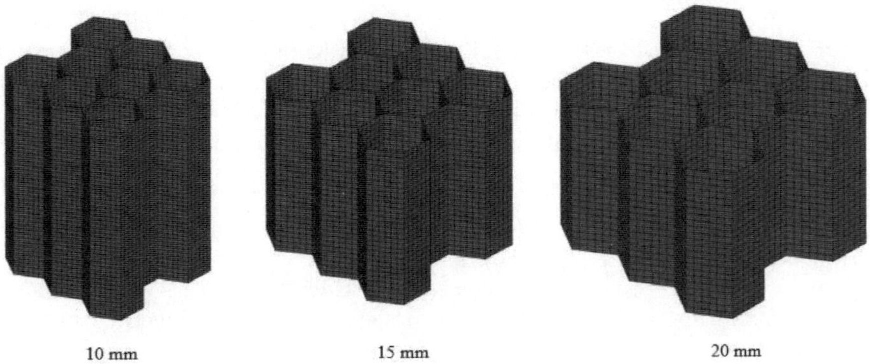

Figure 6. Three different cell size of Kraft paper honeycomb

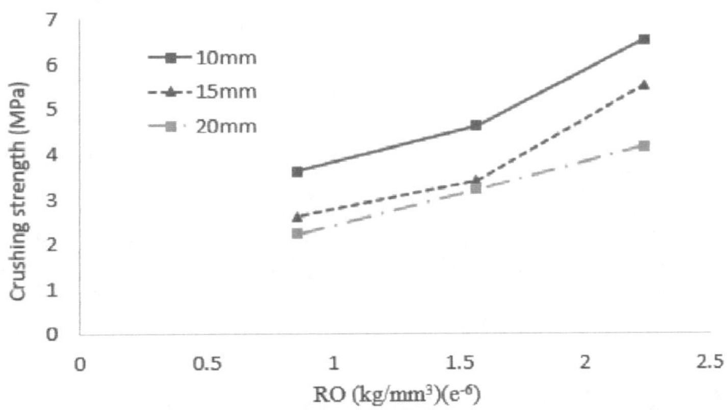

Figure 7. Crushng strength-RO curve for different cell size

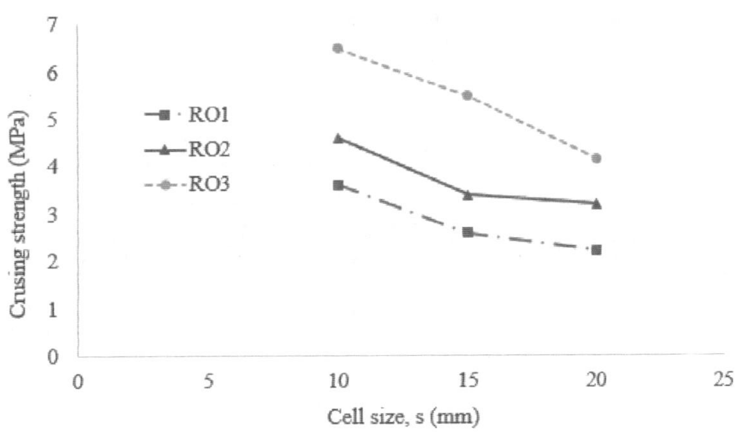

Figure 8. Crushing strength-cell size curve for different mass density

The crushing mechanism of different cell size and mass density in isometric and front view were illustrated in table 2. The crushing mechanism is directly associated with cell size and mass density. It can be seen from the figure that the failure mechanism of Kraft paper honeycomb due to the folding of the cell walls and all cell walls folded at the top surface of the honeycomb. Based on Castenie et al [11] reported in their book that the fold shape differed for each material where rounded shape for aluminum and sharp angle for Nomex. This difference of shape can be attributed to the different plasticity of the materials. In this study, Kraft paper honeycomb deformed in sharp angle shape which same as Nomex.

Besides, it also observed that folding length of the Kraft paper honeycomb are different. Honeycomb with large cell size shows larger folding length compare to small cell size. Xu et al [12] reported that for large cell size of honeycomb, the plastic buckling is more to occur in the middle region of the specimens, indicating that the deformation pattern is influenced by the dimension of honeycomb cell.

Table 2. Deformation of FE model of Kraft paper honeycomb with different cell size (isometric and front view)

	S = 10mm	S = 15mm	S = 20mm
RO1	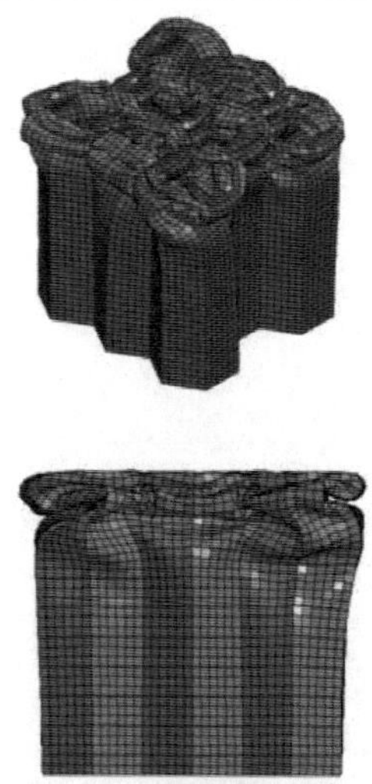		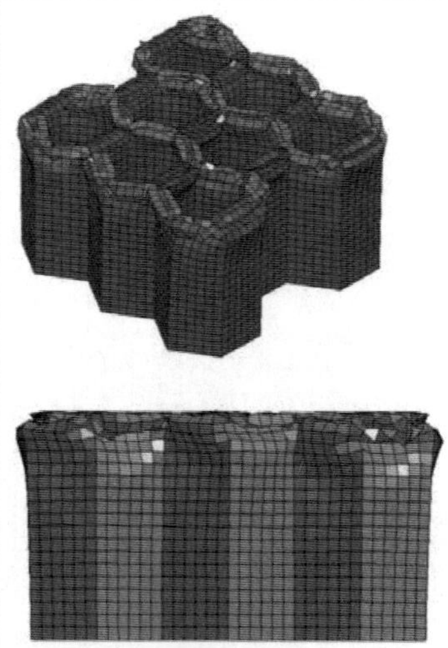
RO2			

		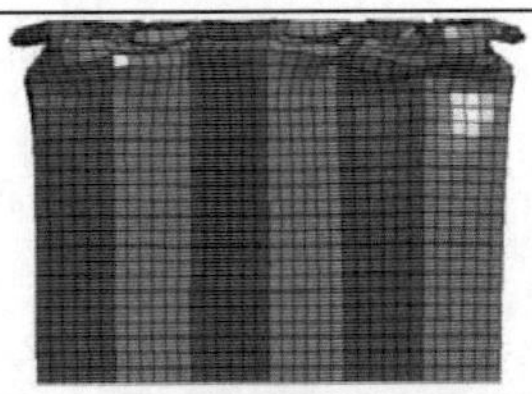	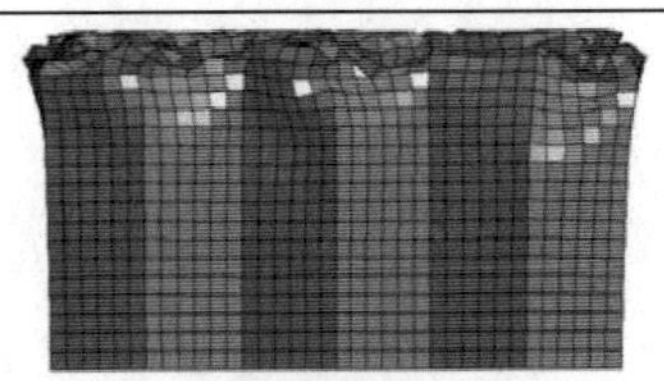
RO3			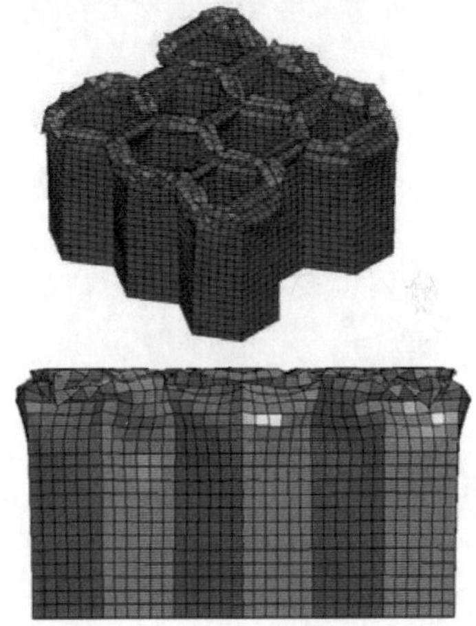

4.3 The effect of thickness of cell wall

The effect of cell wall was studied by changing the thickness of cell wall, *t* between 0.4mm to 0.6mm. The cell size for all specimen is 10mm. Nine FE model was developed in order to study the effect of cell wall thickness. From the simulations, the effect of cell wall on crushing strength were illustrated in figure 9 and 10. Based on the graphs, it observed that the crushing strength increase with increasing the mass density. However, crushing strength shows the optimum value at 0.5mm cell wall thickness. According to simple compression failure mechanism as depict in equation (1), the critical buckling load effected by thickness of cell wall. Where, by increasing the thickness of cell wall, critical buckling load increases and lead to larger energy absorption.

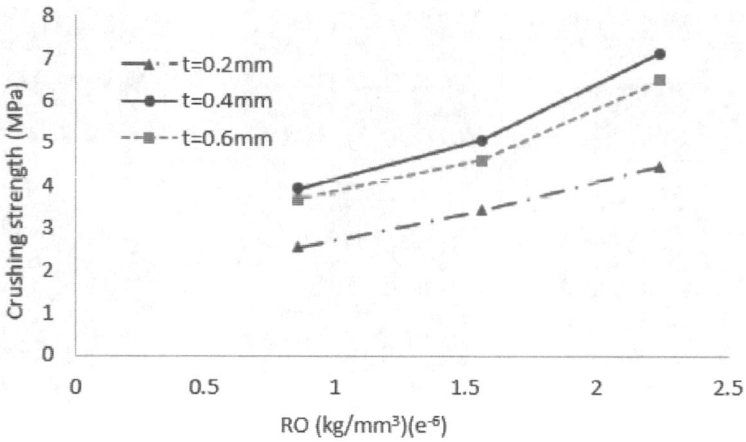

Figure 9. Force -mass density curve for different thickness of cell wall

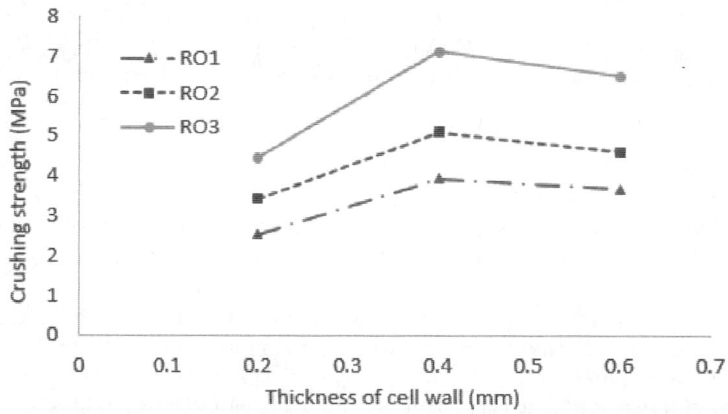

Figure 10. Force -thickness of cell wall curve for different mass density

4. Summary

In this work, the influence of cell size, mass density and thickness of cell wall of Kraft paper honeycomb was investigated numerically. It is shown that mass density of Kraft paper honeycomb affected the crushing force of the honeycomb. Kraft paper honeycomb with higher mass density had higher crushing force where the simulation results shows the good agreement with the experimental result. Both cell size and cell wall thickness also influenced the crushing strength of the honeycomb. 10mm cell size with 0.4 mm cell wall thickness were recommended in increasing the crushing strength of the Kraft paper honeycomb.

References

[1] Chen Z and Yan N 2012 Composites : Part B Investigation of elastic moduli of Kraft paper honeycomb core sandwich panels **43** 2107–14
[2] Wang D 2009 Impact behavior and energy absorption of paper honeycomb sandwich panels *Int. J. Impact Eng.* **36** 110–4
[3] Chen Z, Yan N, Sam-Brew S, Smith G and Deng J 2014 Investigation of mechanical properties

of sandwich panels made of paper honeycomb core and wood composite skins by experimental testing and finite element (FE) modelling methods *Eur. J. Wood Wood Prod.* **72** 311–9

[4] Wang Z, Tian H, Lu Z and Zhou W 2014 High-speed axial impact of aluminum honeycomb – Experiments and simulations *Compos. Part B Eng.* **56** 1–8

[5] Petrone G, Rao S, De Rosa S, Mace B R, Franco F and Bhattacharyya D 2013 Behaviour of fibre-reinforced honeycomb core under low velocity impact loading *Compos. Struct.* **100** 356–62

[6] Castanié B, Bouvet C, Aminanda Y, Barrau J-J and Thevenet P 2008 Modelling of low-energy/low-velocity impact on Nomex honeycomb sandwich structures with metallic skins *Int. J. Impact Eng.* **35** 620–34

[7] Aminanda Y, Castanié B, Barrau J-J and Thevenet P 2009 Experimental and numerical study of compression after impact of sandwich structures with metallic skins *Compos. Sci. Technol.* **69** 50–9

[8] Aminanda Y, Barrau J J and Thevenet P 2011 DISCRETE MODELLING OF THE CRUSHING OF NOMEX

[9] Aktay L, Johnson A F and Kröplin B-H 2008 Numerical modelling of honeycomb core crush behaviour *Eng. Fract. Mech.* **75** 2616–30

[10] Zhang J and Ashby M F 1992 The out-of-plane properties of honeycombs *Int. J. Mech. Sci.* **34** 475–89

[11] Anon *Dynamic Failure of Composite and Sandwich Structures - Google Books*

[12] Xu S, Beynon J H, Ruan D and Lu G 2012 Experimental study of the out-of-plane dynamic compression of hexagonal honeycombs *Compos. Struct.* **94** 2326–36

IOP Conf. Series: Journal of Physics: Conf. Series **914** (2017) 012005 doi:10.1088/1742-6596/914/1/012005

Ground proximity effect on the flow over NACA 4412 multi-element airfoil in clean configuration

Aslam Abdullah, Muhammad Nasreen Yazi, Mohammad Fahmi Abdul Ghafir, Sofian Mohd, Mohammad Zulafif Rahim

Department of Aeronautical Engineering, Faculty of Mechanical and Manufacturing Engineering, Universiti Tun Hussein Onn Malaysia, Johor

Corresponding author: aslam@uthm.edu.my

Abstract. The paper presents the influence of ground distance on aerodynamic characteristics of the flow over NACA 4412 multi-element airfoil in clean configuration. The angle of deflection was kept constant throughout the computations. Finite volume method was used for the grid generation and the numerical calculation. Further investigation suggests the best range of angle of attack where the aerodynamic performance of the airfoil increases in the presence of ground. Furthermore, the condition of flow separation jeopardizing the airfoil performance appeared. It was found that at certain range of angle of attack, the cushion of high pressure at the pressure side of the airfoil is insufficient to compensate the influence of the separation. This work shed insight on the important parameters that need to be taken into account in the operational of an airplane.

1. Introduction

The phenomenon of ground effect becomes apparent in the proximity of the runway, and is beneficial in the control of fixed wing aircrafts during take-off and landing. When an aircraft is flown at approximately one wing span above the runway, the interaction between the airflow around the airfoil and the ground surface modifies the fluid velocity's vertical component and changes its normal pattern which results in change of the lift-to-drag ratio L/D [1,2,3]. In particular, the corresponding rotation of the resultant force vector gives change to the component of lift and drag forces [4,5]. A well trained pilot takes the phenomena into account in order to enhance the safety prior to airborne and touchdown [6].

This paper evaluates the influence of ground on the multi-element airfoil during landing. The problem is analysed with regard to the aerodynamic characteristics of the flow over NACA 4412 airfoils. We take STOL CH750 light sport aircraft (LSA) as a reference for the practical selection of study parameters. In fact, the NACA 4412 airfoils form the constant cross sections of the aircraft wing [7,8]. The main airfoil is equipped with a Junkers flap with constant angle of deflection (AoD). The Junkers flap is slotted below the trailing edge of the main airfoil. The design allows the airflow to pass between the main airfoil and the flap, even when the latter is retracted. Because of the increase in airflow, a significant increase in lift is added to the airfoils, even at low speeds. The flap is employed nowadays in many modern ultralights [9].

The evaluation focuses on various angles of attack α, namely 3°, 7°, 11°, 14°, which is crucial in achieving touchdown point accuracy and preventing stalling of aircraft especially when it is close to the runway during flare; the recovery from such problem would require emergency procedures

[10].The distance between the airfoil and the ground is also taken into account. A steady-state RANS investigation was performed using the commercial CFD code ANSYS. The relationship between α and the nature of ground effect at constant landing speed was observed. The effect was also considered for different distances between the airfoil and the ground.

2. Governing equations and modelling assumptions

A finite volume numerical method based on solving RANS equations describing the case under consideration was used for calculating the flow around the airfoil. In the case of incompressible, steady, 2-D viscous flow, the equations are given by

$$\rho\left(u_1\frac{\partial u_1}{\partial x_1}+u_2\frac{\partial u_1}{\partial x_2}\right)=F_1-\frac{\partial P}{\partial x_1}+\mu\left(\frac{\partial^2 u_1}{\partial x_1^2}+\frac{\partial^2 u_1}{\partial x_2^2}\right)-\rho\left(\frac{\partial\overline{u_1'u_1'}}{\partial x_1}+\frac{\partial\overline{u_1'u_2'}}{\partial x_2}\right), \qquad (3.1)$$

and

$$\rho\left(u_1\frac{\partial u_2}{\partial x_1}+u_2\frac{\partial u_2}{\partial x_2}\right)=F_2-\frac{\partial P}{\partial x_2}+\mu\left(\frac{\partial^2 u_2}{\partial x_1^2}+\frac{\partial^2 u_2}{\partial x_2^2}\right)-\rho\left(\frac{\partial\overline{u_1'u_2'}}{\partial x_1}+\frac{\partial\overline{u_2'u_2'}}{\partial x_2}\right), \qquad (3.2)$$

where u_1, u_2 are time-averaged velocity vectors, P is the pressure, μ is the dynamic viscosity, u_1', u_2' are fluctuating velocity vectors, and F_1, F_2 are body forces. It is necessary that the continuity equation,

$$\frac{\partial u_1}{\partial x_1}+\frac{\partial u_2}{\partial x_2}=0, \qquad (3.3)$$

is satisfied by (3.1) and (3.2).

Since the Reynolds averaging is involved in the derivation of (3.1) and (3.2), then a set of new unknowns called the Reynolds stresses need to be modeled to ensure that the equations can be closed. In particular, standard *k-ε* turbulence model which gives a general description of turbulence by means of two transport equations was applied. In addition, standard wall functions were used for near-wall treatment.

The method was applied with second-order upwind discretization schemes. The algorithm called SIMPLE was employed for the velocity-pressure coupling. The airfoil velocity and the Reynolds number are 25 m/s and 1.7 x 10^6, respectively. The values are within the range of CH750 landing and take-off conditions. The Mach number of flow is equal to 0.07, so then the compressibility effects could be neglected since they should be taken into account for a Mach number greater than 0.3. The investigated α and heights of airfoil flights represented by the relative ratio of the height-to-airfoil chord length (h/c) were introduced.

3. Geometry, grid and computational domain

The flow was studied for the NACA 4412 for both the main airfoil and the Junkers flap. The airfoil has a flat shape of the pressing (bottom) side. The geometry of the airfoil is given in Fig. 1.

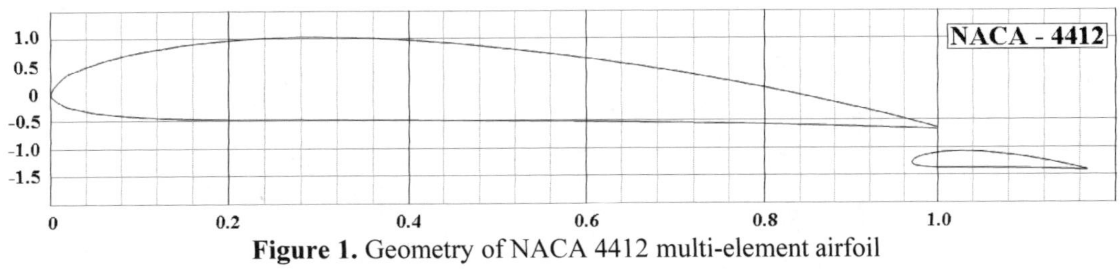

Figure 1. Geometry of NACA 4412 multi-element airfoil

For discretization of the computational domain, an unstructured type of grid with quad elements was selected. The airfoil model moving above the ground and the grid used for calculations are given in Fig. 2.

Inlet and outlet boundary conditions were specified on the outer sides of computational domain with necessary turbulence and flow parameters. The boundary condition on the airfoil is a no-slip condition with zero relative speed enforced. This should produce better representation of the reversing flow assumption.

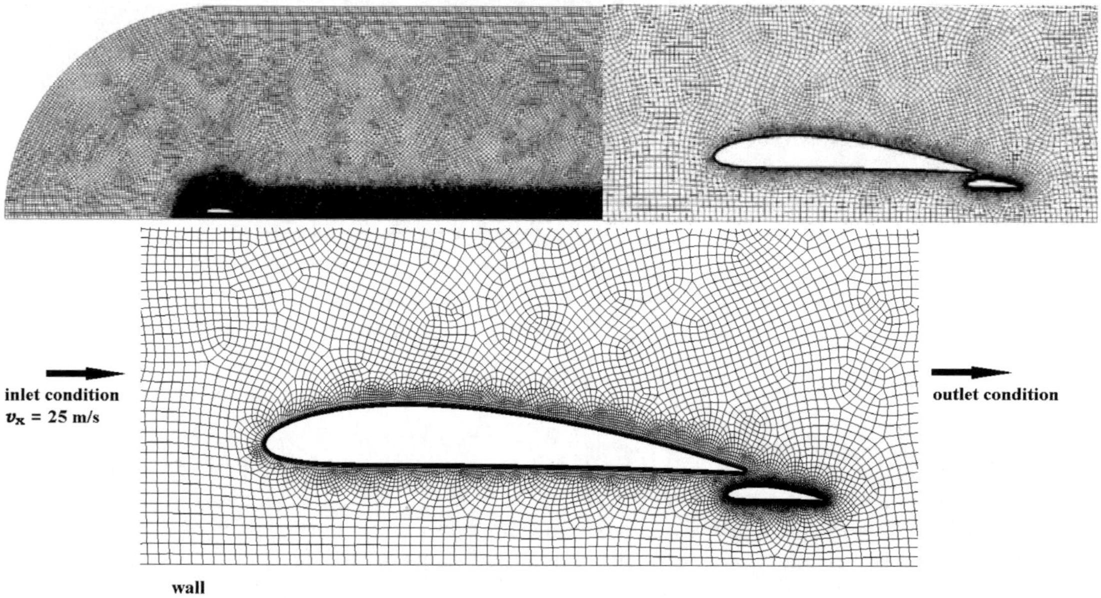

Figure 2. Grid applied for multi-element airfoil moving close to the ground surface

4. Results of calculations

The preliminary result of the velocity field around the NACA 4412 airfoil at zero angle of attack is shown in Fig. 3. Note that the airfoil is in ground proximity. It is obvious that the velocity is relatively low in particular at the pressure side when the airfoil approaches the ground, and thus gives direct influence to *L/D* a moment after the roundout. This indicates that the airfoil floats on a cushion of high-pressure air region above the ground surface. Furthermore, the stagnation point shifts to the lower side of the airfoil due to the ground effect.

(a) (b) (c)

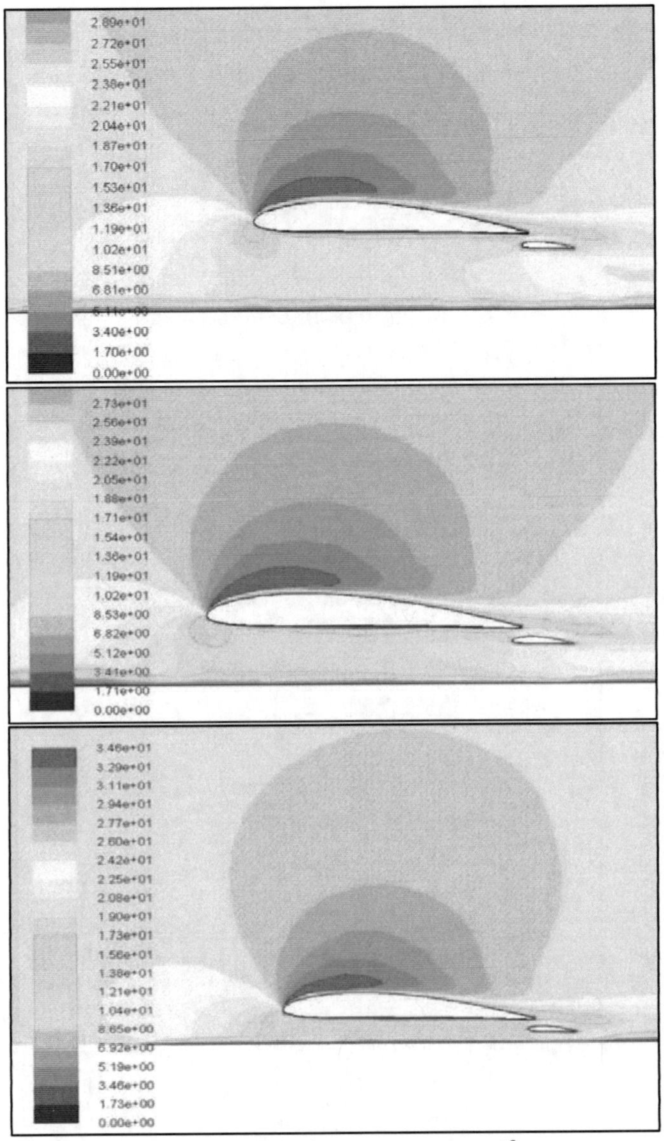

Figure 3. Velocity contour at $\alpha = 0°$
(a) $h/c = 0.25$ (b) $h/c = 0.15$ (c) $h/c = 0.05$

The ground effect on the velocity field around the airfoil at 5% of the chord length from the ground (i.e. $h/c = 0.05$) is shown in Fig. 4 and Fig. 5. Clearly the velocity drop is acute at the pressure side when α changes from 3° to 7°, 11°, and consequently 14°, giving therefore great influence to L/D during the roundout and flare. Apart from the shift of the stagnation point to the lower airfoil side with the increase of α, the onset of flow separation occurs near the trailing edge of the main airfoil at $\alpha = 7°$. The flow separation point then shifts further upstream and finally reaches the leading edge at $\alpha = 11°$ and 14°, respectively. This reveals that the basic aerodynamic characteristics deviate from those for unbounded airflow. Apparently the ground effect advances the flow separation. Note that for unbounded airflow the separation begins at $\alpha = 17\,^{\circ}$.

(a)
(b)
(c)

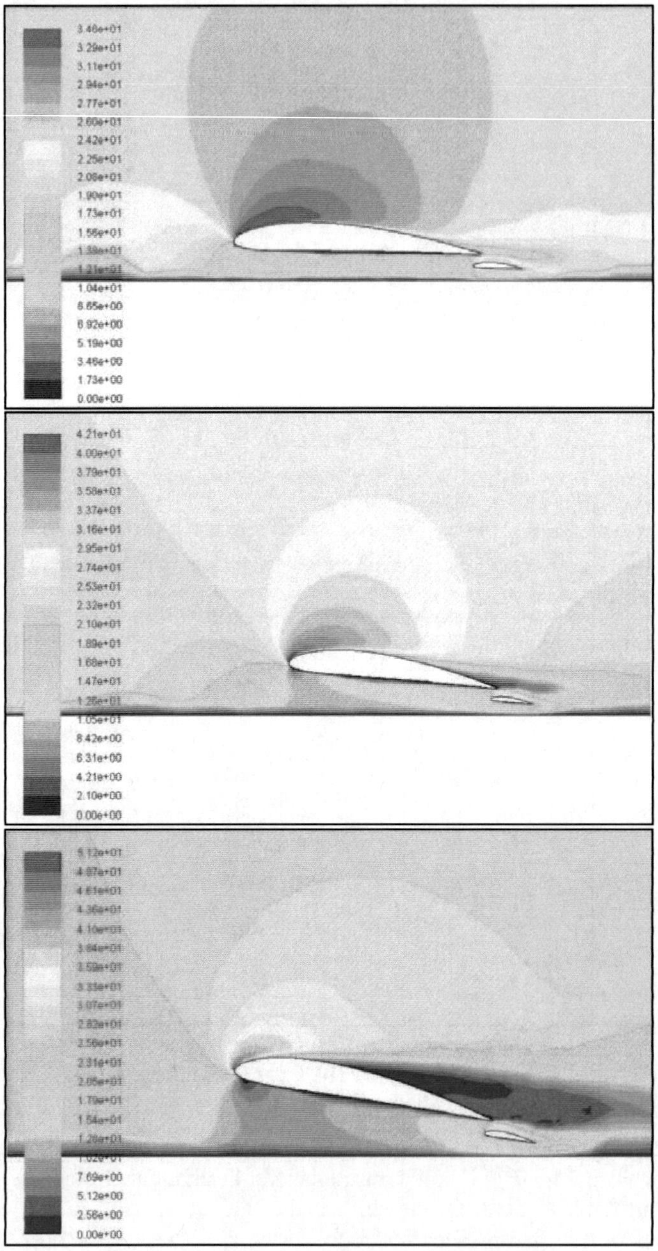

Figure 4. Velocity contour at $h/c = 0.05$
(a) $\alpha = 3^{\circ}$ (b) $\alpha = 7^{\circ}$ (c) $\alpha = 11^{\circ}$

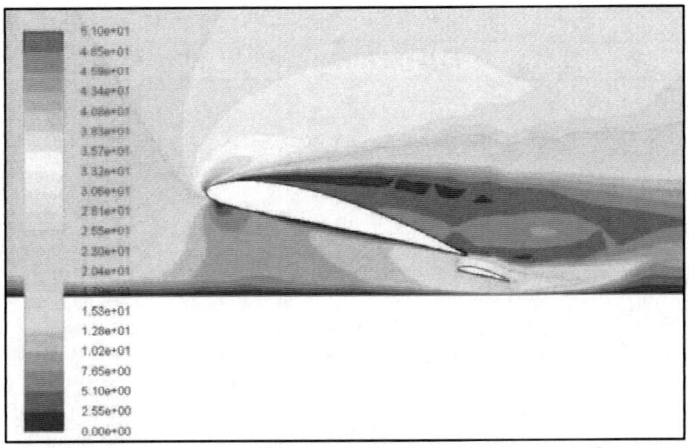

Figure 5. Velocity contour at $h/c = 0.05$, $\alpha = 14^{\circ}$

The early flow separation causes the degradation of the lift-to-drag ratio L/D whose percentage of reduction in the presence of the ground is shown in table 1 for $h/c = 0.05$. The next question is whether the overall aircraft performance can be achieved for $0.05 \leq h/c \leq 0.25$ at a given α.

Table 1. Influence of α on L/D at $h/c = 0.05$

α	*L/D*	Δ *L/D* (%)
3° (ref.)	60.58	--
7°	29.53	-51
11°	11.83	-80
14°	6.00	-90

The computation results based on different settings of the distance between the airfoil and the ground are tabulated in table 2 and table 3. Lift and drag are the forces which act on the geometry of main airfoil and flap as a whole. The tabulated data reveals that at $\alpha = (7^{\circ}, 11^{\circ}, 14^{\circ})$, the average lift-to-drag ratio increment $(\Delta L/D)_{av}$ with respect to h/c is given by a negative value which indicates that the aircraft would sink to the runway relatively faster before the touchdown. Such low overall performance is due to the flow separation which compromises the main airfoil lift (see Fig. 4 and Fig. 5). The airfoil however best performs at $\alpha = 3^{\circ}$ where $(\Delta L/D)_{av}$ is approximately 3%.

Table 2. Influence of h/c on L/D at $\alpha = 3^{\circ}$ and $\alpha = 7^{\circ}$

α	*h/c*	*L/D*	Δ *L/D* (%)	α	*h/c*	*L/D*	Δ *L/D* (%)
	0.25(ref.)	57.14	--		0.25(ref.)	36.17	--
	0.20	59.27	3.73		0.20	33.12	-8.42
3°	0.15	56.69	-0.78	7°	0.15	32.45	-10.27
	0.10	58.66	2.66		0.10	32.46	-10.26
	0.05	60.58	6.03		0.05	29.53	-18.35

Table 3. Influence of *h/c* on *L/D* at $\alpha = 11^\circ$ and $\alpha = 14^\circ$

α	*h/c*	*L/D*	Δ *L/D* (%)	α	*h/c*	*L/D*	Δ *L/D* (%)
	0.25(ref.)	16.19	--		0.25(ref.)	8.05	--
	0.20	15.13	-6.51		0.20	6.65	-17.45
11°	0.15	12.96	-19.93	14°	0.15	6.85	-14.95
	0.10	12.39	-23.47		0.10	6.83	-15.14
	0.05	11.83	-26.90		0.05	6.00	-25.43

5. Final remarks

The study focused on the ground proximity effect on the flow over NACA 4412 multi-element airfoil. In the absence of flow separation (i.e. at $\alpha = 3^\circ$), the ground influence prevails, causing higher pressure at the pressure side of the airfoil, and thus slowing down the touch down of the airplane. Higher α (i.e. $7^\circ \leq \alpha \leq 14^\circ$) will result in sinking of the aircraft at relatively high rate; the cushion of high pressure at the pressure side of the airfoil is insufficient to neutralize the effect of flow separation. It is therefore important to take α and *h/c* into consideration in the operational of an airplane during landing. Further relevant investigation would involve both $\alpha < 3^\circ$ and $3^\circ < \alpha < 7^\circ$.

Acknowledgement

The author would like to thank both Universiti Tun Hussein Onn Malaysia (UTHM) and Ministry of Higher Education of Malaysia for providing space and computing facilities for the research.

References

[1] M Mugunthan, D Pravalika, KMK Babu and SS Kumar 2015 Flow field investigation of flat bottom aerofoil under ground effect *OSR-JMCE* **12** 83-88

[2] MM Tofa, A Maimun, YM Ahmed, SJA Priyanto and Rahimuddin 2014 Experimental Investigation of a wing-in-ground effect craft *The Scientific World Journal* **14** 1-7

[3] MAU Amir, A Maimun, S Mat and MR Saad 2016 Computational analysis of aerodynamic characteristics for wing in ground effect craft in lateral stability *IOP Conf. Series: Materials Science and Engineering* **152** 1-9

[4] SC Luo and YS Chen 2012 Ground effect on flow past a wing with a NACA0015 cross-section *Experimental Thermal and Fluid Science* **40** 18-28

[5] T Abramowski 2007 Numerical investigation of airfoil in ground proximity *Journal of Theoretical and Applied Mechanics* **45** 425-436

[6] KV Rozhdestvensky 2006 Wing-in-ground effect vehicles *Progress in Aerospace Sciences* **42** 211–283

[7] AMD 2009 *STOL 750LS-SLSA with O-200* Airplane and systems description *Engine Pilot Operating Handbook* **2** (Eastman, Georgia) 0-7

[8] Zenair 2017 *Light Sport Utility Aircraft STOL CH750 Specifications and Performance* (Missouri, USA) 1-2

[9] AD Monte, MR Castelli and E Benini 2012 A retrospective of high-lift device technology *International Journal of Mechanical, Aerospace, Industrial, Mechatronic and Manufacturing Engineering* **6** 2561-2566

[10] AERO'Connor Aviation – Flight Training 2014 Emergency procedures *Private Pilot Syllabus* (Frederick, USA) Lesson 5 p 5

IOP Conf. Series: Journal of Physics: Conf. Series **914** (2017) 012006 doi:10.1088/1742-6596/914/1/012006

Flow Simulation of Modified Duct System Wind Turbines Installed on Vehicle

N Rosly[1], S Mohd[2], M F Zulkafli[3], M F Abdul Ghafir[4], S S Shamsuddin[5], W N A Wan Muhammad[6]

[1,2,5,6]Aircraft System and Design Research (ASDR) Group, Faculty of Mechanical and Manufacturing Engineering, University Tun Hussein Onn Malaysia
[3,4]Aerodynamics and Propulsion Research (APR) Group, Faculty of Mechanical and Manufacturing Engineering, University Tun Hussein Onn Malaysia

Corresponding author: [1]nurhayati@uthm.edu.my

Abstract. This study investigates the characteristics of airflow with a flow guide installed and output power generated by wind turbine system being installed on a pickup truck. The wind turbine models were modelled by using SolidWorks 2015 software. In order to investigate the characteristic of air flow inside the wind turbine system, a computer simulation (by using ANSYS Fluent software) is used. There were few models being designed and simulated, one without the rotor installed and another two with rotor installed in the wind turbine system. Three velocities being used for the simulation which are 16.7 m/s (60 km/h), 25 m/s (90 km/h) and 33.33 m/s (120 km/h). The study proved that the flow guide did give an impact to the output power produced by the wind turbine system. The predicted result from this study is the velocity of the air inside the ducting system of the present model is better that reference model. Besides, the flow guide implemented in the ducting system gives a big impact on the characteristics of the air flow.

1. Introduction

There have been many improvements introduced in powered-electric vehicles by using electricity such as hybrid cars and battery-operated electric vehicles. The term hybrid commonly refers to hybrid electric vehicles (HEVs), which combine an internal combustion engine and electric motors to generate extra power [1]. For this study, the wind turbine system has been designed in a way that it can produce external power source that will be applied to the pickup truck. The system mounted on the vehicle is going to charge the batteries of the HEV when it's moving [2].

The capability of wind turbine system depends on few constraints to generate electrical power. The first consideration that needs to be considered is on the appropriate design of wind turbine system that consists of the turbine rotor and ducting system. The rotor chosen to be installed in the turbine system is the Savonius type of turbine rotor. Savonius wind rotor is one of the vertical axis wind turbines that has simple structure, low operating speeds and an ability to capture wind from any directions. The rotor is made up of two semi-cylinders with S-type cross sectional blades in which wind power acts perpendicular on the blades and converts into rotating motion of the central shaft [3]. A studied ran by Alexander and Holownia [4] founds that the performance of the Savonius rotor depends on blades aspect ratio, shielding and end plates. They concluded the increment of aspect ratio lead to better improvement in rotor performance. Supported by other studies [5], they concluded that the rotor blade with shielding and end plates gives better efficiency than that without shielding and end plate. Furthermore, a study covered by Nasef et al. [6] on the aerodynamic performance of Savonius rotor in static and dynamic conditions shows that a suitable model for simulating the flow pattern around the Savonius rotor for both stationary and rotating cases is SST k–ω turbulence model.

This work focused in numerical design and simulation. The design process of the turbine rotor and ducting system is done using a design software, SolidWorks, while the flow field of the air flow inside the turbine system is evaluated by using computer simulation software, ANSYS Fluent. In the simulation, the air is considered as a viscous fluid and is modelled with SST k- ω turbulence model. The purpose of this research is to propose an optimum design of wind turbine system for pickup vehicle.

2. Modeling of Wind Turbine Ducting System

Based on the previous inventions by Mohd et al. [7], this turbine model is chosen as a reference model in order to create a new turbine system model. Based on this turbine model, the appropriate size that is going to be used for the new turbine design is set as 1000 mm length, 1000 mm width and 150 mm height [7]. In the turbine system, there is one part of the system, namely flow guider, was installed. The purpose of the flow guider is to increase the velocity of air and eliminate the moment occurring in an opposite direction for the rotation on the blade of the rotor without making any changes in the basic structure of the conventional Savonius wind rotor. The location of the flow guider has been set for this study is shown in Figure 1.

The second thing to consider in this study is to find the appropriate design and configuration of the rotor. Savonius type is used with respect to the air flow at the roof of the pickup truck body and axis of the wind turbine rotor. The whole body size is set to be 130 mm height and 1000 mm width. Previously, Mohd et al. [7] mentioned that the optimum overlap ratio (*e*/*d*) value has been found as 0.18. This ratio value of Savonius rotor is implemented in this study.

The present design is consists of one Savonius rotor with triple blades configuration. The rotor will be located after the flow guider near the outlet section of the ducting system. A thickness of the rotor has been set as 2 mm with a diameter of d = 528 mm, a rotor height is set as 130 mm and the gap distance of the turbine blade from centre of rotation, e is 10 mm. Figure 2 shows the dimension of the Savonius wind rotor located at the inlet of the wind turbine.

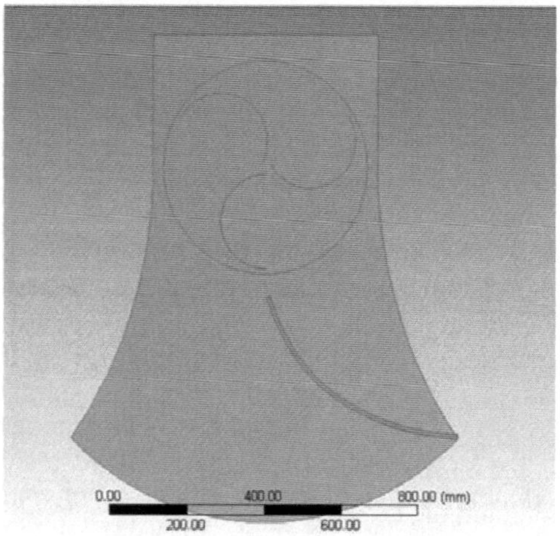

Figure 1. Location of the flow guider in the turbine model.

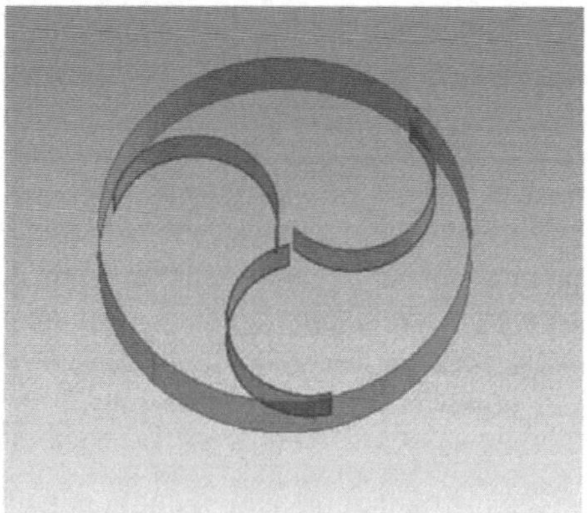

Figure 2. Savonius wind rotor chosen

3. Aerodynamic Performance of the Ducting System

The performance of ducting system between present model and reference model [1] is compared and evaluated in order to validate the present design. The evaluation is based on the air flow inside the turbine's system. Figure 3 shows the characteristic of the air flow inside of the ducting system for both models. During the simulation, the air velocity flows through the inlet is set to be 25 m/s which assume that the pickup truck is moving at 90 km/h. Based on the result obtained from the simulation, each design shows different location for the fastest air flow which proven that the velocity of the air flow is affected by the ducting design.

Based on the result obtained for model shown in Figure 3(b), the highest velocity of air inside the ducting system is at 135.6 m/s. Compared to the previous invention model invented by Nik Samsul Bahari [1], the maximum air velocity flows inside the ducting system is at 82 m/s. Therefore, it can be concluded that the current model is much better compared to the previous invention. In addition, based on the simulation results, it shows that the velocity of airflows inside the present model before it hits the turbine blade is approximately between 99.6 m/s -105 m/s for present model and velocity for Nik Samsul model [1] is only approximately between 78.5 m/s – 85 m/s. From the result, clearly, we can see that the velocity of the present model at neck region is higher than the Nik Samsul model. From the figure shown also, the air velocity inside turbine duct with flow guide before it hits the turbine's rotor was increased four times (25 m/s at the inlet and 99.6 m/s at neck region) with additional flow guide at the inlet of turbine duct. This would justify that the design of the ducting system with application of flow guide increases the velocity of airflows inside the model.

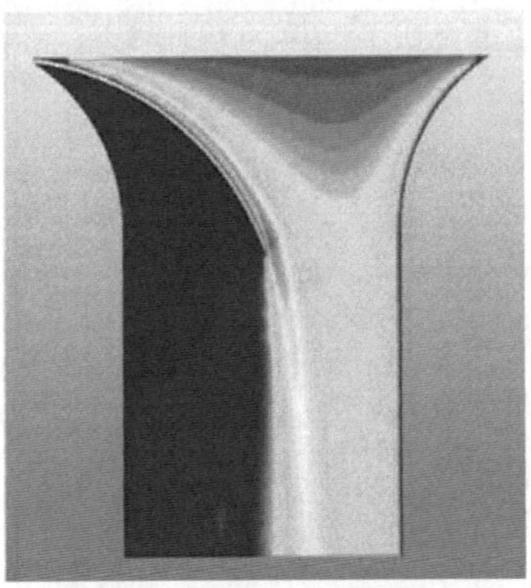

(a) Nik Samsul Model [1]

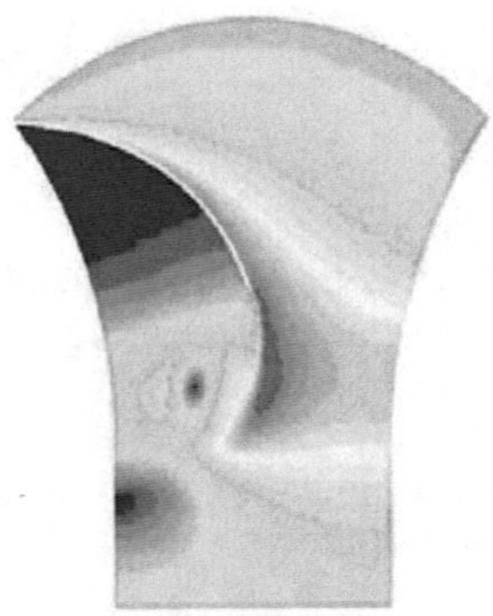

(b) Present model

Figure 3. Velocity characteristic of air flows inside a wind turbine model

4. Analysis of Ducting System With Installed Rotor

Dynamic simulation has been performed to investigate the rotation of the rotor driven by the air that flows into the system. This is important to determine the rotation speed of the rotor which will be used to calculate the power output generate by the system. Figure 4 shows the velocity contour of the ducting system with the installed rotor. The inlet of the air velocity was set at 25 m/s which is equivalent to 90 km/h. Based on the simulation result, the rotor is rotating in clockwise direction at an average angular speed of 405.8 rad/s. The speed of the rotation is calculated from the average linear velocity of the rotor's tip, which is 102.15 m/s.

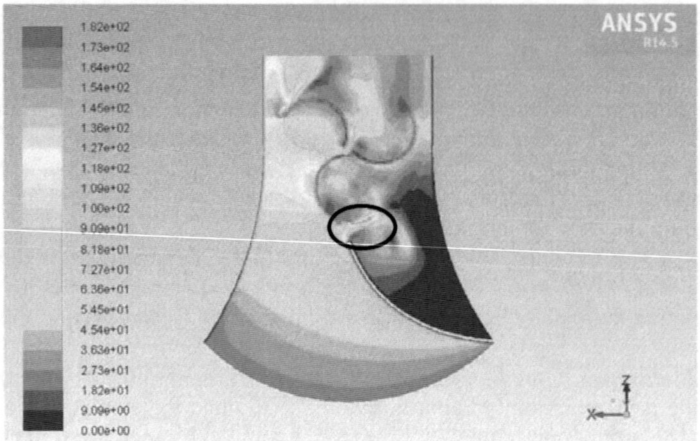

Figure 4. Velocity contour of the air flow inside the ducting system with motor

In addition, the circular section shows that the air does flow in between the gap of the rotor and the flow guide. As the consequences, the air flow will create a negative torque which will cause in the reverse movement direction of the rotor. The rotor will experience slow rotation movement because of the negative torque produced. Therefore, a new model has been produced to counter the problem. The new model was made based on the modification of the flow guider.

5. Analysis of Modified Flow Guide

The modification is made on the flow guide to the ducting system. The flow guide is extended to a new length, i.e. made the gap distance smaller. The reason is to reduce the amount of air flow that moves in between the gap of the flow guide and rotor, which will create negative torque. Figure 5 shows the velocity contour of the modified ducting system installed with a rotor. From the circular section shown in the figure, we can clearly see that less portion of air moves into the gap between the rotor blade and flow guide compared to the previous model. Therefore, the rotation of the rotor will be less affected because of less negative torque created.

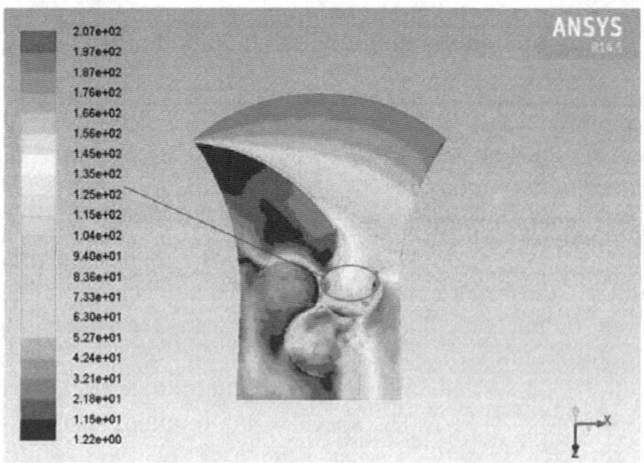

Figure 5. Velocity contour of air characteristic in wind turbine

IOP Conf. Series: Journal of Physics: Conf. Series **914** (2017) 012006 doi:10.1088/1742-6596/914/1/012006

In order to investigate the effect of the flow guide plate and to identify the speed of the air flow inside the ducting system with turbine rotor, simulation is perform on the Savonius wind turbine rotor system. For this simulation, the inlet velocity has been set at 16.7 m/s, 25 m/s and 33.3 m/s. The result (Figure 6) shows that the velocity of air at the neck of duct before it hits the turbine rotor increase about four times for each speed set at the inlet. The figure also shows that air velocity inside the duct system before it hits the turbine blades decreased around 20-25% because the blade acted as a barrier that prevents air flow to enable it to rotate and generate power for wind turbine system.

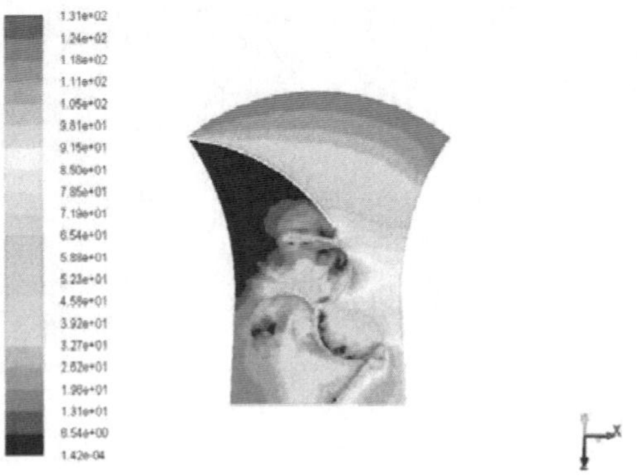

(a) Inlet velocity = 16.7 m/s

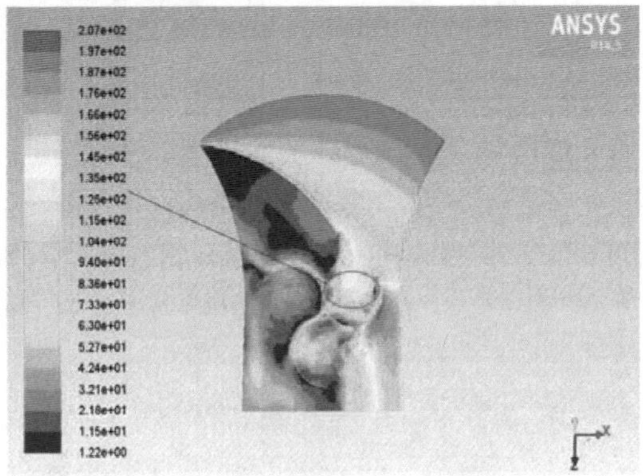

(b) Inlet velocity = 25 m/s.

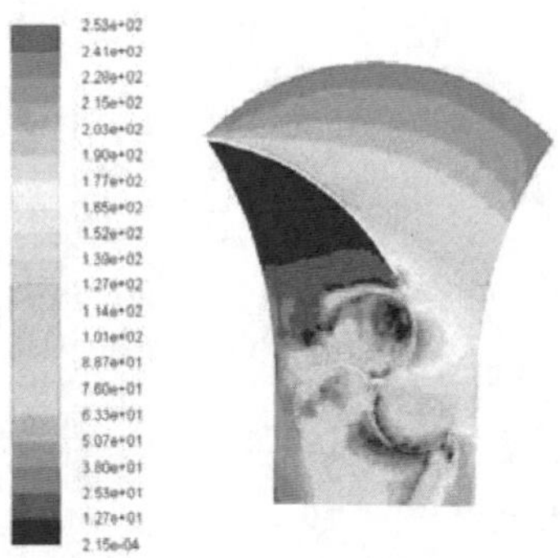

(c) Inlet velocity = 33.3 m/s

Figure 6 . Velocity contours of wind turbine with modified flow guide

Based on the figures shown, the rotation speed of the blade can be estimated based on the maximum air flows that hit the turbine rotor. It is important to evaluate the rotation speed or angular velocity, Ω of the rotor in order to calculate the power generated by the wind turbine system. The calculation of the angular velocity is computed based on the average linear velocity of the tip rotor, V_{tip} obtained from the simulation and the diameter of the rotor, d as shown in equation (1). The calculated value of the angular velocity will be used to calculate power generate by the system.

$$\Omega = 2\frac{V_{tip}}{d} \tag{1}$$

Table 1 shows the summarization of the angular velocity of the rotor calculated based on three different inlet velocity of the system. The relation between the inlet velocity and the calculated angular velocity is depicted in the graph shown in figure 7. Clearly the angular velocity does not increase linearly and slowing down even the inlet velocity is increased. The reason is the flow rate of air that moves through the gap increased and will create significant negative rotation as the inlet velocity increased. Therefore, it can be deduced that the proposed wind turbine design will become less efficient during high speed vehicle movement since the high inlet velocity will produces more air resistance to the rotation of the rotor.

Table 1. Value of calculated angular velocity

Inlet velocity (m/s)	Average tip velocity, V_{tip} (m/s)	Angular Velocity, Ω (rad/s)
16.7	75.2	284.8
25	130	492.1
33.3	145.5	550.7

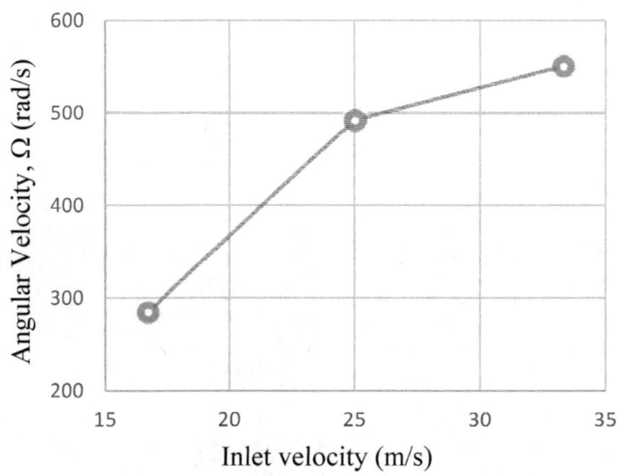

Figure 7. Relation between inlet velocity and angular speed of the rotor

6. Conclusions

The capability of the wind turbine as a machine that converts kinetic energy into electrical energy to produce electrical power could be used in the vehicle. The characteristics of the air flow inside the wind turbine system and output power generated by the system were evaluated.

The air velocity at the inlet of the system was set at 16.7 m/s, 25 m/s and 33.3 m/s which are equivalent to standard vehicle cruising speed of 60 km/h, 90 km/h and 120 km/h. Based on the simulation result, the air velocity was slightly increased at the ducting neck region. The air velocity was significantly increased such as four times of the inlet velocity with the implementation of the flow guide. Therefore, there are few things that can be summarized as the conclusion based on the result obtained:

(a) The air velocity in the ducting system for the present model is much better than reference model [1] as the average highest for present model: (99.6 – 105) m/s compared to reference model: (78.5 -85) m/s.
(b) The air velocity at neck region did increase four times of the inlet velocity: (25-99.6) m/s.
(c) The designation flow guide does give a big impact towards the rotation of the rotor since it can be designed to prevent the negative torque that can encounter the rotor rotation.

IOP Conf. Series: Journal of Physics: Conf. Series **914** (2017) 012006 doi:10.1088/1742-6596/914/1/012006

(d) The rotation speed of the rotor is higher for the modified model compared to the initial model with the rotor.
(e) The modification made on the flow guide did increase the velocity of the air flow inside the ducting system but will become less effective during high speed vehicle movement.

Acknowledgment
This project was supported by Exploratory Research Grant Scheme, ERGS under Grant Vot. No: E034 in order to conduct the research.

References

[1] Mohd S, Mohamad Zainu N S B, Rosly N and Abdullah A 2016 Aerodynamic characteristics evaluation of wind turbine dusting system performance for pick up truck *ARPN J. of Engineering and Applied Sciences* **11** pp 11165-69
[2] Mohd S, Rosly N, Zulkafli M F, Mohamad Zainu N S B, Abdullah A, Shamsudin S S and Wahab A A 2016 Proposed design of wind turbine systems on a pickup truck *ARPN J. of Engineering and Applied Sciences* **11** pp 11160-64
[3] Manwell J F, McGowan J G and Rogers A L 2010 Wind energy explained: theory, design and application second edition *John Wiley & Sons*
[4] Alexander A J and Holownia B P 1978 Wind tunnel tests on a savonius rotor *J. of Industrial Aerodynamics* **3** pp 343-51
[5] Mahmoud N H, El-Haroun A A, Wahba E and Nasef M H 2012 An experimental study on improvement of Savonius rotor performance *Alexandria Engineering Journal* **51** pp 19-25
[6] Nasef M H, El-Askary W A, Abd El-Hamid A A and Gad H E 2013 Evaluation of savonius rotor performance: static and dynamic studies *J. of Wind Engineering and Industrial Aerodynamics* **123** pp 1-11
[7] Mohd S, Rosly N, Rexca A J, Shamsudin S S and Abdullah A 2014 An evaluation of drag coefficient of wind turbine system installed on moving car *Applied Mechanics and Materials* **660** pp 689-93

Experiment on the Effects of Storage Duration of Biodiesel produced from Crude Palm Oil, Waste Cooking oil and Jatropha

Nadiarulah Nanihar[1,a], Amir Khalid[1,2,b], Norrizal Mustaffa[1], Norrizam Jaat[2], Azwan Sapit[2], Azahari Razali[2], Norshuhaila Mohamed Sunar[3]

[1]Automotive and Combustion Synergies Technology Group, Advanced Technology Centre (ATC), Faculty of Engineering Technology, Universiti Tun Hussein Onn Malaysia, Parit Raja, Batu Pahat 86400 Johor, Malaysia.
[2]Centre for Energy and Industrial Environment Studies (CEIES), Universiti Tun Hussein Onn Malaysia, Parit Raja, Batu Pahat, 86400 Johor, Malaysia.
[3]Faculty of Engineering Technology, Universiti Tun Hussein Onn Malaysia, 86400 Parit Raja, Batu Pahat, Johor, Malaysia.

*Corresponding E-mail: nmuhammadnadiarulah@gmail.com[a], amirk@uthm.edu.com[b]

Abstract. Biodiesel based on vegetable oil is an alternative that had various advantage in term of sustainability and environmental attractive compare to others conventional diesel. Biodiesel is product of any fat or oil that derived from any organic sources through a refinery process called transesterification process. This research investigates the effects of storage duration and variant ambient condition on the biodiesel properties and characteristics. In this study, there are three types of blending which is 5vol% blends (5vol% plant oil 95vol% diesel), 10vol% blending (10vol% plant oil and 90vol% diesel) and 15vol% blending (15vol% plant oil and 85vol% diesel) each called CPO5 (crude palm oil 5vol%), CPO10 (crude palm oil 10vol%),CPO15 (crude palm oil 15vol%), JO5 (jatropha oil 5vol%), JO10 (jatropha oil 10vol%),and JO15 (jatropha oil 15vol%) respectively. Biodiesel samples were stored at indoor condition and outdoor condition for a 3 months period. The fuel properties such as acid value, viscosity, density, water content and flash point are observed with the laboratory instrument. Flash point value and water content increased under both of indoor and outdoor condition and a steady data for viscosity and density. However, acid value at indoor condition nearly constant but increased dramatically for outdoor condition over the time.

1. Introduction

Alternative fuels, energy efficiency, energy conservation, environmental protection and management have become important due to depletion of fossil fuel and environmental degradation for last two decade. The alternative fuels or so called biodiesels, can be better used for diesel engine as compared to petrol engines because of the construction of the diesel engine is very robust and can work under high compression ratio along with a significant amount of excess air [1-5]. It can be used as a pure fuel or blended with petroleum in any percentage but the standard storage and handling procedures used for biodiesel are the main issue due to the biodiesel fuel specifications. The specification for this type of file must strictly refer to (ASTM D6751 or EN 14214) in order to insure proper performance. The key issue in using vegetable oil-based fuels is oxidation stability, stoichiometric point, bio-fuel composition, antioxidants on the degradation and much oxygen with comparing to diesel gas oil [2-6]. Some research was done by some researcher to study the effect of storage duration, storage temperature, storage material, and effect of antioxidant to biodiesel properties and biodiesel properties and characteristics. Oxidative degradation undergo the complex secondary reactions further oxidize into acids, leading to an increase in acid value cause by the formation of hydrogen peroxide [2]. Increasing rate of acid value depends on storage duration [3–6]. High temperature of storage can cause the increasing value for the result [5]. Bouaid et al has done the research using different storage condition (expose and not expose to sunlight). Sunlight give higher increasing rate of acid value to biodiesel storage [2,7-8]. M. Shahabuddin et al stated that acid value, density and viscosity is increasing throughout the time [3]. A. Obadiah et al. found that biodiesel without antioxidant gave a significant increase in kinematic viscosity after 50 weeks. The sample with antioxidant still maintain under control for kinematic viscosity. Antioxidant gives more stability to biodiesel to maintain its properties [9-10]. Water content in sample would make the kinematic maintain and decrease because the kinematic viscosity is measure of resistance of substance to flow. The comparison of diesel and biodiesel is described as Table 1.

Table 1: The Comparison of Standards For Diesel And Biodiesel Based On American Socicty For Testing And Material (ASTM) [11].

Fuel Property	Diesel ASTM D975	Biodiesel ASTM D6751
Lower Heating value, BTU/gal	129,050	118,170
Kinematic viscosity @ 40°C, cSt	1.3-1.4	4.0-6.0
Specific gravity @ 60°C, g/cm3	0.85	.88
Carbon, wt%	87	77
Hydrogen, wt%	13	12
Oxygen, by dif. wt%	0	11
Sulfur, ppm	500	0
Boiling Point, °C	180 to 340	315 to 350
Flash point, °C	60 to 80	100 to 170
Cloud point, °C	~15 to 5	~3 to 12
Pour point, °C	~35 to ~15	~15 to 10
Cetane number	40-55	48-65
Lubricity (HFRR), μm	300-600	<300

In this study, six ratio of biodiesel blended from crude palm oil (CPO) based were using for this research, CPO5 (crude palm oil 5vol%), CPO10 (crude palm oil 10vol%),CPO15 (crude palm oil 15vol%), JO5 (jatropha oil 5vol%), JO10 (jatropha oil 10vol%),and JO15 (jatropha oil 15vol%) respectively. All of the samples of one liter were stored in transparences glass bottles under two condition; indoor condition (22-24°C) and outdoor condition (28-30°C). Sample were stored for 1960 hours and tested every week. Five types of properties testing were tested for this study; water content, density, kinematic viscosity, flash point and acid value. For these testing, European standard testing procedures were used.

In this study, biodiesel is be produced by using jatropha oil, was involve a number of processes including titration process, transesterification, and investigate the physical properties of biodiesel. First, the titration was run to determine the content of FFA in the JO. If the value is higher FFA content, JO then must go through to reduce the amount of FFA esterification under 2%. If the FFA content of less than 2%, it will involve this transesterification process.

IOP Conf. Series: Journal of Physics: Conf. Series **914** (2017) 012007 doi:10.1088/1742-6596/914/1/012007

Biodiesel is fuels that can be compose from mono-alkyl esters of long chain fatty acids originates from plant or animals that will need to fulfil requirement of ASTM D6751 fuel specification. The advantages of using extraction biodiesel is it can use either co-products or by-products [8]. Biodiesel is very environmental friendly and non-toxic. If compare with petroleum based diesel, biodiesel is the best in term of combustion emissions. It capable to reduce the emissions of carbon monoxide, particulate matter and unburned hydrocarbons [9]. Carbon dioxide produced form biodiesel can recycled by photosynthesis and will minimize the effects of greenhouse effects [10]. Table 2 shows the fuel properties of jatropha oil, jatropha biodiesel, and diesel. Jatropha (Jatropha carcass) is one of such non-edible oils, which has an estimated annual production potential of 200 thousand metric tons in India and it can be grown in waste land [12]. The oil content for jatropha kernel is 63.16% [13] and higher linseed, soybean, and palm kernel which is 33.33%, 18.35% and 44.6%, respectively [14]. Hence jatropha will be the perfect biodiesel because it more economical in term of oil contains. Table below will show the fuel properties of jatropha oil, jatropha biodiesel, and diesel; Table below will show the fuel properties of jatropha oil, jatropha biodiesel, and diesel [15-16].

Table 2: The fuel properties of jatropha oil, jatropha biodiesel, and diesel [12].

No	Fuel Blended	Density (Kg/m3)	CV (kJ/kg)	Viscosity @ 40 °C (cSt)	Flash Point (°C)
1	Diesel	850	44000	2.87	76
2	JO20	852	43759.5	3.02	88
3	JO50	857	42673	3.59	113
4	JO100	873	43690	4.23	148
Testing Procedure		ASTM D4052	ASTM D240	ASTM D445	ASTM D93

2. Storage Method and Parameter

Biodiesel samples were stored in glass containers under room temperature condition, and ambient temperature condition for 10 weeks in total. These samples were monitored on a weekly basis through the test properties. The tested parameter is acid value, viscosity, density, water content and flash point. This process was be done at a certain temperature and mixing them to produce biodiesel in the presence of methanol and base catalyst. Physical characteristics have been considered including acid value, viscosity, density, water content, and flash point. In this study there are 4 samples are taken which is 5% blends of jatropha oil (JO 5), 10% blends of jatropha oil (JO 10), 15% blends of jatropha (JO 15). A block diagram of blending process and schematic diagram were shown in Figure 1. The purified jatropha oil methyl ester was blended with commercial diesel. The mixer was scale at 60°C and the mixture was stirred at 70°C for 1 hour. The blade speed was maintained at 270 rpm.

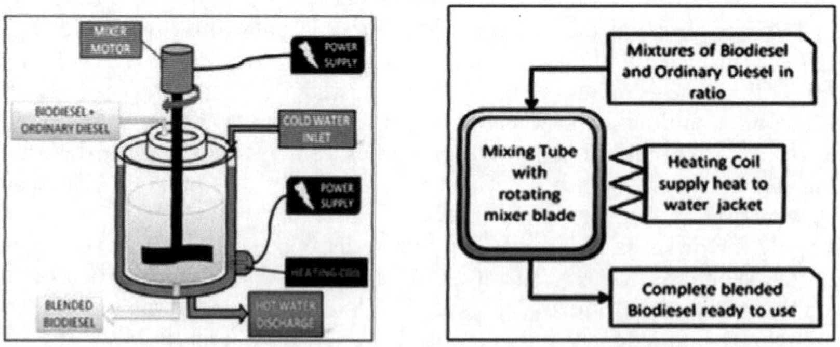

Figure 1: The Schematic Diagram of Blending Process

According to this study, the objectives were to determine the density, viscosity, acid value, water content and flash point of biodiesel blend stored at different storage container material and period. Figure 2 shows the flowchart of the process involved in this study.

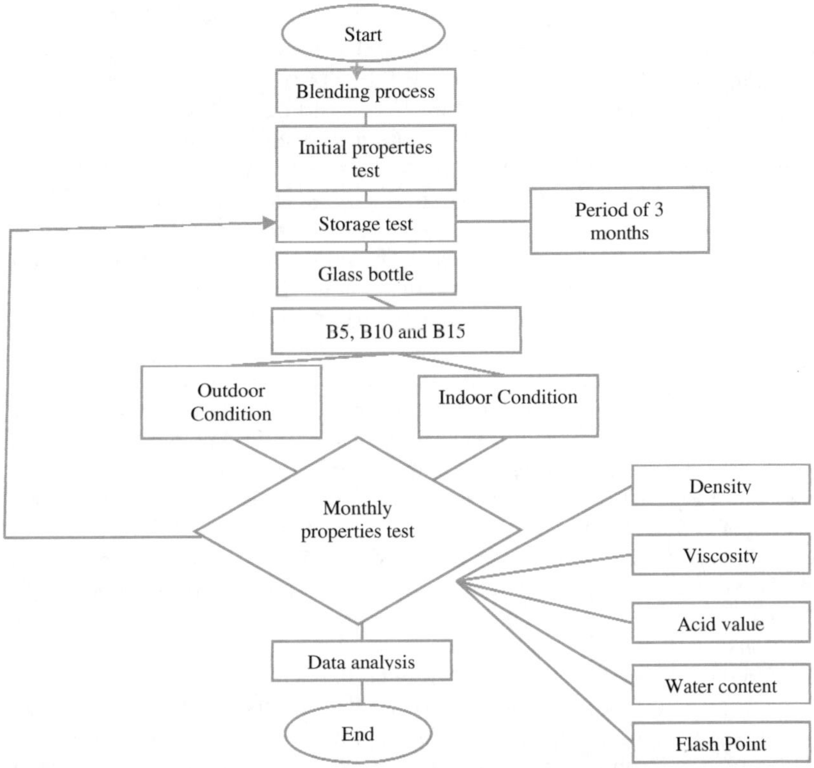

Figure 2: Experiment Flow Chart

3. Procedure of Properties Test

All of the samples is kept on two different conditions, first is inside a cupboard, and another is on a field. Then the sample is undergo five monthly properties test, which is density, viscosity, water content, acid value and lastly flash point test. Density is an important property which influences the overall standard of a biodiesel. The density of biodiesel blend was measured according to, EN ISO 3675 European Standard. By taking approximately 30 ml of sample it was poured into a beaker and left to cool down to 15°C inside a refrigerator. The measured weight of an empty 10.104 ml pycnometer by using a weighing machine and level the weight indicator. The cooled sample was poured into the pycnometer until it was completely filled it. The pycnometer was measured again on a weighing machine and the weight shown this time would be the weight of the sample inside the pycnometer. Viscosity is an important property which indicates the ability of a material to flow. For the viscosity test, EN ISO 3104 European Standard was followed. The sample needed to be heated up to the temperature of 40°C. Heated sample was then poured into a graduated cylinder. A solid stainless steel sensor of the viscometer (Viscolite 700) was then immersed into the sample. The sample level had to be higher than the line on the sensor and the side of the viscometer sensor must not be touching with the graduated cylinder. The reading of the viscometer result was in centipoises (cPs). The left-over of sample on the solid stainless steel sensor was wiped cleanly after used. The acid value analysis was based on the titration process of the sample through the use of sodium hydroxide, phenolphthalein and biodiesel sample. In the process of preparing the alkaline reagent, sodium hydroxide (NaOH) 0.1M, 2 g of sodium hydroxide (NaOH) in powder or pellets form were mixed with 500 ml of distilled water inside a volumetric flask.

The alkaline mixture was mixed well by shaking the flask thoroughly. The mixture was poured into a burette until it was fully filled. In order to ensure no bubble was trapped inside the burette, few drop of solution was titrated out until the bubble was gone. For the acid value test, EN 14104 European Standard was followed. The sample was heated up to around 60 – 70 °C. By taking 4 g of the heated sample was mixed with 50 ml of 2-propanol inside a conical flask. The mixture was heated up again to 40°C and then 5-7 drops of phenolphthalein is added in. It was titrated with the sodium hydroxide mixture until the first pink colour appeared and lasted for around 30 seconds. Water content test, EN ISO 12937: 2000 European Standard was followed. Coulometric Karl Fischer Titrators. It was suitable for small water volumes such as biodiesel blend in ppm unit. A syringe was used to draw approximately 5 ml of biodiesel sample and was weighted by using a weighing machine. The weighing machine was then levelled. The start button on the machine was pressed. The sample was injected into the machine solution through the inlet until a keypad screen appeared at the monitor. It was important that the sample is injected slowly so that no excess sample was injected. Then, the remaining sample in the syringe was weighted again. The weight appeared was the amount of sample injected and was inserted into the machine through the keypad appeared. The result of total water content in the sample was shown in a few seconds. Flash point is a parameter of practical importance because as the flash point increases, the higher is the safety level during handling, transportation and storage. For flash point test, EN ISO 2719: 2002 European Standard was followed. By taking approximately 60 ml of sample was poured into the brass cup until the inner line had reached. Every a part of the flash point tester was assembled back carefully and the estimation of boiling point of the sample was inserted into the machine which is around 80°C. The machine was switched on and the sample was heated up. The overall process was comprised of heating and cooling down of the sample. The duration of the overall process was done in 20 minutes.

4. Results and Discussions

Effect of the storage duration of biodiesel blending ratio from jatropha oil (JO) under room temperature was investigated at the base STD for biodiesel JO 5, JO 10 and JO 15 respectively. In this study, there are three types of blending which is 5vol% blends (5vol% plant oil 95vol% diesel), 10vol% blending (10vol% plant oil and 90vol% diesel) and 15vol% blending (15vol% plant oil and 85vol% diesel) each called CPO5 (crude palm oil 5vol%), CPO10 (crude palm oil 10vol%),CPO15 (crude palm oil 15vol%), JO5 (jatropha oil 5vol%), JO10 (jatropha oil 10vol%),and JO15 (jatropha oil 15vol%) respectively. The average storage temperature for room temperature and outdoor temperature are 24°C and 33°C, respectively.

Figure 3 and Figure 4 illustrates us the changes of physical properties such acid value, viscosity, density and water content for jatropha oil and crude palm oil on different storage condition under room condition and outdoor condition. Overall the graph for JO trends gives the highest result then followed by CPO and standard diesel. For flash point the Crude Palm Oil Biodiesel (CPO) and Standard Diesel showing a steady trend but increasing trend for all of JO 5, JO 10 and JO 15. The highest peak for JO which is store under indoor condition is almost 150°C but for the outdoor condition the value is below 140°C and also the water content increases higher than indoor condition. The peak of the data is at 2190 hours for JO 15 which is 4.4 cTs if we ignore the first data and the lowest is for CPO 10 at 3.0 cTs

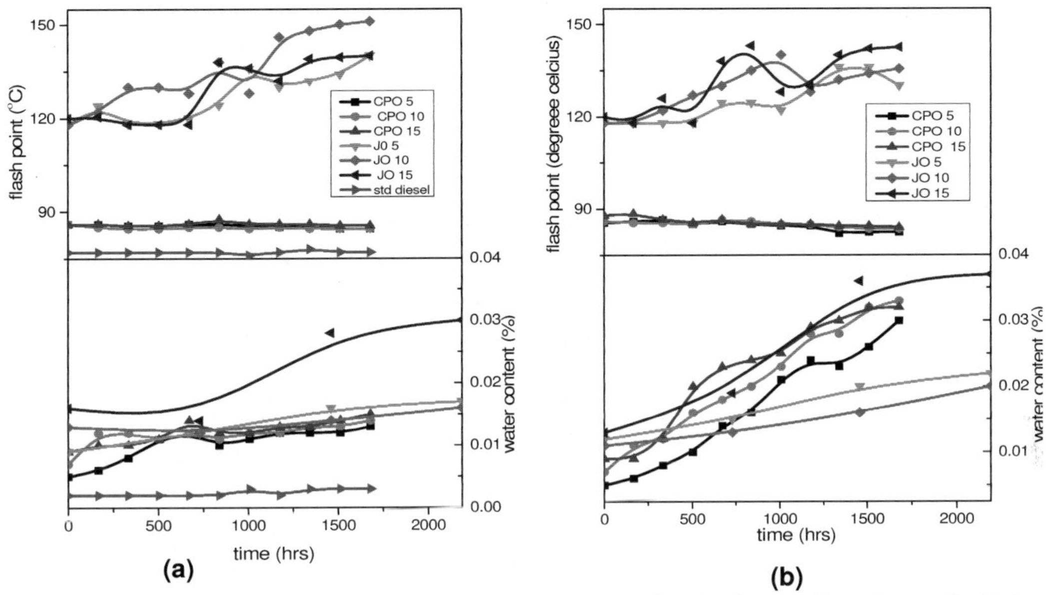

Figure 3: Effect of flash point, water content for storage Duration (a) Storage Duration under Indoor Condition. (b) Storage Duration on Outdoor Condition.

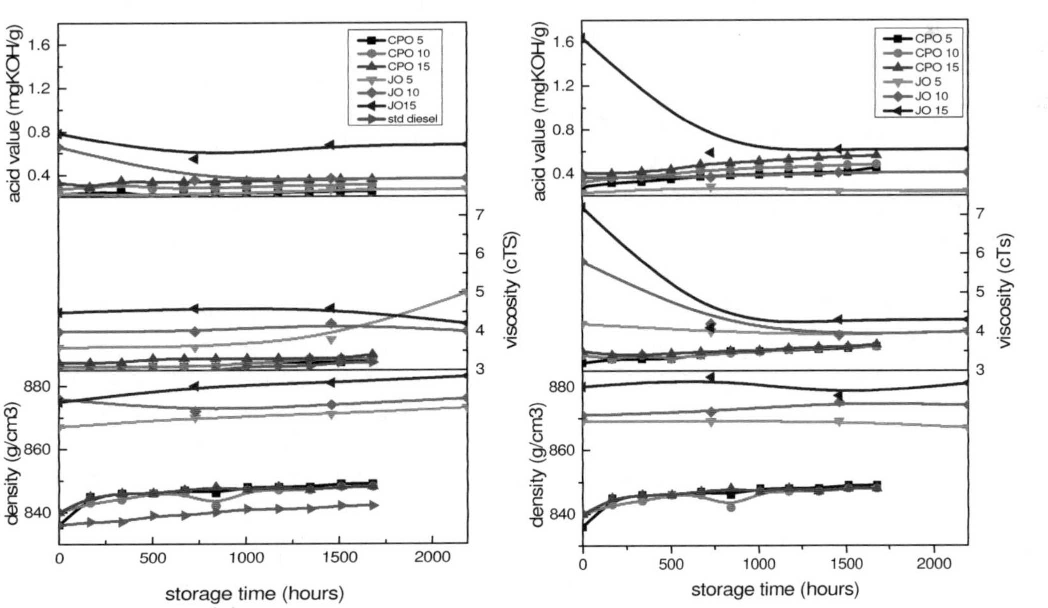

Figure 4: Effect of acid value, viscosity and density for storage Duration (a) Indoor Condition. (b) Outdoor Condition.

The result for density is almost same over the time. The higher of the blending ratio biodiesel influences the higher the density. The maximum value is at 880g/cm^3 for JO 15 and the lowest is for CPO 10 along 840g/cm^3. For the viscosity graph outdoor condition give a surprising result. The JO viscosity decrease throughout the time. The sample JO 10 and JO 15 might stabilize throughout the time. For acid value testing, outdoor condition give us a significant increase for the entire sample. The highest value given by the JO 15 at 0.7 mgKOH/g and the lowest is JO5, 0.2 mgKOH/g. The increasing of acid value was cause by hydrolysis process of methyl ester that induces by the sunlight.

Conclusions

In this research, there are three types of blending which is 5vol% blends (5vol% plant oil 95vol% diesel), 10vol% blending (10vol% plant oil and 90vol% diesel) and 15vol% blending (15vol% plant oil and 85vol% diesel) each called CPO5 (crude palm oil 5vol%), CPO10 (crude palm oil 10vol%),CPO15 (crude palm oil 15vol%), JO5 (jatropha oil 5vol%), JO10 (jatropha oil 10vol%),and JO15 (jatropha oil 15vol%) storage at indoor condition and outdoor condition. The summary as follows:

1. It seems that the small changes in the physical properties of biodiesel when stored at ambient temperature or low temperature. Biodiesel storage at room temperatures is suitable and more advantageous because the impact on the physical properties is minimal and beneficial of slow down the degradation of biodiesel and storage.
2. Indoor storage conditions which are less susceptible to delayed break light chain fatty acids. The physical properties majorly affected by the blending ratio and type of derivation used.

Acknowledgements

The authors also would like to thank the Universiti Tun Hussein Onn Malaysia for the Ministry of Higher Education, Malaysia for supporting this research under COE-MTUN Grant Scheme VOT.C009.

References

[1] Khalid, A., Yatsufusa, T., Miyamoto, T., Kawakami, J., Kidoguchi, Y., "Analysis of relation between mixture formation during ignition delay period and burning process in diesel combustion", SAE Technical Papers, 2009, SAE International.

[2] A. Obadiah, R. Kannan, A. Ramasubbu, and S. V. Kumar, "Studies on the effect of antioxidants on the long-term storage and oxidation stability of Pongamia pinnata (L.) Pierre biodiesel," Fuel Process. Technol., vol. 99, pp. 56–63, 2012.

[3] M. Shahabuddin, M. A. Kalam, H. H. Masjuki, M. M. K. Bhuiya, and M. Mofijur, "An experimental investigation into biodiesel stability by means of oxidation and property determination," Energy, vol. 44, no. 1, pp. 616–622, 2012.

[4] M. Berrios, M. A. Martín, A. F. Chica, and A. Martín, "Storage effect in the quality of different methyl esters and blends with diesel," Fuel, vol. 91, no. 1, pp. 119–125, 2012.

[5] C. Pattamaprom, W. Pakdee, and S. Ngamjaroen, "Storage degradation of palm-derived biodiesels: Its effects on chemical properties and engine performance," Renew. Energy, vol. 37, no. 1, pp. 412–418, 2012.

[6] Khalid, A., Hayashi, K., Kidoguchi, Y., Yatsufusa, T., "Effect of air entrainment and oxygen concentration on endothermic and heat recovery process of diesel ignition", SAE Technical Papers, 2011, Society of Automotive Engineers of Japan, Inc. and SAE International, DOI: 10.4271/2011-01-1834

[7] L. Natalie, "Study of hydrogenation derived renewable diesel as a renewable fuel option in North America," pp. 1–78, 2012.

[8] C. Y. Lin, H. A. Lin, and L. B. Hung, "Fuel structure and properties of biodiesel produced by the peroxidation process," Fuel, vol. 85, no. 12–13, pp. 1743–1749, 2006.

[9] S. M. Hafis, M. J. M. Ridzuan, R. N. Farahana, A. Ayob, and S. Syahrullail, "Paraffinic mineral oil lubrication for cold forward extrusion: Effect of lubricant quantity and friction," Tribol. Int., vol.

60, pp. 111–115, 2013.

[10] Khalid, A., Anuar, M.D., Ishak, Y., Manshoor, B., Sapit, A., Leman, M., Zaman, I., " Emissions characteristics of small diesel engine fuelled by waste cooking oil", MATEC Web of Conferences, Volume 13, 2014, Article number 06006, 2014, DOI: 10.1051/matecconf/20141306006

[11] N. Kaushik, K. Kumar, S. Kumar, N. Kaushik, and S. Roy, "Genetic variability and divergence studies in seed traits and oil content of Jatropha (Jatropha curcas L.) accessions," Biomass and Bioenergy, vol. 31, no. 7, pp. 497–502, 2007.

[12] E. Akbar, Z. Yaakob, S. K. Kamarudin, M. Ismail, and J. Salimon, "Characteristic and Composition of Jatropha Curcas Oil Seed from Malaysia and its Potential as Biodiesel Feedstock Feedstock," Eur. J. Sci. Res., vol. 29, no. 3, pp. 396–403, 2009.

[13] M. H. Gordon, Chemistry of oils and fats – sources, composition, properties and uses, vol. 94. 2006.

[14] P. K. Sahoo and L. M. Das, "Combustion analysis of Jatropha, Karanja and Polanga based biodiesel as fuel in a diesel engine," Fuel, vol. 88, no. 6, pp. 994–999, 2009.

[15] Khalid, A., Azman, N., Zakaria, H., Manshoor, B., Zaman, I., Sapit, A., & Leman, A. M. (2014). Effects of storage duration on biodiesel properties derived from waste cooking oil. doi:10.4028/www.scientific.net/AMM.554.494

[16] Khalid, A., Tamaldin, N., Jaat, M., Ali, M. F. M., Manshoor, B., & Zaman, I. (2013). Impacts of biodiesel storage duration on fuel properties and emissions. Paper presented at the Procedia Engineering, 68 225-230. doi:10.1016/j.proeng.2013.12.172

Analysis of the Effect of Injection Pressure on Ignition Delay and Combustion Process of Biodiesel from Palm Oil, Algae and Waste Cooking Oil

Mohd Irham Anas[1, b], Amir Khalid[1,2,a], Fathul Hakim Zulkifli[1], Norrizam Jaat[1,2], Mohd Faisal Hushim[1], Bukhari Manshoor[2], Izzuddin Zaman[3],

[1]Automotive and Combustion Synergies Technology Group, Advanced Technology Centre (ATC), Faculty of Engineering Technology, Universiti Tun Hussein Onn Malaysia, Parit Raja, Batu Pahat 86400 Johor, Malaysia.
[2]Centre for Energy and Industrial Environment Studies (CEIES), Universiti Tun Hussein Onn Malaysia, Parit Raja, Batu Pahat, 86400 Johor, Malaysia.
[3]Faculty of Mechanical and Manufacturing Engineering, Universiti Tun Hussein Onn Malaysia, 86400 Parit Raja, Batu Pahat, Johor, Malaysia

Corresponding author: amirk@uthm.edu.my[a], an130148@siswa.uthm.edu.my[b]

Abstract. Biodiesel is a domestically produced, renewable fuel that can be manufactured from vegetable oils, animal fats, or recycled restaurant grease for use in diesel engines. The objective of this research is investigation the effects of the variant injection pressure on ignition delay and emission for different biodiesel using rapid compression machine. Rapid Compression Machine (RCM) is used to simulate a single compression stroke of an internal combustion engine as a real engine. Four types of biodiesel which are waste cooking oil, crude palm oil, algae and jatropha were tested at injection pressure of 80 MPa, 90 MPa and 130 MPa under constant ambient temperature at 950 K. Increased in injection pressure resulted shorter ignition delay proven by WCO5 which decreased from 1.3 ms at 80 MPa to 0.7 ms at 130 MPa. Meanwhile, emission for CO_2 increased due to better fuel atomization for fuel-air mixture formation lead to completed combustion.

1. Introduction

There are some factors influences the performance and exhaust emission characteristics for diesel engine. The factors involve quantity injected, injection timing and injection pressure. Diesel engine with direct injection system has a high degree of atomization which undergoes evaporation at a very short time and improves combustion efficiency [1-3]. The spray formation analysis and propagation need to be considered during injection process inside the combustion chamber [4]. High pressure gasses and high temperature due to combustion of fuel take place in the constant volume chamber [5]. Ignition delay was analyzed from the pressure in the combustion chamber when air is compressed and combustion occurs. The pressure change is resulting from competition between endothermic process such as fuel evaporation, decomposition and heat loss to the chamber wall, and exothermic process. Ignition delay is the period between start of fuel injection and end of fuel injection which consists of physical delay process and chemical delay process [5-7]. A free-piston type rapid compression machine (RCM) was used to simulate diesel combustion in a constant volume chamber over a wide range of ambient temperature and pressure conditions similar to actual diesel engines. This system is of a single cylinder, free moving piston and direct injection of fuel and air [8-10].

In this research, the ignition delay and emission are investigated with different injection pressure and constant ambient temperature. The blending machine used for fuel preparation. This experiment was performed in Rapid Compression Machine (RCM). The result from the experiment was compared with results from Palm-Oil biodiesel, Algae biodiesel, Jatropha biodiesel for validation purpose and to find the difference between Waste Cooking Oil biodiesel and different biodiesel.

2. Methodology

Biodiesel is blend with Euro 5 diesel using the blending machine. During blending process, the mixture was stirred at 70ºC for 2 hour and the rotating blade speed was maintaining at 270 RPM. Blending ratios of biodiesel produced were varies at 2% until 40% for preparation of tested fuel. The experimental apparatus can be divided into four main systems such as rapid compression machine, data acquisition system, common rail injection system, emission gas analyser for analysis exhaust emission. Figure 1 illustrate the blending machine and Figure 2 illustrate the schematic diagram of blending machine.

Figure 1: Blending machine.

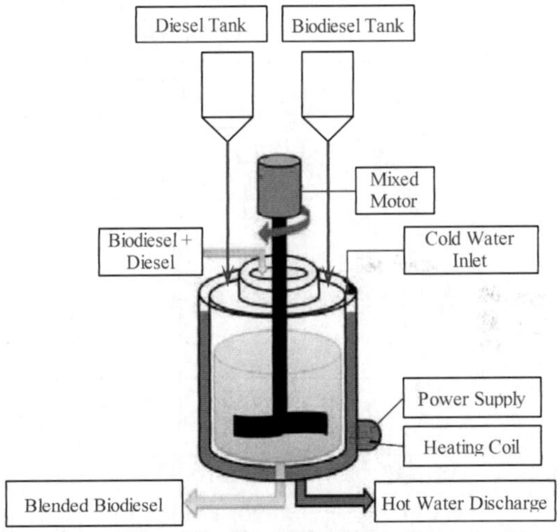

Figure 2: Schematic diagram of blending machine.

Then, the Rapid Compression Machine was used to stimulate the combustion process as real as compression ignition engine. The injection pressure used was 80 MPa, 90 MPa and 130 MPa with constant ambient temperature of 950 K. Figure 3 illustrates the rapid compression machine while Figure 4 shows a schematic diagram of rapid compression machine together with fuel injection system. A free piston is placed inside the cylinder and partitioned by a diaphragm as a start of the experiment. Then, the driver chamber and the pressure reservoir are filled with nitrogen gas which pressures are held below and above the pressure of the diaphragm respectively. Once the valve is opened, the diaphragm bursts and the piston is rammed into a tapered stop ring installed in front of the combustion chamber. Fuel was injected after a fixed interval from the time the piston stopped. The measured air temperature and pressure at the start of injection were used to identify the combustion behaviour by the Pico Scope 3000 series.

Figure 3: Rapid Compression Machine

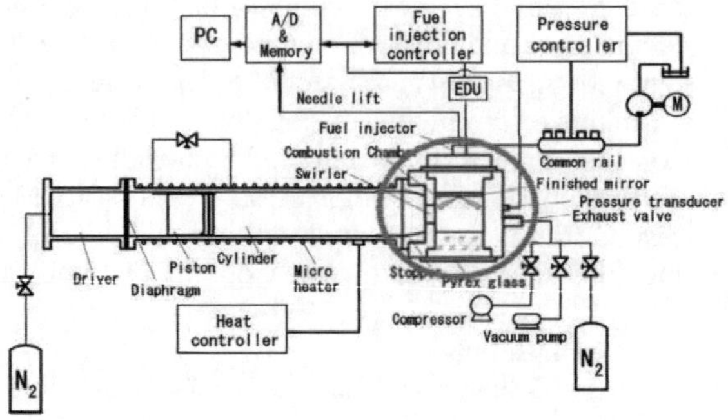

Figure 4: Schematic diagram of RCM

Furthermore, Table 1 shows the fuel properties for different types of biodiesel and Table 2 shows the experimental condition for the experiment.

Table 1: Fuel properties.

Fuel Type	**Properties**	
	Density (g/m^3)	**Kinematic Viscosity (Cp)**
A2	0.860	4.4
J2	0.831	3.6
J5	0.831	3.6
J10	0.878	4.3
J15	0.838	3.6
B5	0.837	3.0
B10	0.838	2.9
B15	0.840	3.0
WCO5	0.889	3.0
WCO10	0.890	3.18
WCO15	0.892	3.2

Table 2: Experimental condition

Fuel	Injector type	6 holes, Ø = 0.16 mm
	Fuel type	Wco, Cpo, Algae and Jatropha
	P_{inj} [MPa]	80, 90 and 130
	q_i [ml]	0.04
Ambient	T_i [K]	950K
	r_s [m/s]	19
	ρ [kg/m^3]	16.6
	O_2 [vol%]	21

3. Results and discussion

The effect of variant injection pressure on ignition delay is investigated for Waste Cooking Oil biodiesel. The injection pressures used were 80 MPa, 90 MPa and 130 MPa with constant ambient temperature of 950 K. Figure 5 illustrate the variant injection pressure on ignition delay for waste cooking oil biodiesel (WCO5, WCO10 and WCO15). The ignition delay decreased constantly when increased the injection pressure. For WCO5, the ignition delay at injection pressure of 80 MPa was 1.3 ms and decreased to 1.1 ms at 90 MPa and to 0.7 ms at 130 MPa. Then, the ignition delay were 1.1ms and 1.5 ms for WCO10 and WCO15 respectively at 80 MPa and this ignition delay continued decreased to 0.8 ms and 0.9 ms at 130 MPa. This situation due to increase in injection pressure effect the increasing of temperature in combustion chamber that helps to improve the air-fuel mixture formation contributes to better premixed combustion which reduced the ignition delay.

Figure 6 illustrate the variation of injection pressure on ignition delay of Waste Cooking Oil biodiesel (WCO5, WCO10 and WCO15) and Palm-Oil Biodiesel (B5, B10 and B15). Injection duration for B5 at 80 MPa was 3.0 ms and decreased to 2.0 ms at 90 MPa and to 1.4 ms at 130 MPa. The similar trend goes to B10 and B15 fuels. The ignition delay was 2.5 ms and 2.6 ms for B10 and B15 respectively at 80 MPa and continued decreased to 2.0 ms and 2.1 ms at 130 MPa. The higher temperature in combustion chamber assists in improvement for better air-fuel mixture formation effect better premixed combustion.

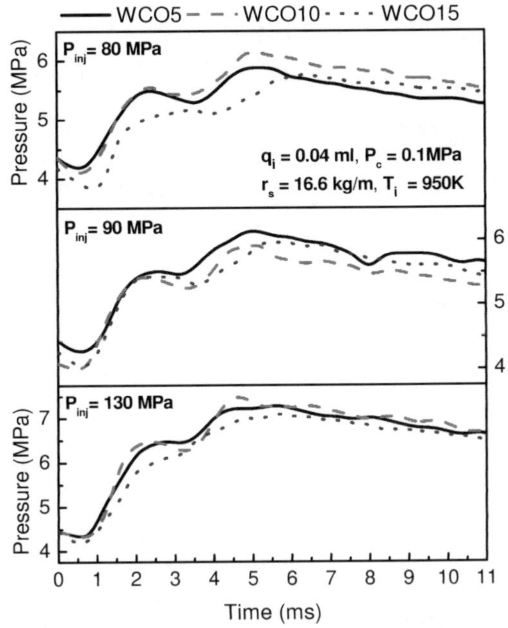

Figure 5: Variant injection pressure of WCO on ignition delay.

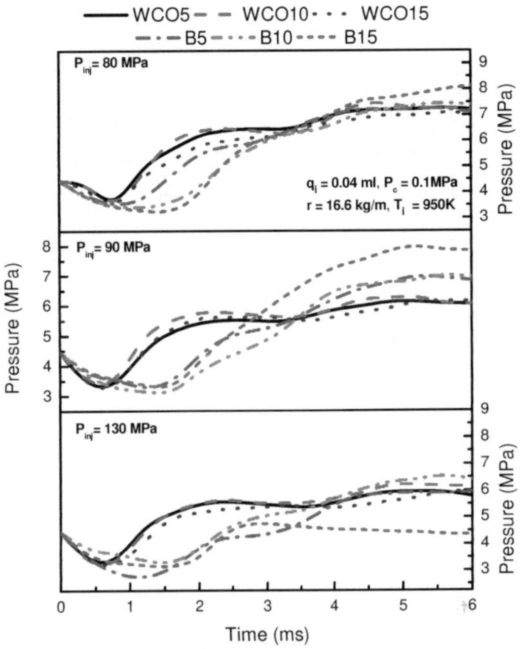

Figure 6: Variant injection pressure of WCO and CPO on ignition delay

Furthermore, the heat release combustion for Jatropha with 5% to 15% blending ratio is shown in Figure 7. The ignition delay for J5 was 2.4 ms at 80 MPa of injection pressure and then decreased to 2.3 ms at 130 MPa. The same trend goes to J10 and J15 fuels. The ignition delay was 2.3 ms and 2.9 ms for J10 and J15 respectively at 80 MPa and continued decreased to 2.2 ms and 2.8 ms at 130 MPa. The higher temperature in combustion chamber helps in improvement for better air-fuel mixture formation effect better premixed combustion.

Comparison of ignition delay for different biodiesel is shown in Figure 8. The ignition delay for Algae biodiesel (A2) decreased when the injection pressure increase. Algae biodiesel has the lower viscosity compared to Palm-Oil Biodiesel (B5, B10 and B15) which help to reduce the ignition delay of the fuel. The ignition delay for Algae Biodiesel (A2) was 1.3 ms at 80 MPa of injection pressure and then decreased to 1.2 ms at 90 MPa and further decreased to 1.0 ms at 130 MPa. In addition, The Jatropha Biodiesel (A2) had compared to previous biodiesel. The ignition delay of J2 decreased with increasing the injection pressure. The ignition delay for J2 at 80 MPa was 0.9 ms and decreased to 0.8 ms at 90 MPa and further reduced to 0.7 ms at 130 MPa. Both J2 and A2 had same blending ratio but the ignition delay for J2 was shorter compared to A2. This was due to Jatropha has higher ignitability when compared with Algae. Then, at injection pressure of 80 MPa the ignition delay for J5 is 1.4ms, J10 is 1.3ms and J15 has ignition delay of 0.9 ms. For injection pressure of 90 MPa, it is observed when the pressure is increased, the ignition delay decreases to 0.9 ms for J5, 1.2 ms for J10 and 1.0ms for J15. The pressure are increased to 130 MPa and its can be seen that ignition delay continued decreased to 0.7 ms and 0.9 ms for J20 and J30 respectively while the delay period for J15 remain unchanged. The ignition delay period of the tested fuel decreases with increasing cylinder pressure and fuel injection pressure. The reduction of ignition delay of Jatropha biodiesel mainly because of cetane number.

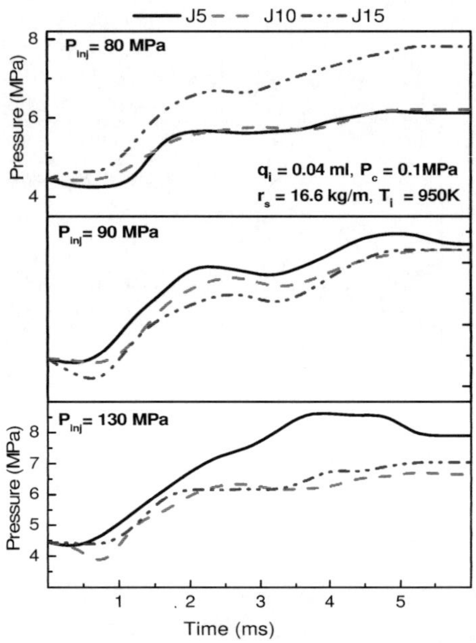

Figure 7: Comparison of ignition delay for Jatropha biodiesel

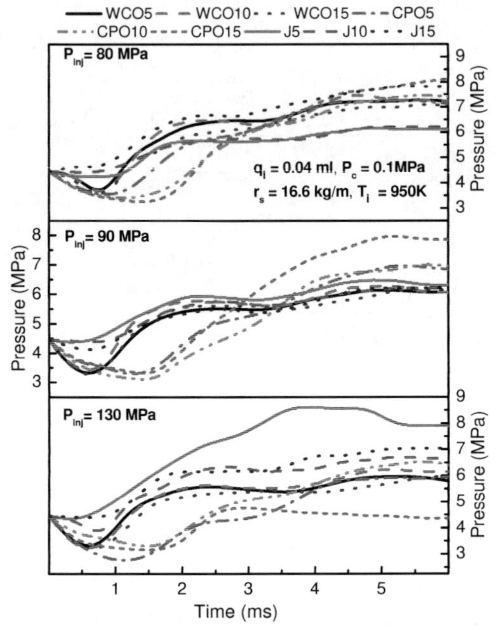

Figure 8: Comparison of ignition delay for different biodiesel.

Then, the effects of injection pressure on emission were discussed. The measured emissions were carbon monoxide (CO), nitrogen oxides (NO_x), carbon dioxide (CO_2) and hydrocarbon (HC). The experiment condition varies at constant ambient temperature of 950 K. Figure 9 illustrates the graph percentage of emission versus injection pressure. Three injection pressures were 80 MPa, 90 MPa and 130 MPa. The emission of carbon monoxide (CO) and hydrocarbon (HC) decreased when increasing the injection pressure for all tested fuels. This was due to better air-fuel mixture formation which leads to complete combustion. In addition, the emission of carbon dioxide (CO_2) and nitrogen oxides (NO_x) seems to increase when increasing the injection pressure. This was due to high temperature in combustion chamber that helps to improve combustion process which produces higher percentage of carbon dioxide emission.

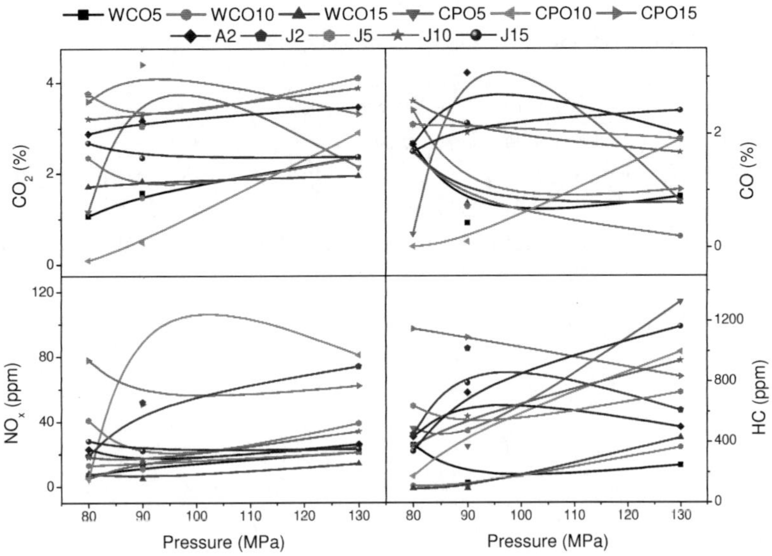

Figure 9: Effect of different injection pressure on emission.

4. Conclusion

Increased in injection pressure attribute to shorter ignition delay. This due to high injection pressure varies with high temperature in combustion chamber which influence better air-fuel mixture formation for better premixed combustion. The formation of CO_2 increased due to complete combustion in combustion chamber.

Acknowledgements

The authors also would like to thank the Ministry of Higher Education, Malaysia for supporting this research under Fundamental Research Grant Scheme (FRGS) vot.1054.

References

[1] A, Valipour,"Experimental Combustion Analysis of Biodiesel Fuel Spray with Hot Surface Ignition," IPASJ Int. J. Mech. Eng, (2014)

[2] Khalid, A., Anuar, M.D., Ishak, Y., Manshoor, B., Sapit, A., Leman, M., Zaman, I., " Emissions characteristics of small diesel engine fuelled by waste cooking oil", MATEC Web of Conferences, Volume 13, 2014, Article number 06006, 2014, DOI: 10.1051/matecconf/20141306006

[3] Khalid, A., Manshoor, B., "Analysis of Mixture formation and Flame Development of Diesel Combustion using a Rapid Compression Machine and Optical Visualization Technique", Applied Mechanics and Materials, Volume 315, 2013, Pages 293-298, Trans Tech Publications, Switzerland, DOI: 10.4028/www.scientific.net/AMM.315.293

[4] Yamashita, H., et al., "Research of the DI Diesel Spray Characteristics at High Temperature and High Pressure Ambient". SAE Paper 2007-01-0665, (2007).

[5] Suk Hong No, et al., "High Temperature Diesel Combustion in a Rapid Compression-Expansion Machine". SAE Paper 911845, (1991)

[6] Ikegami, M., K. Miwa, and M. Inada, " A Study on Ignition and Combustion of a Diesel Spray by Means of a Rapid Compression Machine". Bulletin of the JSME, Vol. **24** No 195, (1981)

[7] Khalid, A., Yatsufusa, T., Miyamoto, T., Kawakami, J., Kidoguchi, Y., "Analysis of relation between mixture formation during ignition delay period and burning process in diesel combustion", SAE Technical Papers, 2009, SAE International.

[8] Wilbur, A. A. Ignition Studies- The Determination of Auto Ignition Temperatures of Hydrocarbon Fuels. Washington DC: Naval Research Laboratory. (1974).

[9] Khalid, A., Hayashi, K., Kidoguchi, Y., Yatsufusa, T., "Effect of air entrainment and oxygen concentration on endothermic and heat recovery process of diesel ignition", SAE Technical Papers, 2011, Society of Automotive Engineers of Japan, Inc. and SAE International, DOI: 10.4271/2011-01-1834

Suction and Blowing Flow Control on Airfoil for Drag Reduction in Subsonic Flow

S S Baljit, M R Saad, A Z Nasib, A Sani, M R A Rahman and A C Idris
Department of Mechanical Engineering, Faculty of Engineering, National Defense University of Malaysia, Kuala Lumpur, Malaysia.

Abstract. Lift force is produced from a pressure difference between the pressures acting in upper and lower surfaces. Therefore, flow becomes detached from the surface of the airfoil at separation point and form vortices. These vortices affect the aerodynamic performance of the airfoil in term of lift and drag coefficient. Therefore, this study is investigating the effect of suction and jet blowing in boundary layer separation control on NACA 0012 airfoil in a subsonic wind tunnel. The experiment examined both methods at the position of 25% of the chord-length of the airfoil at Reynolds number 1.2 x 10^5. The findings show that suction and jet blowing affect the aerodynamic performance of NACA 0012 airfoil and can be an effective means for boundary layer separation control in subsonic flow.

1. Introduction

One of the most vital aerodynamic structure on an airplane is the wing itself which allows the airplane to fly and work efficiently. The airfoil faces heavy frictional and drag force on its surface which further reduces the airplane efficiency. Steady and unsteady blowing is one of the techniques in investigation for separation control on an airfoil. A wide variety of these techniques have been used in the past to delay separation by increasing the momentum of the boundary layer. Blowing has been effective in adjusting and reenergizing the flow to prevent flow separation. A delayed airfoil stall angle due to oscillatory blowing was found and this proved that steady blowing is not as efficient as oscillatory blowing in delaying separation [1]. The separation region forms when the angle of attack of the airfoil is increased, which causes the flow to fail to reattach on the airfoil surface [2]. A pulse blowing technique also was introduced which helped keeping the flow attached to the surface. Further investigations on this topic revealed that blowing also resulted in an increase in lift coefficient with the application of spanwise steady blowing [3]. The blowing system resulted in causing turbulent flow to occur earlier hence causing the flow to come closer to the surface of airfoil near the trailing edge.

The second method opposite to blowing, which is the suction system was then compared [4]. The study proved that the most important factors affecting the outcome are the suction location and the angle. The suction system was also tested on a cylindrical object to reduce drag acting on it by focusing in reducing the fluctuating amplitudes of the dynamic wind loads acting on the test modal [5]. Suction contributes to the stability of the boundary layer by delaying transition so that the flow is laminar rather turbulent [6]. The suction system can also delay the transition occurrence on the upper surface boundary layers compared with the flow without suction control [7]. These methods were then further enhanced by both simulation and experimental studies [8]. Therefore the purpose of this study is to investigate the effectiveness of the suction system and jet blowing in reducing the coefficient of drag and also in increasing the lift coefficient of NACA 0012 airfoil.

IOP Conf. Series: Journal of Physics: Conf. Series **914** (2017) 012009 doi:10.1088/1742-6596/914/1/012009

2. Methodology

2.1. Airfoil CAD Model

The airfoil model was designed using SolidWorks 2016 as shown in Figure 1. NACA 0012 was chosen as the model. The chord length of the airfoil is 150mm and a span of 240mm.

Figure 1. Isometric view of airfoil.

2.2. *Airfoil Model Fabrication*

The airfoil model was then fabricated using a 3D printer. The material used was made out of ABS plastic. The material was first melted in high temperature to create the base which usually is larger than the model size as shown in Figure 2. After the base was complete, the fabrication process of the model was initiated and took approximately 7 hours until completion. The airfoil model was then drilled from the base with holes of diameter of 3mm to allow the insertion of the suction and blowing tubes.

Figure 2. Base of the airfoil.

2.3. *Air Distributor Fabrication*

Aluminum was chosen as the material to fabricate the air distributor. This distributor was used in both the suction sytem and jet blowing for suction and pressurised air delivery. The air distributor was then connected to the airfoil model using five tubes as shown in Figure 3.

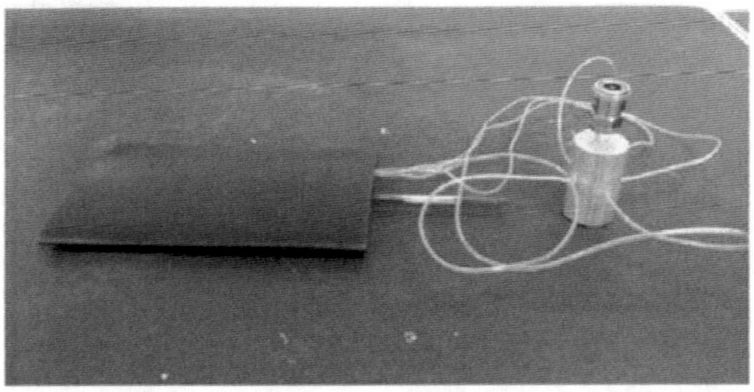

Figure 3. NACA 0012 Airfoil connected to the Air Distributor.

2.4. *Subsonic Wind Tunnel Testing*

The airfoil model was then tested in the LW-9300R Subsonic Wind Tunnel as shown in Figure 4, at the National Defence University of Malaysia. The velocity speed used was 20m/s. Temperature and static pressure were set at 293.20 K and 101.3 kPa. The angle of the airfoil was increased by 3 degrees for every data collected. The wind tunnel is equipped with a 3-component force balance, smoke flow visualization and pressure transducers for pressure measurements.

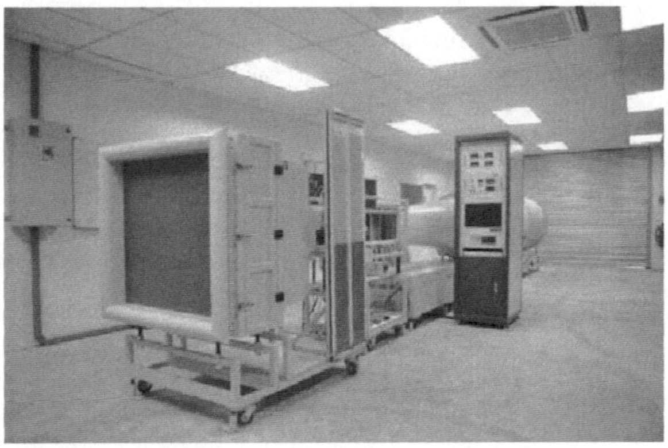

Figure 4. LW-9300R Subsonic Wind Tunnel.

3. Results and Discussions

3.1. *Validation of experimental study for suction system*

The validation process for suction system was carried out by comparing results experimentally against the results produced by Huang et al. [4]. An experiment was carried out in the wind tunnel to obtain results of drag and lift coefficients against various angle of attacks. The airfoil model was mounted inside the test section and connected to the 3-component force balance. The results produced were similar to Huang et al which is shown in Figure 5 and Figure 6. The graph produced was of similar trend as Huang et al. for both cases of lift and drag coefficients.

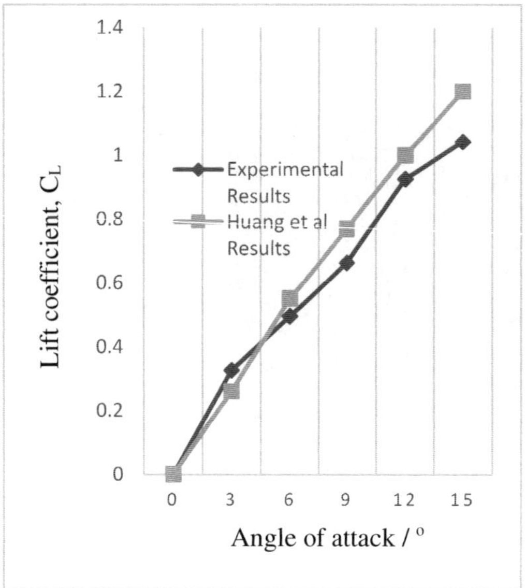

Figure 5. C_L vs Angle of attack.

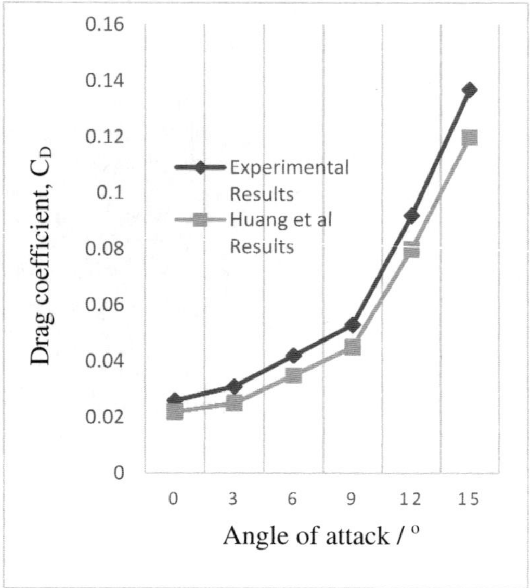

Figure 6. C_D vs Angle of attack.

3.2. Effectiveness of suction system

Figure 7 shows the drag coefficient obtained against various angle of attacks between airfoil with and without the suction system. The drag coefficient produced by the airfoil with suction system has a relatively lower value compared to airfoil without the suction system at all angle of attacks. This proves that the suction system on airfoil can indeed reduce drag acting on the airfoil at almost all angle of attack. Lift coefficient also has a vital role in determining the efficiency of an airplane. Figure 8 shows the graph of lift coefficient against various angle of attacks. It can be seen that the C_L value for airfoil with suction system is higher comparing to the C_L value for airfoil without the suction system.

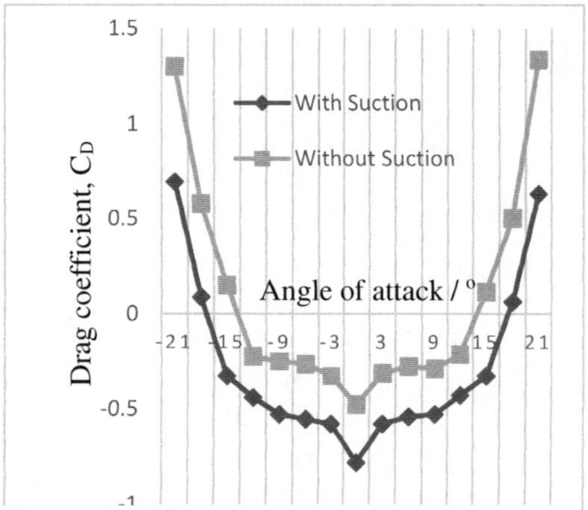

Figure 7. C_D vs Angle of attack with and without suction.

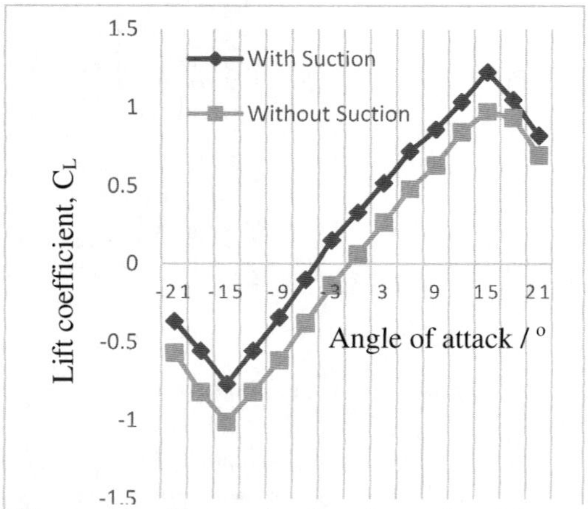

Figure 8. C_L vs Angle of attack with and without suction.

Table 1 shows the difference of C_L/C_D for airfoil with and without suction system at various angle of attacks. The table clearly shows that the airfoil with suction system has a higher lift over drag ratio compared to the airfoil without the suction system. Theoretically it is said that a higher lift over drag ratio is more favorable has it helps lower the glide ratio which allows a longer gliding distance and further increase the airplane efficiency.

Table 1. C_L/C_D ratio comparison.

Angle of Attack	No suction C_L/C_D	Suction C_L/C_D
3	-0.500	-0.261
6	-1.115	-0.766
8	-2.133	-1.119
12	-3.583	-1.618
15	-4.900	-2.269
18	-5.556	-3.529

3.3. *Effectiveness of jet blowing*
In Figure 9, the lift coefficient value of airfoil with jet blowing has no changes at negative angle of attack but differences can be seen at positive angle of attack comparing to the lift coefficient without jet blowing. This proved that blowing device has an effect on the the aerodynamic performance of the airfoil. The lift coefficient increases as the angle of attack increases with blowing device. The lift coefficient is highest at 18° angle of attack with blowing device due to the delayed flow separation. Based on drag coefficient, Figure 10 shows an improvement for airfoil with blowing device compared to airfoil without blowing device for almost all angle of attack. Looking at 21° angle of attack, the C_D for airfoil with blowing device is slightly lesser comparing to the C_D for airfoil without blowing device.

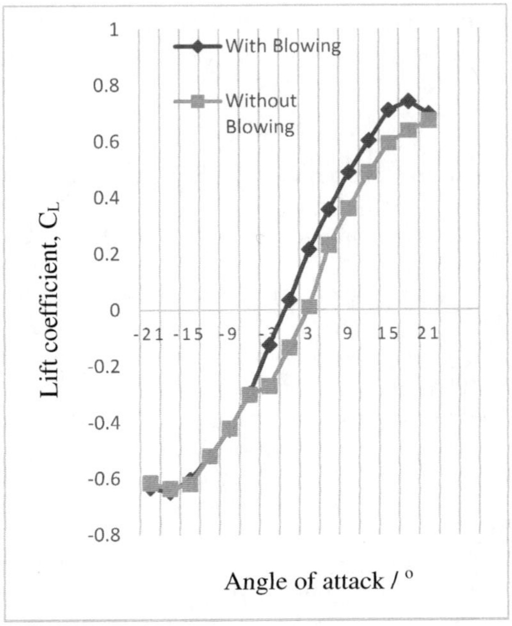

Figure 9. C_L vs Angle of attack with and without blowing.

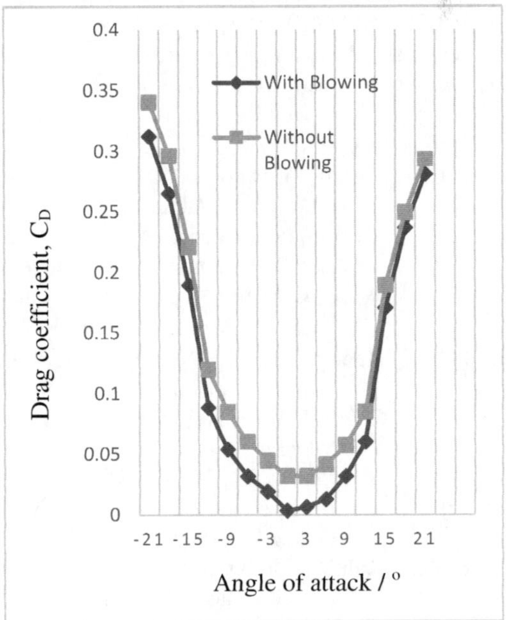

Figure 10. C_D vs Angle of attack with and without blowing.

Table 2 shows that lift and drag coefficient ratio for airfoil with blowing device has higher value compared to lift and drag coefficient value for airfoil without blowing device. Hence it proves that blowing indeed has an improvement effect in aerodynamics performance on airfoil since it is known that higher value of lift and drag coefficient ratio has higher efficiency. It can be seen that the blowing device effect is at its highest at 0° angle of attack.

Table 2. C_L/C_D ratio comparison.

Angle of Attack	No Blowing C_L/C_D	Blowing C_L/C_D
3	7.300	34.000
6	8.769	28.250
9	8.611	15.500
12	6.963	10.052
15	3.366	4.166
18	2.696	3.133

4. Conclusion

Based on the validation results, it is valid to say that the subsonic wind tunnel indeed produces decent results. The suction system and jet blowing also has proven its capability in producing positive results in lift and drag coefficients acting on NACA 0012. Both the devices further delay the separation region by keeping the flow attached on the skin surface of the airfoil.

Acknowledgement

The authors would like to thank the Ministry of Education Malaysia for the financial support under the RAGS (RAGS/1/2014/TK09/UPNM/1) and RACE (RACE/F3/PK6/UPNM/13) research grant schemes. Also an appreciation to the technical staffs of the Faculty of Engineering, Universiti Pertahanan Nasional Malaysia for providing the assistances to the equipment.

References

[1] Seifert, A., and L. G. Pack. "Oscillatory control of separation at high Reynolds numbers." *AIAA journal* 37.9 (1999): 1062-1071.

[2] Mccullough, Goerge B., and Donald E. Gault. "Examples of three representative types of airfoil-section stall at low speed." (1951).

[3] Wong, Chi, and Konstantinos Kontis. "Flow control by spanwise blowing on a NACA 0012." *Journal of aircraft* 44.1 (2007): 337-340.

[4] Huang, L., et al. "Numerical study of blowing and suction control mechanism on NACA0012 airfoil." *Journal of aircraft* 41.5 (2004): 1005-1013.

[5] Chen, Wen-Li, Hui Li, and Hui Hu. "An experimental study on a suction flow control method to reduce the unsteadiness of the wind loads acting on a circular cylinder." *Experiments in fluids* 55.4 (2014): 1707.

[6] Liu, P. Q., et al. "Numerical study of suction-blowing flow control technology for an airfoil." *Journal of aircraft* 47.1 (2010): 229-239.

[7] Kühn, Timo, et al. "Active Flow Separation Control on a High-Lift Wing-Body Configuration Part 1: Baseline Flow and Constant Blowing." *29th AIAA Applied Aerodynamics Conference*. 2011.

[8] Ghoddoussi, Armin, and L. Scott Miller. "A conceptual study of airfoil performance enhancements using CFD." *AIAA Atmospheric Flight Mechanics Conference*. 2011.

Effect of Milling times and Carbon content on Structural and Magnetic properties of Fe-Mn Alloys

K. Tarigan[1], Y. Kusuma[1], D. Sebayang[1] and D. S. Yang[2]

[1]Department of Mechanical Engineering, Mercu Buana University, Jakarta 11650, Indonesia.
[2]Physics Division, School of Science Education, Chungbuk National University, Cheongju 361-763, Korea

Corresponding author: kontan.tarigan@mercubuana.ac.id

Abstract: The structural and the magnetic properties of nanocrystalline $Fe_{47.5}Mn_{47.5}C_5$ alloys were prepared by using a mechanical alloying technique, used the commercial *Fe, Mn,* and *C* powders as precursors. It was studied in detail as function of the milling times of *1*- to *48* hrs. The structural analysis based on X-ray diffraction and extended X-ray absorption fine structure spectroscopy revealed that the alloying process took place after 36 hrs milling. Concerning the magnetic behavior, the data obtained from a vibrating sample magnetometer showed that both the magnetic saturation and the coercivity depended strongly on the milling time and the crystallite size. With these results, by adjusting the milling time shows that an appropriate structural transformation and appropriate magnetization value.

1. Introduction

In recent years, nanocrystalline and amorphous magnetic materials have been studied for many applications in industrial products. The wide range of applications arises from the versatile nature of these materials which can provide fast magnetization reversal with minimal magnetic losses. Mechanical alloying (MA) is able to produce nanostructure materials with unique chemical, structural, electrical and magnetic properties, due to type of disorder created by the high density of defects and the small crystal size [1-3].

MA has been shown to be capable of synthesizing a variety of equilibrium and non-equilibrium alloy phases starting from blended elemental or pre-alloyed powders. The non-equilibrium phases synthesized include supersaturated solid solutions, meta-stable crystalline and quasi-crystalline phases, nanostructures, and amorphous alloys. MA produces nanostructured materials by the structural disintegration of coarser-grained structure as a result of severe plastic deformation. MA is able to produce nanostructure materials with unique chemical, structural, electrical and magnetic properties, due to type of disorder created by the high density of defects and the small supersaturated solid solution, amorphous phases and nano powders, starting from a crystal size. Nowadays, MA has been used to prepare metastable phases such as mixture of elemental components or inter-metallic compounds [4-5].

In fact, MA process is an effective way to fabricate nanocrystalline alloys, and their physical properties are related to structural variations. Some regularity in atomic arrangement in solids can be classified by the short-range order (SRO) and long-range order (LRO). Among these, LRO is frequently examined by X-ray diffraction studies while SRO could be examined by extended X-ray absorption fine structure (EXAFS). EXAFS give useful information related to the local structure around specific atoms [6-8]. In recent years, nanocrystalline and amorphous magnetic materials have been studied for many applications in industrial products, including transformers, motors, and a wide variety of magnetic components in sensors, power electronics, electrical energy control/management systems, telecommunication equipment and pulse power devices. The wide range of applications arises from the versatile nature of these materials which can provide fast magnetization reversal with minimal magnetic losses [9-10].

Recently, research shows that additional of C to FeCo lead to decrease in both the magnetic anisotropy and coercivity by reducing the crystallite size [11]. Xing Lu, et al. [12] reported that the C is the most effective element among Mn, Al, Cr and C to reduce magnetic saturation. So, effects of C in Fe-Mn alloys, especially very important to analysis.

2. Experimental Method

$Fe_{47.5}Mn_{47.5}C_5$ metastable alloys were prepared by mechanical alloying using SPEX 8000 mixer with stainless steel balls and vial. The starting material was a mixture of pure Fe(53 um, 99 %), Mn(75 um, 99 %) and C(105 um, 99 %) powders which were used commercial powders as the precursors. The weight ratio of balls-to-powder mixture was 5:1. $Fe_{47.5}Mn_{47.5}C_5$ alloys were mixed and ground for different times 1-, 6-, 12-, 24-, 36- and 48 hrs. The process was performed in Ar ambient to prevent oxidation during the alloying process. Magnetic measurements were carried out on VSM in magnetic field of 10 kOe. Structure data were obtained by using the X-ray diffractometer (XRD) Cu-K_α radiation. The data were analyzed using Material Data Inc. (MDI) software. EXAFS data were collected with energy of 2.5 GeV, and a maximum current of 200 mA. EXAFS spectra were obtained at Fe K-edge (7112 eV) in the transmission mode at room temperature. The sample chamber was filled with pure nitrogen gas. Then the EXAFS data were analyzed using IFEFFIT software, an interactive program for XAFS analysis.

3. Results and Discussion

Fig. 1 shows XRD patterns of mechanically alloyed $Fe_{47.5}Mn_{47.5}C_5$ powders. In the beginning of process, its shows the presence of Fe (Im-3m), Mn (I-43m) and C (P63mc) phases. Their diffraction peaks become weaker and broader when the milling time is increased. This is due to the structure deformed. Carbon peaks disappear from beginning of milling, it diffused as interstitial to Fe and Mn structures as host structure, also of variation in crystallite size. Started from 36 hrs milling where the main peak is shifted to smaller angle and some peaks disappeared. The single phase occurred after 36- and 48 hrs milling. It looks like the structures of 36- and 48 hrs milling are fixed; the peaks are tended to form a space group of fm3m. The structure is changed to manganese structure after 36 hrs and more completed after 48 hrs milling.

The EXAFS give direct information about the variation of local structure. Its use to examine the local structure around the Fe ions in the $Fe_{47.5}Mn_{47.5}C_5$ alloys. **Fig. 2** shows the normalized near edge spectra for the processed samples were similar to each other but above the edge the spectra gradually changed. This suggests that the electronic configuration for the Fe central atoms was unchanged but the surrounding around the Fe atoms was changed during the MA processing.

Fig. 3 shows the EXAFS spectra of $Fe_{47.5}Mn_{47.5}C_5$ alloys for 1- to 48 hrs. The reduction of amplitude is related to the disorder of local structure, and the variation of the phase is related to the change of chemical order [13]. In Figure shows the significant change in the amplitude and the phase took place after 36 hrs milling. Its mean there is a huge changed in local structure. The systematic variations of the amplitude and the phase in the EXAFS spectra confirmed that alloying at atomic scale occurred during the MA process. The EXAFS spectra were obtained from the absorption spectra by removing the background with AUTOBK.

IOP Conf. Series: Journal of Physics: Conf. Series **914** (2017) 012010 doi:10.1088/1742-6596/914/1/012010

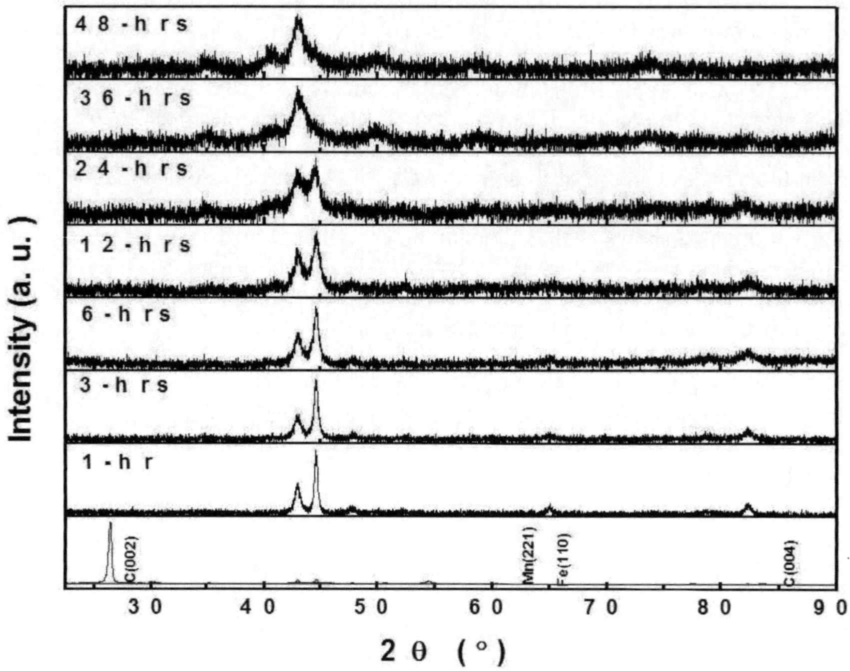

Fig.1: XRD profiles of $Fe_{47.5}Mn_{47.5}C_5$ mechanically alloyed as a function of milling times.

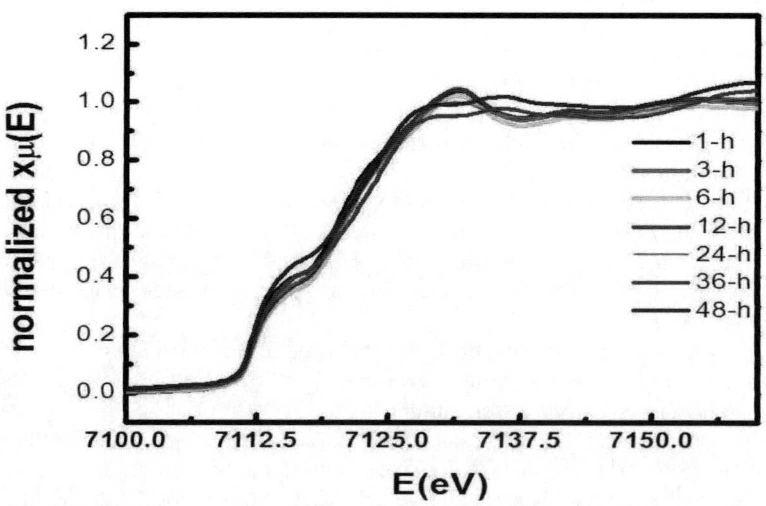

Fig.2: The EXAFS spectra of $Fe_{47.5}Mn_{47.5}C_5$, each line represents the processing time.

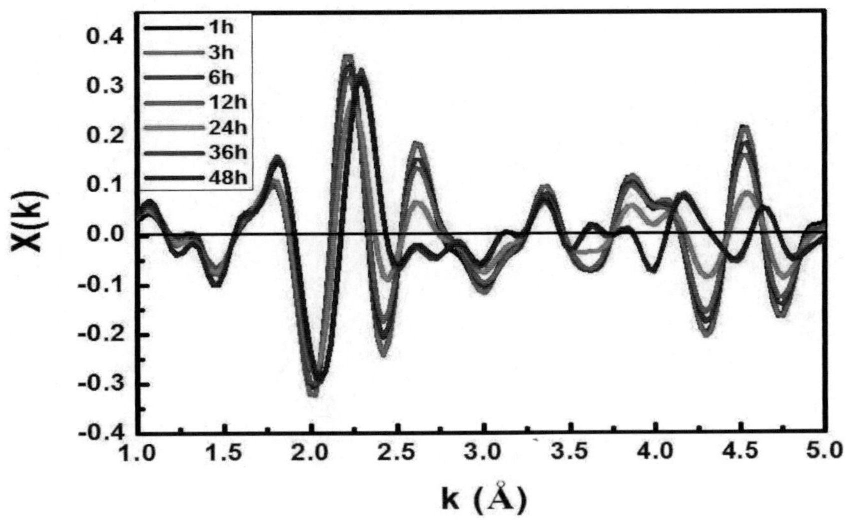

Fig. 3: The k-weighed EXAFS spectra of 1-hr to 48 hrs milling.

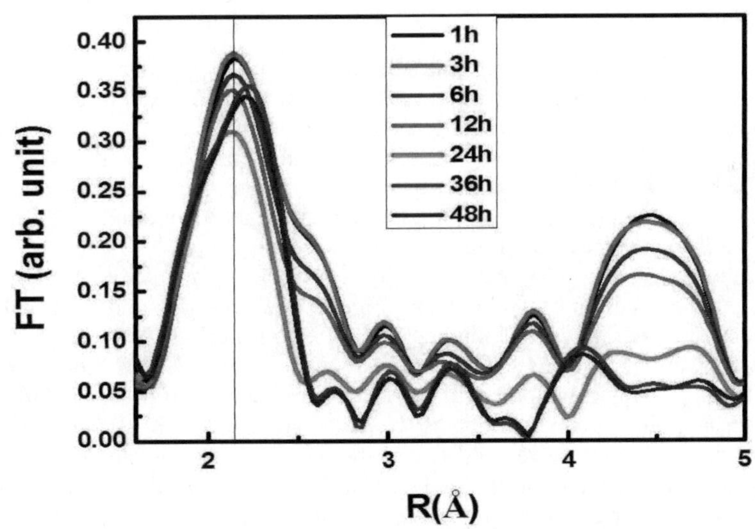

Fig. 4: Fourier transform of EXAFS spectra for $Fe_{47.5}Mn_{47.5}C_5$ alloys measured at Fe K edge as a function of milling times.

The decrease of the amplitude before 36 hours indicates that the fracture was dominant. After 36 hours milling, the phase was shifted significantly. This indicates that the Mn and C atoms were diffused into the Fe structure and new materials were produced during this process. The amount of the new alloy increased as the processing time 36 hrs and afterword, but for 36- and 48 hrs milling the alloys were look like stabled.

Fig. 4 shows that the Fourier (FT) of EXAFS spectra measured at Fe K-edge. The radial atomic density has been seen in spectrum FT. In the Figure, the magnitude of the first peak in the

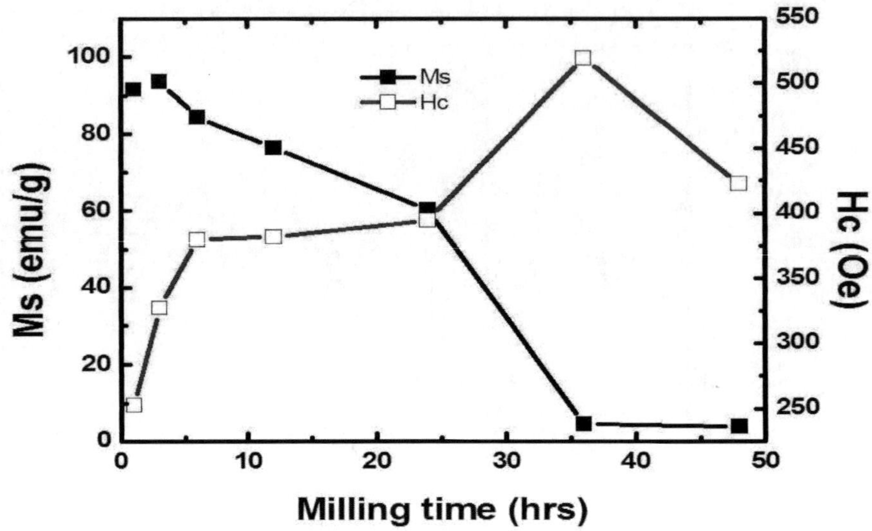

Fig. 5: Variation of magnetization and coercivity for $Fe_{47.5}Mn_{47.5}C_5$ alloys as a function of milling times.

Fourier transformed spectrum decreased gradually as the processing time increased. This suggests that the number of Fe-Fe direct bond decreased due to both the fracture of crystalline and the alloying with other kind atoms. The short and long range orders were reduced gradually. The spectrum around of 2.2 A have changed for 36- and 48 hrs, it mean for the milling time have occurred the new phase. At 4.5 Å, the spectrum have changed for 36-, and 48 hrs, the new phase have occurred. In 36 hrs milling and after, both short and long range order are changed in amplitudes and phases. It means there is a huge change in local structure.

The magnetic saturation (*Ms*), and the Coercivity (*Hc*) are found from the hysteresis loop depend strongly on milling times and the C content. **Fig. 5** shows that *Ms* is decreased continuously as the milling times increased. Magnetization rapidly decreased until 24 hrs, but after 36 hrs milling the graph is flat to 48 hrs with *Ms* is 1 emu/g, it means there is transition from Fe grains to the $Fe_{47.5}Mn_{47.5}C_5$ grain. This variation could come from the dilution of magnetic lattice of Fe by Mn and C with increasing milling time. On the other hand, the coercivity (*Hc*) is increased until 36 hrs, then at 48 hrs the coercivity (*Hc*) is decreased because the sample tend go to single domain, so the coercivity drop to smaller. The increase of H_c could be attributed to crystallite size reduction for Fe. The coercivity (H_c) is reaching a maximum value of approximately 550 Oe after 36 hrs, then it reduce to 400 Oe after 48 hrs milling.

4. Conclusions

The relatively new phase of Fe-Mn-C alloy is explicitly shown in the EXAFS spectra by the variation of amplitude and phase between 36- and 48 hrs milling times. The significant change of the structural phase revealed that new atom neighbors, the atom central Fe substituted by Mn and C atoms were increased during the MA process. The magnetic saturation (*Ms*) is decreased in long milling times as effect of crystallite size decreased. The coercivity (*Hc*) is increased until 36 hrs milling as effect of decreased of grain then it is decreased to 48 hrs milling because of single domain is growth as an effect of the carbons.

Acknowledgments
The research was supported by the Community Service Centre - Mercu Buana University (Grant No: 02-5/01/B-SPK/VIII/2016).

References

[1] A. Djekoun et. al., Structure and magnetic properties of Fe-rich nanostructured Fe_{100-X} Ni_X powders obtained by mechanical alloying, Physics Procedia **2** (2009) 693–700.
[2] C. Suryanarayana, C. C. Koch, Nanocrystalline materials–Current research and future directions, Hyperfine Interactions **130** (2000) 5–44.
[3] Gema González et. al., Mechanical Alloying of FeCoCr, Revista Latinoamericana de Metalurgia y Materiales 2011; **31** (1): 64-70.
[4] C. Suryanarayana, Mechanical alloying and milling, Progress in Materials Science **46** (2001) 1-184
[5] I. Chicinaş, Soft magnetic nanocrystalline powders produced by mechanical alloying routes, Journal of Optoelectronics and Advanced Materials **8**, No. 2, (April 2006) 439-448.
[6] M. M. Rico et al., Magnetic and Structural Properties of Mechanically Alloyed $Fe_xMn_{0.70-x}$ $Al_{0.30}$ (x = 0.40 and 0.45) Alloys, Physica Status Solidi (a) **189**, No. 3, (2002) 811–816.
[7] Yong-Goo Yoo, Bingzhi Jiang, J.M. Greneche, Dong-Seok Yang and Seong-Cho Yu, Local structures of nanostructured $(Fe_xCo_{1-x})_{75}C_{25}$ alloys, J. Magn. & Mag. Mat., **304** (2006) e715~e717.
[8] Hongwei Shi, Debo Guo, Yifang Ouyang, Structural evolution of mechanically alloyed nanocrystalline FeAl intermetallics, J. Alloys Comp. **455** (2008) 207-209
[9] R. Hasegawa, Present status of amorphous soft magnetic alloys, Journal of Magnetism and Magnetic Materials **215–216** (2000) 240-245.
[10] Michael E. McHenry, Matthew A. Willard, David E. Laughlin, Amorphous and nanocrystalline materials for applications as soft magnets, Progress in Materials Science **44** (1999) 291-433.
[11] V. Edon et al, Journal of Applied Physics 107, 09A321 (2010) 09A321-1-3.
[12] Xing LU, Zuoxiang QIN, Yansheng ZHANG, Xingyu WANG and Fengxian LI, J. Mater. Sci. Technol., Vol. 16 No.3, 2000 pp. 297-301.
[13] D. S. Yang, et al., EXAFS Study for a Magnetic Shape Memory Alloy Ni-Mn-Ga, Journal of the Korean Physical Society, Vol. 50, No. 4, April 2007, pp. 1062~1067.

Morphological and Physical Behaviour of LSCF-SDCC-Ag Composite Cathode with the Incorporation of Ag as an Additive Element

M.S.A. Bakar[1], H.A. Rahman[1*], S.F. Mohammad, S. Ahmad[1] and A. Muchtar[2]

[1]Faculty of Mechanical and Manufacturing Engineering, Universiti Tun Hussein Onn Malaysia, 86400 Batu Pahat, Johor, Malaysia.
[2]Fuel Cell Institute, Universiti Kebangsaan Malaysia, 43600 UKM Bangi, Selangor, Malaysia

Corresponding author: hamimah@uthm.edu.my

Abstract. The composite cathode, lanthanum strontium cobalt ferrite-samarium-doped ceria carbonate-argentum (LSCF–SDCC–Ag), was developed and investigated for low-temperature solid oxide fuel cell (LTSOFC) applications. This research studied the effect of Ag on the development and properties of the composite cathode. The composite cathode powders, LSCF–SDCC–Ag with an Ag addition of 1–5 wt% were developed via high energy ball milling technique, and all prepared pellets were sintered at 500, 550, and 600 °C. The morphological properties of the composite cathode powders were observed via FESEM micrograph, and the average particle sizes of the composite powders were measured via SmartTiff Software. The total porosity (%) of the LSCF–SDCC–Ag composite cathode pellets was determined using the Archimedes method. The FESEM micrograph revealed that the obtained composite cathode powder is homogeneous, with particle sizes of 100–200 nm. The particle size of the LSCF–SDCC–Ag cathode increased with the increase of Ag content. Meanwhile, the porosity decreased in value with the increase of the Ag content and the sintering temperature. The porosity percentage range from 26.6% to 33.8%.

1. Introduction

The solid oxide fuel cell (SOFC) is an electrochemical energy conversion device that delivers high electrical efficiency and significant environmental friendliness in terms of fuel flexibility. It is also clean and efficient with more than 70% with fuel regeneration electric power generation. A SOFC produces useful electricity from the reaction via the diffusion of oxide ions (or protons) through an ion acting between solid electrolyte layers [1–2]. Initial research in SOFCs focused mostly on high temperatures between 800 and 1000 °C. Reliability issues at high temperature have caused a shift in SOFC research towards operation at low temperatures (LT) to intermediate temperatures (IT), that is, from 400 °C to 750 °C. However, a reduction in operating temperature reduces the efficiency of the cathode elements, as it leads to the increase in cathode polarization resistance causes a substantial performance decline [3]. Designing a high-performance electrode with low polarization resistance (such as by using a composite cathode) is a potential solution for the aforementioned problems [4]. One of the method to improve low-temperature cathode performance involves the development of composite cathode materials. Recently, numerous SOFC researchers have studied issues associated with operational temperature. They have presented significant solutions regarding cathode material selection, synthesis, and processing.

Perovskite-type materials have been widely used as cathode materials in SOFCs. A key characteristic of the perovskite structure is its ability to conduct both electrons and ionic species, thus leading to them being designated as mixed electronic and ionic conductors (MIECs) [5]. MIEC materials from rare earth cobalt oxide groups include $La_{0.6}Sr_{0.4}Co_{0.2}Fe_{0.8}O_{3-\delta}$ (LSCF), one of the promising cathode materials with good electrical conductivity, a good thermal expansion coefficient, and a good melting point [6]. The universal properties of MIECs include good ionic and electronic conductivity. Their excellent conductivity property enhances the triple phase boundary (TPB), which holds the reactive site at the cathode. This phenomenon may account for the improved electrochemical performance of the SOFC [7].

Additionally, an electrolyte composite known as samarium doped ceria (SDC) carbonate (SDCC) was found to perform well at low temperatures (400–600 °C). The incorporation of SDCC into LSCF was reported to dramatically affect the electrochemical potential of IT–LTSOFCs cathode. The use of binary carbonate ($(Li/Na_2)CO_3$) together with SDC is to overcome ionic conductivity issues, in which the LSCF–SDCC composite cathode was reported to be efficient as LTSOFC. It has been proven to yield outstanding properties [8–10]. Nevertheless, ceria-based electrolyte systems have contributed to an increase in the power outputs of LTSOFCs. The use of SDCC electrolytes is believed to improve the performance of LSCF–SDCC composite cathode because carbonates universally exhibit good compatibility with many composite cathode powders.

The evaluation of SOFCs has significantly enhanced the performance of composite cathode powders through the incorporation of metallic elements such as silver (Ag), platinum (Pt), paladium (Pd), and cuprum (Cu). Given its superior electronic conductivity, high catalytic activity, substantial oxygen solubility, and mobility, Ag is considered the most suitable candidate as a cathode material in LTSOFC applications [11–12]. Moreover, the use of this metal also improves the cathode microstructure, as the melting point of Ag is 961 °C. This advantage may aid the sintering behavior of the LSCF–SDCC–Ag composite cathode. This research work focuses on developing the composite cathode, LSCF–SDCC–Ag, to suit LTSOFC applications. Further investigations are on the effect of Ag addition to the morphological and characteristic of the composite cathode LSCF–SDCC materials.

2. Materials and Methods

A commercial LSCF powder (LSCF6428; KCeracell, Korea), self-fabricated SDC, and a binary carbonate (SDCC) powder were used. A mixture of SDC and binary carbonate was ball-milled for 24 hours before being dried overnight in an oven and undergoing calcination at 680 °C. The composite cathode powder of LSCF–SDCC was prepared with a weight percentage ratio of 50:50 using high energy ball milling (HEBM) techniques with propanol as the solvent medium. The obtained powder was then ground using an agate mortar and calcined in a furnace at 750 °C. For the preparation of LSCF–SDCC–Ag, a dry mixing method of 100 rpm was employed with the addition of 1–5 wt% Ag powder. The LSCF–SDCC–Ag composite cathode pellets were fabricated via a uniaxial press using a constant pressure of 46 MPa followed by a sintering process at temperatures ranging from 500–600 °C. To obtain the desired porosity ranging from 20%–40%, the sintering temperatures used in this study were lower than the temperature used for calcinations.

Next, the morphology of the composite cathode LSCF–SDCC without Ag content and the LSCF–SDCC–Ag (1–5 wt%) were observed via FESEM. The average particle size measurements were obtained using SmartTiff Software (Version V02.01, Carl Zeiss Microscopy Limited) based on the exact scale on the FESEM image. By using the Mettler Toledo machine, the Archimedes method was employed to measure the porosity of the composite cathode sintered pellet. The porosity measurement methods were based on the standard ASTM C20-00 (Standard test Method for Apparent Porosity, Water Absorption, Apparent Specific Gravity, and Bulk Density of Burned Refractory Brick and Shapes by Boiling Water), with several modifications for compatibility with the composite cathode pellet samples. Composite cathode pellets were dried at 150 °C for 24 hours, followed by cooling at room temperature, before the porosity test was conducted.

3. Results and Discussion

3.1 Morphology of LSCF-SDCC-Ag Composite Cathode Powders

The images of the microstructures of LSCF–SDCC and the LSCF–SDCC–Ag composite cathode powder were captured using a FESEM micrograph (Figures 1(a) to 1(e)). The FESEM micrograph showed that the particles tended to agglomerate as Ag content increased. As observed, the particle size was in nanoscale. The particle size of the composite cathode powder LSCF–SDCC produced via HEBM was less than 200 nm. The HEBM has also been proven to produce a composite cathode powder measuring below 200 nm [13]. A composite cathode powder with a small size (between 100–200 nm) was desirable because of the reduced particle size and increased surface area. This phenomenon enlarged the TPB area for oxygen reduction reaction to occur in the cathode, which might directly improve the cathode performance [14].

The average particle size was taken from five different spots of the FESEM images with 100 measurements for each sample. Table 1 lists the average particle sizes of LSCF–SDCC without Ag and LSCF–SDCC–Ag with Ag content ranging from 1–5 wt%. The average particle size ranged from 80.0–89.2 nm. The composite cathode of LSCF–SDCC without Ag had the lowest average particle size (80.0 nm), whereas the composite cathode LSF–SDCC with a 5% addition of Ag content had the highest average particle size (89.2 nm). The particle size tended to increase as the amount of added Ag was increased. The average particle size of the composite cathode powder did not change drastically with the amount of Ag content (1–5 wt%). As the average particle sizes ranged from 80–89.2 nm, all of the composite cathode powders met the required characteristics and were thus suitable for composite cathode powder materials [15–16]. This conclusion is based on the criterion that particle sizes below 1 μm can provide an appropriate surface area absorption in the ceramic body. Hence, the average particle size of all composite cathode powder of LSCF–SDCC–Ag (1–5 wt%) in the present study satisfied the requirement of particle size for a SOFC's cathode powder. The nano-particles powder with large surface areas will extend the TPB and supply an increased catalytic activity toward the oxygen reduction reaction (ORR), consequently improving electrochemical performance [17].

Figure 1. FESEM micrographs of composite cathode LSCF–SDCC (a) without Ag, (b) Ag 1 wt.%, (c) Ag 2 wt.%, (d) Ag 3 wt.%, (e) Ag 4 wt.%, and (f) Ag 5 wt.% at 50K magnification

Figure 2. Measurement of average particle size from FESEM image of LSCF–SDCC–Ag composite cathode powder

Table 1. Average particle size of composite cathode powders LSCF–SDCC–Ag based on different amounts of added Ag

LSCF–SDCC–Ag (wt%)	Particle size (nm)
0	80.0 ± 17.4
1	81.5 ± 12.4
2	83.7 ± 14.0
3	86.8 ± 14.7
4	87.8 ± 15.3
5	89.2 ± 16.4

3.2 Porosity of LSCF-SDCC Composite Cathode Pellets

The porosity values of the cathode pellets for different sintering temperatures and various amounts of Ag content are shown in Figure 3 below. The increasing amount of Ag clearly decreased the porosity of the LSCF–SDCC–Ag composite cathode pellets. The particle size of LSCF–SDCC–Ag also dramatically affected the porosity value of the composite cathode pellet, such that the increased size of the particle with Ag content significantly reduced the porosity. Nevertheless, high sintering temperatures produced dense bodies, which affected the porosity of the LSCF–SDCC–Ag [8, 18].

Researchers believe that the LTSOFC achieves optimal ionic conductivity when porosity is approximately 30%. Several researchers suggested that the porosity of the cathode should range from 20% to 40%. Cathode components should have acceptable porosity to allow for a sufficient amount of oxygen to flow through the cathode, reaching the cathode–electrolyte region for ORR to occur. Optimal cathode porosity also produces the best power output from the LTSOFC [19]. Therefore, the implementation of a sintering temperature higher than 600 °C for LSCF–SDCC–Ag was avoided to ensure that the value of porosity would not be lower than 20%.

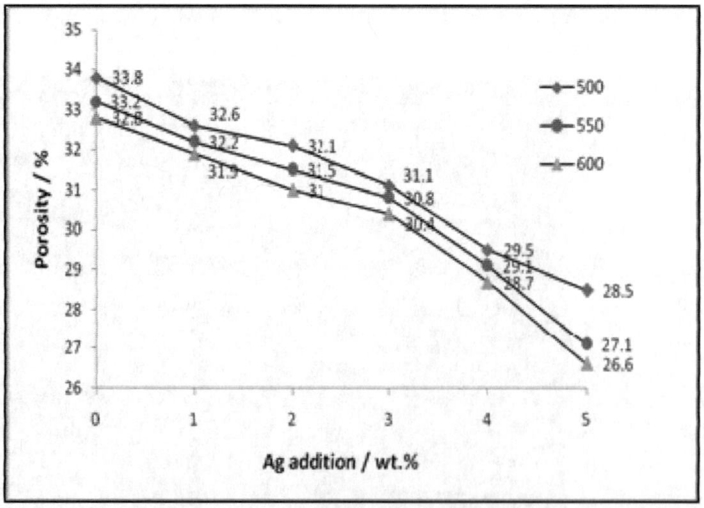

Figure 3. Average porosity of sintered pellet LSCF-SDCC-Ag at 500, 550 and 600°C

4. Conclusions

The present work investigated the effect of Ag as an additive element on a composite cathode, LSCF–SDCC–Ag (mixed using the HEBM). The morphological and physical characteristics of the composite were examined in relation to the 50% weight fraction of the SDCC electrolyte. The FESEM micrographs indicated that adding Ag to the composite cathode powder made the powder homogeneous and produced particle sizes ranging from 100–200 nm. The particle sizes of LSCF–SDCC–Ag increased with the increase of the Ag content. The porosity value decreased as the Ag content and sintering temperature increased. However, the porosity percentage was within the acceptable range of 26.6% to 33.8%. Finally, the influence of the addition of Ag on the morphological and physical properties of LSCF–SDCC composite cathodes was discussed briefly.

5. Acknowledgments

The authors would like to acknowledge the financial supports from the Universiti Tun Hussein Onn Malaysia (UTHM) and the research sponsorship from Ministry of Higher Education (FRGS Vot 1585) for supporting this work.

6. References

[1] Mahato N *et al* 2015 Progress in Material Selection for Solid Oxide Fuel Cell Technology: A Review *Progress in Materials Science* **72** 141–337

[2] Bidrawn F 2010 Investigation and Improvement of SOFC Composite Cathodes.

[3] Liu Z *et al* 2013 High-performance, ceria-based solid oxide fuel cells fabricated at low temperatures. *Journal of Power Sources*, **241** 454–459

[4] Muhammed Ali *et al* 2011 A review on preparation of SDC-carbonate as composite electrolyte material for intermediate temperature Solid Oxide Fuel Cells (IT-SOFC). *2011 IEEE 1st Conference on Clean Energy and Technology, CET 2011*, (March), 394–399

[5] Ahmad S *et al* 2014 Brief Review: Electrochemical Performance of LSCF Composite Cathodes - Influence of Ceria-Electrolyte and Metals Element. *Applied Mechanics and Materials*, **695** 3–7.

[6] Rahman H.A *et al* 2012 Structure and thermal properties of $La_{0.6}Sr_{0.4}Co_{0.2}Fe_{0.8}O_{3-\delta}$–SDC carbonate composite cathodes for intermediate- to low-temperature solid oxide fuel cells. *Ceramics International*, **38(2)** 1571–1576

[7] Bakar M *et al* 2015 XRD and EDS Analysis of Composite Cathode Powders LSCF-SDCC-Ag for Low Temperature Solid Oxide Fuel Cells (LTSOFC). *Advanced Materials Research*, **1087** 207–211

[8] Agun L *et al* 2014 Influence of Binary Carbonate on the Properties of Low Temperature Cathode Composite SOFC, *2014* (ICXRI 2014) 1–2

[9] Rahman H.A *et al* 2016 Ba- and La- Strontium Cobalt Ferrite Carbonate Composite as Cathode Materials for Low Temperature SOFC, **694** 125–129

[10] Ng K H *et al* 2016 Influence of Calcination on the Properties of Nickel Oxide-Samarium Doped Ceria Carbonate (NiO-SDCC) Composite Anodes. *Procedia Chemistry*, **19** 267–274

[11] Ahmad S *et al* 2014 Brief Review: Electrochemical Performance of LSCF Composite Cathodes - Influence of Ceria-Electrolyte and Metals Element. *Applied Mechanics and Materials*, **695** 3–7.

[12] Simner S *et al* 2007 Silver-perovskite composite SOFC cathodes processed via mechanofusion. *Journal of Power Sources,* **168** 236-9

[13] Jing Y *et al* 2013 An improved synthesis method of ceria-carbonate based composite electrolytes for low-temperature SOFC fuel cells, *Int. J. Hydrogen Energy 38,* **36** 16532-16538

[14] Sun Y *et al* 2014 The effect of calcination temperature on the electrochemical properties of $La_{0.3}Sr_{0.7}Fe_{0.7}Cr_{0.3}O_{3-x}$ (LSFC) perovskite oxide anode of solid oxide fuel cells (SOFCs). *Sustainable Energy Technologies and Assessments*, **8** 92–98

[15] Fergus J *et al* 2009 *Solid Oxide Fuel Cells: Material Properties and Performance.* Boca Raton, London, New York: Taylor & Francis Group

[16] Fan L *et al* 2012 Pr_2NiO_4–Ag composite cathode for low temperature solid oxide fuel cells with ceria-carbonate composite electrolyte. *International Journal of Hydrogen Energy*, **37(24)** 19388–19394

[17] Rahman H.A *et al* 2014 Influence of Calcination on the Properties of $La_{0.6}Sr_{0.4}Co_{0.2}Fe_{0.8}O_{3-\delta}$-Samarium Doped Ceria Carbonate. *4th Mechanical and Manufacturing Engineering, Pts 1 and 2*, **465–466** 949–953

[18] Niwa E Uematsu C and Hashimoto T 2013 Sintering temperature dependence of conductivity, porosity and specific surface area of LaNi0.6Fe0.4O3 ceramics as cathode material for solid oxide fuel cells - Superiority of Pechini method among various solution mixing processes. *Materials Research Bulletin*, **48(1)** 1–6

[19] Raharjo J *et al* 2011 Fabrication of Porous LSCF-SDC Carbonates Composite Cathode for Solid Oxide Fuel Cell (SOFC) Applications, **472** 179–184

Review of CO_2 Reduction Technologies using Mineral Carbonation of Iron and Steel Making Slag in Malaysia

Siti Norhana Selamat[1], Nik Hisyamudin Muhd Nor[1,*], Muhammad Hanif Abdul Rashid[1], Mohd Fauzi Ahmad[1], Fariza Mohamad[1], Al Emran Ismail[1], Mohd Fahrul Hassan[1], Faiz Mohd Turan[2], Mohd Zamzuri Mohd Zain[3], Elmi Abu Bakar[4], Yokoyama Seiji[5]

[1]Department of Materials and Design Engineering, Faculty of Mechanical and Manufacturing Engineering, Universiti Tun Hussein Onn Malaysia , 86400, Parit Raja Batu Pahat Johor , Malaysia.
[2]Faculty of Manufacturing Engineering, Universiti Malaysia Pahang, 26600 Pekan, Pahang, Malaysia.
[3]School of Manufacturing Eng., Universiti Malaysia Perlis, Kampus Tetap Pauh Putra, Jln Arau-Changlun, 02600, Arau, Perlis, Malaysia.
[4]School of Aerospace Engineering, Universiti Sains Malaysia, 14300 Nibong Tebal, Seberang Perai Selatan, Pulau Pinang Malaysia.
[5]Faculty of Mechanical Engineering, Toyohashi University of Technology, 1-1 Hibarigaoka Tempaku Toyohashi Aichi Japan.

Corresponding author: sitinorhanaselamat@gmail.com

Abstract. Climate change, greenhouse gas effect, and global warming is envisioning to turn more awful and more terrible by year. Since the leading cause of global warming is uncontrolled CO_2 in atmosphere. The amount of unused steel slag is expected to increment later on, steel industries is one of the mechanical industries that contribute the CO_2 emission. That because this businesses deliver carbon in light of powers reductant and substantial volume of steel. The changes of atmosphere these day is truly developing concern and that make steel creator are confronted with test of discovering methods for bringing down CO2 emission. Malaysia is working decidedly in the diminishment of CO_2 gas. There are a few techniques in decreasing the amount of CO_2 in the air as underlined by the Intergovernmental Panel of Climate Change (IPCC), an organization under the United Country however CCS is an extremely encouraging innovation to moderate CO_2 emission in air. Mineral carbonation is another technique to store carbon dioxide permanently, long term stability and vast capacity.

1. Introduction

As per Malaysian Iron and Steel Industry Federation (MISIF), around 7.5 million tons of steel had been created in year 2014 [1]. In Malaysia around 750 tons of steel were delivered every day and roughly 7.5 tons of slag were produced, from the exploration completed in the Asian nation, roughly 10% of steel slag was produced from the industries of steel [2]. There are a few strategies in lessening the amount of CO2 in the air underlined by the Intergovernmental Panel of Climate Change (IPCC), an office under the United Country. Those are CO2 capture and storage (CCS), renewable carbon free energy (RE), Sources (solar),high efficiency of energy conversion and savings (ECS), policies and legislation (PL) and High Efficiency of Energy Conversion & Savings (ECS) are all the more concentrating on the lessening of CO2 produced, not to the diminishment of CO2 radiated [3-4]. Policies and legislation (PL) is covering extensive variety of degrees and advances. Carbon Capture and Storage (CCS) is extremely encouraging innovation to relieve CO2 emission in air. CO2 capture and storage (CCS) are being examined broadly everywhere throughout the world.

The CCS technologies include capturing the CO2 from the preparation of petroleum products and in addition the combustion of non-renewable energy sources, as in power generation. The CCS can likewise be connected to different procedures with high measure of CO2 emitted [4].The major components of a CCS process include separation of CO_2 from the exhausted gas, compression of CO_2 by high pressure, transportation of CO_2 from plant to stock yard and storage [9]. CO_2 storage, even with small amount of CO_2, would make an useful contribution to lowering emissions. Lately, the specialized writing on this field has extended quickly.

There are several CCS technologies namely biological storage (photosynthesis, soil carbon stock, ocean life storage etc), geological storage (underground or underwater), oil and gas reservoirs, enhanced oil recovery, chemical fixation, membrane flash process, artificial weathering, mineral carbonation and so on. The main concern of these techniques, besides applicability of them, is the temporary storage capability. For example, the CO_2 gas will stored in the sea, however about hundreds to thousands of years it will come back to the atmosphere. W J J Huijgen et al state, even though it has been brought out that the storage is not essentially be perpetual [10]. The inquiry stays with respect to whether re-discharge of CO2 impacts future era. Extra shortcomings of ocean storage are the nearby change of pH of the water and its instigated consequences for the earth. In comparison to other CCS technologies, mineral carbonation have some advantages in term of storage stability, long term continuous monitoring not required, and economically feasible.

In term of material used for the carbonation, carbonation can be done either using natural resources or industrial waste. However, the utilization of natural resources such as olivines, mafic, peridotite, wollastnite and some others can contribute to some other environmental and public acceptance issues [11]. Thus, the use of industrial waste is very much preferable in CCS. Finding a suitable industrial waste for the carbonation process is very much essential for this work. Since the re-use of steel slag in Malaysia is all more concentrating on the development area such as construction sector, the reutilization of steel slag in carbon capture and storage technologies is especially imperative. In mechanical procedures, ball milling process is most likely the best options because of the low energy utilization. There are no much literatures on CCS technology in Malaysia. Most of the literature are focusing on the CCS in Malaysian forest soils, power plant generation, renewable energy and policies [12-15] and the use of steel slag in carbon capture and storage that may be extremely useful for Malaysia particularly and all different nations in the world generally

2. Steel making manufacturing and CO_2 emission

Steel industries is one of the huge mechanical industries in Malaysia that donate the CO_2 gas in atmosphere which danger the environment and earth . It is on account of this industries deliver carbon in view of powers reductance and expansive volume of steel. The change of climate these days is truly developing unease and that make steel producer are confronted with problem of discovering methods for bringing down CO_2 emission without genuinely influencing the process of productivity or adding to cost [6].

The steel industries and the carbon intensity of iron have diverse differs impressively production courses, unrefined steel for scrap/electric have going from 0.4 t CO_2/t circular segment heater (EAFs), 1.7-1.8 tCO_2/t rough steel for the integrated blast furnace(BF) basic oxygen furnace (BOF) to 2.5 tCO_2/t rough steel for coal based reduced iron directly (DRI) processes. [41, 52]

The indirect carbon capture and storage discharge from the all steel plant would bring down their CO_2 outflow. This outflow source from the utilization of power delivered. It could be change to power produced from sustainable power source, hydroelectric, or nuclear power plant [26]. Henceforth this outcome will end up plainly close to zero of CO_2 emission for a piece or small scale process. Power decarbonised will frame if indirect CO_2 outflow from steel industries is diminish in future.

\

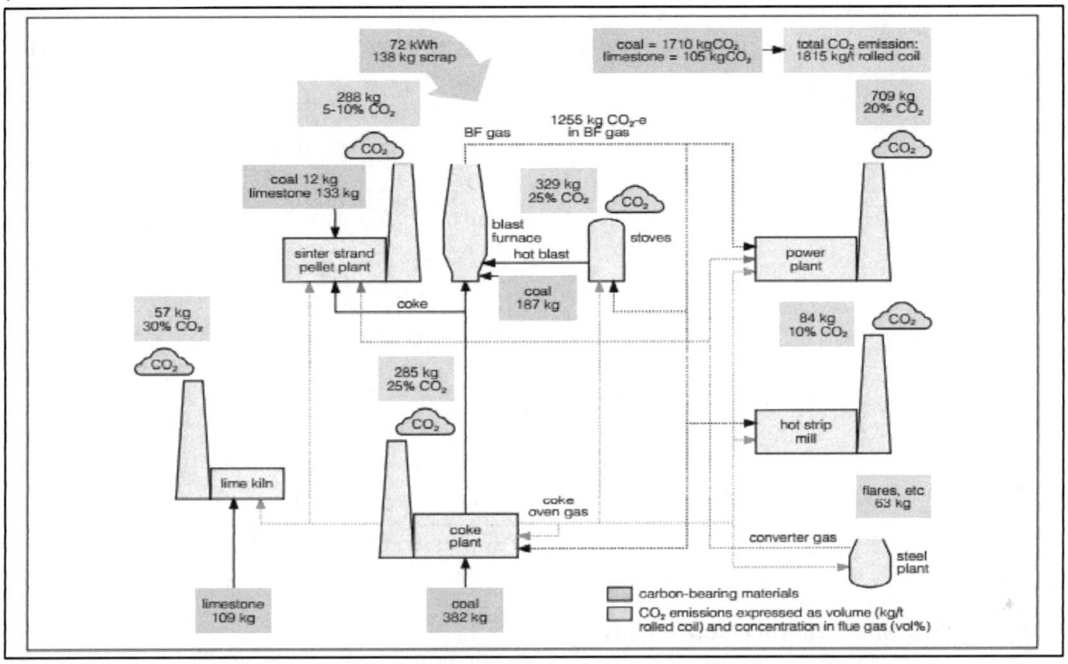

Fig. 1 CO_2 emission of typical steel mill [6]

The Fig. 1 above demonstrates the important CO_2 emission of typical steel mill process delivering hot rolled coil. The biggest wellspring of CO_2 is the on location control plant. The BF taken after by the sinter/pellet and coking plants are the biggest producers of CO_2 on the iron making side. Consequently to limit this impact, the innovation alleviation should be considered. Production of iron and steel likewise prompt the production of steel making slag as result. Essential utilization of slag are, as a bond material of cement, road base course material, material for structural designing and development work, silicate-calcium compost, crude material for high-level glass and numerous more [25]. However, the utilization of steel slag has been constrained by its propensity to self-corruption, caused by nearness of free lime in the slag and its hydration capacity joined by volume changes, the lessening of open work ventures and other reused materials, for example, reused roadbed materials and fly fiery debris, additionally progress toward becoming contender of the slag in the fields. Fig. 2 and Fig. 3 shows the total production of crude steel in Malaysia and the total production of crude steel in Asia and world. Both figure shows that the production of crude steel in Malaysia and Asia are worrying and necessity to overcome to lowest CO_2 emission.

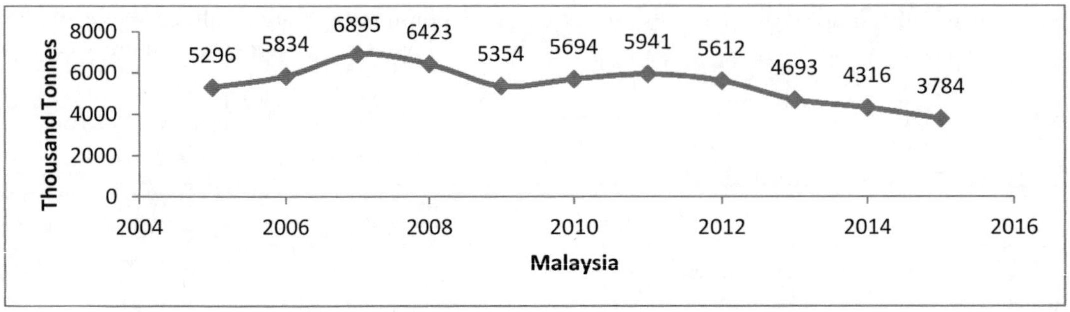

Fig. 2 Total production of crude steel in Malaysia

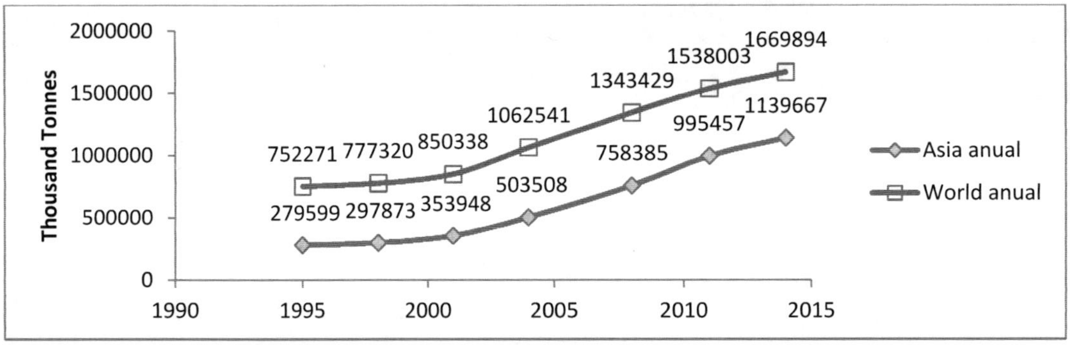

Fig. 3 Total production of Crude steel in Asia and World

3. Physical and chemical characteristics of steel slag

Steel making industry can take control of its vaporous CO_2 outflow with the strong of something created alongside steel slag, which is a rich wellspring of basic oxides, for example, CaO and MgO.The first parts of iron and steel slag are limestone (CaO) and silica (SiO_2) ,CaO, Fe_2O_3, FeO, Al_2O_3, P_2O_5 , SiO_2, MgO and MnO [17]. 17Utilizing mineral processing innovation, the part of iron and steel slag can be isolated and reused for sintering, blast furnace and steel making. Steel slag contain high of CaO, MgO, and Mno which can diminish iron and steel making cost on the grounds that can be utilized as substitution for part of limestones, dolomite and manganese mineral. The chemical part of steel slag diverse with the furnace sort, steel evaluations, and pre-treatment technique [19-21]. Some steel slags contain higher measure of P_2O_5 and S than iron concentration. This may impact the quick reusing of the steel slags to the iron and steel making process. Various applications using the physical and compound characteristics of slag have been created and are being put to use in a wide extent of fields. The chemical composition of the steel slags from different sources is showed up in Table 1. Table 2 records the link between the characteristics and application fields of steel slag.

Table 1. Chemical composition of steel slag (oxides %)

COUNTRY	Type Of Slag	CaO	SiO_2	Fe_2O_3	Al_2O_3	MgO	K_2O	Na_2O	SO_3
Malaysia(Ahmedzade & Sengoz, 2009)[22]	EAF	55.2	11	17	4	3.1	0.16	0.75	0.85

Thailand(Penpolcharoen, 2005)[23]	BFS	10.5	24	47	3.4	1.3	0.48	4.12	-
China(Xuequan, Hong, Xinkai, & Husen, 1999)[24]	Steel Slag	48.2	12	8.5	2.6	6.4	-	-	0.43
Japan(Shen & Forssberg, 2003)[25]	BOF	44.3	14	18	1.5	6.4	-	-	-
	EAF	38	19	15	7	6	-	-	-

Table 2. Characteristic and applications of steel slag [26]

Characteristics	Applications
Hard,wear resistant,Adhesive, rough	Aggregates for road and hydraulic construction Waste
Porous, alkaline	Waste water treatment
FeOx, Fe components	Iron reclamation
CaO,MgO,FeO,MgO,MnO componets	Fluxing agent
Cementitious components (C_3S, C_2S and C_4AF)	Cement and concrete production
CaO, MgO components ,FeO,	CO2 capture and flue gas desulfurization
FeO, CaO, SiO_2 components ,Fertilizer	Raw material for cement clinker
Fertilizer components (CaO, SiO_2, MgO, FeO)	Fertilizer and soil improvement

4. Mineral carbonation

Mineral carbonation has been proposed as one of the central procedures for CO_2 sequestration. Mineral carbonation Technology (MCT) is a system whereby CO_2 is chemically reacted with calcium or possibly magnesium-containing minerals to form stable carbonate materials which don't cause any whole deal hazard or watching obligations [5]. Mineral carbonation contemplates are huge with respect to utilizing CO_2 for mechanical purposes [6-7].One imperative preferred standpoint of CO_2 sequestration by mineral carbonation include the actually liberal and in every way that really matters enduring catching of CO_2 as carbonated minerals by using rich mineral assets, for instance, Mg-silicates [27].There are a few potential materials that can be utilized as a sorbent. These materials are ordered into minerals and waste materials. The major idea driving the mineral sequestration is the guideline of principle of natural rock weathering [6]. Throughout weathering, Ca and Mg silicate mineral-bearing rock materials respond actually with the encompassing CO_2 to change into strong carbonates which are then put away in the ground forever [7-8]. As of now the most explored mineral assets are olivine, serpentine and wollastonite. The carbonation of common minerals normally requires a high temperature and high weight since regular minerals more often than not requires a high temperature and high weight since characteristic minerals are thermodynamically steady. For squander material and additionally certain mechanical waste, steel slag has gotten a great deal of consideration, for example, electric arc furnace (EAF) dust, blast, furnace slag and waste cement have been assessed for use as crude materials for mineral carbonation. (28-37).

4.1 Methods

Catching carbon dioxide in carbonates can be accomplished through different process. Under the mineral carbonation, there are a few systems, for example, natural carbonation, heat and treatment, aging, aqueous carbonation and mechanical activation. Mineral carbonation prepare ideas that have been contemplated can generally be isolated into direct and indirect methods as shows in Fig. 4 and Fig. 5

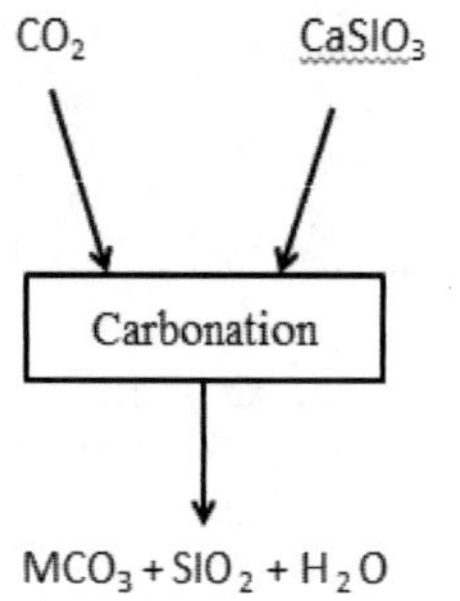

Fig. 4 Direct carbonation

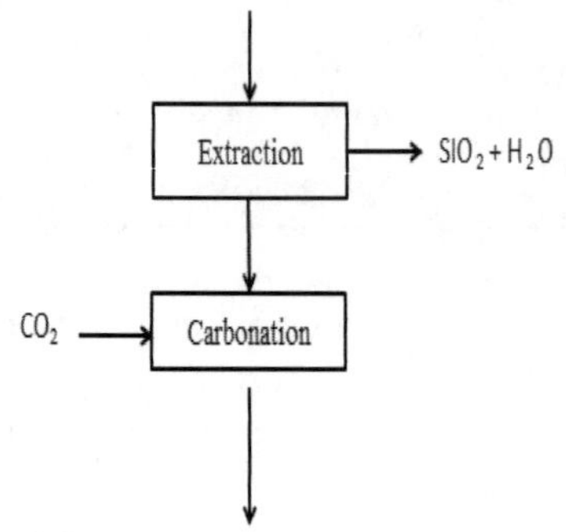

Fig. 5 Indirect carbonation

4.1.1 Direct gas solid carbonation

In the common carbonation process, the material that goes about as a sorbent will be presented to the CO_2 with no added substance or extra treatment for the response to occur. Gas-solid carbonation is a straightforward approach towards mineral carbonation. Shockingly, the response rates of such a procedure have been too moderate and the procedure experiences thermodynamic impediments and along these lines, not reasonable for the productive CO_2 sequestration. In Malaysia, mineral carbonation of red gypsum as a wellspring of calcium particle was concentrated to assess the possibility of utilizing it for CO_2 sequestration [31]. Direct carbonation technique is utilized to deliver solid carbonates in the exploration. Fresh red gypsum tests utilized as a part of this examination were gotten from landfill of Huntsman Tioxide, Kemaman, Terengganu, Malaysia. The composition arrangements of the red gypsum test were analyzed which demonstrates that the red gypsum is plainly ruled by CaO (32.2%), SO_3 (31.6%), and Fe_2O_3 (28.99%), and in addition TiO_2 (3.4%) together with some immaterial polluting influences. Plainly, the red gypsum is rich with Ca, a standout amongst the most ideal minerals for mineral carbonation. However experiment result demonstrated that the calcium and iron carbonates could be delivered through direct carbonation of red gypsum has low item immaculateness and carbonation productivity [32]. Past research from Reddy et al 2008, Baciocchi et al 2009, and Prigiobbe et al, 2009 additionally express that the direct gas–solid response of strong Ca/Mg silicate in equation 1 experiences moderate response rates and has for all intents and purposes been relinquished. Reaction kinetics, are moderate for Ca/Mg-silicates [25].

$$\text{Ca/Mg} - \text{silicate}_{(s)} + CO_{2(g)} - (\text{Ca/Mg})CO_{3\,(s)} + \text{SiO}_{2(s)} \quad (1)$$

4.1.2 Direct aqueous carbonation

Aqueous carbonation process is the procedure that utilizing strong or weak acid, (HCL, CH_3COOH), base solution(NaOH), or chelating agents (EDTA, citrus extract), basic aqueous solutions or chelating operators to break down metal oxides, at that point respond it with vaporous CO_2 which at long last would prompt the form of carbonates. The direct aqueous mineral carbonation-route alluding to carbonation preformed in a solitary stride in a aqueous solution, gives off an impression of being the most encouraging CO2 mineralization other option to date [38]. Direct aqueous mineral carbonation can be additionally divided into two subcategories, contingent upon the kind of arrangement utilized. Be that as it may, it has been accounted for [39] that there are still upgrades to be made in regards to the previously mentioned arrangement. K song at el has examined impact effect of polyacrilyc acid on direct aqueous mineral carbonation of flue gas desulfurization gypsum and expressed that the expansion of PAA brought down the general carbonation yield through the emission of $CO_2(g)$, which hampered CO_2 disintegration [32]. The lower carbonation yield within the sight of PAA would thus be able to be ascribed to the loss of carbonate through the emission of $CO_2(g)$. The direct aqueous carbonation of FGD gypsum can be expressed as far as the condition of every species, as in the following equation 2:

$$\mathrm{CaSO_4\,.\,2H_2O}_{(s)} + \mathrm{2NH_3(aq)} + CO_2\mathrm{(g)} \rightarrow CaCO_3(s) + (NH_4)_2\mathrm{SO_4(aq)} \quad (2)$$

4.1.3 Indirect carbonation

At the point when the mineral carbonation process is partitioned into a few stages it is delegated as indirect carbonation. For example, the extraction of Ca as well as Mg from the feedstock, the dissolution of CO_2 (on account of aqueous carbonation) and the precipitation of carbonate materials occur as isolated strides and additionally in various reactors. In light of review of mineral carbonation utilizing indirect carbonation [ ], pH swing mineral carbonation is one of indirect carbonation. The research state that pH swing mineral carbonation have been fundamentally looked into, the cost is too high than others mineral carbonation technique, generally high carbonation effectiveness is accomplished through pH swing mineral carbonation and high purity carbonated are delivered by pH swing mineral carbonation [28].

4.2 Mechanical activation

Numerous methods have been utilized to treat the mineral metals that are environmentally worthy among which mechanical activation is an inventive procedure to make new surfaces and basic imperfections in solid state [21–33]. Mechanical activation was observed to be a positive pre-treatment process adding to altogether upgrade metal recuperations or enhanced material performances [34]. Think about others strategy, mechanical activation required less energy consumption [5, 37]. One of the previous study that utilizing mechanical activation is mechanical activation of chemically activated high phosphorous slag content concrete.

Mechanical activation has noteworthy enhancing impact on the advance of hydration responses of chemically activated high phosphorous slag content cement. This thus significantly influences the microstructure and the compressive quality improvement of the cement [34]. Compare others method, mechanical activation required less energy consumption [5, 37]. One of the previous study that using mechanical activation is Mechanical activation of chemically activated high phosphorous slag content cement. Mechanical activation has significant improving effect on the progress of hydration reactions of chemically activated high phosphorous slag content cement. This in turn greatly affects the microstructure and the compressive strength development of the cement [34].

In Malaysia there is not much study and method using mechanical activation however, in industrial processes especially steel making industry, a traditional ball process is most likely the best option because of the the low energy consumption [37].Various techniques for mineral CO_2 sequestration have been suggested in many literatures such as the membrane flash process, pressurized gas/solid process, aqueous acid and alkaline solution process, mechanical activation and so on. However under mechanical activation technique, the use of mechanical grinding machines plays a significant role in enhancing the carbonation process. There are several types of grinding machines that typically used, for examples centrifugal mill, attrition mill, nutating mill, tumbling mill, planetary mill and vibrating mill. [40-48].

4.2.1Effect of Co₂ absorption into steel slag using mechanical milling

As a progression of major investigation of CCS by the mechanical grinding technique, the reaction between CO_2 and numerous ground metal powders; Fe, Ni, Pb, Sn and Cu, are most component in steel slag were done at the first place to investigate the to investigate the reaction behavior and mechanism, and to prove the possibility of utilization of this strategy to CCS. The specimen were chosen since they had moderately low proclivity of oxygen, as such, they were effortlessly diminished by hydrogen and generally simple to be gotten. An investigation using mechanical activation on past research studied the absorption of carbon dioxide onto metals and compound reagent calcium oxide under dry grinding which express that CO_2 not decomposed during the grinding but rather it happened on the new uncovered surface of the Fe powder produced by the grinding [6].

The absorption of CO_2 into Fe powder was quickened and upgraded with the grinding procedure by around 26 times in contrast with that without grinding process. Increment in the weight of Fe and rotational speed appeared to altogether influence the the rate of CO_2 absorption. However the underlying weight of CO_2 seemed to have no impact on the rate of CO_2 absorption into Fe powder. The absorption likewise seen on other specimens with the amount of CO_2 absorption tends to increment with the standard free energy of formation of each metal oxide. The grinding process by centrifugal mill did not crush the whole sample to finer particle, but only grazed the surface of the sample. The use of other type of grinding system that has higher impact intensity might lead to the enhancement of the co2 absorption into the metal powders. Fig. 6

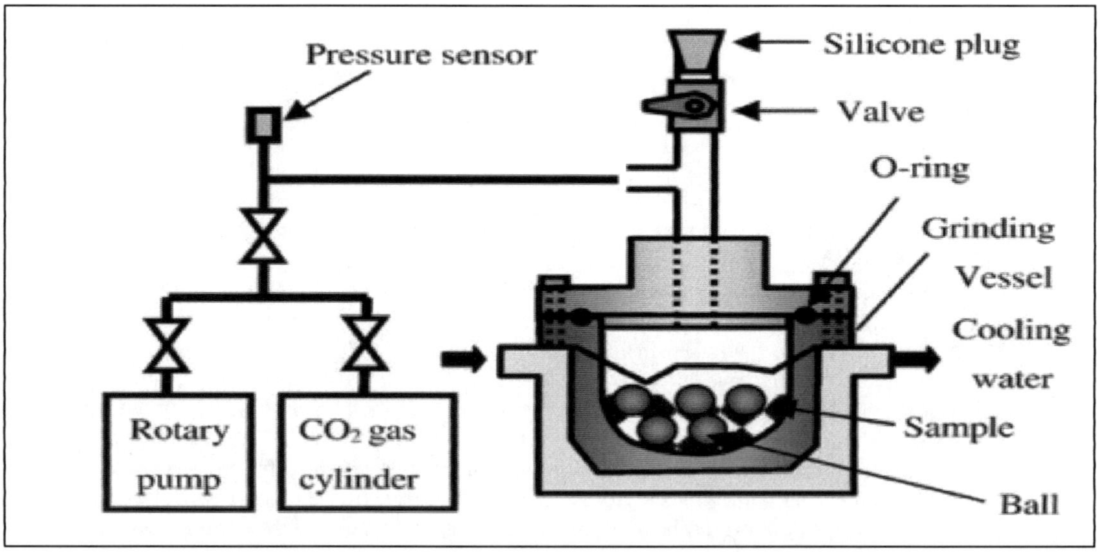

Fig. 6 Schematic Diagram of Experimental apparatus of dry grinding [5]

Fig. 7 shows the example of schematic diagram of the experimental apparatus in Malaysia. It is mainly composed of stirrer, pressure sensor (SDL 700), pressure gauge, tank, motor, CO_2 gas cylinder, and rotary pump, this machine required low energy to operate because using simple stirrer to mixture the CO_2 and slag sample. Both experiments show how mechanical activation process can absorb CO_2 gas without using much energy.

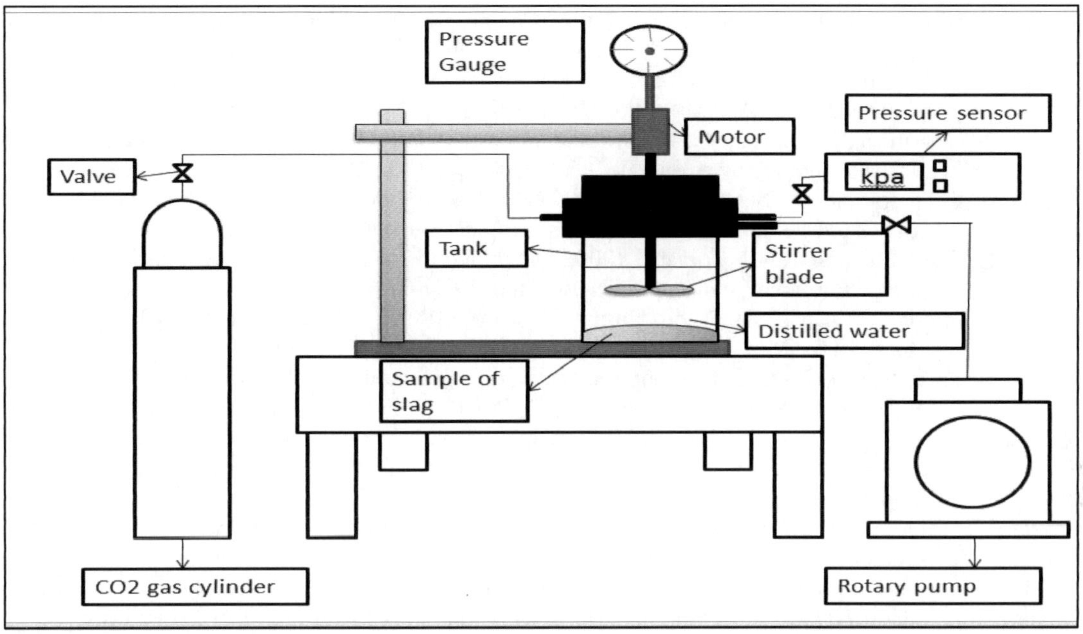

Fig. 7 Schematic Diagram of Experimental Apparatus of Mechanical Stirrer in Malaysia [52]

Conclusion

Malaysia is working positively in the reduction of CO_2 gas in line with the commitment for the 40% reduction in CO_2 intensity by GDP compared to 2005 [2,5]. Carbon storage technology must be increased in order to achieve further reduction of CO_2 gas emission. It can be concluded that the among techniques using additives for reactivity rising gave better results than those not using additives and for that, studies were performed with the purpose of a better understanding on the reactions complexity regarding the carbonation techniques. Compared to the developed countries, the utilization rate of steel slag is still very low in Malaysia. Therefore, large-scale utilization is a substantial resolution to the environmental problems arisen by steel slag dump, thus, CCS is a very promising technology to mitigate CO_2 concentration in atmosphere. Mineral carbonation is a potentially attractive sequestration technology for the permanent and safe storage of CO_2.

Acknowledgement

The authors would like to thank the financial support from Centre for Graduates Studies UTHM, Universiti Tun Hussein Onn Malaysia (UTHM).

References

[1] Malaysian Iron and Steel Industry Federation (MISIF), Performance of the Iron and steel Industry in 2014 information on http://www.misif.org.my/.

[2] Sharliza, N., Safaai, M., Noor, Z., Hashim, H., & Ujang, Z. (2010). Projection of CO_2 Emissions in Malaysia, *00*(00), 1–8.

[3] Intergovernmental Panel on Climate Change (IPCC), (1966) Technologies, Policies, and Measures for Mitigating Climate change – IPCC Technical Report.

[4] Intergovernmental Panel on Climate Change (IPCC), (2005). Carbon Capture and Storage

[5] Nor, N. H. B. M., Yokoyama, S., Kawakami, M., & Umemoto, M. (2009). Reaction between CO_2 and CaO under dry grinding. *Powder Technology*, *196*(2), 156–162.

[6] W.J.J Huijgen, G.J. Witkamp, R.N.J. Comans, (2006). Mechanisms of aqueous wollastonite carbonation as a possible CO_2 sequestration process, Chem. Eng. Sci. 61 (13) 4251.

[7] W.J.J. Huijgen, R.N.J. Comans,, G.J. Witkamp, (2007) . Cost evaluation of CO_2 sequestration by aqueous mineral carbonation, Energy Convers. Manage. 48 (7) 1923-1935.

[8] Ukwattage, N. L., Ranjith, P. G., & Li, X. (2017). Steel-making slag for mineral sequestration of carbon dioxide by accelerated carbonation, *97*, 15–22.

[9] R. Steeneveldt, B. Berger, T.A. Torp, (2006).CO_2 Capture And Storage: Closing the Knowing-Doing Gap, Chemical Engineering Research and Design 84(A9) p. 739-763

[10] W.J.J. Huijgen, R.N.J. Comans, (2003) Carbon dioxide sequestration by mineral carbonation, Energy Research Center of the Netherlands ECN-C--03-016

[11] Intergovernmental Panel on Climate Change (IPCC), (1966). Technologies, Policies, and Measures for Mitigating Climate change – IPCC Technical Report.

[12] Tick Hui Oh, (2010).Carbon capture and storage potential in coal-fired plant in Malaysia A review. RENEWABLE AND SUSTAINABLE ENERGY REVIEWS 14(9) 2697-2709.

[13] N. Y. G. Lai, Lee Chan Wai and E. H. Yap.(2011).Carbon capture and storage for developing economies: Yhe case for Malaysia, Clean Energy and Technology (CET), IEEE First Conference on, Kuala Lumpur pp. 182-186

[14] Md. Maniruzzaman B. A. Aziz, (2015).Feasibility of Carbon Capture, Storage Technology and Biofuels in Malaysia. 3rd International Symposium on Energy Challenges and Mechanics .

[15] Ebenezer Akin Oluwasola, Mohd Rosli Hainina, Md. Maniruzzaman A. Aziz, Characteristics and Utilization of Steel Slag in Road Construction. Engineering) 70:7 (2014), 117–123

[16] F. Goff, K.S. Lackner (1988). Carbon dioxide sequestering using ultramafic rocks, Environ. Geosci. 5 89–102.

[17] Motz H, Geiseler J. (2001) Products of steel slags an opportunity to save natural resources. Waste Manage ; 21:285-293.

[18] Liu SZ. (1944). Application of slag in steelmaking. Steelmaking (in Chinese) ;6:54-9.

[19] Wu XQ, Zhu H, Hou XK, Li HS. (1999). Study on steel slag and fly ash composite Portland cement. Cement Concrete Res ; 29: 1103-6

[20] Huang YH, Liu CJ. (2008). Analysis on comprehensive utilization of electric furnace slag. Ind Heat (in Chinese) ;37(5):4-6.

[21] Kourounis S, Tsivilis S, Tsakiridis PE, Papadimitriou GD, Tsibouki Z. (2007). Properties and hydration of blended cements with steelmaking slag Cement Concrete Res ; 37: 815-822.

[22] Ahmedzade, P., & Sengoz, B. (2009). Evaluation of steel slag coarse aggregate in hot mix asphalt concrete, *165*, 300–305

[23] Penpolcharoen, M. (2005). Utilization of secondary lead slag as construction material,*35*(August2003),1050-1055.

[24] Xuequan, W., Hong, Z., Xinkai, H., & Husen, L. (1999). Study on steel slag and fly ash composite Portland cement, *29*, 1103–1106.

[25] Shen, H., & Forssberg, E. (2003). An overview of recovery of metals from slags, *23*, 933–949.

[26] Yi, H., Xu, G., Cheng, H., Wang, J., Wan, Y., & Chen, H. (2012). An Overview of Utilization of Steel Slag. *Procedia Environmental Sciences*, *16*, 791–801.

[27] Metz, B., O. Davidson, H. C. de Coninck, M. Loos, and L. A. Meyer (eds.) (2005). IPCC special report on carbon dioxide capture and storage, Prepared by Working Group III of the Intergovernmental Panel on Climate Change, Cambridge University Press, Cambridge, United Kingdom and New York, NY, USA, IPCC.

[28] Lackner KS, Wendt CH, Butt DP, Joyce EL, Sharp DH. (1995). Carbon dioxide disposal in carbonate minerals. Energy ;20(11):1153e70.

[29] Lee MG, Ryu KW, Jang YN, Kim W, Bang J-H. (2011). Effect of oxalic acid on heat pretreatment for serpentine carbonation. Mater Trans ;52(2): 235e8.

[30] Ryu KW, Lee MG, Jang YN. (2011). Mechanism of tremolite carbonation. Appl Geochem ;26:1215e21

[31] O'Conner WK, Dahlin DC, Rush GE, Dahlin CL, Collins WK. (2002). Carbon dioxide sequestration by direct mineral carbonation: process mineralogy of feed and products. Miner Metall Proc ;19(2):95e101

[32] Eloneva S, Teir S, Salminen J, Fogelholm C-J, Zevenhoven R. (2008). Fixation of CO2 by carbonation calcium derived from blast furnace slag. Energy ;33: 1461e7

[33] Huijgen WJJ, Witkamp G-J, Comans RNJ. (2006). Mechanisms of aqueous wollastonite carbonation as a possible CO2 sequestration process. Chem Eng Sci ;61: 4242e51

[34] Huijgen WJJ, Comans RNJ, Witkamp G-J. (2007). Cost evaluation of CO2 sequestration by aqueous mineral carbonation. Energ Convers Manage ;45: 1923e35

[35] Iizuka A, Fujii M, Yamasaki A, Yanagisawa Y. (2004). Development of a new CO2 sequestration process utilizing the carbonation of waste cement. Ind Eng Chem Res ;43:7880e7.

[36] Teir S, Eloneva S, Fogelholm C-J, Zevenhoven R. (2007). Dissolution of steelmaking slags in acetic acid for precipitated calcium carbonate production. Energy ;32:528e39.

[37] Jang YN, Chae SC, Jung SB, Kim HS, Ryu KW. (2009). Fundamental study on sequestration of CO2 by mineral carbonation, Korea ministry of knowledge economy, KIGAM research report GP-2009-002-2009(1),

[38] Lackner, K.S., CLIMATE CHANGE: A guide to CO2 sequestration, Science. (300), pp. 1677-1678, 2003.

[39] McKelvy, M.J., Chizmeshya, A.V.G., Squires, K.D., Carpenter, R.W., Béarat, H. (2005). A novel approach to mineral carbonation: Enhancing carbonation while avoiding mineral pretreatment process cost, Arizona State University, Center for Solid State Science, Science and Engineering of Materials Graduate Program, and Department of Mechanical and Aerospace Engineering, June 21-22.

[40] K.E Kurrer, E. Gock, V. Vogt, R. Florescu, (2004).centrifugal tube mill for finest grinding, International Journal of Mineral processing 74S, S75-S8.

[41] D.I. Hoyer. (1992). Power consumption in centrifugal and nutating mills, Minerals Engineering 5(6) p. 671- 684.

[42] S.H.C. Cho, K.H. Kim, H. Lee, D.J. Kim. (2011). Study of residence time distribution and mill hold up fpr a ontinious centrifugal mill with various G/D ratios in a dry- grinding environment, minerals engineering 24 (1) p. 77-81.

[43] H. Choi, W. Lee, s. Kim. (2009). Effect of grinding aids on the kinetics of fine grinding energy consumed of calcite powders by a stirred ball mill, advanced powder Technology, 20(4) p. 350-354.

[44] V.mannhein, empirical and scale- up modelling in stirred ball mills. (2010). Chemical Engineering research and design, In Press, corrected Proof, available online 11 August.

[45] H. watanabe. (1999). Critical rotation speed for ball-milling, Powder technology 104 (1) p. 95-99.

[46] B.K. Mishra. (2003). A Review of computer simulation of tumbling mills by the discrete element method: Part II – Practical application, International Journal of mineral processing 71 91-4) p. 95-112.

[47] D,Chen, s. ni, z. chen, synthesis of Fe_3O_4 nanoparticle by wey millingiron powder in a planery ball mill hina Particuology 5(5) p. 357-358.

[48] T. Itsukaichi, S. Shiga, K. Masayama, M. Umemoto, I. Okane,. (2008).Materials Sc. Forum, vol. 88–90, 1992, pp. 631–638.

[49] A. J. B. Dutra, P. r. p. paiva, L. M. Travares. (2006). Alkaline leaching of zinc from electric arc furnace steel dust, mineral engineering 19 p. 478 – 485.

[50] C. Suryanaryana. (2001). Programme material science. 46 (1–2) (2001) 1–184.

[51] B.S. Murthy, S. Ranganathan. (1998). International material revision. 43 1–40

[52] S. N. Selamat, N. H. Muhd Nor, F. Mohamad, M. H. A. Rashid, A. K. Abd Rahim, A. E. Ismail, M. Z. M. Zamzuri, S. Yokoyama. (2017). Effect of Pressure on the CO_2 Absorption into

Ladle Furnace Slag, Materials Science Forum, Vol. 888, pp. 508-512.

IOP Conf. Series: Journal of Physics: Conf. Series **914** (2017) 012013 doi:10.1088/1742-6596/914/1/012013

The Influence of Porosity on Corrosion Attack of Austenitic Stainless Steel

Z Abdullah[1], A Ismail[2] and S Ahmad[3]

[1]PhD Student, Department of Material Engineering and Design, Faculty of Mechanical Engineering and Manufacturing, Universiti Tun Hussein Onn Malaysia, Batu Pahat, Johor

[2,3]Senior Lecturer, Department of Material Engineering and Design, Faculty of Mechanical Engineering and Manufacturing, Universiti Tun Hussein Onn Malaysia, Batu Pahat, Johor

Corresponding author: ikauthm88@gmail.com

Abstract. Porous metals also known as metal foams is a metallic body having spaces orpores through which liquid or air may pass. Porous metals get an attention from researchers nowadays due to their unique combination of properties includes excellent mechanical and electrical, high energy absorption, good thermal and sound insulation and water and gas permeability. Porous metals have been applied in numerous applications such as in automotive, aerospace and also in biomedical applications. This research reveals the influence of corrosion attack in porous austenitic stainless steel 316L. The cyclic polarization potential analysis was performed on the porous austenitic stainless steel 316L in 3.5% NaCl solution. The morphology and the element presence on the samples before and after corrosion attack was examined using scanning electron microscopy (SEM) and energy dispersive x-ray (EDX) respectively to determine the corrosion mechanism structure. The cyclic polarization potential analysis showed the result of (E_{corr}) for porous austenitic stainless steel type 316L in the range of -0.40v to -0.60v and breakdown potential (E_b) is -0.3v to -0.4v in NaCl solution.

1. Introduction

Porous metals and cellular structure are artificial metals that have pores in their bodies and having large volume of porosities. The interesting combination properties of porous metals are availability to maintain their mechanical properties at higher temperatures, lightweight while be able to be strong and stiff, capability of absorbance in energy, vibration and sound, and great thermal conductivity. Porous metals are essentials for a wide range of industrial technologies includes filters, electrodes, catalysts, heat exchanger and excellent for medical devices such as implants[1,2]. With an advanced in medical technology nowadays, porous metals has been applied in biomedical application such as in orthopaedic, dentistry, surgical instruments, cardiovascular surgery, and neurosurgery[3]. Thus, it is important to produce the porous metals that have high porosity and excellent in mechanical properties to provide synergy effect between implant and the bones to avoid implant loosening to happen[4].

Metals such stainless steel, cobalt-base alloys, and titanium-base alloys have been successfully used as biomaterials implants due to their excellent characteristics such as ability to provide good mechanical properties, high corrosion resistance and biocompatibility. Research has proved that stainless steel was widely used to fabricate implants due to their ability to withstand significant loads, fatigue resistance, high fracture toughness and corrosion resistance [5]. SS316L have the low carbon content in their composition and able to produce passive layer (chromium oxide) to protect from corrosion attack. Since, corrosion attack on stainless steel become a serious problem due to possibility of implant to sudden falling continuous exposed of extracellular tissue fluid after implanted may exposed the human body to contamination of corrosion. The surface of the implant will interact with the surrounding environment of human body which contains water, complex organic compounds, dissolved gasses will trigger chemical reactions to cause implant degradations. In addition, the reaction between SS316L implant will release Fe, Cr and Ni ions and lead to allergens and carcinogens which harmful to human body[6].

It has been reported that pitting and crevice corrosion attack are the main reasons of the failure of SS316L implant. Pitting corrosion happened when passive film on SS316L breached and exposed the surface for another corrosion attack due to existing surrounding of tissue fluid and serum. Not only pitting, crevice corrosion may also happen between bone plate and screws made from SS316L such as between the heads and the countersink holes [6]. Previous researchers has established that; fabricate porous metal by using stainless steel type 316L exhibited poor corrosion resistance because of porous surface geometry[7]. However, there are lacks of study in corrosion attack of porous metal for biomedical applications especially for implant. Advanced in medical technology clearly demand for porous metal as an implant. Therefore, continuing research are essential for better understanding of how corrosion mechanism on porous metal and could continue the manageable implant failure in medication[6].

The aim of this research is to study the effect of corrosion attack on porous SS316L. The cyclic polarization potentiodynamic analysis has been performed to the porous SS316L in the 3.5% NaCl solution. The morphology of the porous SS316L before and after performing cyclic polarization potentiodynamic analysis was investigated by using scanning electron microscopy (SEM). The element presence to the porous SS316L before and after cyclic polarization potentiodynamic analysis was is determined by energy dispersive x-ray (EDX).

2. Methodology

As per received porous SS316L was used represent as implant material. The preparation starts with cutting the samples into size 2cm length with an exposed area 2.5cm^2. The cyclic polarization potentiodynamic analysis was carried out with the working electrode (SS316L) in 3.5% NaCl solution at room temperature. By using potentiostat apparatus the scan rate has been set to 0.25 mV/s. The NaCl solution was prepared by diluting 35 gram of NaCl particles with 1000ml distilled water. The potentiodynamic cyclic polarization analysis was repeated to get statically acceptable results. The characterization of the morphology before and after performing cyclic polarization potentiodynamic analysis of the porous SS316L samples was evaluated by using Hitachi Scanning electron Microscopy while the energy dispersive x-ray was performed to study the element present to the samples.

3. Result and Discussion

3.1 Potentiodynamic cyclic polarization analysis

Figure 1 presents the result of CPP of porous SS316L. The cyclic polarization potential curve in all solutions are shifted in positive hysteresis may be subjected to the susceptibility to pitting corrosion happen to the samples in NaCl solution. The corrosion potential (E_{corr}) of porous austenitic stainless steel type 316L are in the range of -0.40V to – 0.60V in NaCl solutions. Previous studies claim that, the austenitic stainless steels with higher chromium, molybdenum, nickel, and nitrogen exhibit higher E_{corr} value and improved the pitting corrosion resistance[8]. The breakdown potential (E_b) of porous SS316L are in the range of -0.30V to -0.40V in NaCl solution. E_b is where potential starts to slow down at increasing current. This is where the passive film starts to breach and exposed stainless steel for another corrosion attack.

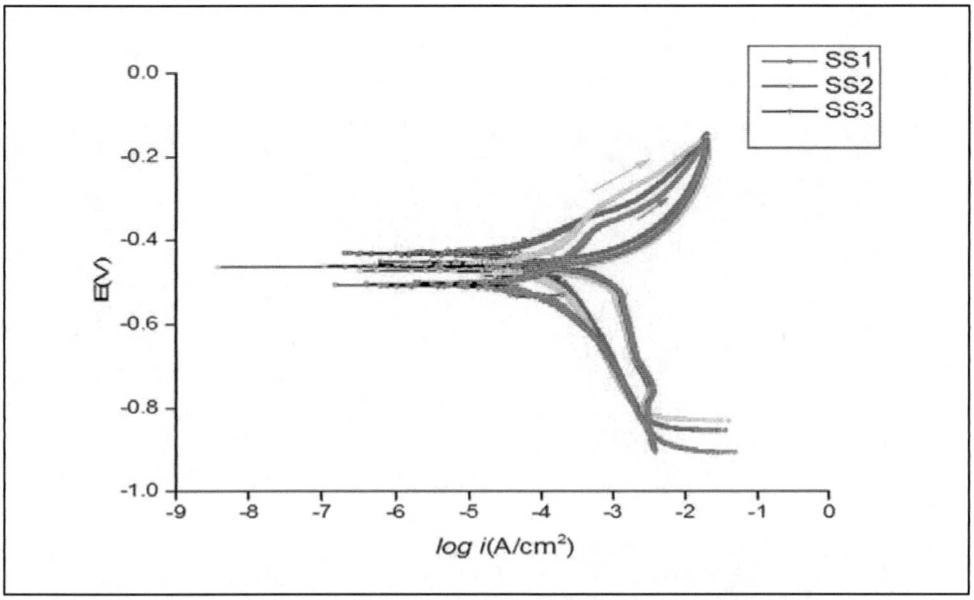

Figure 1. Anodic polarization curves of porous austenitic stainless steel type 316L

3.2 Morphology analysis of porous austenitic stainless steel type 316L

Figure 2. shows the SEM micrograph with the magnification of 500x and 1000x of the samples surface morphology before performing cyclic polarization potentiodynamic test. The surface of porous SS316L were polished using sand paper with the grit size of 600, 800 and 1000 in order to obtain the clear view of pores before performing corrosion test. From Figure 2(a) and 2(b), it can be seen that the fine pores already exist in sample before the corrosion test was performed. The size of the pore approximately to 6μm and can be categorized as micropores and the structure of the pore can be see clearly at the SEM micrograph with magnification 1000x.

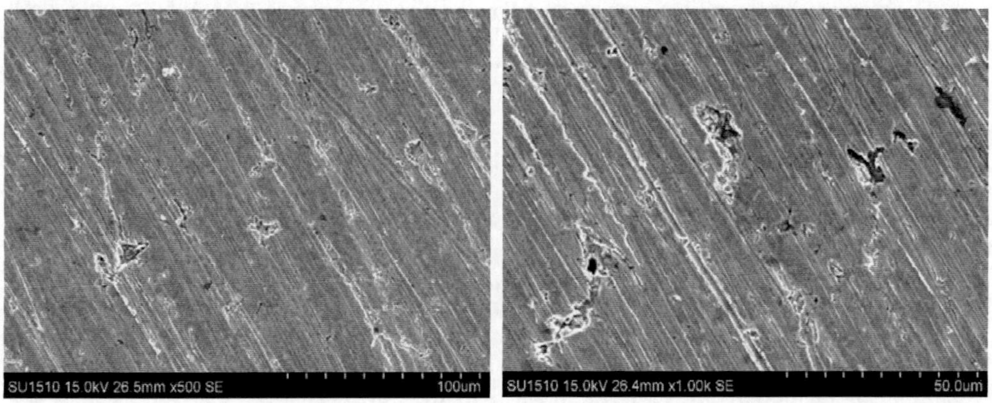

(a) (b)

Figure 2. SEM micrograph with the magnification of a) 500X, and b) 1000X of the samples surface morphology before performing potentiodynamic cyclic polarization test

IOP Conf. Series: Journal of Physics: Conf. Series **914** (2017) 012013 doi:10.1088/1742-6596/914/1/012013

Figure 3 shows the SEM micrograph of the samples before and after performing cyclic polarization *potentiodynamic* analysis. *It* can see *from the Figure 3(b) that parallel* corrosion pits form on the surface of the samples after corrosion test. Pitting holes starting to enlarge and become local anodic which creates another corrosion to happen.

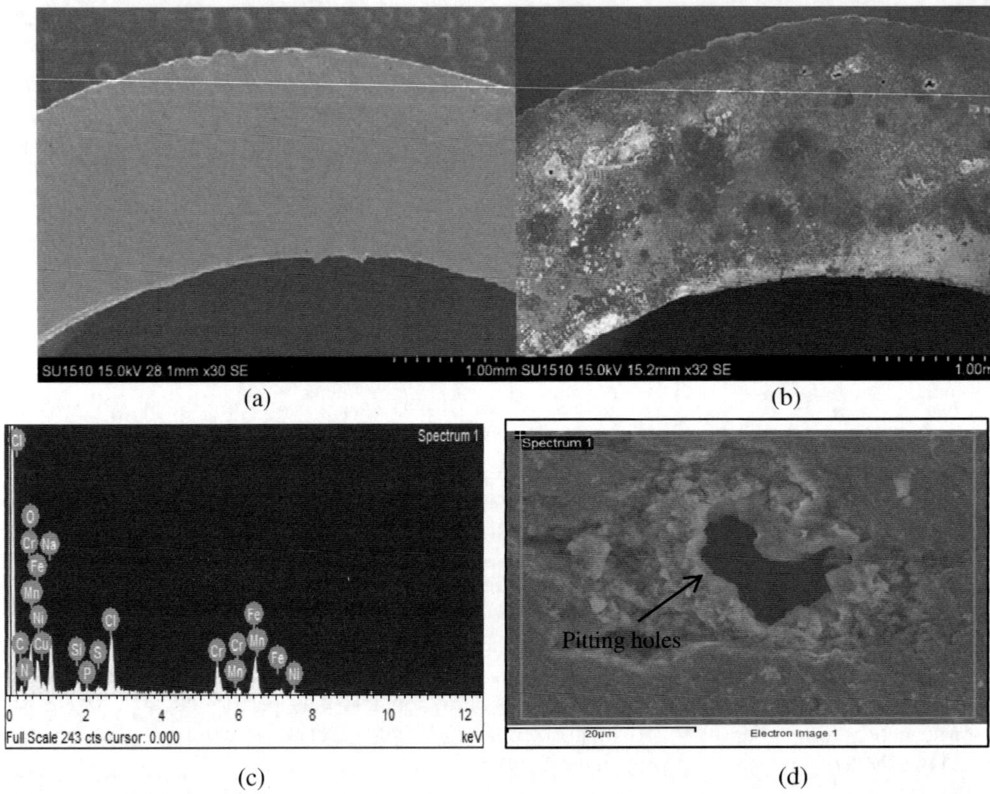

Figure 3. SEM micrograph of the sample a,b)before and after performing potentiodynamic cyclic polarization test, c)EDX spectrum of the samples, d) Pitting holes on the sample surface

Figure 3(c) shows the EDX spectrum of the samples after performing corrosion test. Figure 3(d) displays the focus area of EDX analysis which pitting holes on the samples surface formed. As mentioned, SS316L rich with the element iron (Fe) and from the EDX analysis, the element of iron (Fe) was dominated on SS316L surface samples. The element of carbon (C) recorded highest value and signify the standard of carbon content in SS316L. However, the high amount of carbon element can be neglected due to the use of carbon electrode during performing potentiodynamic cyclic polarization analysis.

Figure 4 shows the SEM micrographs with the magnification of 100x surface of *SS316L* before performing corrosion test and after corrosion attack *in* 3.5% NaCl solution. Figure 4(b) displays the formation of branching and clustered crystallite structures on the surface of the porous *SS*316L. The corrosion product form in irregular structure at the SS316L surface after the testing. The same structure also obtained by previous researcher that study the corrosion behaviour of self-ligating and conventional metal bracket [10]. Figure 4(c) presents the EDX spectrum for the focus area that displays on Figure 4(d) which branching and clustered crystallite structure was formed in the surface of the sample after corrosion test. The element of iron is the highest element in the composition followed by sodium (Na), carbon and chloride (Cl).

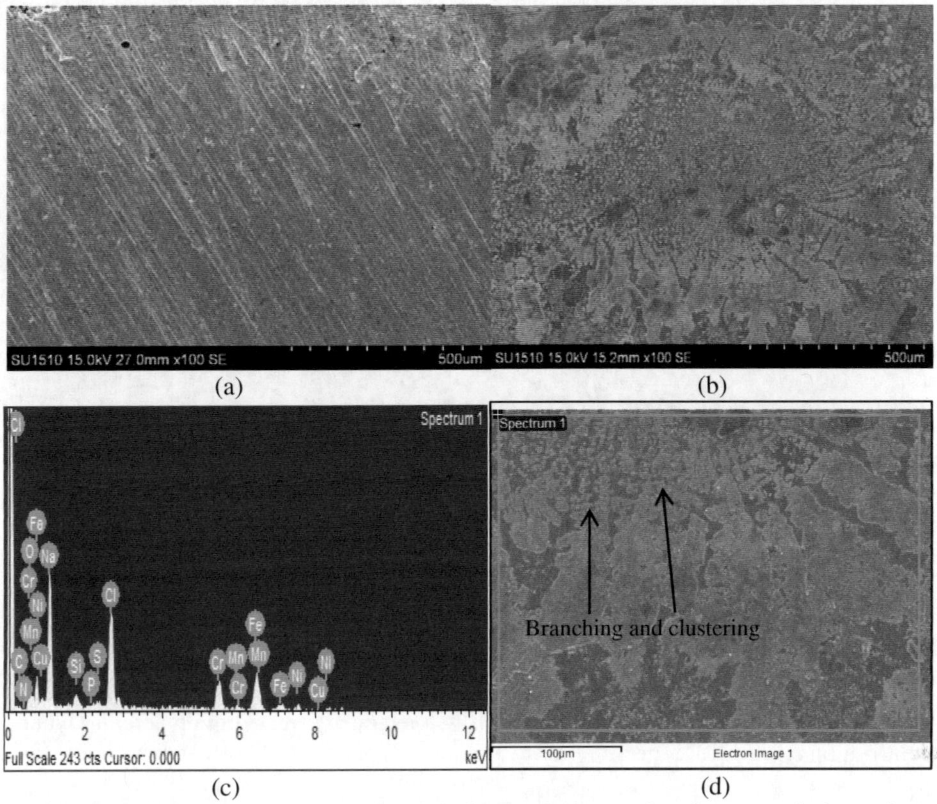

Figure 4. SEM micrograph of the sample a,b)before and after performing potentiodynamic cyclic polarization test, c)EDX spectrum of the samples, d) branching and clustered crystallite structures on the sample surface

Figure 5 compares the SEM micrograph for the surfaces structures of the samples before and after performing potentiodynamic cyclic polarization analysis. As shown in Figure 5. the particles of corrosion product appear on the surfaces of the samples after corrosion test. Figure 5(c) provides the EDX spectrum of the samples that consist of the particles. The iron element dominates the composition of the samples compare to other element followed by chromium (Cr) content that sufficient into the composition of the samples. Generally, austenitic stainless steel consists of different composition such as chromium, molybdenum, nickel, and etc. These contribute to galvanic effect. Galvanic is the effect from the two different of metal combine together, and exposed to the oxidize atmosphere and then start to corrode and produce the galvanic effect. Moreover, the other elements such nickel and chromium that consist in austenitic stainless steel will react and produce the oxide layer with the aim of increasing resistance to stainless steel [10].

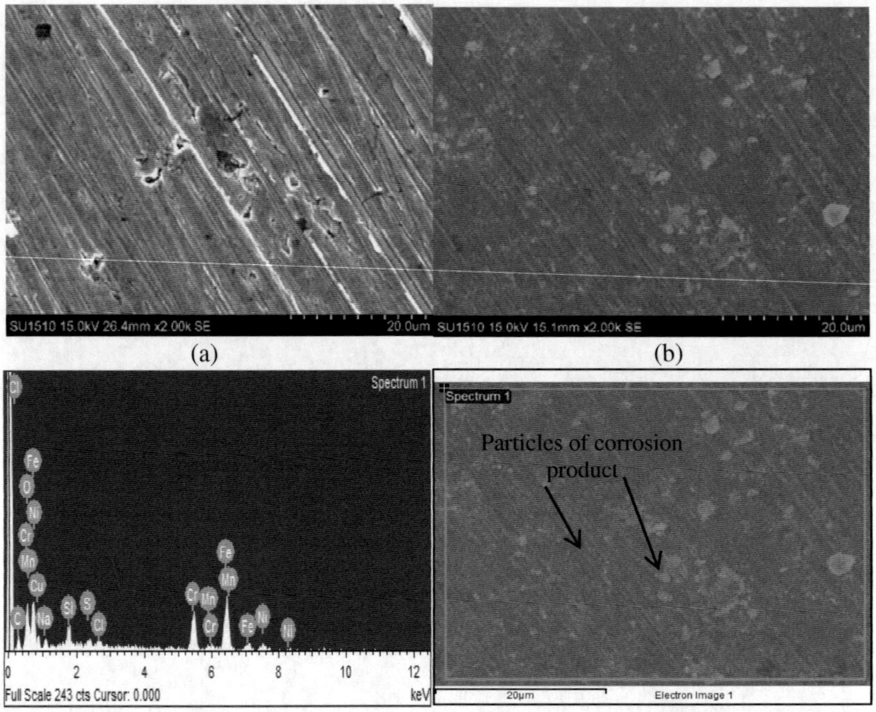

Figure 5. SEM micrograph of the sample a,b)before and after performing potentiodynamic cyclic polarization test, c) EDX spectrum of the samples, d) particles of corrosion product appear on the sample surface

4. Conclusion

Cyclic polarization potentiodynamic analysis successfully performed in order to study the corrosion attack to the porous SS316L. From the result, the anodic polarization curve recorded the result of (E_{corr}) for porous SS316L in the range of -0.40v to -0.60v and breakdown potential (E_b) is -0.3v to -0.4v in NaCl solution. From the SEM observation, there are three types of formation happen to the surface due to corrosion includes the in line formation of corrosion pits, the formation of branching and clustered crystallite structure, and the formation of the particles of corrosion product. Based on EDX analysis, the dominant element in the composition of the porous SS316L after performing cyclic polarization potentiodynamic analysis is iron and followed by chromium and carbon. It can be conclude that, the pore structure and morphology of the porous austenitic stainless steel type 316L influence the corrosion attack.

5. References

[1] A. Kennedy, "Porous Metals and Metal Foams Made from Powders," in *Powder Metallurgy*, Katsuyoshi Kondoh, Ed. InTech, 2012, pp. 31–46.

[2] L. P. Lefebvre, J. Banhart, and D. C. Dunand, "Porous metals and metallic foams: Current status and recent developments," *Adv. Eng. Mater.*, vol. 10, no. 9, pp. 775–787, 2008.

[3] G. Manivasagam, D. Dhinasekaran, and A. Rajamanickam, "Biomedical Implants: Corrosion and its Prevention - A Review," *Recent Patents Corros. Sci.*, vol. 2, no. 1, pp. 40–54, 2010.

[4] M. Guden, E. Celik, E. Akar, and S. Cetiner, "Compression testing of a sintered Ti6Al4V powder compact for biomedical applications," *Mater. Charact.*, vol. 54, no. 4–5, pp. 399–408, 2005.

[5] J. . Davis, *Handbook of Materials for Medical Devices*. ASM International, 2003.
[6] D. C. Hansen, "Metal corrosion in the human body: The ultimate bio-corrosion scenario," *Electrochem. Soc. Interface*, vol. 17, no. 2, pp. 31–34, 2008.
[7] K. H. W. Seah, R. Thampuran, and S. H. Teoh, "The influence of pore morphology on corrosion," *Corros. Sci.*, vol. 40, no. 4–5, pp. 547–556, 1998.
[8] A. Szewczyk-nykiel, "The Influence of Molibdenum on Corrosion Resistance of Sintered Austenitic Stainless Steels," *Tech. Trans. Mech.*, vol. 26, no. 4, pp. 131–142, 2015.
[9] E. Salahinejad, M. J. Hadianfard, D. D. Macdonald, S. Sharifi-Asl, M. Mozafari, K. J. Walker, A. T. Rad, S. V. Madihally, and L. Tayebi, "In Vitro Electrochemical Corrosion and Cell Viability Studies on Nickel-Free Stainless Steel Orthopedic Implants," *PLoS One*, vol. 8, no. 4, pp. 1–8, 2013.
[10] L. H. E. G. Maia, H. Lopes Filho, A. C. de O. Ruellas, M. T. de S. Araújo, and D. S. Vaitsman, "Corrosion behavior of self-ligating and conventional metal brackets," *Dental Press J. Orthod.*, vol. 19, no. 2, pp. 108–114, 2014.

IOP Conf. Series: Journal of Physics: Conf. Series **914** (2017) 012014 doi:10.1088/1742-6596/914/1/012014

Natural and synthetics nanomaterials: comparative study on their mechanical and thermal properties as nanofiller in polymer composite

Ferial Ghaemi, Luqman Chuah Abdullah, Paridah Md Tahir

Institute of Tropical Forestry and Forest products (INTROP), Universiti Putra Malaysia, 43400UPM, Serdang, Selangor, Malaysia

Email: ferialghaemi@yahoo.com

Abstract. In this research, natural nanomaterials including cellulose nanocrystal (CNC) and nanofiber cellulose (NFC) and synthetic nanomaterials such as carbon nanofiber (CNF) and carbon nanotube (CNT) with different structures, sizes and surface areas were produced and analyzed. CNC and NFC were fabricated by use of acid hydrolysing and mechanical methods, respectively and synthetics nanomaterials were produced by chemical vapor deposition method (CVD). Here, in order to evaluate the effect of these nanomaterials on the mechanical and thermal properties of the polymer, use them as nanofiller in the polymer matrix. The most important challenge in this study is to evaluate and compare these nanofibers based on the effects of their structures and compositions on the mechanical and thermal properties. The characteristics of these nanofibers such as morphology, structure, and composition and also the properties of these nanomaterials including thermal stability and mechanical strength were investigated, comprehensively. Based on the results, CNT with high thermals stability and strong structure has the best filler for using in polymer composite.

1. Introduction

Synthetics nanomaterials including carbon nanomaterials has many applications in industries and sciences. Carbon nanomaterials such as CNT and CNF are made from sp^2 carbon atoms with one-dimension (1D) structures. The structure of pure CNT can be visualized as a single sheet of graphite rolled to form a tube. Besides, CNFs have cylindrical nanostructures with different stacking arrangements of graphene sheets, such as stacked platelet, ribbon or herringbone [1]. CNF and CNT with low-density and high aspect ratio, as well as extraordinary mechanical, thermal, electrical and electrochemical properties have produced a huge amount of activity in most areas of science and engineering [2]. There are many techniques to synthesis carbon nanomaterials but chemical vapor deposition as the most effective method has been employed to fabricate carbon nanomaterials [3,4]. Besides, in order to obtain different types of carbon nanomaterials with different structures, some parameters such as temperature, time, flow rate of carrier gas and carbon source amount should be changed [5,6].

To prepare natural nanomaterials there are many natural sources. Kenaf is a natural tropical plant that has been grown commercially to generate a secondary source of income for developing countries including Malaysia. Its high cellulose content, ranging between 44 and 63.5% in kenaf, [7] has generated interest in exploiting the material as nanofillers in composites. The cellulose nanocrystals (CNC) and nanofibers (NFC) can be obtained via acid hydrolysis and mechanical treatment, respectively. CNC and NFC have been used in many applications such as filler in polymer composite to make a wide range of other functional materials [8-11].

In this research we produced natural and synthetics nanomaterials and analyzing their properties based on their characterization. Besides, the produced nanomaterials were used as nanofiller in polymer matrix to evaluate the effects of these nanomaterials on the thermal and mechanical properties of different composites and compare together.

2. Experimental Part

2.1. Synthesis of synthetic and natural fibers

In this part the nickel nitrate hexahydrate powder ($Ni(NO_3)_2.6H_2O$; Sigma-Aldrich), as Ni precursors, was put in a quartz crucible locate in the CVD reactor and dried in the reactor at 150°C to remove water for 1 h and then increased the temperature to 400°C for 1h to remove nitrate. The reaction temperature was set at 650°C and 800°C to produce high quality CNF and CNT, respectively. The process was done by the decomposition of the acetylene at 50 standard cubic centimeters per minute (sccm) flow rate on the Ni particles at 100/100sccm H_2/N_2 flow rates for 30min.

Here, CNC and NFC were produced by use of acid hydrolysing and mechanical methods, respectively. For Preparation of CNC, firstly, acid hydrolysis was conducted under mechanical stirring using 65wt% H_2SO_4 (pre-heated) for 45 min. Then the suspension dilute with cold distilled water to stop the reaction. The suspension then centrifuge for 10 min at 10,000 rpm. The aqueous suspension subsequently dialyzes against distilled water until a constant pH attain. The resulting suspension was subsequently dried and stored in the refrigerator.

For fabricating NFC, firstly, Kenaf bast fibers retted in water was cut and then cooked in a rotatory digester contains 25wt% NaOH and 0.1 wt% anthraquinone solution (liquor to fiber ratio was 7:1) at 160°C for 2 h. After that, AQ was added to the solution to not only increase the delignification rate but also protect the fibers from alkali degradation called end-wise degradation of cellulosic chains. Extraction of nanofibers were done by further mechanical destruction using super-masscolloider. Aqueous suspensions with the concentration of 3 wt% was prepared and prepared and blend until formation of a homogeneous mixture. To some up, the produced suspensions were fed into the grinder and the process repeated until a gel was formed. And then it was dried and formed into the powder shape.

2.2. Preparing of polymer composite

Initially, 5 wt.% nanomaterials was added into the polypropylene matrix in to the melting mixing machine at 200°C of screw temperature and 60 rpm of screw rotation for 5 min. Then, the produced composite was cooled to 25°C in a mold. After that, it was pressed by Hot Press Machine at a 200◦C under 150kg/cm^2 pressure for 10 min and then cooled.

2.3. Characterization of the composite

The prepared polymer composite was cut into the dog-bone shape based on the ASTM D638 standard. To fulfill a tensile test, an Instron Universal Testing Machine was used at room temperature to analysis the modulus of elasticity and the strength of PP, CNC/PP, NFC/PP, CNT/PP and CNF/PP. Moreover, to determine the thermal resistance and stability of the produced nanocomposites, a thermal gravimetric analysis (TGA) test was employed. Here, TGA was fulfilled on a Mettle Stare SW 9.10 thermal gravimetric analyzer.

3. Results and Discussion

The FESEM images of the produced nanocellulose were shown in Figure 1. CNCs present a simple needle-like structure with an average length of 200nm, a diameter of 20nm. NFCs exhibit a complex, highly entangled, web-like structure. Twisted/untwisted, curled/straight, and entangled/separate nano-

IOP Conf. Series: Journal of Physics: Conf. Series **914** (2017) 012014 doi:10.1088/1742-6596/914/1/012014

fibrils and their bundles with diameters ranging from 50 to 200 nm in diameter can be identified from the micrograph.

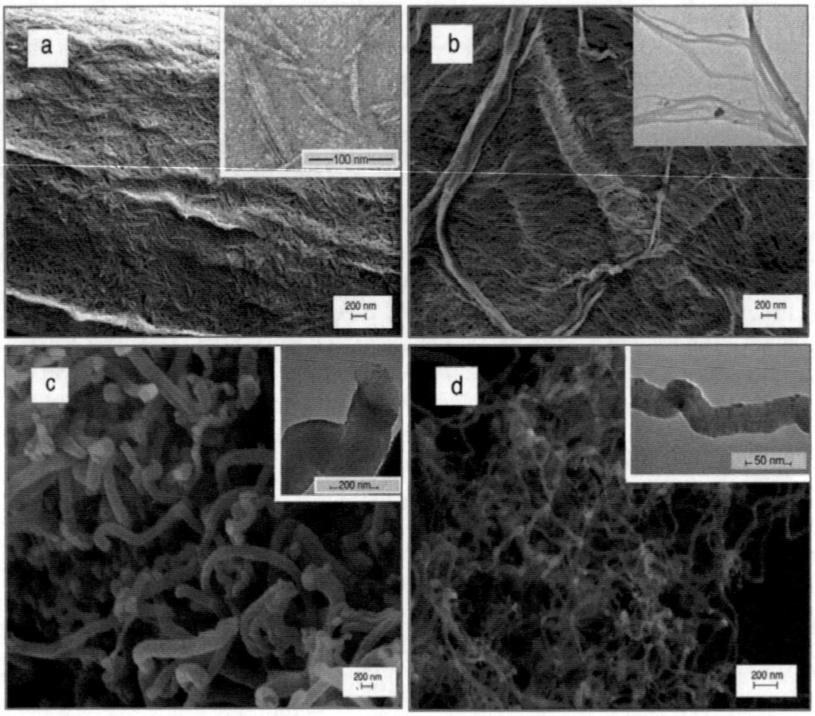

Figure 1. SEM/TEM images of (a) CNC, (b) NFC, (c) CNF and (d) CNT

The TGA curves in Figure 2 illustrate the degradation process of different nanocomposite filled with natural and synthetics nanofibers based on the weight loss (wt.%) versus temperature (°C). the curves related to CNC/PP and NFC/PP reveal that the initial weight loss related to the evaporation of moisture. Further loss in weight may be attributable to pyrolysis of NFC/PP and CNC/PP decomposed at 300°C–400°C and 250°C–370°C, respectively. The lower degradation temperature of CNC/PP could be due to the smaller dimensions compared to the NFC/PP, which leads to higher specific surface area.

On the other hand, it is obvious that the decomposition temperature of the CNF/PP was started about 400°C and complete at 550°C and for CNT/PP, the decomposition temperature of sample was increased to about 500° C and completed at 610° C. since the composition of CNT/PP and CNF/PP is same, therefore, the thermal stability of CNT is more than CNF due to its structure and morphology. Briefly, the thermal analysis showed that the thermal degradation phenomenon of natural nanofibers (CNC and CNF) is much lower than synthetics nanofibers (CNF and CNT).

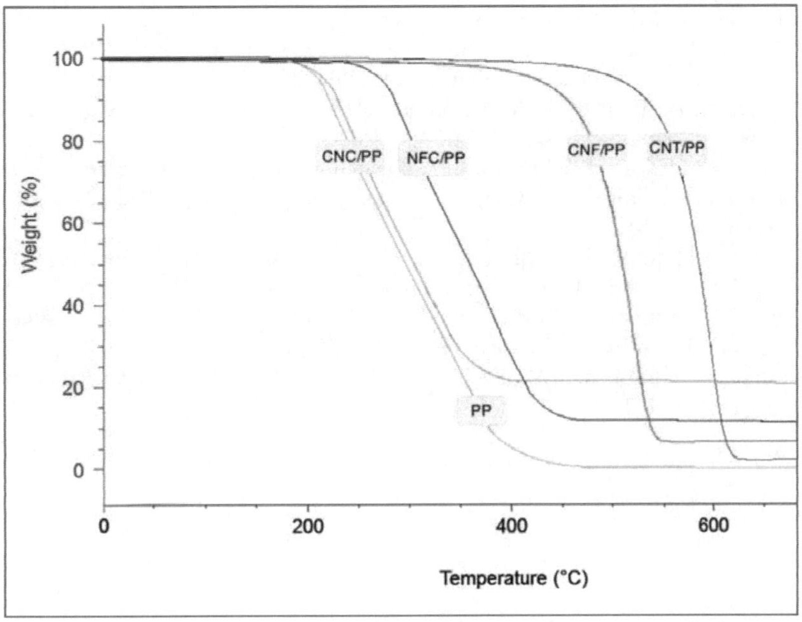

Figure 2. TGA curves of PP, CNC/PP, NFC/PP, CNF/PP and CNT/PP

According to Table 1, presence of nanofibers enhanced the mechanical properties of the composite. Comparison of the elasticity and strength of them illustrates a significant difference in the tensile modulus that related to the different nanofillers. The tensile modulus and strength of synthetics fiber are significantly more than natural nanomaterials related to the hard structures and bigger length of carbon nanomaterials.

Table 1. Mechanical results of different polymer composites

Sample	Tensile modulus (GPa)	Tensile strength (GPa)
PP	1.5	0.02
CNC/PP	620	19
NFC/PP	405	10
CNF/PP	650	21
CNT/PP	780	27

4. Conclusion

The most important contribution of this research is related to apply the CNC and NFC as natural fibers with CNT and CNF as synthetics fibers in order to apply in polypropylene matrix and analysis the thermal and mechanical properties of them. In order to evaluate the effects of the types of nanofillers on the polymer matrix in terms of thermal and mechanical properties, the TGA and tensile test were performed on the produced nanocomposites. Since the CNT has highest thermals stability and strongest structure, the thermal stability, tensile stress and Young's modulus of this composite are higher than others.

References

[1] P. Serp, M. Corrias and P. Kalck, Appl. Catal., A, 2003, 253, 337–358.
[2] V. Vamvakaki, K. Tsagaraki and N. Chaniotakis, Anal. Chem., 2006, 78, 5538–5542.

[3] S. H. Liu and S. C. Zhao, Adv. Mater. Res., 2014, 875, 1590.
[4] H. Shu, X. M. Tao and F. Ding, Nanoscale, 2015, **7**, 1627.
[5] Z. Fan, C. Wu, J. Chen and S. Yi, Carbon, 2008, 46, 378.
[6] F. Ghaemi, R. Yunus, M. A. M. Salleh, H. N. Lim and S. A. Rashid, Fullerenes, Nanotubes, Carbon Nanostruct., 2015, 23, 669.
[7] Jonoobi, M., Harun, J., Mishra, M., & Oksman, K. (2009). *BioResources*, *4*(2), 626-639.
[8] Fukuzumi, H., Saito, T., Iwata, T., Kumamoto, Y., & Isogai, A. (2008). *Biomacromolecules*, *10*(1), 162-165.
[9] Kelly, J. A., Shukaliak, A. M., Cheung, C. C., Shopsowitz, K. E., Hamad, W. Y., & MacLachlan, M. J. (2013). *Angewandte Chemie International Edition*, *52*(34), 8912-8916.
[10] Mendez, J., Annamalai, P. K., Eichhorn, S. J., Rusli, R., Rowan, S. J., Foster, E. J., & Weder, C. (2011). *Macromolecules*, *44*(17), 6827-6835.
[11] Walther, A., Timonen, J. V., Díez, I., Laukkanen, A., & Ikkala, O. (2011). *Advanced Materials*, *23*(26), 2924-2928.

An Evaluation of Mechanical Properties on Kenaf Natural Fiber/Polyester Composite Structures as Table Tennis Blade

Amin, M. H. M., Arifin, A. M. T.*, Hassan, M. F., Haq, R. H. A., Rahman, M. N. A., Ismail, A. E., Rahim, M. Z., Ibrahim, M. R., Yunos, M. Z., Ismail, R.

Engineering Design Research Group (EDRG), Department of Material and Engineering Design, Faculty of Mechanical and Manufacturing Engineering, Universiti Tun Hussein Onn Malaysia, Batu Pahat, 86400 Johor, Malaysia.

Corresponding author: *mubarak@uthm.edu.my

Abstract: Nowadays, natural fibers getting attention from researchers and industries to optimize the use it, with combination of polymers as composite structure, due to environmental awareness. Furthermore, it show a few advantage, such as biodegradability, light in weight and non-toxic characteristic. In this study, kenaf natural fibers was used as reinforcement material, with combination of polyester as matrix material, known as polymer matrix composites. The main purpose of this study is to analysis the mechanical properties of kenaf natural fiber/polyester composite structure, in order to know the suitability of kenaf natural fibers as replacement material for table tennis blade structure, instead of using wood. The structural panel of composite laminates has been produced using hand lay-up technique. The experimental works are performed in tension, impact (Charpy) and shear condition. The characteristic of different condition on kenaf composite structure was studied. Based on the result, it found the properties of kenaf composite structure, and it will used as a benchmark, to compare with initial properties of table tennis blade made by wood. In addition, the strength and a weakness of that particular materials and lamination structure will be identified.

1. Introduction

Over the years, people have found various applications by using natural fibers. An example of natural fibers, such as kenaf, hemp, flax, sisal and ramie. The initial characteristic of natural fibers, are emerging as low cost, lightweight and environmental friendly [1]. Therefore, natural fibers become one of the most famous material selected as a reinforcement materials, in order to produce composite structures. Whereby, a composite material is a combination of reinforcement and matrix material, and it can have better properties than the individual materials used alone [2,3]. So, a reinforcement and matrix material need to be chosen according to the application and desired characteristics. As reported by, Davoodi et. al. (2010) natural fiber composites have an advantage, which is better formability, renewable, cost effective and safe toward health [4]. Furthermore, by comparing glass fiber composites, natural fiber composites are likely to be enviromentally superior [1].

In this study, kenaf natural fibers or also known as *Hibiscus Cannabinus*, are consider as a material to be investigated, in term of suitability to become reinforcement materials for table tennis blade structures. Kenaf fibers is obtained from the bast/stems of plants and it requires less water to grow because it growing cycle around 150 to 180 days with avarage yield of 1700 kg/ha [5]. As mentioned by Saba et. al. (2015), kenaf is commercially available and economically cheap amongst other natural fibers. It also hardy, strong and tough plant with a fibrous stalk. Figure 1 shows the kenaf plantation [6] ; (a) cross-section of Kenaf stalk [7]. Due to the primary stage on development of natural plant fibers reinforced composites, researcher need to continously investigate, especially the basic experimental, such as the mechanical properties [8]. According to previous research, the kenaf bast fibers showed better mechanical properties than other part, that make them as a replacement to glass fibers in polymer composites as reinforcement materials [9]. On mechanical properties, as reported by Wambua et. al. (2003), it was shown kenaf fibers specifically, increased the ultimate tensile strength, tensile modulus, and impact strength with increasing fiber weight fraction. However, the composites tested showed low impact strengths compared to glass material composites [10].

Figure 1. Kenaf plantation; (a) cross-section of Kenaf stalk [5,6]

An existing of table tennis blade, consist of a wooden blade which incorporates the handle and table tennis rubber which is affixed to each side of the blade using water-based glue, as shown in Figure 2, the side view of a blade showing the wooden blade with table tennis rubber affixed to each side of the blade. The wood structure, will be replaced with other materials. Therefore, the aim of this study is to replace the wood structure with kenaf natural fiber. However, the most important thing to understand is on the characteristic of table tennis blade itself. The characteristics need to investigate and evaluate are the energy absorption, hardness, weight, size and shape of table tennis blade. These characteristics contribute on the performance of table tennis blade [11]. An understanding of the factor affecting the performance of table tennis blade is required to ensure replacement material, contribute a better performance, or at least at the similarity level.

Reviewing on related cases show, a need to investigate the characteristics and behaviours of new designated structures, which is kenaf natural fiber/polyester composite structures, in order to know the strengths and weaknesses of structures. Thus, the objective of this study is to investigate the suitability of kenaf natural fibers used as replacement material for table tennis blade, in term of tension (strength), impact charpy (energy absorption) and shear test (layer strength). The hand lay-up technique is used to fabricate composite materials, with is combination of kenaf natural fibers and polyester resin. An expected results of this study are to evaluate the initial mechanical properties, based on basic experimental that reflected on characteristic need for table tennis blade.

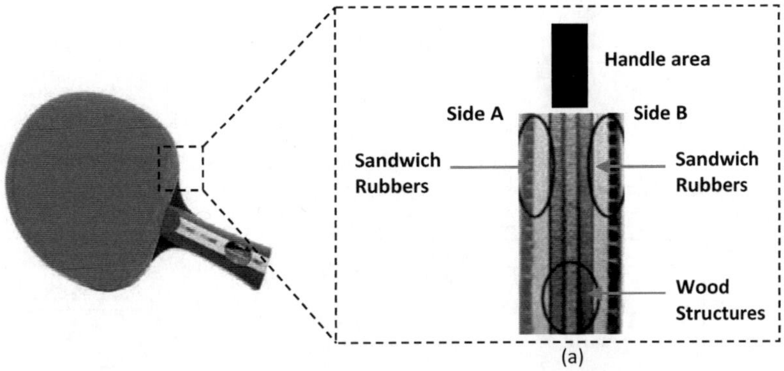

Figure 2 Table tennis blade; (a) Side view of table tennis blade

2. Methodology

2.1 Specimen Preparation

Kenaf natural fiber/polyester composite structures were fabricated using hand lay-up techniques. Each specimen consist of kenaf fiber as reinforcement and polyester resin as matrix materials, on five layers lamination structure. On fabrication process, kenaf natural fibers was impregnated with polyester resin until five layers configuration. The resin was applied to the surface of material, after the material was laid down at the mould. Brush rollers was used to remove void in the lamination structure. After lamination process, the specimen was cured at room temperature, and it was cut out according to experimental specimen standard. Figure 3 shows the flowchart of experimental and followed by Table 1 represents the characteristics of laminate specimen.

Table 1. Characteristics of laminate specimen

Characteristics Kenaf Composite Structures		
Kenaf Natural Fiber	5	t= 4mm } Kenaf
Stacking Sequence / Orientation	Kenaf [0/90]	
Ply	5	
Resin	Polyester	
Average Thickness	$\pm$4 mm	

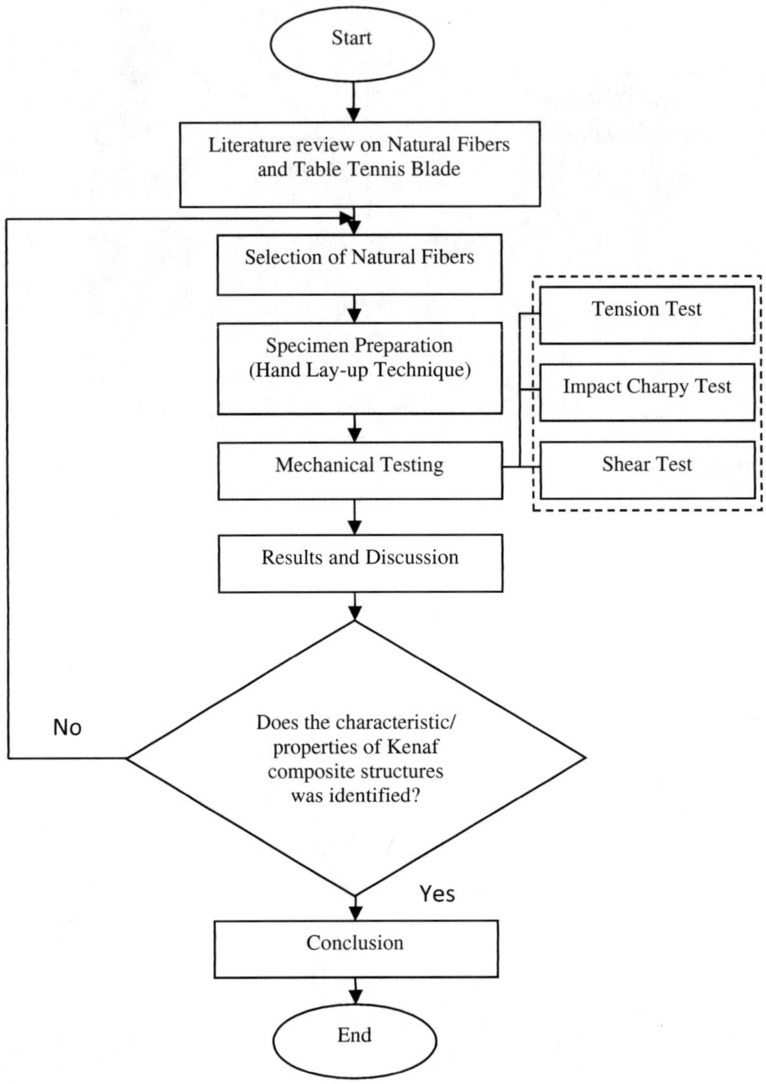

Figure 3. Flowchart of experimental

2.2 Mechanical Testing

In this study, tension test, impact charpy test and shear test were performed, in order to investigate the characteristic and behaviour of composite structures. All experiments follow difference standard, which is; ASTM D638 (Tension Test), ISO 179 (Impact Charpy Test) and ASTM D2344 (Shear Test). Specifically, tension test was conducted to determine the ability of composite structures when subjected to the axial load on the specimen until failure. Then, followed by impact charpy test for examine the potential of the kenaf composite structure to withstand the impact force at high rates. The output from impact charpy test was energy absorption, whereby the test determines the amount of energy absorbed by a material during fracture. At last, shear test was carried out in this study. This experiment examines the shear strength of composite structures. Due to the difference of standard, each experiment have different of specimen geometry, as shown in Figure 4.

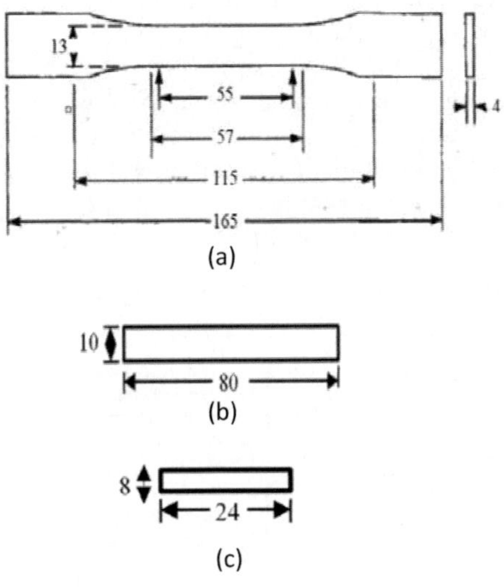

Figure 4. Specimen geometry (mm); (a) Tension test, (b) Impact charpy test, (c) Shear test

3. Result and Discussion

Three sets of experimental results were obtained from the three types of experimental, which is tension test, impact charpy test and shear test. The experimental were demonstrate the characteristics of composite lamination structures, in the field of stress, modulus young, energy absorption and shear strength. As shown in Figure 5, tensile strength and modulus young with different of specimen. It shows, different test specimen exhibit different strength behaviour. This is due to, different of condition on lamination layer, which is on interface condition [12,13]. The result indicate that the highest value for the tensile strength is for specimen 4, approximately 19 MPa. An average of tensile strength for kenaf composite structures at 15 MPa. As reported in [14], the tensile strength for kenaf/ unsaturated polyester is 27 MPa and higher than other natural fibers, which is jute and bamboo. However, due to difference of geometry specimen, it shows difference of behaviour. With respect to the modulus young, average modulus young is 708 MPa. Based on the results, it shows the low modulus of elasticity, will provide high stability when table tennis blade hit the ball. In addition, the structure shows capability for absorption the vibration.

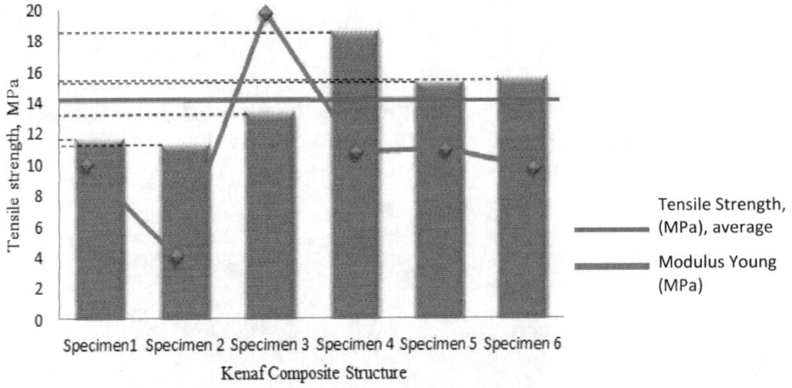

Figure 5. Tensile strength (MPa) and Modulus young (MPa) of kenaf composite structures

IOP Conf. Series: Journal of Physics: Conf. Series **914** (2017) 012015 doi:10.1088/1742-6596/914/1/012015

In this study, an average of impact energy absorption for kenaf composite structures, approaching 0.55 J, as shown in Figure 6. The energy absorption also affected due to orientation angles, as reported by Arifin et. al. (2014). The result shows that the kenaf composite structures has higher capability to absorb energy during impact force imposed on the structures [12]. On manufacturing table tennis blade, this criterion is most important aspect to be investigated. It is essential to note that, this initial study shows the possibility of kenaf natural fibers used as reinforcement material on table tennis blade, due to capability of absorption.

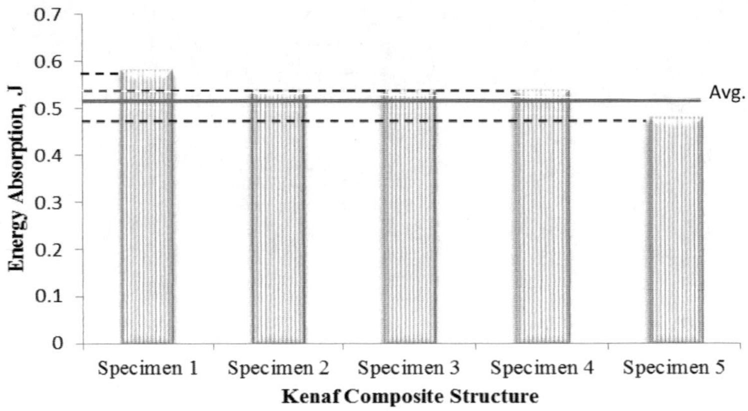

Figure 6. Energy absorption of kenaf composite structures

As shown in Figure 7, the scanning electron migrographs of kenaf hybrid composites, as reported by Ghani et. al. (2012), shows the surface failure due to matrix fracture, fiber-matrix debonding, fiber pull-out and fiber fracture [5]. This phenomenon represents, the composite materials can exhibit multiple types of damage, if it exceeds the accepted load. In the actual condition of table tennis blade, the load is not high, therefore based on the results, kenaf composite structures can sustain and have capability to absorb the load. Next, the effect of shear condition on the kenaf composite structures are shown in Figure 8. Shear strength represent the capabilities of structures to withstand from structural failure, where the material fails in shear. Additionally, shear load is a force that produce the sliding failure on material along the plane. In this study, shear test was performed on bending condition, in order to investigate the strength of the lamination layer. The result shows that the average shear strength of the kenaf composite structures, in the range 420 N– 450 N. Moreover, it should be noted that the delamination between lamination layers will affect the shear strength of structures. In composite material structures, if delamination of plies occurs between each lamination structure, it means a progressive failure mode has occurred. The failure mode, can be divided into three stages: damage initiation, damage propagation and total perforation (fracture) [12,15,16]. The results obtained in this experimental shows in the range of accepted, but need more investigation related on the actual condition of table tennis blade condition.

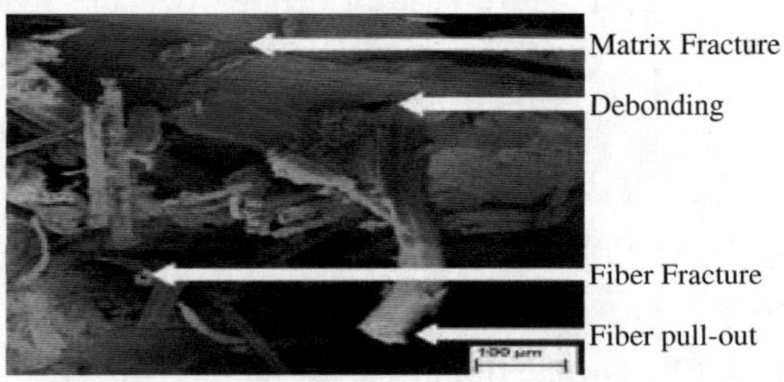

Figure 7. SEM micrograph on fracture surface of kenaf haybrid composites [5]

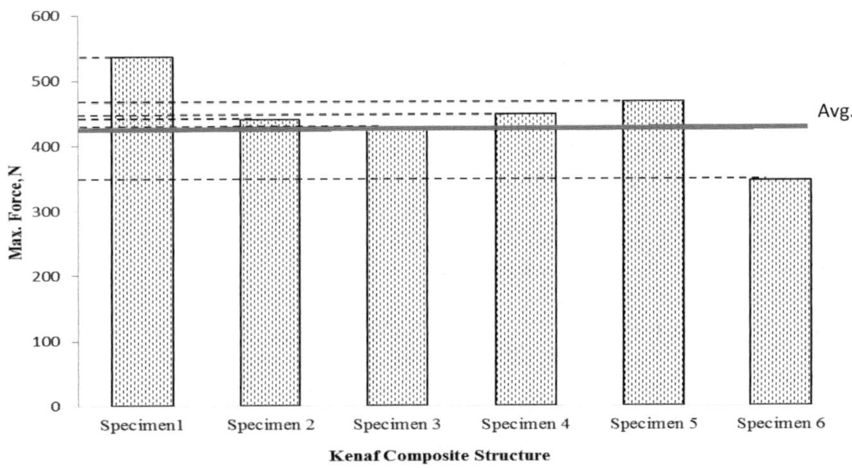

Figure 8. Shear strength of kenaf composite structures

4. Conclusion

In this study , kenaf composite structures were made using hand lay-up technique and tested to evaluate the characteristic of lamination structures, in term of strength and impact condition. The novel ideas from this investigation is to produce new table tennis blade, that have natural fibers in its structures, to replace existing materials used, which is wood. The attraction and motivation using natural fibers as replacement materials in table tennis blade structures, due to the biodegradability, light in weight and relatively stronger [6]. Therefore, this study was conducted, on a basic experimental for mechanical characteristic, in order to know the capability of new structures. Based on the result, it shows the possibility to replace wood, as a main material for the new table tennis blade. However, it should be noted that, this study need to be extended on other scope of study, each aspect need to investigate, to ensure it can be realized, created a same characteristic behaviour with an existing table tennis blade and will be commercialized.

Acknowledgement

The author would like to express their gratitude to Universiti Tun Hussein Onn Malaysia for supporting these research activities.

References

[1] Joshi SV, Drzal LT, Mohanty AK and Arora S. Are natural fiber composites environmentally superior to glass fiber reinforced composites 2004 *Composites Part A* **35** 371-6
[2] Arifin AMT, Abdullah S, Wahab DA and Rafiquzzaman M Behavior of chopped strand mat and woven roving under bending 2014 *Materials Testing* **56** 1-7
[3] Nazeth da Silva AL, Teixeira SCS, Widal ACC and Coutinho FMB Mechanical properties of polymer composites based on commercial epoxy vinyl ester resin and glass fiber 2001 *Int. J. Polymers Testing* **20** 895-9
[4] Davoovi MM, Sapuan SM, Ahmad D, Ali A, Khalina A and Jonoobi M Mechanical properties of hybrid kenaf/glass reinforced epoxy composite for passenger car bumper beam 2010 *Material Design* **31** 4927-32
[5] Ghani MAA, Salleh Z, Hyie KM, Berhan MN, Taib YMD and Bakri MAI Mechanical properties of Kenaf/Fiberglass polyester hybrid composite 2012 *Procedia Engineering* **41** 1654-59
[6] Saba N, Paridah MT, Jawaid M Mechanical properties of kenaf fibre reinforced polymer composite: A review 2015 *Construction and Building Materials* **76** 87-96

[7] Andrew S. 2014 Preliminary evaluation of kenaf as a structural material. Master Thesis Graduate Scholl of Clemson University
[8] Bledzki A and Gassan J Composites reinforced with cellulose based fibers 1999 *Prog Polym Sci* **24** 221–74
[9] Paridah MT, Basher AB, SaifulAzry S and Ahmed Z Retting process of some bast plant fibers and its effect on fiber quality: a review 2011 *Bio Resour* **6** 5260-81
[10] Wambua P, Ivens J and Verpoest I Natural fibres: can they replace glass in fibre reinforced plastics 2003 *Composites Science and Technology* **63** 1259 –64
[11] Seemiler D and Holowchak M. 1997 Winning Table Tennis: Skill, Drills and Strategies (United States of America: Human Kinetics)
[12] Arifin AMT, Abdullah S, Rafiquzzaman M, Zulkifli R, Wahab DA and Arifin AK Investigation of the behaviour of a chopped strand mat/woven roving/foam-klegecell composite lamination structure during charpy testing 2014 *Materials and Design* **59** 475-85
[13] Arifin AMT, Abdullah S, Rafiquzzaman M, Zulkifli R and Wahab DA Failure characterisation in polymer matrix composite for un-notched and notched (open –hole) specimens under tension condition 2014 *Fibers and Polymers* **15** 1729-38
[14] Hojo T, Xu Z, Yang Y, Hamada H Tensile properties of bamboo,jute and kenaf mat reinforced composite 2014 *Energy Procedia* **56** 72-9
[15] Othman A, Abdullah S, Ariffin AK and Mohamed NAN Investigating The Quasi-Static Axial Crushing Behavior Of Polymeric Foam-Filled Composite Pultrusion Square Tubes 2014 *Materials and Design* **63** 446-59
[16] Rafiquzzaman M, Abdullah S and Arifin AMT Behavioural observation of laminated polymer composite under uniaxial quasi-static and cyclic loads 2015 *Fibers and Polymers* **16** 640-49

Characterization and Thermal Properties of Nitrate Based Molten Salt for Heat Recovery System

Mohd Faizal Tukimon[1], Wan Nur Azrina Wan Muhammad[1] , Md Nor Annuar Mohamad[1] , Farazila Yusof[2]

[1]Faculty of Mechanical and Manufacturing Engineering, Universiti Tun Hussein Onn Malaysia, 86400 Parit Raja, Batu Pahat, Johor, Malaysia.
[2]Faculty of Mechanic, University of Malaya, 50603 Kuala Lumpur, Wilayah Persekutuan Kuala Lumpur, Malaysia.

Corresponding Author: wazrina@uthm.edu.my

Abstract. Molten salt can acts like a storage medium or heat transfer fluid in heat recovery system. Heat transfer fluid is a fluid that has the capability to deliver heat this one side to another while heat recovery system is a system that transfers heat to produce energy. This studies shows about determining the new formulation of different molten nitrate/nitrite salts consisting of $LiNO_3$, KNO_2, KNO_3 and $NaNO_2$ that give a low temperature of melting point and high average specific heat capacity. Mixed alkaline molten nitrate/nitrite salt can act as a heat transfer fluid due to their advantageous in terms of its properties that feasible in heat recovery system such as high specific heat capacity, low vapour pressure, low cost and wide range of temperature in its application. The mixing of these primary substances will form a new line of quaternary nitrate salt ($LiNO_3$ - KNO_2 - KNO_3- $NaNO_2$). The quaternary mixture was heated inside the box furnace at 150°C for four hours and rose up the temperature to 400°C for eight hours to homogenize the mixture. Through heating process, the elements of nitrate/nitrite base were mixed completely. The temperature was then reduced to 115°C for several hours before removing the mixture from the furnace. The melting point of each sample were testified by using thermal gravimetric analysis, TGA/DTA and experiment of determining the specific heat capacity were conducted by using Differential Scanning Calorimeter, DSC. From the result, it is found that the melting point Sample 1 with percentage of weightage (25.4wt% of $LiNO_3$, 33.8wt% of KNO_2, 20.7wt% of KNO_3 and 20.1wt% of $NaNO_2$) is 94.4°C whereas the average specific heat capacity was 1.0484/g°C while for Sample 3 with percentages of weightage (30.0wt% of $LiNO_3$, 50.2wt% of KNO_2, 3.1wt% of KNO_3 and 16.7wt% of $NaNO_2$), the melting point is 86.1°C with average specific heat capacity of 0.7274 J/g°C. In the nut shell, the quaternary mixture salts had been a good mixture with good thermal properties that low in melting point and have high specific heat capacity which could be a potential heat transfer fluid in heat recovery application.

1. Introduction

Molten salt technology is a combination of different technologies such as electro-chemistry, heat transfer, chemical oxidation or reduction baths and nuclear reactors. All of these applications are related by the common characteristics of molten salts which it can act as a mixture that has good specific heat capacity. Basically, the molten salt can produce an energy to the system while can withstand a high temperature course material [1]. According to Bradshaw *et al.*, [2] molten salt heat

IOP Conf. Series: Journal of Physics: Conf. Series **914** (2017) 012016 doi:10.1088/1742-6596/914/1/012016

transfer is a compound that may enhance the thermal energy storage for a designated system. For further development according to this molten salt mixture that containing an alkaline nitrate and nitrite salts, both mixture have the advantageous properties that relevant in the application of energy recovery system. Bradshaw also state that molten salt have a very low vapour pressure and most inactive substances in the air [2].

Molten salt is both a non-toxic and inert liquid. It is frequently used as a heat transfer fluid as well as the thermal energy storage in different industries such as organic rankine cycle, concentrated solar power and electricity. According to Lu-di Zhang *et al.,* [3] the usage of molten salts as a working fluid in industries might help in reducing the global warming by using solar system. Molten salts was very commercial in solar thermal power plant as heat transfer fluid and as a medium of thermal energy storage due to its advantages such as low viscosity, low vapor pressure, good chemical stability, low cost, wide range of operating temperature and environmental friendly [3]. According to US patent by Cordaro [4] and Bradshaw [5] molten salt is practically harmless substances in term of environmental surrounding. Also, it is stated that molten salt have a high density and high heat capacity that boosting the heat transfer capability. Besides, molten salt heat transfer fluid is a stable with contact with air and not a flammable substance.

One of the applications of molten salt is act as a heat transfer fluid. From previous researcher, the heat transfer fluids contain a high melting point and limit the practical application in concentrating solar power application [1, 6-11]. Quaternary molten salts are a mixture of four different salts that mixed well to be a new molten salt. To investigate the properties of thermal inside the system, quaternary salt is the most preferable mixture for previous researcher that can bring the best result of melting point, heat capacity and heat storage [6]. According to Wang *et al.,* [6] a new quaternary mixture of $LiNO_3$ – $NaNO_3$ – KNO_3 – $Ca(NO_2)_3$ gives a better physical and chemical characteristic compare to binary Solar salt. At the end of these studies, the new compositions of quaternary nitrate salt that contain low melting point with high average specific heat capacity will be proposed for heat recovery system application.

2. Experimental Procedure

In order to find the best mixture of the quaternary nitrate salts, several composition were made by using the range of composition as shown in Table 2.1.

Table 2.1 Range of Compositions (wt%).

Molten Salt	Range of composition
$LiNO_3$	20-30 wt %
KNO_2	30-52 wt %
KNO_3	0-20 wt %
$NaNO_2$	15-20 wt %

The mixtures of primary molten salt were prepared by following the range of composition listed on Table 2.1 to develop a new quaternary heat transfer fluid. By using the range of composition shown in Table 2.1, the new line of quaternary molten salt that gives low melting point has been determined. From this study, the low melting point of quaternary mixture is defined from 0 to 100°C whereas medium melting point is defined from 101 to 500°C. Meanwhile, high melting point is defined from 501°C and above. The nitrate based alkali was very famous among previous researcher since its gives a low temperature and high specific heat capacity in their application [3, 6, 12-14]. Table 2.2 shows that the mass ratio of sample 1, 2 and 3.

Table 2.2 The composition of molten salts in milligrams, (mg).

No. sample	Mass in milligram (mg)			
	$LiNO_3$	KNO_2	KNO_3	$NaNO_2$
1	25.4	33.8	20.7	20.1
2	26.2	46.3	10.0	17.5
3	30.0	50.2	03.1	16.7

The mixing of quaternary molten salt was heated inside a box furnace. Firstly, the mixture of salt was heated at the temperature of 150°C for four hours to complete the process of removing existing moisture in the mixture. Then, the temperature was increased up to 400°C for eight hours to let the element well mixed. After the heating process was done for 8 hours, several hours was needed for the sample to go through cooling process until the temperature reached 115°C. Then, a new quaternary mixture was produced after exposing the mixture for several minutes at the room temperature. The phase of new quaternary molten salt after being exposed to the surrounding was solid phase. Then, the salt mixtures have been crushed into powder to proceed for the next step. The melting point of quaternary mixture was determined by using TGA/DTA machine. Meanwhile, we determined the specific heat capacity of the molten salt by using Differential Scanning Calorimeter, DSC.

3. Results and Discussion

In order to determine the melting point of each quaternary mixture, all sample were undergoes TGA/DTA testing. The result of melting point was detected on each peak of endothermic or exothermic peak that observed at DTA graph. The operating parameter that used for quaternary molten salt were, mass in milligram, (120-130mg), maximum temperature, (500°C), and temperature rise per minutes, (10°C/min). As for the quaternary molten salt result, Figure 3.1 shows the results of melting point of each mixture through TGA/DTA testing.

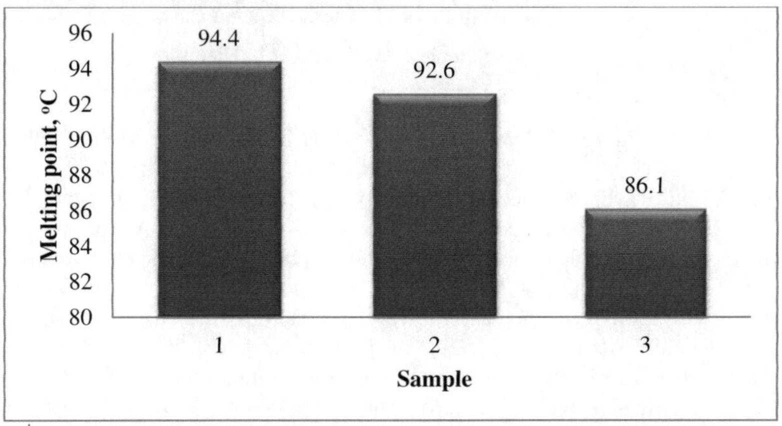

Figure 3.1 Graph of Melting Point Versus Sample.

From Figure 3.1, all three quaternary samples achieved the low temperature of melting point which is under 100°C. Based on the table 2.2, as the value of $LiNO_3$ increase, the melting point will decrease. According to previous researchers, by adding more $LiNO_3$ into the mixture will reduce the melting point of the molten salt [15] since $LiNO_3$ melts at lower temperature compared to other three primary elements. So, it is important to increase the number of weight percentages, (%wt) of $LiNO_3$ in achieving a low temperature of melting point. Meanwhile, as the value of weightage KNO_3 decrease, the melting point will also decrease. According to Nan Ren *et al.,* the sample gives a higher temperature due to the melting point of KNO_3 itself which was higher compared to other three elements [16]. Also, another researcher stated that the melting point of KNO_3 is higher than $NaNO_3$ which are 337°C and 307°C respectively [12]. By replacing or adding $NaNO_3$ into the mixture will help the molten salt to get a lower melting point. With this point, it is strong enough to be evidence that addition of KNO_3 will increase the melting point of the mixture. Next, as the weight of KNO_2 increase, the melting point will decrease whereas the melting points decrease as the mass of $NaNO_2$ decrease. By changing the nitrate solution to nitrite solution also can diminish the melting point [12] since its help molten salt to enhance the range of thermal stability of molten salt.

By completing the testing of DSC, the average results of specific heat capacity were then calculated. Differential scanning calorimeter (DSC) was conducted to provide the value of melting point and its specific heat capacity in the present of quaternary nitrate salt mixture [6]. Based on previous researcher, it is shows that the peak of the graph plotted was the value of specific heat capacity with its melting point [3, 6, 10, 13, 16-18]. From these results, the determination of formulation that contains low temperature and high specific heat capacity can be done. Figure 3.2 shows the graph of average specific heat capacity for three samples respectively.

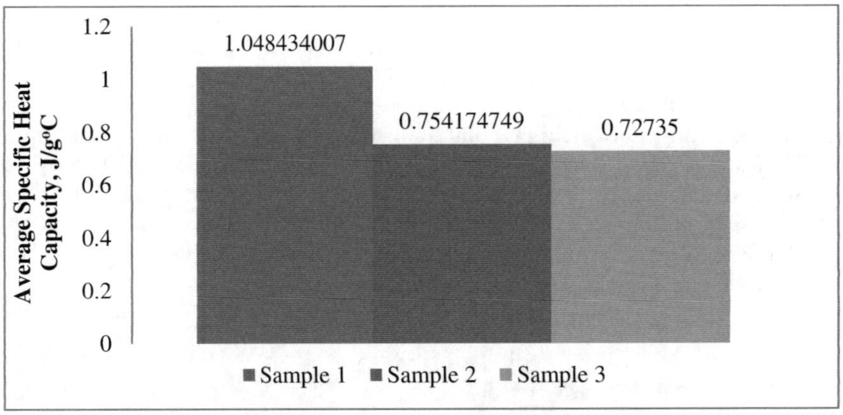

Figure 3.2 Average Specific Heat Capacities for Molten Salts.

From Figure 3.2, the average specific heat capacity for Sample 1, Sample 2 and Sample 3 are 1.0484J/g°C, 0.7542J/g°C and 0.7274J/g°C respectively. Sample 1 shows that the average specific heat capacity was the highest while sample 3 shows the lowest value of average specific heat capacity. By referring Table 2.2, as the value of $NaNO_2$ decrease, the value of average specific heat capacity will also decrease. It is important to increase the value of $NaNO_2$ since it gives a higher value of average specific heat capacity. Also, as the value of $LiNO_3$ increase, the average specific heat capacity decrease. This is because $LiNO_3$ melts at a very low temperature [12] and has low specific heat capacity. Hence, by decreasing the value of $LiNO_3$ will increase the value of average specific heat capacity. In contrast, the quaternary mixture produce high average specific heat capacity compared to primary, binary and ternary mixture. Based on these three average specific heat capacities, Sample 1 gives the highest value compared to Sample 3 that gives the lowest value average specific heat capacity. It is mentioned that the higher the specific heat capacity, the greater the thermal stability that satisfied the application in thermal energy storage [6].

4. Conclusion

From result TGA/DTA testing, sample 3 of quaternary mixture give the lowest melting point which was 86.1°C but give the lowest value of specific heat capacity which was 0.7274 J/g°C. Besides that, sample 1, also give a low value of melting point which was 94.4°C but gives the highest value of specific heat capacity, 1.0484 J/g°C. For a conclusion, the best quaternary mixture molten salt that will be proposed was sample 1 that shows its properties in low melting point but high value in specific heat capacity. Hence, the aim of this study was fulfilled by finding the better formulation of molten salts in low temperature condition which are 25.4wt% of $LiNO_3$, 33.8wt% of KNO_2, 20.7wt% of KNO_3 and 20.1wt% of $NaNO_2$.

Acknowledgement

This work is supported by University Tun Hussein Onn Malaysia (UTHM) under Postgraduate Research Grant (GPPS) Vot No U721 and Fundamental Research Grant Scheme (FRGS) Vot No 1588.

References

[1] Peng. Q, Yang, X., Ding, J., Wei, X., & Yang, J.(2013). Design of new molten salt thermal energy storage material for solar thermal power plant. Applied Energy, 112., 682-689.

[2] Bradshaw RW, Cordaro JG, Siegel NP. Molten nitrate development for thermal energy storage in parabolic through solar power system. In: Conference Molten Nitrate salts development for thermal energy storage in parabolic through solar power system, vol. 2 ;2009.

[3] Zhang, L., Chen, X., Wu, Y., Lu, Y., & Ma, C. (2016). Solar Energy Materials & Solar Cells Effect of nanoparticle dispersion on enhancing the specific heat capacity of quaternary nitrate for solar thermal energy storage application, *157*, 808–813.

[4] Cordaro JG., "Low-Melting Point Heat Transfer Fluid" U.S. Patent 7,828,990 issued November 9, 2010.

[5] Bradshaw RW., "Low Melting Point Inorganic Nitrate Salt Heat Transfer Fluids" U.S. Patent 7,588,694 issued September 15, 2009

[6] Wang, T., Mantha, D., & Reddy, R. G. (2013). Novel low melting point quaternary eutectic system for solar thermal energy storage. *Applied Energy*, *102*, 1422–1429.

[7] Desai, N. B., & Bandyopadhyay, S. (2015). Optimization of concentrating solar thermal power plant based on parabolic trough collector. Journal of Cleaner Production, 89, 262–271

[8] Tian, Y., & Zhao, C. Y. (2013). A review of solar collectors and thermal energy storage in solar thermal applications. Applied Energy, 104, 538–553.

[9] Martin, C., Bauer, T., & Müller-steinhagen, H. (2013). An experimental study of a non-eutectic mixture of KNO 3 and NaNO 3 with a melting range for thermal energy storage, 56.

[10] Herrmann, U., Kelly, B., & Price, H. (2004). Two-tank molten salt storage for parabolic trough solar power plants, 29, 883–893.

[11] Valverde, J. L., & Sánchez, P. (2013). Characterization of different heat transfer 5lu ids and degradation study by using a pilot plant device operating at real conditions. Energy, 54, 240–250.

[12] Redzuan N.A., Wan Muhammad W.N.A., & Mohamad M.N.A. (2016). Synthesis of molten salt as heat transfer fluid for waste heat recovery, *11*(10), 6589–6593.

[13] Wu, Y., Li, Y., & Ren, N. (2017). Solar Energy Materials & Solar Cells, *160*(April 2016), 263–268.

[14] Wu, Y., Li, Y., Lu, Y., Wang, H., & Ma, C. (2017). Solar Energy Materials & Solar Cells Novel low melting point binary nitrates for thermal energy storage applications, *164*(February), 114–121

[15] Bradshaw, R. W., & Siegel, N. P. (2008). ES2008-54174, 1–7.

[16] Ren, N., Wu, Y., Ma, C., & Sang, L. (2014). Solar Energy Materials & Solar Cells Preparation and thermal properties of quaternary mixed nitrate with low melting point. *Solar Energy*

Materials and Solar Cells, *127*, 6–13.

[17] Li, R, Zhu, J., Zhou, W., Cheng, X., & Li, Y. (2016). Thermal compatibility of Sodium Nitrate/ Expanded Perlite composite phase change material. Applied Thermal Energy, 103, 452-458.

[18] Zhao, C. Y., & Wu, Z. G. (2011). Solar Energy Materials & Solar Cells, Thermal property characterization of a low melting-temperature ternary nitrate salt mixture for thermal energy storage systems, *95*, 3341–3346.

[19] Herrmann, U., Kelly, B., & Price, H. (2004). Two-tank molten salt storage for parabolic trough solar power plants, 29, 883–893.

[20] Ã, B. L., Chien, K., & Wang, C. (2004). Effect of working fluids on organic Rankine cycle for waste heat recovery, *29*, 1207–1217.

IOP Conf. Series: Journal of Physics: Conf. Series **914** (2017) 012017 doi:10.1088/1742-6596/914/1/012017

PCL/PLA Polymer Composite Filament Fabrication using Full Factorial Design (DOE) for Fused Deposition Modelling

Haq, R. H. A.[1,2], Rd. Khairilhijra, K.[1,2], Wahab, M.S.[1,2], Sa'ude, N.[1,2], Ibrahim, M.[1,2], Marwah, O.M.F.[1,2], Yusof, M. S.[1,2], Rahman, M. N. A.[1], Ariffin, A. M. T.[3], Hassan, M. F.[3], Yunos, M. Z.[3], Adzila, S.[3]

[1] Department of Manufacturing and Industrial Engineering, Faculty of Mechanical and Manufacturing Engineering, Universiti Tun Hussein Onn Malaysia, 86400 Parit Raja, Batu Pahat, Johor, Malaysia
[2]Additive Manufacturing Research Group (ADDMAN). Department of Manufacturing and Industrial Engineering, Faculty of Mechanical and Manufacturing Engineering,, Universiti Tun Hussein Onn Malaysia, 86400 Parit Raja, Batu Pahat, Johor, Malaysia
[3]Department of Material and Design Engineering, Faculty of Mechanical and Manufacturing Engineering, Universiti Tun Hussein Onn Malaysia, 86400 Parit Raja, Batu Pahat, Johor, Malaysia

Corresponding author: reazul@uthm.edu.my

Abstract: In this study, Polycaprolactone / Polylactice Acid (PCL/PLA) composite are used to fabricate filament wire with the specific diameter, which is in the range of 1.75 to 1.80 mm. Full factorial experimental design technique was used to study the main effects and the interaction effects between operational parameter which is (A) die temperature, (B) roller puller speed, (C) spindle speed and (D) inlet temperature. Besides that, there are two levels (-1 and +1) and the response are filament wire diameter. There are 16 numbers of runs and plus 8 centre points per blocks which makes the runs into 24 runs. From the experiment it shows that there are four factor that are significant effects on the filament wire diameter which is A, B, C and BC. The optimum parameter setting are also determined and there are 10 suggestions to achieve the target with different setting of parameter. The margin error for confirmation run is below than 15% when the parameter set at 6 Hz spindle speed, 4.99 rpm roller puller, 100.31 °C die temperature and 79.65 °C inlet temperature which can be noted that the confirmation run result is acceptable. The optimization parameter setting can use to continue in Fused Deposition Modelling (FDM). Filament wire from PCL/PLA are succesfully fabricated with acceptable diameter size and ready to be used for Fused Depotion Modelling process (FDM).

1.0 Introduction

Additive manufacturing has grown in the manufacturing process, where 3D printing is giving the ability to prototype and manufacture process, for example to make the difficult shape and it compromises important advantages over the traditional manufacturing process. In FDM, one of the most important parts is the filament. There are several types of filaments available in today's market which depends on the part to produce and its functionality. There are many types of standard filaments such as Polylactic Acid (PLA), Acrylonitrile Butadiene Styrene (ABS), Polyamide (Nylon) and

IOP Conf. Series: Journal of Physics: Conf. Series **914** (2017) 012017 doi:10.1088/1742-6596/914/1/012017

Polyethylene Terephthalate (PET) which is commonly used for producing prototypes. However, there are recent studies have been carried out to produce filament from composites either from Ceramic Matrix Composite (CMCs) or Metal Matrix Composite (MMCs). The challenges developing composite filament is to produce this filament in a specific size and strength in order for the filament to be extruded through the printer successfully [1].

The most significant in the filament is that the diameter size must be uniform and constancy in order to used in 3D printing during the fabrication process and capability to optimize the right diameter to suit in the 3D printer [2]. The advantages of small diameter filament are that the filament is easy to be produce through the extruder machine besides, the printing process will be more accurate and easier.

There are a lot types of materials that are in market to produce implants such as titanium, steel, ceramic and aluminium, but nowadays, most of implants are made from polymers. There are a lot of benefits to produce implants from olymer where it reduces cost and it has a good physical properties. Polycaprolactone/Polylactic Acid (PCL/PLA), both of these polymers are most used in medical applications. These polymers are very famous in biomaterials because they are complementary on their physical properties and biodegradability. The PCL / PLA composite has been used as a tissue engineering scaffolds for regeneration, bone, ligament, orthopedic, skin, nerve and vascular tissue [3][4].

Usually, Fused Deposition Modelling (FDM) uses Acrylonitrile Butadiene Styrene (ABS) as its base materials, however, there are possibilities to produce new composites, especially from the mixture of Polycaprolactone/ Polylactic Acid (PCL/PLA) which can be an alternative material for the implant.

2.0 Methodology of research

2.1Materials

Type of polymer used in this experiment are biodegradable polycaprolactone (PCL), medical grade of BGH600C meanwhile polylactic Acid (PLA) that used in this research is a medical grade of PTG600C. Both of this polymers are in pellet form and supplied by Shenzen Bright China Industrial Co, China.

Table 1: Blend formulation of PCL and PLA

Designation	PCL (wt%)	PLA(wt%)
PCL30PLA70	70	30

2.2 Sample preparation

The PCL and PLA which is in resin form are mix manually. These uniformly dry-mixed batches are melt-blended in Brabender Plastograph machine where the temperatures were set at 160^{0}C before it is crush into pellet form. All the compounded materials were then fed into injection section. The tensile and flexural specimens were produced on an injection molding machine with an injection temperature between 40 ^{0}C – 80 ^{0}C. Injection pressure was set to 140 MPa.

2.3 Filament preparation

The Polymer nano-composites that crushed into pallet size later is fed into single screw extruder machine which have been set along with a water bath and also a roller puller machine as shown in Figure 1.

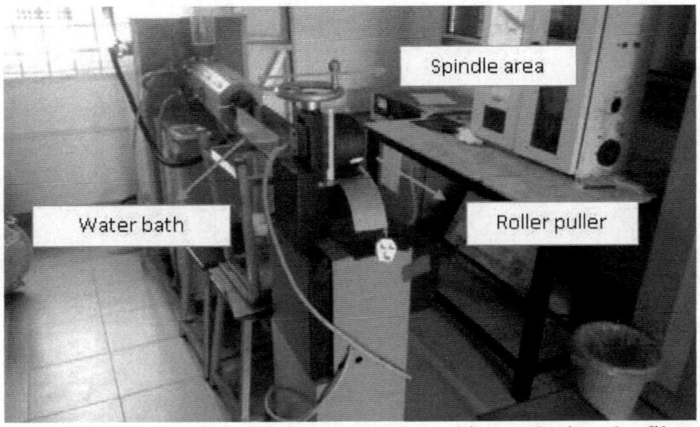

Figure 1 : Layout of the experiment conducted in producing the filament.

2.4 Full Factorial Design (DOE)

The Design Expert Software version 6 is used in this experiment to discover the main effects for the parameters that involve in the filament fabrication process. Analysis of Variance (ANOVA) remained proficient in finding the conditional on variables that affect the machining appearances and to create the optimum situation. There is four parameter use in this experiment, Table 2 shows the range of parameter factor and level were carried out in two level high and low. The high was design in (+1) and for low was designed in (-1). The response is very important because from that the result of this experiment will be obtained. The incline of the factor that continued in demand to recognized the response which is the filament wire size. Two level full factorial design have four factors (A, B, C and D) and each factor to be studied at 2-level. There are altogether 24 runs with added 8 center points per blocks. The full factorial experiment run is shown in Table 3.

Table 2: The optimized factor using DOE

Factor code	Name	Units	Low (-1)	High (+1)
A	Die temperature	°C	100	160
B	Roller puller speed	rpm	2	6
C	Spindle speed	Hz	2	6
D	Inlet temperature	°C	60	100

IOP Conf. Series: Journal of Physics: Conf. Series **914** (2017) 012017 doi:10.1088/1742-6596/914/1/012017

Table 3: Full factorial experiment runs

Std	Run	Block	**Factor (A)** Die temperature (°C)	**Factor (B)** Roller puller speed (RPM)	**Factor (C)** Spindle speed (Hz)	**Factor (D)** Inlet temperature (°C)	**Response** Filament size (mm)
9	1	PCL70PLA30	100	2	2	80	
15	2	PCL70PLA30	100	6	6	80	
13	3	PCL70PLA30	100	2	6	80	
21	4	PCL70PLA30	130	4	4	65	
18	5	PCL70PLA30	130	4	4	65	
8	6	PCL70PLA30	160	6	6	50	
10	7	PCL70PLA30	160	2	2	80	
7	8	PCL70PLA30	100	6	6	50	
3	9	PCL70PLA30	100	6	2	50	
5	10	PCL70PLA30	100	2	6	50	
22	11	PCL70PLA30	130	4	4	65	
4	12	PCL70PLA30	160	6	2	50	
16	13	PCL70PLA30	160	6	6	80	
24	14	PCL70PLA30	130	4	4	65	
2	15	PCL70PLA30	160	2	2	50	
19	16	PCL70PLA30	130	4	4	65	
17	17	PCL70PLA30	130	4	4	65	
20	18	PCL70PLA30	130	4	4	65	
1	19	PCL70PLA30	100	2	2	50	
14	20	PCL70PLA30	160	2	6	80	
6	21	PCL70PLA30	160	2	6	50	

12	22	PCL70PLA30	160	6	2	80	
11	23	PCL70PLA30	100	6	2	80	
23	24	PCL70PLA30	130	4	4	65	

3.0 Result & Discussion.

3.1 Two Level Full Factorial Design.

The complete result of all 24 runs of experiment are shown in Table 4. From the experiment, it can be seen that there are two result that produce the filament with diameter size in range of 1.75mm-1.80mm which are run number 6 and 13. The parameter setting is when the die temperature at 100°C, the roller puller speed at 6 RPM, the spindle speed at 6 Hz. The differences is only at the inlet temperature where at run 6 the temperature is at 50°C meanwhile at run 13 the temperature is at 80°C.

Table 4: The respond result for filament size diameter

Run	**Factor (A)** Die temperature (°C)	**Factor (B)** Roller puller speed (RPM)	**Factor (C)** Spindle speed (Hz)	**Factor (D)** Inlet temperature (°C)	**Response** Filament size (mm)
1	100	2	2	80	1.18
2	100	6	6	80	1.63
3	100	2	6	80	1.98
4	130	4	4	65	1.61
5	130	4	4	65	1.61
6	160	6	6	50	1.77
7	160	2	2	80	1.36
8	100	6	6	50	1.53
9	100	6	2	50	0.98
10	100	2	6	50	2.49
11	130	4	4	65	1.63
12	160	6	2	50	1.12
13	160	6	6	80	1.76
14	130	4	4	65	1.57
15	160	2	2	50	1.70
16	130	4	4	65	1.59
17	130	4	4	65	1.59
18	130	4	4	65	1.59
19	100	2	2	50	1.43
20	160	2	6	80	2.73
21	160	2	6	50	2.81
22	160	6	2	80	1.11
23	100	6	2	80	0.93
24	130	4	4	65	1.53

3.1.1 Analyzed in of Variance (ANOVA) for Filament Wire.

All the data were analyzed in variance (ANOVA). Figure 2 shows the half normal plot graph. From the figure, all the specimens are located far from line that have a significant effect which are point A, B, C, and BC. Table 5 below shows the ANOVA summarize for the next interpretation.

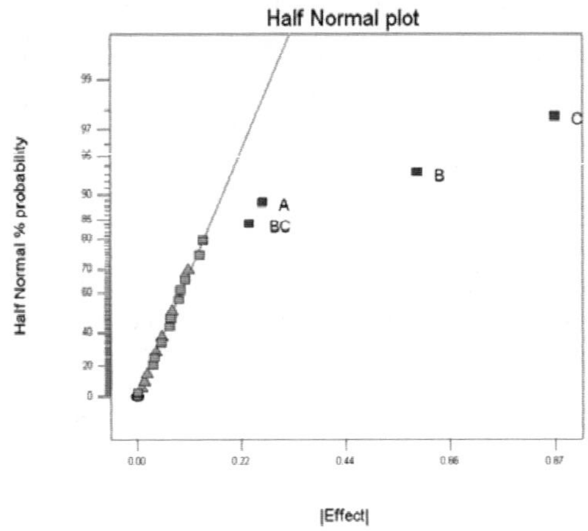

Figure 2: The half normal plot graph for filament wire diameter PCL70PLA30

Table 5: ANOVA summarize for filament wire diameter

Response: Filament wire diameter					
ANOVA for selected factorial model					
Analysis for variance table (Partial sum of square)					
Source	Sum of squares	DF	Mean Square	F value	Prob > F
Model	4.94	4	1.24	36.75	< 0.0001*
A	0.28	1	0.28	8.28	0.0100
B	1.39	1	1.39	41.25	<0.0001
C	3.05	1	3.05	90.85	<0.0001
BC	0.22	1	0.22	6.64	0.0190
Curvature	0.018	1	0.018	0.54	0.4735**
Residuals	0.61	18	0.034		
Lack of Fit	0.31	11	0.028	0.67	0.7318**
Pure error	0.29	7	0.042		
Cor Total	5.56	23			

Significant*
Not-significant**

The model of F-value is 36.75 indicates the model is significant. There are only a 0.01% chance that a model F- value this larger could occur due to noise. The value of Prob > F less than 0.0500 indicates that the model is significant. In this case, only point A, B, C, BC are significant model terms because the value is below than 0.0500. The Curvature F- value is 0.54 implies the curvature. The curvature measure of difference between the average of the center point and average of the factorial point. This design is significant relative to noise. The Curvature Prob F> value has chance 47.35%, this is the large value and might occur due noise. The Lack of Fit F- value is 0.67, this is not significant relative to the purse error. Prob > F there is a 73.18% chance, this is large could occur due to noise. Non- significant is the best result because when reject the factory are not involved the value is close to 1.

Table 6 below shows the ANOVA and a regression summary of filament wire diameter fabrication. The Predicted R- Squared, Adjusted R-square and R- Square value is 0.8127, 0.8867, and 0.8909 respectively. This values are in reasonable because the value were close to 1. For the Adequate Precision the value must greater than 4 is desirable. In this adequate result, the value is 18.831 indicating an adequate signal. The Adequate Precision is measure the signal to the noise ratio and this model can be used to navigate the design space.

Table 6: ANOVA and regrassion summary filament wire diameter for PCL70PLA30

Response model	Quadratic model
R-squared	0.8909
Adjusted R-Squared	0.8667
Predicted R-squared	0.8127
Adequate Precision	18.831

The quadratic regression equation might be expressed as:

Final Equation in terms of Coded Factor:
*Filament wire diameter = 1.65 + 0.13*A – 0.29*B + 0.44*C – 0.12*B*C*

Final Equation in terms of Actual Factor
*Filament wire diameter = o.3191 + 4.39583E*A – 0.029062*B + 0.33656*C – 0.029531*B*C*

3.1.2 Graphic Plots Analysis Filament wire Diameter
The analysis continues with the graph plots. The factor among spindle speed, roller puller, inlet temperature and die temperature were observed better in the normal plot of residuals graph, residuals vs. predicted graph and Outlier T graph. The graphs plot are shown in Figure 3, 4 and 5 respectively.

Figure 3 shows that the residuals point lies on a straight line indicating that the errors are distributed normally. Meanwhile, the plots in Figures 4 and 5 shows that no obvious pattern and unusual structure and all the results fall in the acceptable range. Therefore, it can be concluded that the model proposed was adequate and could proceed for further analysis. At this point, only significant effects were analyzed.

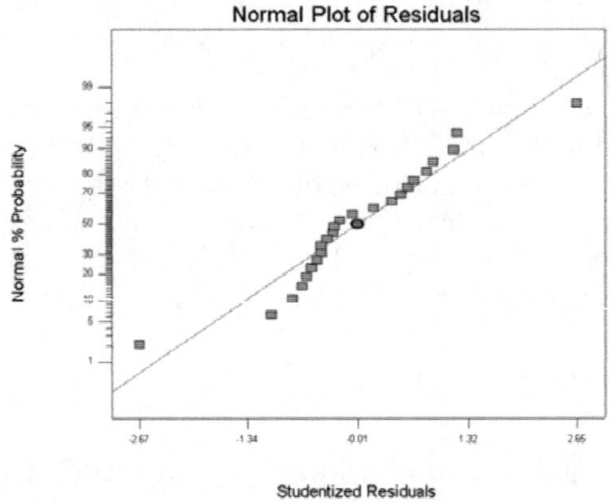

Figure 3: The normal plot of residuals graph of filament wire diameter PCL70PLA30

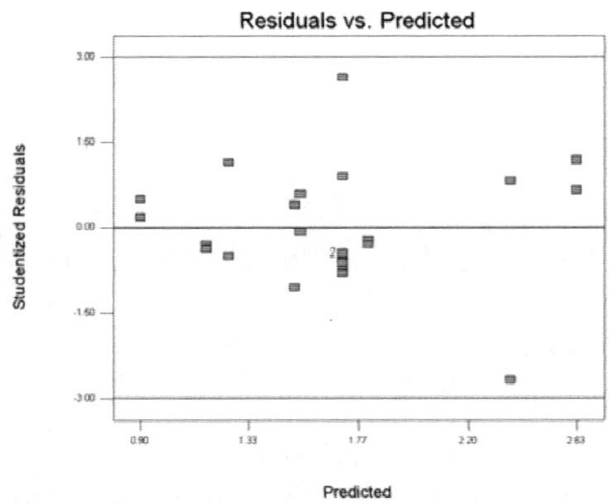

Figure 4: The residuals vs. predicted graph of filament wire diameter PCL70PLA30

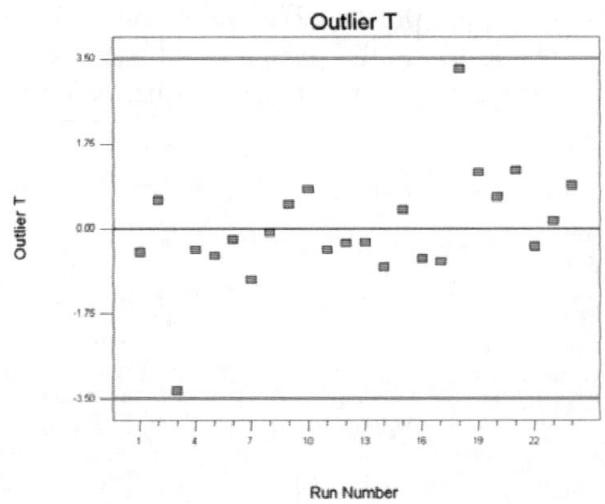

Figure 5: The outline T graph of filament wire diameter PCL70PLA30

IOP Conf. Series: Journal of Physics: Conf. Series **914** (2017) 012017 doi:10.1088/1742-6596/914/1/012017

Figure 6 shows the contour and interaction graph among (B) roller puller speed and (C) spindle speed. It have a strength interaction between two factors to fabricate the filament wire diameter. When the spindle speed was increased and the roller puller speed was decreased, the diameter size of filament wire produced is in a large size. When the spindle speed was decrease, the reverse result was found and the diameter of filament wire that produce in a small size.

This design of effect were contributed through the die swell outcome which is significant miracle defining to size or shape and quality performance of screw speed of extruder polymer. Meanwhile, the roller puller speed is important in fabricating of filament wire diameter in required range. If the roller puller setting was in a higher speed, it is possible that the wire that pull out from the die with a high speed will create the size of the wire that came out is in a small diameter. In order to achieve the required size of wire diameter, both of this parameter need to adjust in a perfect manner as both are correlated [5].

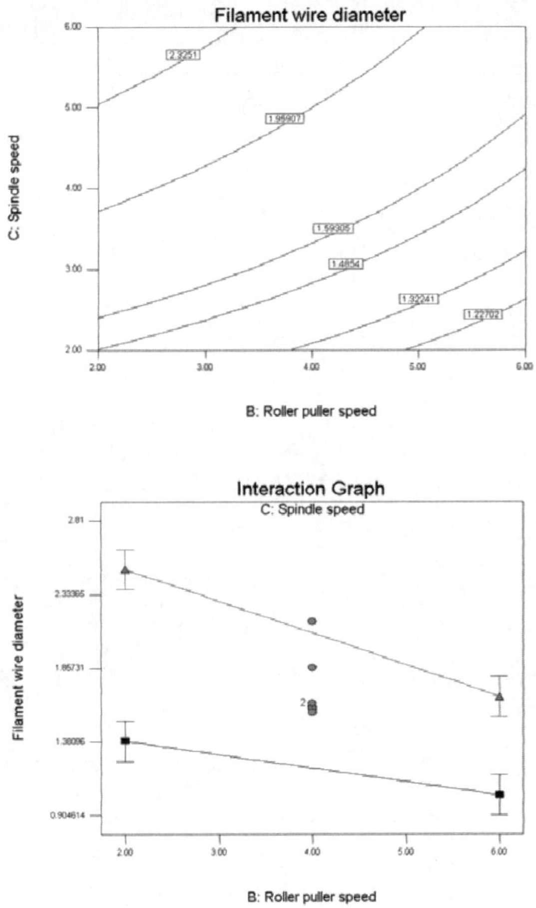

Figure 6: Contour and interaction graph among (B) roller puller speed and (C) spindle speed on filament wire diameter PCL70PLA30

3.1.3 Optimization Condition

The optimum condition for the filament wire diameter were obtained from the 2 level factorial design. The combination from few parameters were combined which is to achieve the filament wire diameter in the range 1.75 mm . From the result, there are 10 suggestions to achieve the target with different setting of parameter. Meanwhile, the combination of all suggestions were in specific desirability value because all value is closed to 1. Table 7 shows the suggested setting parameter for 1.75mm filament wire diameter.

Table 7: Suggested setting for filament wire diameter

Number	Die temperature (°C)	Roller puller speed (rpm)	Spindle speed (rpm)	Inlet temperature (°C)	Filament wire diameter (mm)	Desirability
1	100	4.97	6.00	79.96	1.75316	0.987
2	100.31	4.99	6.00	79.65	1.75	0.986
3	160.00	2.00	2.83	50.00	1.74992	0.984
4	124.99	5.52	6.00	80.00	1.75	0.972
5	113.96	5.24	5.95	80.00	1.75	0.972
6	100.00	4.98	6.00	74.71	1.75	0.966
7	160.00	2.00	2.82	52.62	1.74662	0.956
8	148.56	6.00	6.00	60.66	1.75417	0.939
9	160.00	4.19	3.99	50.00	1.75	0.907
10	100.00	4.98	6.00	60.12	1.75137	0.901

Figure 7 shows the ramps plots for suggestion number 1, where the parameter setting value must set on optimization conditions to achieve goals. Where the die temperature, roller puller speed, spindle speed, and inlet temperature value must set at 100.00 °C, 4.97 rpm, 6.00 Hz, 79.96 °C respectively.

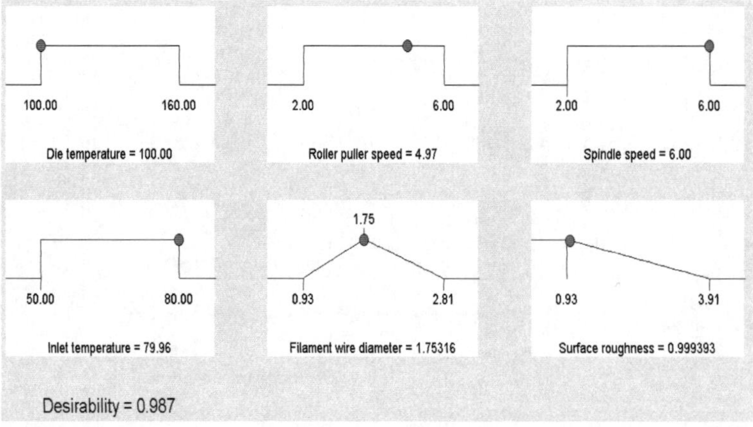

Figure 7: Ramps plots for suggestion number 1

3.2 Confirmation run for Filament Wire Diameter

The confirmation run are needed to compare the result between the theoretical and experimental value for the size of wire in the range 1.75 mm-1.80mm. From the optimization 2 level factorial design, there are 10 runs suggested to achieve the target. From the parameter range, run 2 were selected for the confirmation run to evaluate the value of margin error. Table 8 shows the confirmation run selected from the 2 level factorial design.

Table 8: Confirmation run selected (run 2)

Number	Die temperature (ºC)	Roller puller speed (rpm)	Spindle speed (rpm)	Inlet temperature (ºC)	Filament wire diameter (mm)	Desirability
2	100.31	4.99	6.00	79.65	1.75	0.986

By referring to the filament size in Table 8, the filament wire diameter for run 2 is 1.75mm. The confirmation run was repeated for 5 times, in order to identify the best setting on single screw extruder machine and roller puller machine. The actual test result in confirmation run is 1.75mm diameter size, the comparison of theoretical prediction result and the actual test result can be a final consideration for the optimal parameter predicted allowed in range. The margin error value of actual and predicted result was set less than 15%.

Table 9 shows the result of confirmation run. The parameter were set at 6 Hz spindle speed, 4.99 RPM roller puller, 100.31 ºC die temperature and 79.65 ºC inlet temperature. This optimizes setting for a confirmation run in a 95% prediction interval (PI) are reasonably accurate and the error value for filament wire diameter in the range 0 - 15%. The percentages are below than 15% and the confirmation run result is verified as valid to the existence of an optimal point. From this experiment, PCL/PLA filament wire succesfully fabricated with exact size, strength and properties to be used in Fused Deposition Modelling (FDM) machine.

Table 9: Confirmation run result for filament wire diameter

Run	Wire diameter (PCL70PLA30)		
	Actual (mm)	Predicted (mm)	Error (%)
1	1.76	1.75	0.57
2	1.77		1.14
3	1.75		0
4	1.76		0.57
5	1.78		1.69
	Average error (%)		0.79

4. Conclusion

Filament wire from Polycaprolactone/ Polylactic Acid (PCL/PLA) composites are succesfully fabricated with the aid of two level full factorial design using Design of Expert version 6 software. From the study, it can concluded that:

1. From the experiment there are 2 out of 24 runs that gives a promising result as the filament produce falls into an acceptable range. The parameter setting is when the die temperature at 100°C, the roller puller speed at 6 RPM, the spindle speed at 6 Hz. The differences is only at the inlet temperature where at run 6 the temperature is at 50°C meanwhile at run 13 the temperature is at 80°C.

2. The Analyzed of Variance (ANOVA) for filament wire diameter indicate that the significant effect were the die temperature (A), roller puller speed (B), spindle speed (C) and roller puller speed with spindle speed (BC). For the curvature and lack of fit value, there were not significant for the response. In curvature, if the value is significant, an additional run was needed. The lack of fit value must be closes to 1.
3. In order to achieve the target required diameter size of the filament which is 1.75 ± 0.05 mm, there are 10 suggestion of parameter setting for the all the 4 factors.
4. The confirmation run are conducted using the value from optimal condition that is suggested. The setting for confirmation was 6Hz for spindle speed, 4. 99 RPM for roller puller, 100.31 ºC for die temperature and 79.65 ºC for inlet temperature. From the confirmation run result, the percentage average error are below 15% and its is acceptable.

Acknowledgement

The authors would like to deliver their greatest acknowledgements to Universiti Tun Hussien Onn (UTHM) for funding this research – Short Term Grant (STG) U642.

References

[1] Rajendra P Pawar, 2014 'Biomedical Applications of Poly (Lactic Acid)', 40–51.

[2] Arrieta M P and Peponi L 2017 Polyurethane based on PLA and PCL incorporated with catechin: Structural, thermal and mechanical characterization. *European Polymer Journal* **89** 174–184.

[3] Arijit Gandhi and others, 2015 'Studies on Thermoresponsive Polymers: Phase Behaviour, Drug Delivery and Biomedical Applications', *Asian Journal of Pharmaceutical Sciences*, 10.2, 99–107

[4] Manfred F. Maitz, 2015 'Applications of Synthetic Polymers in Clinical Medicine', *Biosurface and Biotribology*, 1.3, 161–76.

[5] Haq, R.H.A., Wahab, M.S., Jaimi, N.I. 2013 Fabrication Process of Polymer Nano-Composite Filament for Fused deposition Modelling. *Journal of Applied Mechanics and Material* 465-466, pp. 8-12.

Neural-Fuzzy model Based Steel Pipeline Multiple Cracks classification

[1]*Hatem Mostafa Elwalwal*, [1]*Shahruddin Bin Hj. Mahzan*, [2]*Ahmed N Abdalla*

[1]Faculty Of Mech. And Man. Eng., Univeriti Tun Hussein Onn Malaysia, Johor 86400, Malaysia

[2]Faculty of Electr. Info. Eng., Huaiyin Institute of Technology, Huai'an 223003, Jiangsu, P. R. China

Email: sharudin@uthm.edu.my

Abstract. While pipes are cheaper than other means of transportation, this cost saving comes with a major price: pipes are subject to cracks, corrosion etc., which in turn can cause leakage and environmental damage. In this paper, Neural-Fuzzy model for multiple cracks classification based on Lamb Guide Wave. Simulation results for 42 sample were collected using ANSYS software. The current research object to carry on the numerical simulation and experimental study, aiming at finding an effective way to detection and the localization of cracks and holes defects in the main body of pipeline. Considering the damage form of multiple cracks and holes which may exist in pipeline, to determine the respective position in the steel pipe. In addition, the technique used in this research a guided lamb wave based structural health monitoring method whereas piezoelectric transducers will use as exciting and receiving sensors by Pitch-Catch method. Implementation of simple learning mechanism has been developed specially for the ANN for fuzzy the system represented.

1. Introduction

Pipelines are used in virtually every nation around the globe to transport oil and gas from the fields to the market. Oil spills, gas leaks and their associated environmental problems has become a serious and major concern in the oil and gas industry; and consequently, this has led to significant losses in revenue, severe disruption of operations, persistent threat to marine life and the ecosystem. This accidental discharge of petroleum products on/offshore has hitherto caused untold and unimaginable environmental hazards and economic loss that requires urgent remedial action and attention [1,2].

Many researchers have studied inspection of this pipes line is critical, not only to maintain normal functions of various life support facilities, but also to prevent fires, explosions, water damage, and contamination from broken gas, water, or sewage lines. All pipes whether made of coated steel or plastic composites will

IOP Conf. Series: Journal of Physics: Conf. Series **914** (2017) 012018 doi:10.1088/1742-6596/914/1/012018

corrode and form some kind of a defect in the form of a crack or small pit on the surface of the pipe. In power industries any leakage due to these cracks might be hazardous to both man and nature. Also, such leaks will eventually cause economic losses since the resources are wasted. Periodic inspection of pipes is needed to satisfy regulatory requirements. The inspection characterizes the condition of the pipes and directs strategic planning on whether and when to replace, repair or continue operation of the pipe [2-4].

Starting from the fuzzy inference system and the combination of artificial neural networks has led to increased interest in research and the application of a variety of different applications [5-9]. There are an understanding between the development of this system due to the attitude of hole complement each other rather than for competition. In one part the FIS offering method of calculation that emphasizes the interpretability of results arises, imprecision and uncertainty data for system capability of ANN calculation. Additionally, ANN has the ability to understand, adapt to change is an advantage to the FIS. In-depth study has shown merger between FIS and ANN has been created [10-13]. Mamdani FIS is one of the neuro-fuzzy that have multiple branches, which includes fuzzy network adjustment control study [14], the controller of fuzzy neurons [15] and neuro-fuzzy function estimate [16]. In addition to the integrated neuro-fuzzy made by Takagi-Sugeno FIS is adaptive neuro-fuzzy inference system (ANFIS), and dynamically evolving fuzzy neural network [17].

This study will focus on the multiple cracks and holes detection of steel pipelines using PZT-based guided waves /monitoring of critical pipeline system utilizing embedded piezoelectric guided waves. The statistical modelling procedure aids in the proper classification of extracted features for various damage levels. The first class is supervised learning where a training set of known damage states is used to help classify extracted features. The second class is unsupervised learning where feature classification is based upon statistics of a given process.

2. Research Background

2.1 Typical defects of pipeline structures

The investigations result on the incidents of pipeline leakage introduce the main cause of this incident is the degradation mechanisms on pipeline structure. The degradation mechanism of the pipeline is due to various factors such as mechanical damage, corrosion, cracking caused by the environmental and the original manufacturer defect. Therefore, continues assessment is require ensure the integrity of pipeline structure and to prevent the incident of gas and oil leakage. Many defects may result in lowering the security of the pipeline working, and eventually lead to leakage, even explosion accidents as mention. Therefore, it is of importance to develop the defect detection method for pipeline structures by using an effective way such as the PZT-based guided waves. The typical defects of pipeline structures are shown in Figure 1[18,19].

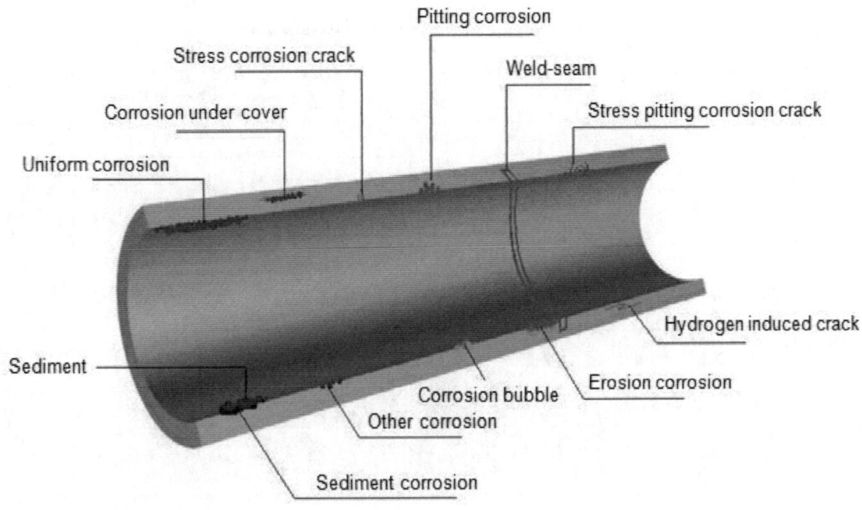

Fig. 1. The pipeline structure schematic of typical defects.

2.2 Wave Propagation in Hollow Cylinders

Guided waves in cylindrical structures may travel in the circumferential or axial direction. Based on boundary conditions, material properties, and geometric properties of the hollow cylinder, the wave behaviour can be described by solving the governing wave equations with appropriate boundary conditions.

Wave propagation in cylindrical waveguides has been studied on the theoretical and experimental levels by researchers at academic institutions and the industry. Such studies can be found in the literature [20-23].

There are basically two categories of guided waves that propagate in a hollow cylinder, the circumferential and the axial guided waves. Under the axial category, there are three types of modes generated when waves are propagating along a cylindrical structure. These are the longitudinal, torsional and flexural wave modes as seen in Figure 2.

The longitudinal and torsional waves inflict symmetrical displacement of particles in the axial direction across the structure while the flexural waves impose a non-symmetric particle displacement along the structure.

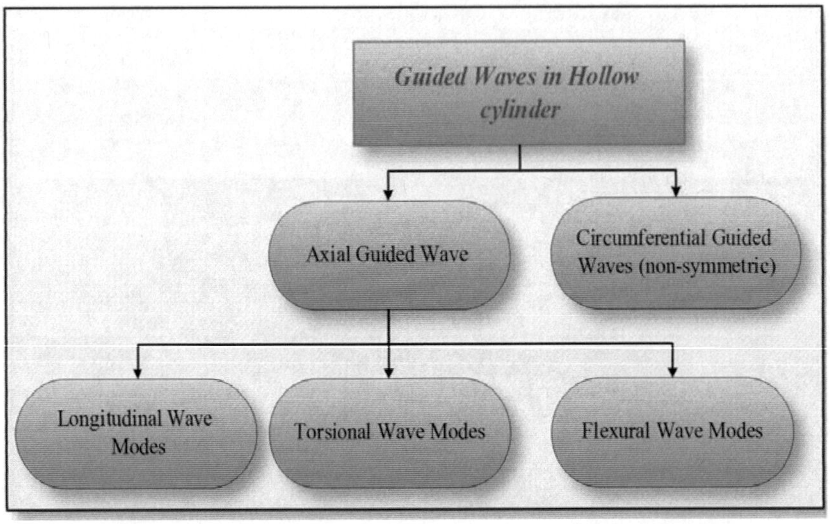

Fig. 2. Flow chart showing the types of guided waves in pipes.

2.3 Neuro-Fuzzy Systems

Various methods have been discover in combining FIS and ANN which is called neuro-fuzzy system. Based on studies has conducted by Lin et al [24] and Hsiao [25] showing a combination of FIS and ANN system can be categorized into three forms, namely integrated (hybrid) neuro-fuzzy systems, cooperative and simultaneous.

Based on neuro-fuzzy cooperative system shows that FIS and ANN have a method for free initial phase. The determining the role of FIS membership or fuzzy rules which follow the learning method based on training data. After the completion of the first phase of the ANN is no longer used in the system and the FIS activated. This was shown by Abraham [26], based on the example of neuro-fuzzy system of cooperatives that have been created. It also been made by Kosko [27] in addition to fuzzy memories unite fuzzy rule extraction is applied to the map. This is also shown by Nomura [28] by means learning fuzzy set parameters.

The uses of ANN to FIS continuously in neuro-fuzzy systems simultaneously not only occur at the beginning just like in the cooperative system. The cooperative system is equivalent to that normally exists with a combination of ANN and FIS inherent in neuro-fuzzy system simultaneously and independently. The simplest example can be provided on the situations in which the input process is difficult to measure directly. The ANN serves as a pre-processor for the input parameter in the determination of requirements that need to be used (for example, fuzzy sets and fuzzy rules). Additionally, ANN also serves as a post-processor for fuzzy and its related output [26]. ANN application as a pre-processor or post-processor for neuro-fuzzy system simultaneously serves as a performance enhancer for FIS as shown in Figure 3.

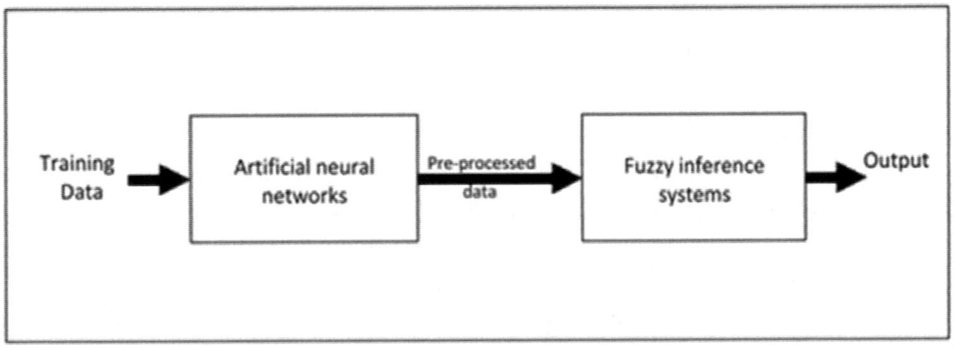

Fig. 3. Concurrent neuro-fuzzy systems

3. Proposed Model

The combination of reduced power method has been implemented by Takagi-Sugeno as hybrid neuro fuzzy systems (for the determination the parameters of linear coupled partly as a result of the rules) and the learning algorithm of back-propagation (to determine the parameters of the membership function in the antecedent rules) as shown in Figure 4. The Anfis model include six inputs present Harmonics in collected data and three outputs present crack Size, Number, and Types. (CS, CN, CT). The results were obtained using the software ANSYS 17.1 is used to simulate the Lamb wave propagation in the steel pipe. For the FEA, the 42 sample had been selected form 128 possible sample based on Design of Experiment (DOE). The pipe model section radius is 60mm and the wall thickness is 3mm. To simulate the defect, the degree of central angle is used to measure the size of defect, and two cracks and holes are simulated in the same way. The feature extraction process involves the extraction of damage sensitive features from sensor data collected during the data acquisition procedure to determine the presence of damage.

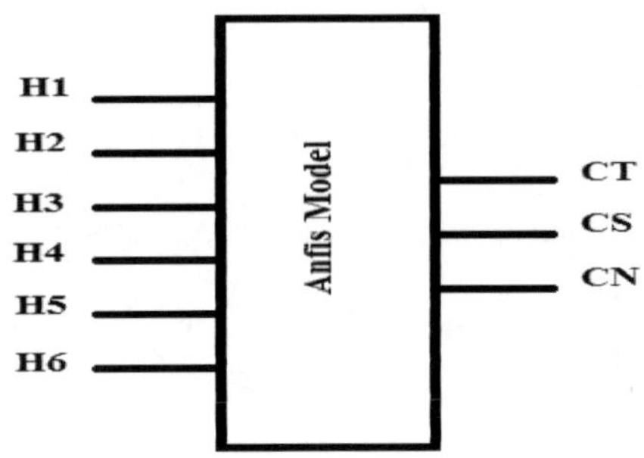

Fig.4. Proposed Model

Where:- crack type's (CT); Crack Number (CN); and crack size (CZ)

The extraction of features is done through signal processing of sensor data, which includes the processes of data normalization, compression and fusion. Data normalization reduces sensitivity to environmental variability such as temperature fluctuations. The data compression process decreases the dimensionality of the

IOP Conf. Series: Journal of Physics: Conf. Series **914** (2017) 012018 doi:10.1088/1742-6596/914/1/012018

acquired data. Data fusion involves combining data from multiple transducers to aide in feature identification.

4. The Numerical Model

It is possible to model pipes structures by ANSYS/ Multi physics product. The integration of control actions to the ANSYS solution is realized. Each of pipeline design should considered frequency response transient, and the model of the system into account. The material and dimensions of the pipe and types of the signal source and the placement measurement points. To analyse the transient dynamics problem full method will choose to simulate guided the pipe. The simulation test parameter is chosen as the excitation signal a 6-cycle 70 KHz sine wave, the step length is 1.52 μ/s, the load action is 100 μ/s, and propagating time is 1.52 m/s.

Figure 7.a&b, show sample of simulated pipeline and the concept of active guided wave focusing where time delay-amplitude phasing is applied to the segmented guided wave transducer array to cause constructive interference of the guided wave energy to occur at a given focal point along the pipe.

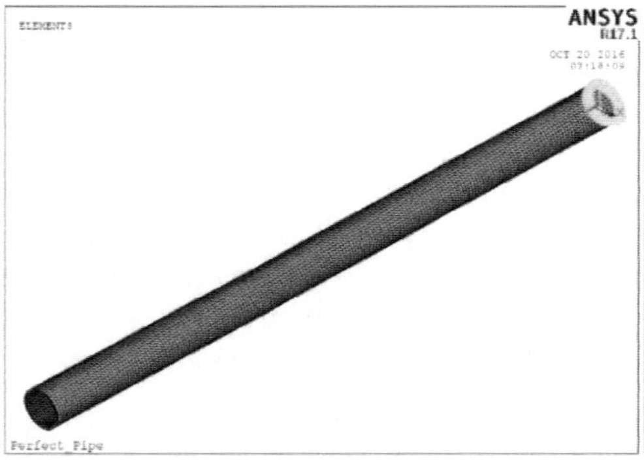

(a)

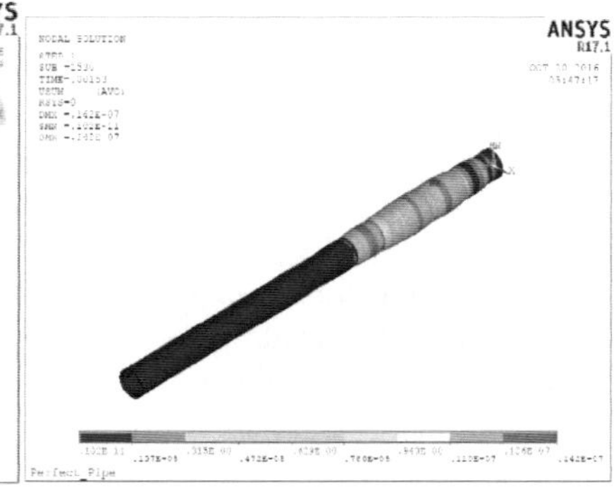

(b)

Fig.5 The meshing model of perfect pipe using FEM simulation demonstrating the concept of axisymmetric guided wave excitation

Figure 6, show the averaged signal and shows the induced signal from the transmitter to the receiver which directly refers by guided wave propagation.

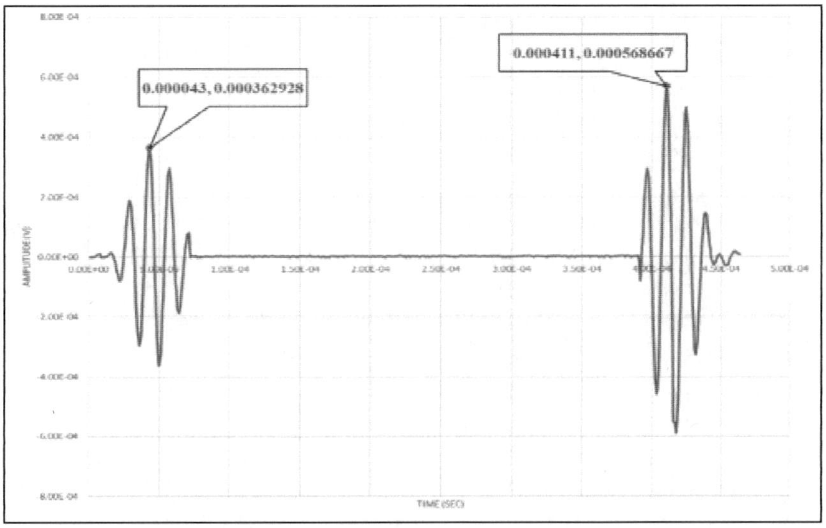

Fig. 6 Steel pipeline displacement-time curves and propagation along 1m of perfect pipe.

5. Results and Discussion

The signal harmonic extracted from converting time domain response to Frequency domain using Fast Fourier Transform (FFT) method for responses of the circumference, slop and holes defect in the pipe as shown in Table 1. Forty two samples from ANSYS simulation data was used to construct to verify the performances of ANFIS models. In order to improve the ANFIS model, about 85% of data are used for training and the remainder for testing performance. In Table 1, the training and testing Experimental available data are specified at different run number. Figure 7 show the training optimized structure has been selected as ANFIS, based on the percentage error minimum at 25 epochs.

Table 1. Training and testing data

Sample No.	ANFIS input						ANFIS output		
	H1	H2	H3	H4	H5	H6	CS	CN	CT
Training Sample									
1	11.48139	0.4511	0.26651	0.20033	0.16621	0.15146	1.5	3	0.25
2	12.89888	0.11014	0.07199	0.05419	0.04677	0.0411	2	3	0.25
3	15.89129	0.06144	0.04194	0.03127	0.02688	0.02441	2.5	3	0.25
4	9.45601	0.13304	0.06977	0.05143	0.04307	0.03375	1.5	1	0.25
5	10.08956	0.14195	0.07444	0.05487	0.04595	0.03601	2.5	1	0.25
6	10.03205	0.14114	0.07402	0.05456	0.04569	0.0358	2.5	1	0.25
7	9.46022	0.13309	0.0698	0.05145	0.04308	0.03376	2	1	0.25
8	9.31359	0.13103	0.06871	0.05065	0.04242	0.03324	2	1	0.25
10	10.5165	0.24362	0.11708	0.07767	0.05862	0.05739	2	2	0.25
11	10.58317	0.24516	0.11782	0.07816	0.059	0.05775	2.5	2	0.25
12	10.04755	0.23276	0.11186	0.07421	0.05601	0.05483	1.5	2	0.25
13	10.13676	0.23482	0.11285	0.07486	0.05651	0.05532	1.5	2	0.25

14	10.21542	0.23665	0.11373	0.07545	0.05695	0.05575	2	2	0.25
16	1.27668	0.01518	0.01169	0.00886	0.00718	0.00297	2	3	0.5
17	2.27802	0.01869	0.00764	0.00452	0.00484	0.00441	2.5	3	0.5
18	1.11069	0.02408	0.01399	0.0097	0.00837	0.00586	1.5	1	0.5
19	1.1851	0.02569	0.01492	0.01034	0.00893	0.00625	2.5	1	0.5
20	1.17835	0.02554	0.01484	0.01029	0.00887	0.00621	2.5	1	0.5
22	1.09396	0.02371	0.01377	0.00953	0.00824	0.00571	2	1	0.5
23	1.10435	0.02394	0.01391	0.00964	0.00832	0.00582	1.5	1	0.5
24	1.14845	0.01898	0.01299	0.01046	0.01263	0.01057	2	2	0.5
25	1.20662	0.01994	0.01365	0.01099	0.0133	0.01152	2.5	2	0.5
27	1.15573	0.0191	0.01307	0.01052	0.01274	0.01063	1.5	2	0.5
28	1.1647	0.01925	0.01318	0.01061	0.012843	0.01072	2	2	0.5
30	53.59083	0.27824	0.15317	0.10365	0.09158	0.08365	2	3	0.75
31	70.96711	0.36506	0.21695	0.17027	0.13646	0.13254	2.5	3	0.75
32	41.62189	0.06814	0.02106	0.00971	0.01118	0.00888	2	1	0.75
33	43.03703	0.07045	0.02177	0.01004	0.01156	0.00918	2.5	1	0.75
35	40.27142	0.06592	0.02037	0.00934	0.01081	0.00859	2	1	0.75
36	39.55862	0.06476	0.02016	0.00922	0.010625	0.00843	2	1	0.75
37	41.33886	0.06767	0.02091	0.00964	0.01139	0.00881	1.5	1	0.75
38	43.94270	0.32027	0.12356	0.07493	0.0623	0.0525	2	2	0.75
39	46.13162	0.33638	0.12971	0.07866	0.06541	0.05511	2.5	2	0.75
40	43.79685	0.31921	0.12315	0.07468	0.06212	0.05232	1.5	2	0.75
42	44.52860	0.32455	0.12521	0.07593	0.06314	0.0532	2.5	2	0.75
Testing data									
9	9.40211	0.13228	0.06937	0.05113	0.04282	0.03355	1.5	1	0.25
15	1.26707	0.03858	0.02656	0.02215	0.02325	0.01652	1.5	3	0.5
21	1.11851	0.02409	0.01396	0.00974	0.00837	0.00586	2	1	0.5
26	1.14556	0.01893	0.01296	0.01043	0.01263	0.01054	1.5	2	0.5
29	45.44689	0.39232	0.17486	0.11701	0.10421	0.0866	1.5	3	0.75
34	42.70564	0.06914	0.0216	0.00962	0.011471	0.00916	2.5	1	0.75
41	44.18572	0.32205	0.12424	0.07534	0.06265	0.05279	1.5	2	0.75

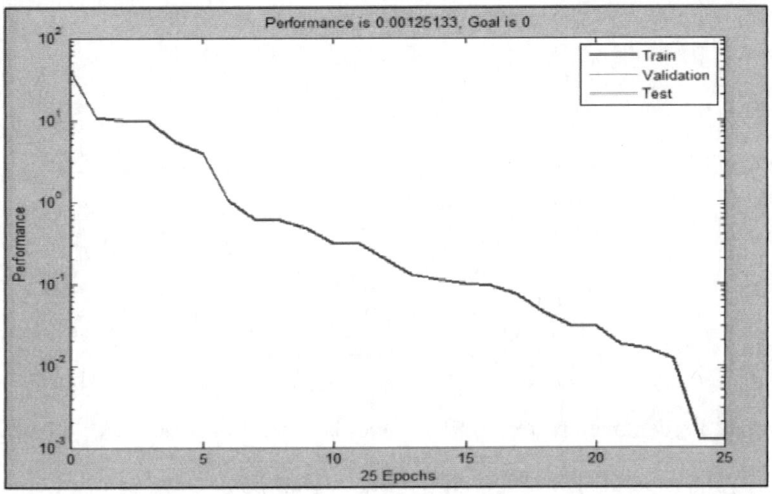

Fig.7. ANFIS training

The relative error results of ANFIS model are shown in Figure 8 for training data, where the percentage Error (ER) for input test variable and the average error (AER) are estimated as:

$$ER\% = \frac{|Y_{Ex} - Y_{pre}|}{Y_{Ex}} * 10 \quad (1)$$

$$AER\% = \frac{1}{N}\sum_{i=1}^{N}(ER\%) \quad (2)$$

Where (ER%), (YEX) and (Ypre) stand for error percentage values, experimental test data and ANFIS predicted values respectively.

The ANFIS models for training and testing data are shown in Figure 8. It is observed that the maximum error among the numerical and predicted for Slop, circumference, and Hole defect using ANFIS model are in good agreement with value of 99.83%.

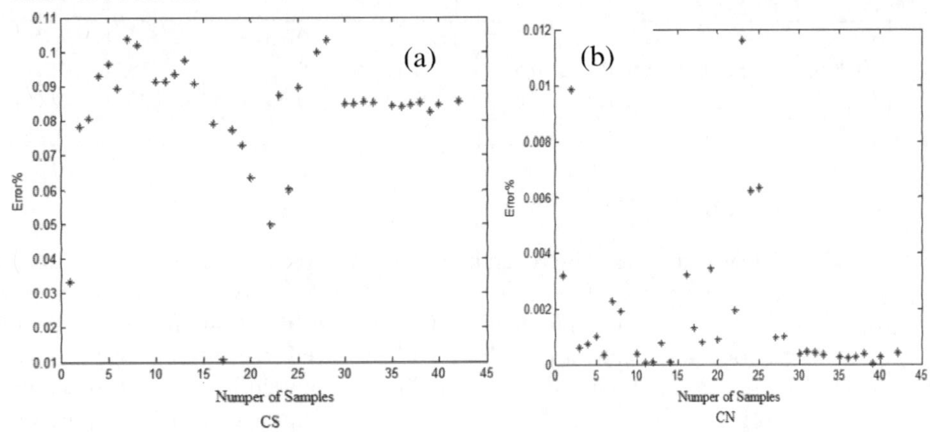

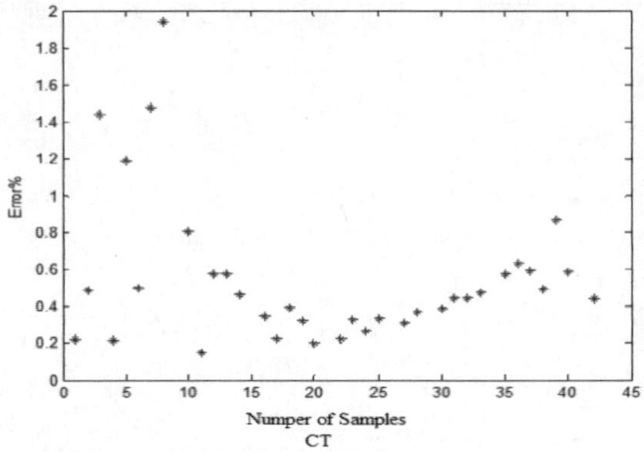

c)

Fig. 8. Percentage error of ANFIS models for training data: a) CS b) CN c) CT.

From Table 2, it is also observed that the higher percentage error of crack size was approximately 0.0.3239 and the average error is 0.01898. The higher values of percentage error of crack Number was approximately 0.04383and the average error is 0.0259. The higher values of percentage error of crack Types was approximately 0.00559 and the average error is 0.00288. From this result it can approve that Anfis predication model has same superior as ANSYS simulation result.

Table 2. Comparison between the numerical result and ANFIS models for testing data.

Sample No.	9	15	21	26	29	34	41
CS	1.5	1.5	2	1.5	1.5	2.5	1.5
ANFIS	1.520845	1.548579	2.004002	1.510003	1.557444	2.525992	1.543851
ER	0.0139	0.03239	0.002	0.00667	0.0383	0.0104	0.02923
AER	0.01898						
CN	1	3	1	2	3	1	2
ANFIS	1.043832	3.027557	1.060253	2.00163	3.043815	1.035197	2.034869
ER	0.04383	0.00919	0.06025	0.00082	0.0146	0.0352	0.01743
AER	0.0259						
CT	0.25	0.5	0.5	0.5	0.75	0.75	0.75
ANFIS	0.251398	0.500111	0.501819	0.501753	0.75337	0.751879	0.750163
ER	0.00559	0.00022	0.00364	0.00351	0.00449	0.0025	0.00022
AER	0.00288						

6. Conclusion

The GW-SHM technique, which can be regarded as the integration of actuators and sensors into structural components combined with automated advanced signal processing procedure for implementing a crack identification strategy, makes this challenging demand realizable. This work showed the effectiveness and the generality of Integrated nural Fuzzy (ANFIS) with guided lamb wave to detect and localize crack in pipes in the presence of significant operational. A data-driven framework based on pattern recognition and machine learning is then developed to differentiate between different types of cracks. Several simulation trials were performed using ANYSIS software to validate the feasibility and applicability of the modified model. Under the simulation effort, various crack types with different sizes and number along a pipeline were analyzed. It was found out that the model can accurately estimate the extent and number and size of crack. These findings indicate that the model can accurately estimate the location and the size of the flaw as the location of the flaw changes.

7. References

1. D.Rifai, A. N. Abdalla, R.Razali, K. Ali, & M. A. Faraj. Sensors, **17**, 3, 579. (2017)
2. I. D.Adewale, *Multiple parameters based pulsed eddy current non-destructive testing and evaluation*. (2015).
3. K. B. Ali, A. N. Abdalla, D. Rifai, & M. A. Faraj, IET Cir., Dev. & Sys., (2017).
4. M. A. Faraj, S., A. N. Abdalla, Fahmi, D.Rifai, & K. B. Ali, Ind. J. of Sci. & Tech., **10**, 2, 1-5, (2017).
5. S.Christodoulou, A.Deligianni, P., Aslani, & Agathokleous, Comp., Env. & Urban Sys., **33**, 2, 138-149, (2009).
6. B. Khoshnevisan, S. Rafiee, & H. Mousazadeh. Measurement, 47, 903-910. (2014).
7. Aryza, Solly, A.N. Abdalla, Zulkeflee Khalidin, and Zulkarnain Lubis. Procedia Engineering 15: 4188-4193. (2011)
8. H. F. Rashag, S. P. Koh, A. N.Abdalla, N. M. Tan, & K. H.Chong, Journal of Power Electronics, **13**, 3, 369-380. (2013)
9. A. M Haidar, I. A. Ahmed, & A. N. Abdalla, Int. J. of Soft Comp., **5,** 2, 62-66, (2010).
10. J. S. Jang, IEEE trans. on sys. man, and cybernetics, **23**, 3, 665-685, (1993).
11. J. S. Jang, Fifth IEEE International Conference on, **2**, 1493-1499, (1996).
12. L.Naderloo, R. Alimardani, M. Omid, F. Sarmadian, P. Javadikia, M. Y. Torabi & F. Alimardani, Measurement, **45**, 6, 1406-1413, (2012).
13. M. Hayati, A. Rezaei, M. Seifi, & A. Naderi, Microele. J., **41**, 7, 381-387. (2010).
14. C. T. Lin, & C. S. Lee, IEEE Trans. on comp., **40**, 12, 1320-1336. (1991).
15. D. Nauck, & R. Kruse, Fuzzy sets and systems, **101**, 2, 261-271. (1999).
16. D. Nauck, & Kruse, R. 4th International Workshop Fuzzy-Neuro Systems, (1999).
17. N. K. Kasabov, & Song, Q. Depar. of Info. Sci., University of Otago. (1999).
18. Yan, Y. Measurement Science and Technology, **7**, 12, 1687, (1996).
19. D. Rifai, A. N. Abdalla, K. Ali, & R. Razali, Sensors, **16**, 3, 298. (2016).
20. D. C. Gazis, The Journal of the Acoustical Society of America, **30**, 8, 786-794. (1958).
21. Achenbach, J. **Wave propagation in elastic solids** Elsevier, Book, **16**. (2012).
22. J. L. Rose, *Ultrasonic waves in solid media*. Cambridge university press. (2004).
23. Towfighi, S., & T.Kundu, International journal of solids and structures, **40**, 20, 5495-5510. (2003).
24. C. T. Lin, & C. G. Lee, *Neural fuzzy systems*. PTR Prentice Hall. (1996).
25. S. W. Hsiao, & H. C. Tsai, International Journal of Industrial Ergonomics, 35(5), 411-428. (2005).

26. A. Abraham, Fuzzy systems engineering, 914-914. (2005).
27. B. Kosko, *Neural networks and fuzzy systems: a dynamical systems approach to machine intelligence*. Prentice hall, **1** (1992).
28. H. Nomura, I. Hayashi, & N. Wakami, In Proc. IFSA (Vol. 91, pp. 155-158). (1991).

Mixture of natural fiber with gypsum to improve the fire resistance rating of a fire door: The effect of kapok fiber

Azieyanti N A[1] and Alif Hakim[2] and Hasril Hasini[3]

[1]Mechanical Engineering Department, College of Engineering, Universiti Tenaga Nasional

Corresponding author: Azieyanti@uniten.edu.my

Abstract. A composite mixture of gypsum and natural fibers has been considered in this study to enhance the fire resistance rating of a fire door. Previously the materials used to make a fire door are gypsum and fiber wool where it acts as a protective coating. Normally this fire door must be compact and able to close on its own. Natural fibers have the ability to replace glass fiber cotton because of its features that are available in fiber glass wool. When using fiberglass, it can cause health problem once it is swallowed and inhaled, and may remain in the lungs indefinitely. It also can contribute to lungs cancer. Kapok fiber has been used in this experiment as natural fibers. The objective of the experiment is to analyze the fire resistant rating of the composite mixture of gypsum with kapok fiber. The scopes of the experiment consist of a preparation of composite mixture samples of gypsum with kapok fiber with different composition and thickness, and the fabrication of a fire resistant testing furnace. A testing of samples which were conducted in accordance with the standard MS 1073: PART 2:1996.

1. Introduction

Nowadays, the materials used to make a fire door consist of materials that are able to withstand fire for a long time. Buildings, offshore structures, ships, etc. are equipped with doors. Generally, it is a movable structure which is used to close off an entrance, typically comprises a panel that swings on hinges or that slides or rotates. Past experiences have taught humans a good lesson to use a fire door instead a conventional wooden or steel door. Usually a fire door is provided with a rating of 30 or 60 minutes. It is related to the duration of time the door can resist fire before becoming unstable [1]. Fire resistant rating has been traditionally determined standardized fire testing performed in accordance with ASTM E-119, Standard Method for Fire Tests of Building Construction and Material. However, in accordance to the ASTM standard which is the American standard, other countries to have adopted tentative standard accordance to the national Institute for standard in the respective countries, among them are BS476 which is used in Britain, MS 1073 (1996) which is used in Malaysia under Standard and Industrial research Institute of Malaysia.

Natural fibers are increasingly being used as reinforcement due to their low cost, high specific properties and renewable nature. Kapok with its scientific name is (Ceiba Pentandra) is the best insulation material and readily available. Now, it is grown worldwide in several plantations in Southeast Asia. Kapok fiber poses an excellent hydrophobic-oleophilic characteristic that we can label it as a natural plant fiber. It has a penetrable hollow lumen structure and waxy material on its surface.

IOP Conf. Series: Journal of Physics: Conf. Series **914** (2017) 012019 doi:10.1088/1742-6596/914/1/012019

Kapok has high cellulose content and has the potential to replace glass fiber. Kapok fiber has a unique characteristic which does not get wet easily with water. This characteristic shows that kapok fiber is a hydrophobic [2-3]. The raw kapok is yellow to brownish or whitish in color. Kapok is an excellent thermal and acoustic insulation because it contains 70%-80% of air. Kapok cannot withstand a strong stress due to its brittleness. Nowadays kapok is used as a stuffing and for insulation of sound due to good thermal insulating properties. For kapok fiber, it can improve the crystalline structure of fibers and will induce topographic changes of the fibers surface due to the alkalization of kapok. It has a good resistance at high temperatures but sensitive to ultraviolet rays. Nowadays, for a small refrigerating system, kapok is used as an insulating material. It also can be used for a lining in sleeping bags and as gloves for dry ice handling. This is because kapok is equipped with its excellent thermal insulating properties. Beneficially, due to its good thermal insulation, it also can be used for thermal insulation of conduit pipes and as an acoustic insulation of dwelling rooms [4].

Gypsum is a natural resistance of fire and its scientific name is hydrated calcium sulphate (CaSO4 2H2O). It has thermal conductivity of 0.17 W/(m.k) and has a monocline structure. Pure gypsum is a white rock. Generally gypsum can be found in laminated, granular or compact masses. Gypsum is also known as a product of reducing the emission levels of coal-fired power stations. The United States of America leads the world in gypsum production. Synthetic gypsum is generated worldwide with over 110 million metric tons [5].

Fire resistance is a performance of a given complete system and also as a construction when it is exposed to the standard heating conditions [6]. It also can be illustrated as a measurement of the building element ability to resist a fire. Besides that, fire resistance also is assessed as a time when the element exposes to a standard fire resistance test, and must meet certain criteria [7]. It also can be described as an element of a building construction property and as an ability of measurement for a stated period. There are some criteria for the fire resistance which are resistance to collapse, resistance to flame and hot gases penetration, and resistance to temperature rise on the unexposed face which is up to a maximum of 180oC [8]. Furthermore, it is also as a length of time for a structural element that performs well in the standard furnace test. It can be specified for the exposure of a fire on side of walls, partitions, and floors [9].

2. Details Experimental

2.1. Materials and Experimental Procedures

The experiment was performed by setting up an experimental furnace and the specimen for the testing method. The testing was carried at ambient condition with reference to the Malaysian Standard MS 1073: PART 2: 1996. The essential materials required were stated as follows:

1. Gas Stove (Ignition Source)
   - It is required to yield the heat up to 1200°C. A high pressure of gas stove was used to produce the required heat and to start the ignition for the experiment. Therefore the ignition source was initiated to supply the required energy to carry out the combustion process.
2. Bricks
   - It is used to make up the furnace linings.
3. LPG Gas Cylinder (Fuel)
   -It is used as a combustion fuel.
4. Thermometer
   - The ambient temperature was found to be in the range from 27°C to 33°C.
5. Stopwatch (Timing Device)
   - It is used to measure the time taken for the specimen to refuse the fire.
6. Type K Thermocouple
   - Thermo K (chromel-alumel) thermocouples were used to obtain the temperature measurement of exposed and unexposed specimen surface. The standard specification was ASTM C 177. Its

function under the principle of resistance changes linearly with respect to the temperature changes. Five thermocouples were placed on unexposed side of the specimen to measure its temperature changes. The bare wire thermocouples were used for exposure process.

2.2 Sample Preparation

The steps for preparation of the sample as follows:

Table 1. Thickness of composite mixture of gypsum and kapok fiber

Sample	Thickness (mm)
1	6.4
2	9.5
3	12.7

Table 2. Composition ratio of composite mixture of gypsum and kapok fiber

Ratio (mixture)	
Gypsum (gram)	Kapok (gram)
1000	10
1000	20
1000	30

Step:

(1) Framing and mold base: The mold was designated and fabricated with dimension of 225x225x30 mm.it was made of wood and panel board.
(2) The kapok fiber must be in dry condition and the samples were dried under the hot sun for seven days.
(3) A balance weight is used to measure the weight of raw materials
(4) Kapok fiber and gypsum powder were mixed to get the slurry of the mixture
(5) The slurry mixture were poured into the mold and then compressed to get the required thickness of the sample. The mixture then placed under the sun to let them dried and hardened.
(6) Placement of thermocouples: Fixed the healed samples with five thermocouples at the selected points of an unexposed surface.

Testing:

To evaluate the fire resistance of the healed samples, each of the samples was tested using the experimental furnace. The non-combustibility method was used in this experiment because it was cheaper and simpler than the former method. During the experiment, the temperature of the furnace was kept constant in the range of 750-950°Celsius and flame temperature was recorded using a high temperature type K, and the temperature readings were also taken to face the unexposed samples at five different locations using a type of lower temperatures K. Readings were taken at each time interval of 3 minutes. Fire resistance rating for overall sample tested are shown in Table 3:

Table 3. Fire Resistance Ratings

Composition	Thickness (mm)	Rating (minutes)	Maximum Mean Temperature (°C)	Density, ρ (kg/m³)
G = 1000g KF = 10 g	6.4	71	111.6	2469.14
G = 1000g KF = 10 g	9.5	90	117.3	1663.42
G = 1000g KF = 10 g	12.7	99	99.4	1244.29
G = 1000g KF = 20 g	6.4	81	119.6	2160.49
G = 1000g KF = 20 g	9.5	93	106.5	1455.49
G = 1000g KF = 20 g	12.7	105	126.3	1088.75
G = 1000g KF = 30 g	6.4	102	110.6	1098.75
G = 1000g KF = 30 g	9.5	126	136.4	984.91
G = 1000g KF = 30 g	12.7	153	146	836.39

- G = Gypsum, KF = Kapok Fiber

3. Results and Discussion

The fire resistant test was conducted on nine samples with different composition ratios with three different thickness of composite mixture of gypsum and kapok fiber. The composition ratio were gypsum 1000 g: kapok fiber 10g; gypsum 1000 g: kapok fiber 20 g; gypsum 1000 g: kapok fiber 30 g; and those ratios were prepared for each of these three different thicknesses of 6.4 mm, 9.5 mm and 12.7 mm.

3.1. Composition of Gypsum 1000 g: Kapok Fiber 10 g

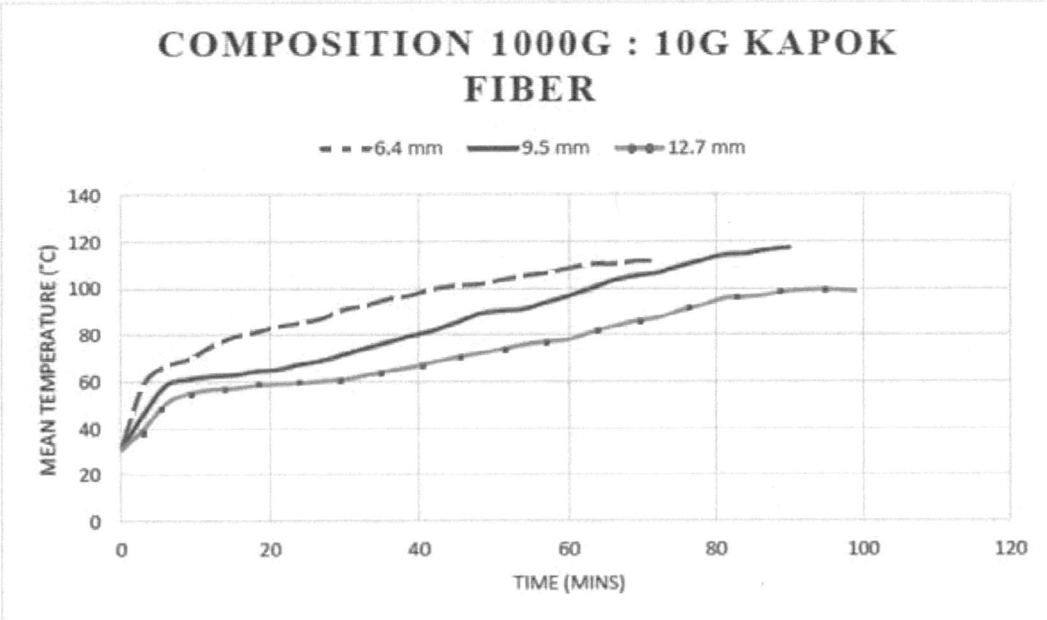

Figure 1. Fire resistant test results of gypsum 1000 g: kapok fiber 10 g for different thicknesses.

3.2. Composition of Gypsum 1000 g: Kapok Fiber 20g

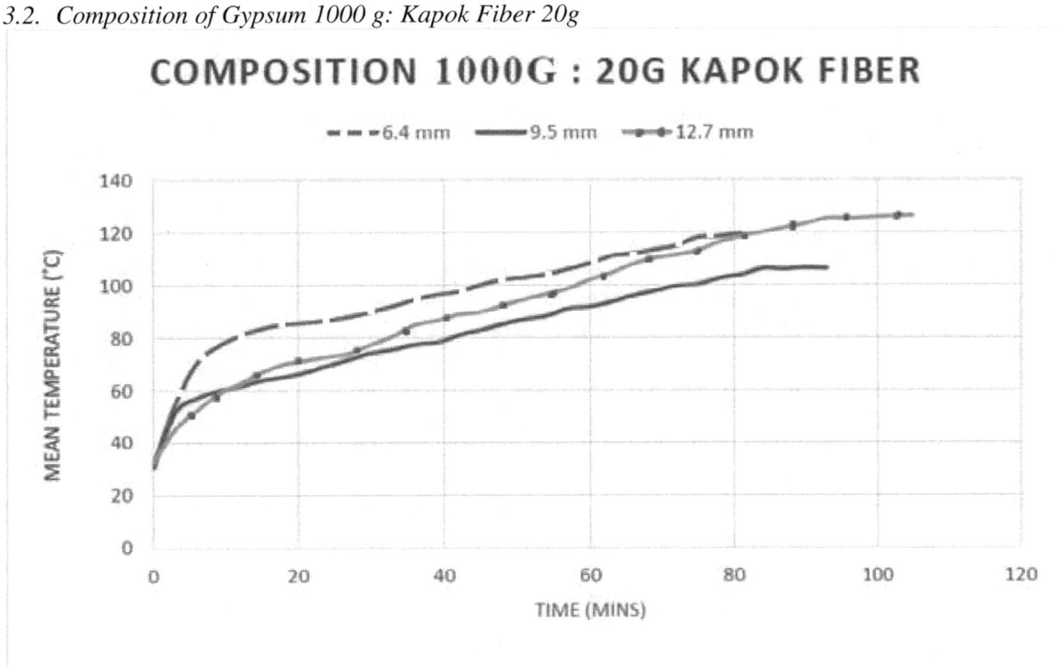

Figure 2. Fire resistant test results of gypsum 1000 g: kapok fiber 20 g for different thicknesses

3.3. Composition of Gypsum 1000 g: Kapok Fiber 30 g

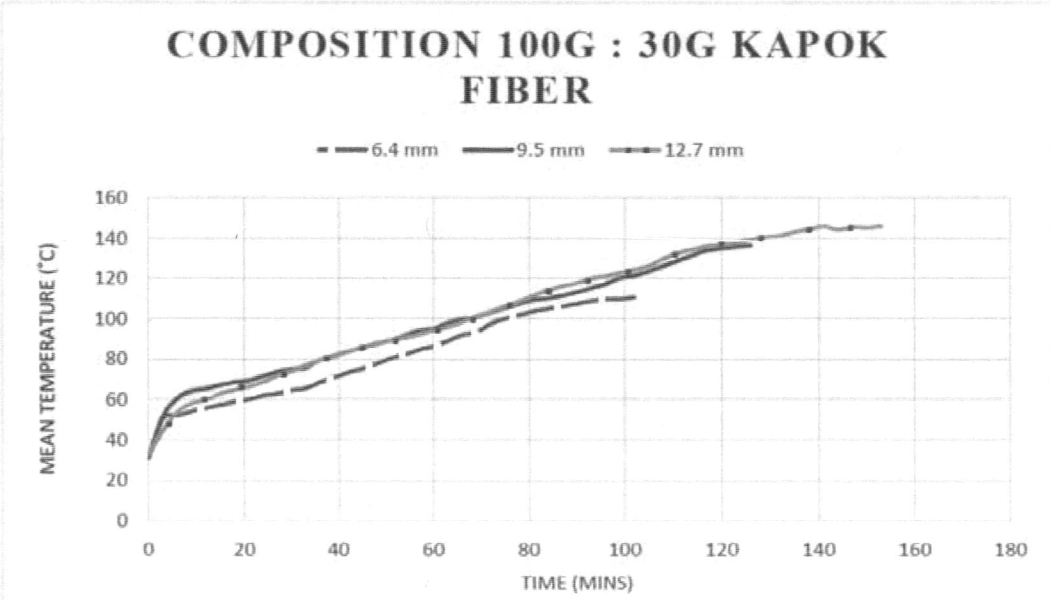

Figure 3. Fire resistant test results of gypsum 1000 g: kapok fiber 20 g for different thicknesses

Figure 1 shows that the composite sample of 12.7 mm thickness gave the highest fire resistant rating which 99 minutes. The second highest was the composite sample of 9.5 mm thickness which recorded 90 minutes. Lastly, the composite sample of 6.4 mm thickness produced the fire resistant rating of 71 minutes only.

Figure 2 shows the plots for gypsum 1000 g: kapok fiber 20 g for different thicknesses. The best fire resistant performance was shown by composite sample of 12.7 mm thicknesses which could last for 105 minutes before it failed. This was followed by the composite samples of 9.5 mm for 93 minutes and 6.4 mm for 81 minutes.

Figure 3 shows that the highest fire resistant rating recorded was 153 minutes, which was achieved by the composite sample of 12.7 mm thickness. This was followed by the composite samples of 9.5 mm thickness for 126 minutes and 6.4 mm thickness for 102 minutes.

These findings were based on the integrity failure. After a failure, cracks and smaller gaps were found on the samples. Generally it is found that the performance of fire resistant rating achieved by these composite samples increases with thickness. It means that the increasing content of kapok fiber gave better performance due to its structure and properties. The experimental works and observations that have been carried out revealed that a mixture of natural fibers and gypsum has the potential to be used in manufacturing fire resistant doors commercially. Based on the standards, scope, objective, and procedures of this experiment, the composite sample of natural fiber and gypsum could last for 30 minutes when it was burnt, so it can be said that the suitable good sample is for making fire resistance doors.

4. Conclusion

Natural fibers are a good insulating material which has the potential to be used in a fire door manufacturing. The highest fire resistant rating achieved was 153 minutes for the composition of 1000 g gypsum : 30 g kapok fiber with a thickness of 12.7 mm. The lowest fire resistant rating was 71 minutes for the composition of 1000 g gypsum : 10 g kapok fiber with a thickness of 6.4 mm. From the observation, it clearly showed that the thickness of the samples has a significant impact on the fire resistant rating. In addition, the different compositions of kapok fiber in the composite samples also were found to have an effect on the fire resistant rating. Therefore, kapok fiber should be considered as an insulated construction material in the manufacturing of fire resistant doors, ceilings, and wall panels in order to provide a better safety protection from fire.

References

[1] Jackson S 2006 *Development of Wireless Reproducts* (Durham Theses Durham University).
[2] Hori K, Kuga M E F S, et al. 2000 Excellent oil Absorbent Kapok [Ceibapentandra (L.) Gaertn.] Fiber: Fiber Structure, Chemical Characteristics, and Application *The Japan Wood Research Society.*
[3] Kartina S 2012) Oil Sorption Capacity of Kapok Fiber IEEE *Colloquium on Humanities, Science & Engineering Research.*
[4] Tekstilstvo N F and Ljubljani U V 2008 Kapok in Technical Textiles *PoslanoJunij.*
[5] Donald W O *by Gemstone Domestic survey data* and tables were prepared by Christine K. Pisut, statistical assistant, and the world production table was prepared by Glenn J W, international data coordinator.
[6] Promat 2008 The Passive Fire Protection Handbook *Etex Group Company.*
[7] Bernhart D 2004 The Effect of Support Conditions on the Fire Resistance of a Reinforced Concrete Beam *Thesis.Department of Civil Engineering University of Canterbury.*
[8] MS 1073 PART 2 1996 *Methods for Determination of the Fire Resistance-General Principle* SIRIM, Malaysia.
[9] Cooper L Y 1996) Methodology for Developing and Implementimg Alternative Temperature-Time Curves for Testing the Fire Resistance of Barriers for Nuclear Power Plant Applications *Building and Fire Research Laboratory National Institute of Standards and Technology Gaithersburg, MD 20899.*

Acknowledgement

The authors would like to thank Universiti Tenaga Nasional (UNITEN) for the funding of this research under the Fundamental Research Grant Sceme (FRGS) and UniversitiTeknologi Malaysia Laboratory for the support.

IOP Conf. Series: Journal of Physics: Conf. Series **914** (2017) 012020 doi:10.1088/1742-6596/914/1/012020

Influence of Alkali Resistant (Ar) Fibreglass in Porcelain Clay for Manufacturing Vitrified Clay Pipes

Muhammad Ikhmal Hanapi, Sufizar Ahmad, Hariati Taib, Al Emran Ismail, Mohd Nasrull Abdol Rahman, Salihatun Md Salleh, Azmahani Sadikin, Shahruddin Mahzan

Faculty of Mechanical and Manufacturing Engineering, Universiti Tun Hussein Onn Malaysia, 86400 Parit Raja, Batu Pahat, Johor

Corresponding author: sufizar@uthm.edu.my

Abstract. The aim of this work is to determine the characteristics of porcelain ceramic with influence of milled Alkali Resistant (AR) fiberglass for manufacturing vitrified clay pipes. In this study, raw materials consist of porcelain clay and AR fiberglass were refined into powders less than 90μm. Subsequently, these samples were compacted into cylindrical pellet for chemical analysis using X-Ray Fluorescence (XRF). The ceramic sample was produced by mixing different weight percentage of AR glass to porcelain ceramic with 3 wt%, 6 wt%, 9 wt% and 12 wt%. Subsequently, the sample was compacted with 3 ton of pressure load and sintered at 900 °C, 1000 °C, 1100 °C and 1200 °C. The phase identification by using X-Ray Diffraction (XRD) and microstructural analysis were performed for the sintered sample. Chemical analysis revealed that the significant element for all raw material are SiO_2, Al_2O_3, Na_2O and K_2O. Phase identification analysis shown that sample sintered at 1000 °C produces quartz (SiO_2), berlinite ($AlPO_4$), albite ($NaAlSi_3O_8$) and calcium-magnesium-aluminum-silicate (CaMgAlSiO). The formation of primary mullite was observed in sample sintered at 1100 °C. The image of microstructural morphology denoted that the formation of glassy phase with decreasing amount of void when sintering temperature and addition of AR glass were increased, which correspond well to phase identification analysis.

1. Introduction

Ceramic is essential for advanced application because of its properties towards thermal and high strength. The coalition of three materials that consists of silica, clay, and feldspar, form a main component in structure of traditional ceramic. Silica function as filler whereas clay provide plasticity in ceramic body. As for fluxing agent in ceramic material, feldspar was significant and commonly used in formulation of ceramic material [1]. It is previously acknowledged that the fluxing agent were included in ceramic material to assist the adherence of different particle and has the tendency to affect the formation of glassy form at lower melting temperature to ceramic bodies. Therefore, nowadays there are more endeavours has been put into increasing the fluxing agent in ceramic material [2]. Notably, ceramic has been commonly used in manufacturing pipes for sewage application such as vitrified clay pipes. The purpose of this work is to identify the suitability of porcelain ceramic in manufacturing of vitrified clay pipes. This is crucial as the conventional ceramic powder formulation is not feasible for mixing the main components together, namely clay, feldspar, and grog. Therefore, porcelain has been chosen in this study due to the fact that the tri-axial ceramic material comprises in porcelain produced highly glassy finish which make up 70% of glass and 30% crystalline after sintering process [3].

Consequently, ceramic was a dubious metal and polymeric materials with high plastic deformation properties. Thus, the main drawback of ceramic material there is a tendency for sudden or catastrophic failure when thermal or mechanical loading were subjected due to restricted displacement of stress induced crack propagation. To resolve this, a small fraction of short fibre is commonly added into the ceramic body to feasibly govern the crack growth and to reduce impact of early shrinkage, in which result in dimensional control [4]. Moreover, low porosity of ceramic material can be achieved by high fluidity of glass forming throughout the sintering process by constructing ceramic composite with glass ceramic system such as Alkaline Resistant (AR) fiberglass material. The complex matrix reinforcement interaction can also resulted in absorbing fracture energy due to crack deviating [5].

Sintering process is another crucial factor in manufacturing of ceramic product because it involves physicochemical reactions that significantly affect the final properties of ceramic material. Besides that, it is important to properly contemplate the choice of raw material preparation, drying condition, and sintering cycle to develop a better phase and microstructure intricacy in producing quality product. Thus, time consuming and high energy consumption can be made optimal by examining the preliminary properties [6]. In this study, Alkali Resistant (AR) fiberglass was incorporated in porcelain material via simple planetary ball milling method. AR fiberglass is expected to enhance the properties of porcelain material with optimum sintering temperature. AR fiberglass was added at the ratio of 0-12 wt.% into porcelain. The samples were sintered at various sintering temperatures of 900-1200 °C. The phase formation and microstructure of sintered samples were investigated.

2. Experimental Methods

The raw material consists of porcelain clay and Alkali Resistant (AR) Fiberglass (18mm in length) was milled into powder with the average particle size of 90 μm. These powders were later milled in different weight percentage of 3 wt%, 6 wt%, 9 wt% and 12 wt% via planetary ball milling at 150 rpm as shown in Figure 1.

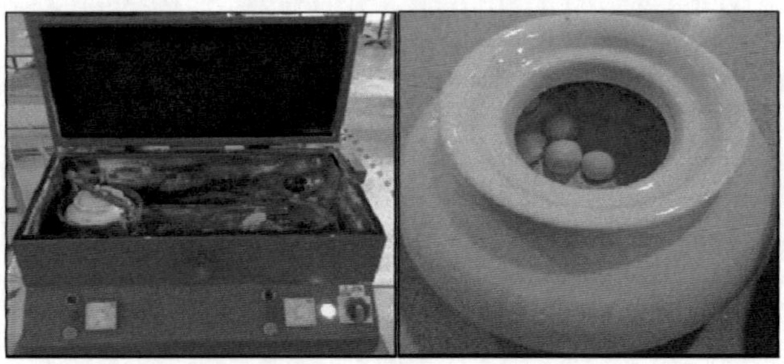

Figure 1. Planetary ball milling

The chemical analysis was performed via X-Ray Fluorescence spectrometry (Xsupreme8000, Oxford Instrument, United Kingdom). The sample is pressed to bar shape at 3 tons and sintered at various temperatures at 900 °C, 1000 °C, 1100°C and 1200 °C for 2 h in air. The sintering rate was controlled at 5°C/min. The phase identification analysis was conducted using X-Ray Diffraction (D8 Advance, Bruker, England). The microstructural analysis was performed for sintered sample using Scanning Electron Microscope (Hitachi, Japan). The magnification of 1000X were selected at an accelerating voltage of 15kV.

3. Results and Discussions

3.1. Chemical analysis

It is fundamental to identify chemical composition of all ceramic material as it provides the result of weighing oxides required to determine the better formulation for final properties of ceramic product. Table 1 shows the chemical composition analysis of porcelain ceramic, AR glass and mixture of porcelain and AR glass.

Table 1. The chemical composition of the porcelain powder and mixture by using XRF

Element	Raw Material		Mixture (Porcelain + AR Glass (%))			
	Porcelain	AR Glass	3 wt%	6 wt%	9 wt%	12 wt%
SiO_2	70.9290	59.4350	70.351	70.055	69.997	69.841
Al_2O_3	24.319	6.1370	24.206	23.597	23.046	22.540
Na_2O	1.7500	24.4780	2.251	2.954	3.339	3.864
K_2O	1.5680	1.2180	1.529	1.497	1.526	1.487
Fe_2O_3	0.4450	0.1240	0.428	0.405	0.396	0.387
MgO	0.3580	0.2090	0.365	0.368	0.345	0.308
CaO	0.2260	1.8780	0.286	0.339	0.407	0.469
SO_3	0.1780	0.1220	0.165	0.159	0.150	0.150
TiO_2	0.1170	1.6380	0.183	0.237	0.296	0.359
P_2O_5	0.0870	4.7140	0.215	0.365	0.478	0.575
Mn_2O_3	0.0152	0.0019	0.0141	0.0132	0.0116	0.0112
CI	0.0054	0.0384	0.0057	0.0058	0.0064	0.0049
ZnO	0.0029	0.0000	0.0025	0.0026	0.0021	0.0021
Cr_2O_3	0.0000	0.0056	0.0000	0.0016	0.0019	0.0019
SrO	0.0000	0.0000	0.0000	0.0000	0.0000	0.0000

Based on the XRF result in Table 1, silica (SiO_2), alumina (Al_2O_3), sodium oxide (Na_2O) and potassium oxides (K_2O) produced higher concentration and can be indicated as major element in producing ceramic product. Silica produced the highest concentration among all elements, which is more than 59% for all material. Silica is essential to the final properties of ceramic as it function as quartz and affect the major crystalline phase of ceramic body. Whereas, the concentrations of alumina (Al_2O_3) was higher for porcelain and all mixture, that is significant for finer mechanical properties of ceramic. This is due to its function in increasing the viscosity of glassy phase to produce primary mullite and reducing recrystallization to form secondary mullite [7].

Simultaneously, Sodium Oxide (Na_2O) manifested an outstanding concentration percentage in AR glass yet yielded in small amount in porcelain and all mixture. Sodium is commonly favorable in the mixture of ceramic material as it gave the strongest flux in sintering process and slightly stiffer than potassium. [8]. Sodium and potassium is vital to aid different particle because it will ensure high fluidity of glassy phase to form eutectics at low melting temperature causing the network structure of ceramic particle breakdown during sintering process. [9]. Thus, the addition of AR glass to porcelain has slightly decreased the concentration of alumina and slightly increase the sodium oxides concentration for all mixtures depending on the elevation of AR glass addition.

3.2 Phase identification analysis

Phase identification was conducted to identify the crystalline structure formed in ceramic material after the sintering process. The crystalline structure can be correlated with the amount of glassy phase formation. Figure 2 and Figure 3 display the of X-Ray Diffraction patterns for sample sintered at 1000°C and 1100°C, respectively.

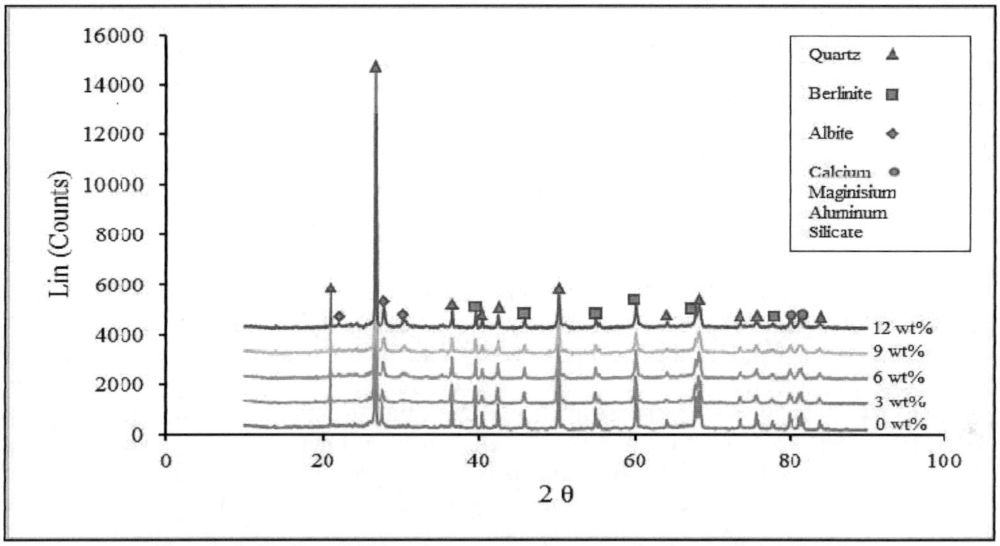

Figure 2. X- Ray Diffraction pattern for sample sintered at 1000°C

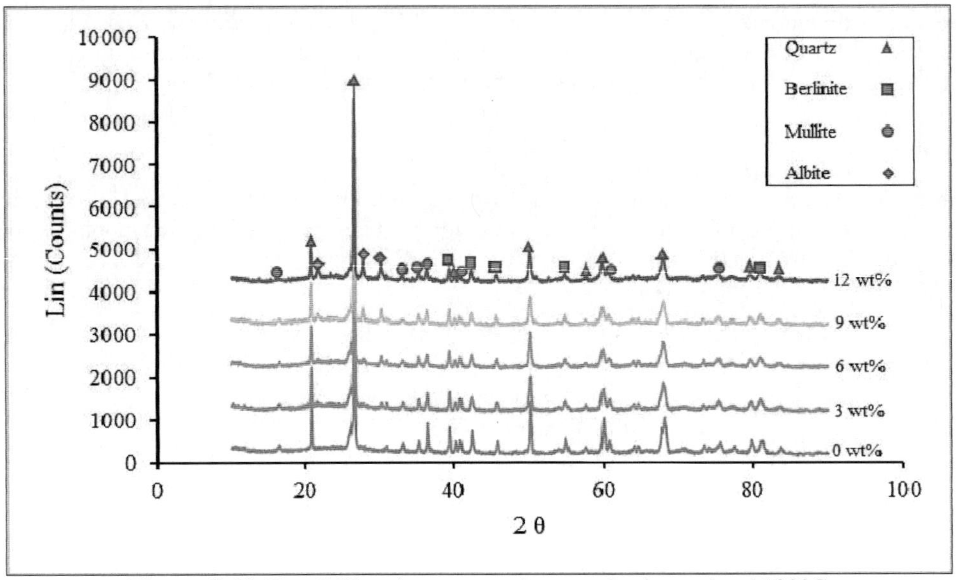

Figure 3. X- Ray Diffraction pattern for sample sintered at 1100°C

Figure 2 reveals that the formation of crystalline structures such as quartz, berlinite, calcium magnesium aluminum silicate and albite have been identified for sample sintered at 1000°C. Quartz was found to be the major crystalline phase based on the peaks produced. This can be attributed to the highest content of silica, which is in good agreement with the XRF results in Table 1. Quartz possesses refractory property and commonly maintained its shape in which imparts rigidity to ceramic bodies [10]. The formation of berlinite occurs due to the equivalent size of aluminum and phosphorus ions in comparison to silicon. Moreover, quartz is isostructural with berlinite that has different mineral chemistries with although with the same crystal structure [11]. The small intensity for albite ($NaAlSi_3O_8$) and calcium-magnesium-aluminum-silicate (CaMgAlSiO) shift produces is shown in Figure 2. The formation of albite has proven the process of particle consolidation where liquid phase takes place while it is to known that calcium magnesium aluminum silicate visible as sealing and wetting material [12].

Despite that, Figure 3 displays almost comparable results with Figure 2. However, the formation of mullite starts to appear at 1100°C based on the shift produce. Other than that, the formation of berlinite was reduced and no constituent of CaMgAlSiO was detected. The formation of albite and mullite proved the formation of a much glassy phase that adhere with quartz. Mullite appear at 1100°C due to the formation of liquid phase that infiltrated the interconnected pores, which had derived from breaking reaction of sodium and potassium [13]. The evidence of mullite crystal is vital because it provide high tenacity and better mechanical properties to ceramic material [14]. By looking at the overall result, the degree of crystallinity obtained was the same and the elevation of temperature lowering the peak intensity.

3.3 Microstructural morphology

Microstructural studies were conducted to observe quartz relict, void formation, and glassy phase formation that were correlated with previous result such as phase identification analysis. Figure 4 and Figure 5 display the SEM microstructure for sample sintered at for sample sintered at 900°C and 1200°C.

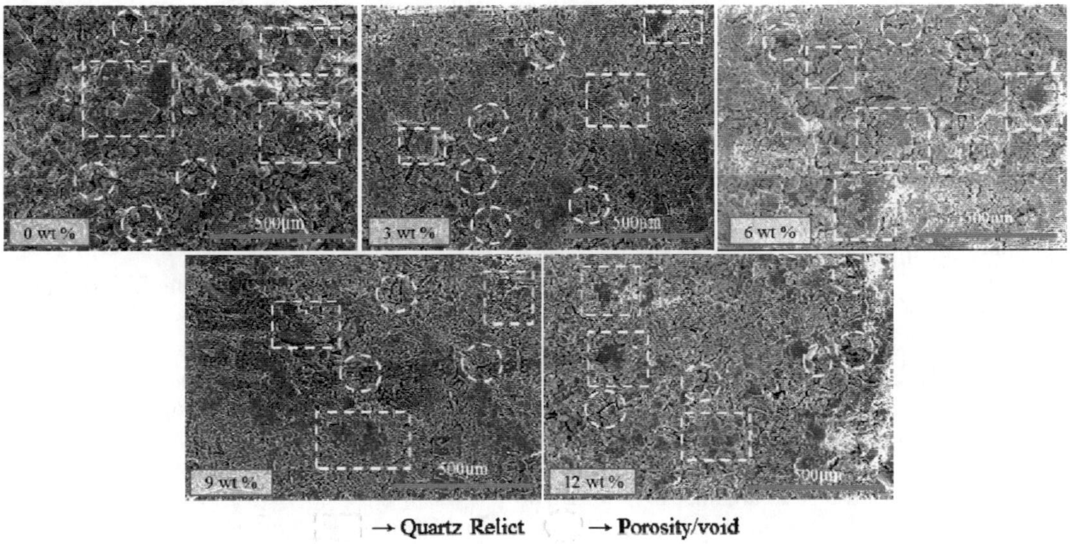

Figure 4: SEM images of sample sintered at 900°C

By looking at Figures 4, scatter formation of quartz and the incomplete grain growth can be observed which consequently resulting in large void formation and uncountable porosity. Moreover, the sample shows less dense where lack of glassy phase produce to the sample sintered at 900 °C. Referring to M. Salehi & A. Salem (2010) [15], a low sintering temperature will construct separation gap between particle due to the reduction of particle interaction. This result corresponded well with XRD results in Figure 2, whereby no formation of mullite or less glassy phase and high formation of quartz were found at low temperature of 900 °C, which possibly produced a less dense ceramic body.

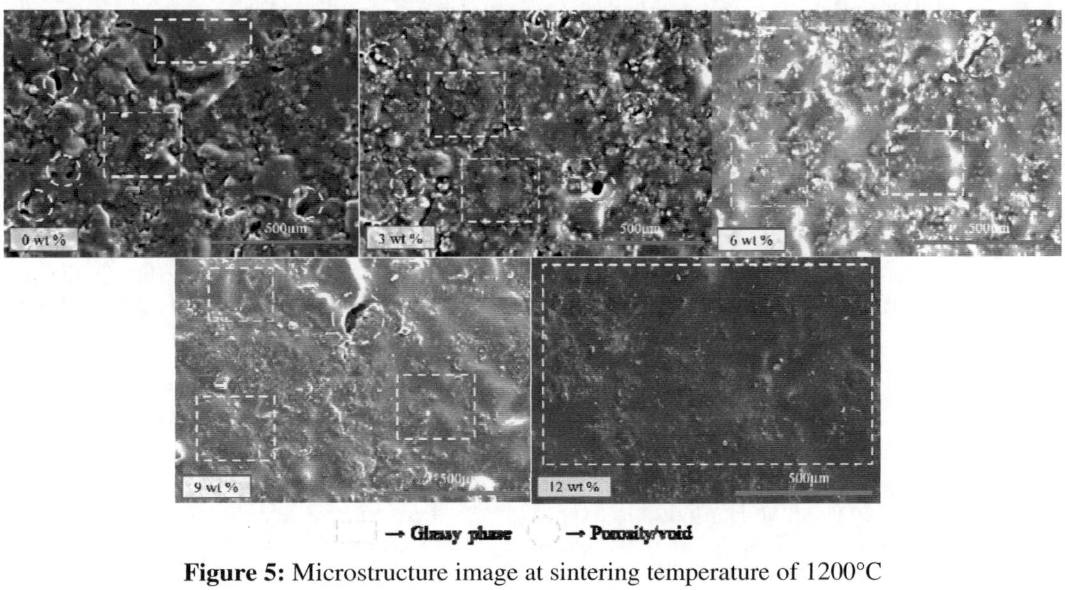

Figure 5: Microstructure image at sintering temperature of 1200°C

The image of microstructure morphology shown in Figure 5 revealed that the dense sample was produce with increasing temperature and addition of AR glass. It can be seen that the formation of void or porosity is eliminated due to the formation of mullite (glassy phase) that has covered and filled pores within the quartz. In addition, the presence of the mullite as glassy phase was also coherent with the result shown in Figure 3. Based on C.M. F. Viera & S. N. Monteiro, (2007) [16], it is stated that the liquid phase from AR will intensify as the sintering temperature is increasing, whereby the inclusion of fluxes aids the particle diffusion, thus producing a dense solid ceramic body. Therefore, the microstructure produced at the highest AR glass content of 12 wt.% manifest a more densify body with the least occurrence of porosity as compared to the rest of the samples.

4. Conclusions

In this study, the effects of sintering temperature and addition of AR glass into porcelain were investigated on porcelain ceramic. A formation of mullite as glassy phase is important for development of a dense ceramic body. A sintering temperature of 900 °C demonstrated absence of mullite phase on the phase analysis results, which will possibly lead to the formation of more voids and lesser dense ceramic structure regardless of AR glass content. A sintering temperature of 1100 °C coupled with the addition of AR glass were favourable to induce the formation of primary mullite, as observed on XRD analysis. Moreover, the microstructural analysis also demonstrated a denser and lesser void ceramic body at sintering temperature of 1200 °C, as compared to 900 °C and 1100 °C. Therefore, the elevation of sintering temperature and addition of AR glass are crucial factors that can enhance the properties of porcelain ceramic for vitrified clay pipes application. Further studies regarding the mechanical properties with respected to AR glass content and sintering temperature will be deliberated for future research.

Acknowledgement

The authors gratefully acknowledge Universiti Tun Hussein Onn Malaysia (UTHM) for funding this research under Geran Kontrak UTHM (Vot-U553).

References

[1] R. D. Rawlings, J. P. Wu, and A. R. Boccaccini, "Glass-ceramics: Their production from wastes-A Review," *J. Mater. Sci.*, vol. 41, no. 3, pp. 733–761, 2006.

[2] Y. N. El-shimy, S. K. Amin, S. A. El-sherbiny, and M. F. Abadir, "The use of cullet in the manufacture of vitrified clay pipes," *Constr. Build. Mater.*, vol. 73, pp. 452–457, 2014.

[3] F. Keith and D. Y. Goswani, *The CRC HANDBOOK of Mechanical Engineering*, 2nd ed. CRC Press LLC, 2005.

[4] G. Barluenga and F. Hernández-olivares, "Cracking control of concretes modified with short AR-glass fibers at early age. Experimental results on standard concrete and SCC," vol. 37, pp. 1624–1638, 2007.

[5] A. V. Gorokhovsky, J. I. Escalante-Garcia, E. Sanchez-Valdes, I. N. Burmistrov, and D. V. Kuznetsov, "Synthesis and characterization of high-strength ceramic composites in the system of potassium titanate – Metallurgical slag," *Ceram. Int.*, vol. 41, no. 10, pp. 13294–13303, 2015.

[6] S. R. Bragança and C. P. Bergmann, "Traditional and glass powder porcelain: Technical and microstructure analysis," *J. Eur. Ceram. Soc.*, vol. 24, no. 8, pp. 2383–2388, 2004.

[7] C. . Prasad, K. . Maiti, and R. Venugopal, "Effect of rice husk ash in whiteware compositions," *Ceram. Int.*, vol. 27, no. 6, pp. 629–635, 2001.

[8] T. Hansen, "Na_2O (Sodium Oxide, Soda)," *Digitalfire References Database*, 2015. [Online]. Retrieved from: https://digitalfire.com/4sight/oxide/na2o.html. at July 2016

[9] D. Askeland, P. Fulay, and W. Wright, *The Science and Engineering of Materials*, 6th ed. Cengage Learning, Inc, 2010.

[10] M. S. Spiller and M. Jameson, "Dental Ceramics," Academy of Dental Learning and OSHA Training 2015.

[11] G. Dhanaraj, K. Byrappa, V. Prasad, and M. Dudley, *Springer Handbook of Crystal Growth*.

Springer Berlin Heidelberg, 2010.

[12] A. Amarnath, A. Goel, D. U. Tulyaganov, S. Kapoor, K. Pradeesh, M. J. Pascual, and J. M. F. Ferreira, "Study of calcium magnesium aluminum silicate (CMAS) glass and glass-ceramic sealant for solid oxide fuel cells," *J. Power Sources*, vol. 231, pp. 203–212, 2013.

[13] M.-M. Jorge, J. M. Rincón, and M. Romero, "Mullite development on firing in porcelain stoneware bodies," *J. Eur. Ceram. Soc.*, vol. 30, no. 7, p. 1599, 2010.

[14] T. Manfredini and M. Hanuskova, "Natural Raw Materials in ' Traditional ' Ceramic Manufacturing," *J. Univ. Chem. Technol. Metall.*, vol. 47, no. 4, pp. 465–470, 2012.

[15] M. Salehi and A. Salem, "Porosity-strength correlations in ceramic Raschig ring: Effects of sintering temperature and water content," *Int. J. Appl. Ceram. Technol.*, vol. 7, no. 6, pp. 918–924, 2010.

[16] C. M. F. Vieira and S. N. Monteiro, "Evaluation of a plastic clay from the state of Rio de Janeiro as a component of porcelain tile body," *Matéria (Rio Janeiro)*, vol. 12, no. 1, pp. 1–7, 2007.

IOP Conf. Series: Journal of Physics: Conf. Series **914** (2017) 012021 doi:10.1088/1742-6596/914/1/012021

A study on tensile deformation at room temperature and 650 °C in the directional solidified Ni-base superalloy GTD-111

AA Pauzi[1], MH Abdul Ghaffar[2], SY Chang[3], GP Ng[4], S Husin[5]

[1,2,4,5]Materials Engineering Group, TNB Research Sdn Bhd, Kawasan Institusi Penyelidikan dan Latihan Bangi, 4300 Kajang, Selangor, Malaysia.
[3] Korea Electric Power Research Institute (KEPRI), 105 Munji-Ro, Yuseong-gu, Daejeon, 34056, South Korea.

Corresponding author:afiqpauzi@tnb.com.my[1], hafiz.ghaffar@tnb.com.my[2], sychang@kepco.co.kr[3], guatpeng@tnb.com.my[4], shuib.husin@tnb.com.my[5]

Abstract. GTD-111 DS generally used for gas turbine blades is a high performance Ni-base superalloy. This alloy, with high volume of γ' phase, has excellent tensile properties at high temperature. The effect of temperature on the tensile deformation of GTD-111 DS was investigated by using tensile test and microstructure evaluation of the fractured specimens. The tensile behaviour of GTD-111 DS was studied in the room temperature (RT) and 650 °C. From the yield strength results, the yield strength decreases from the average of 702.72 MPa to the average of 645.62 MPa with the increase of temperature from RT to 650 °C. The scanning electron microscope (SEM) results on fractured specimens confirmed that the tensile behaviour affected by deformation of the surface at 650 °C compared to fractured surface at RT. Based on the laboratory testing results, the correlation between tensile deformation of fractured surface and yield strength were discussed.

1. Introduction

Nickel base superalloys are widely used as the materials of gas turbine blades. These materials are selected for this application because of their excellent properties at elevated temperature. Directional solidified (DS) Ni-base GTD-111 was selected for this study. GTD-111 DS has a multi-phase microstructure containing of γ matrices, γ' precipitates, γ-γ' eutectics and a small amount of phases and carbides [1]. The fine shape of γ' that provides this directional solidified GTD-111with excellent high temperature properties in comparison with other Ni-base alloys. The total amount of γ' in GTD-111 DS is 60 – 70%. For this reason, the behavior of this alloy at high temperature depends on the amount γ'. With the increase of firing temperature for future advanced gas turbines, it is important to investigate and understand the behavior of GTD-111 DS as the superalloy mainly operates all of its lifetime at high temperatures during the operation [2]. Other than its capability to maintain excellent resistance at high temperature, this superalloy also has good oxidation resistance. These characteristics are the main reason for using this superalloy in severe and harsh operating condition of hot section gas turbine [3].

There are numerous studies have been carried out to investigate the deformation of GTD-111 at high temperatures. For materials researchers, it is important to study the mechanical properties of this alloy for high temperature gas turbine applications. One of the main laboratory life assessments of this alloy is by performing tensile test to investigate the strength of the materials at high temperature [4].

Previous work carried out by Sajjadi et al. [5] revealed that the behavior of this Ni-base superalloy is mainly depended by phase of γ'. The study also found that the yield strength decreased rapidly with the increase of temperature. In a study conducted by Yoon et al. [6], it showed that the yield strength increased with the increase of temperature up to 800 °C. The yield strength then dropped with the increase of temperature more than 800 °C. Lipski and Mrozinski [7] revealed that the value of yield strength normally decreases for high temperature, but the yield strength indicated different value in different temperature ranges. Another study on the temperature dependence of Ni-based alloys conducted by Zhang et al. [8], it stated the yield strength and elongation of this alloy attribute the variation to different deformation mechanisms at different elevated temperature.

The aim of this study was to investigate the temperature dependence of tensile properties of directional solidified Ni-based GTD 111at temperature 650 °C. While previous literatures mainly focused on the temperature 800 °C and more, this study has been focusing on strength of this material at the region between 25 °C and 650 °C. It is important to investigate the correlation between yield strength and tensile deformation structures at these temperatures. For this study, the room temperature at 25 °C has been used as a reference.

2. Experimental procedures

The elemental composition of the Ni-base superalloy was verified by using X-ray fluorescent. The chemical composition of the directional-solidified Ni-base superalloy is listed in Table 1.

Table 1: Chemical composition of GTD 111 (in wt %)

Ni	Cr	Co	Mo	Fe	W	Ta	Ti	Hf
60.65	14.60	9.93	1.78	0.78	4.48	2.77	4.37	0.20

The test specimens were fabricated from a rod according to ASTM E21. Tensile test were performed at two different temperatures, 25 °C and 650 °C at a constant strain rate of $10^{-4}s^{-1}$. At least five samples were tested at each temperature to obtain precise results. Microstructures of the tested samples were examined in a scanning electron microscope to analyse the degraded microstructure and investigate the deformation mechanisms.

3. Results and discussion

The fractured surfaces of the GTD-111 DS superalloy from the tensile test were analysed to investigate the damage characteristics.

3.1. Variation of yield strength (YS) with temperature

Tensile test results for both temperatures are shown in Figure 1. For comparison, the yield strength of the alloys GTD-111 at room temperature (RT) and 650 °C are included. From the results, the yield strength decreases from the average of 702.72 MPa to the average of 645.62 MPa with the increase of temperature from RT to 650 °C. According to the previous works conducted by Sajjadi et al. [9] and Zhang et al. [8], the yield strength of directional solidified GTD-111 at RT should be higher than the yield strength at 650 °C. The detail explanations of this situation will be discussed in the next sections.

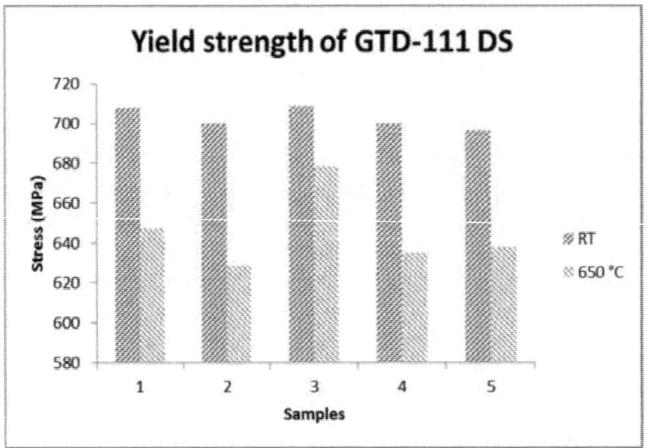

Figure 1: Yield strength of GTD-111 DS at RT and 650 °C

3.2. Tensile deformation

3.2.1. At low temperature. Figure 2 shows the fractured surface of one of the tensile samples at RT. The microstructure of the fractured sample is shown in Figure 3. From the micrograph image, the γ' phase appeared in cuboidal shape. The shapes were uniform and the average size of the γ' phase is between 0.5 µm to 1 µm. The γ' phase shapes in the micrograph image exhibit similar deformation pattern at RT. In a study by Sajjadi et al. [5], there was no evidence of voids on the fractured surface. The author concluded that the cracks on the fractured surface are due to the brittleness of grain boundary. With the proof of no voids on the fractured surface, it also concluded that the cracks are not due to the void growth on the surface.

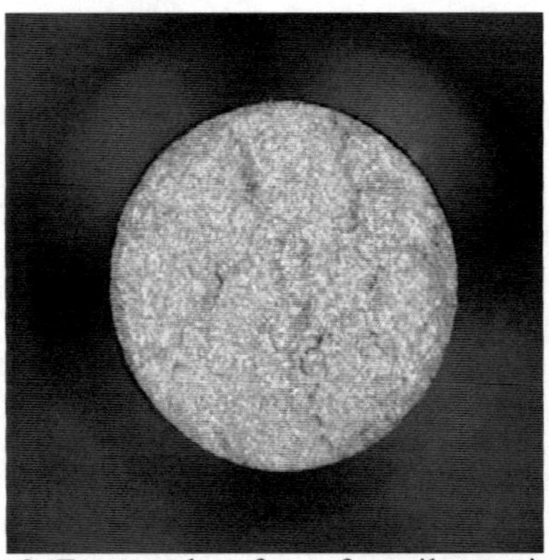

Figure 2: Fractured surface of tensile specimen at RT.

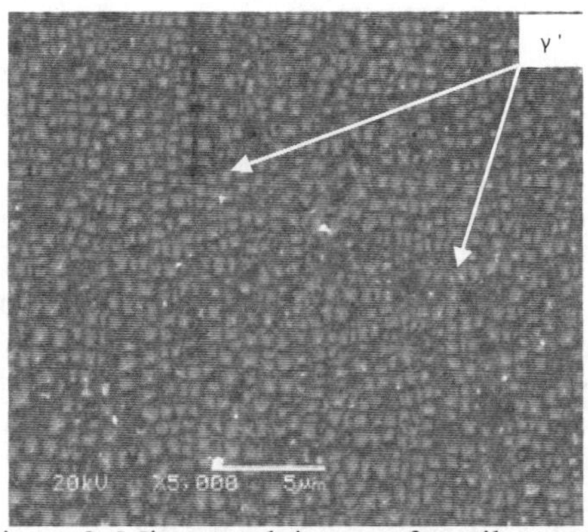

Figure 3: Micrograph image of tensile specimen at RT (X5000 magnification).

3.2.2. At high temperature. Figure 4 shows the fractured surface of one of the tensile samples at 650 °C. The microstructure of the fractured sample is shown in Figure 5. From the micrograph image, the γ' phase appeared in elliptical shape. The γ' began to degrade with the increase of temperature. The average size of γ' at 650 °C increased between 1.5 µm to 3 µm. The cuboidal shape of γ' that has been shown at RT totally elongated and showed non-uniform elliptical shape. Several previous studies by Sajjadi and Nategh[10] stated that, when γ' began to degrade due to the exposure at high temperature, the size of γ' increased, the γ' began to elongate and changes the shape from cuboidal to elliptical or spherical.

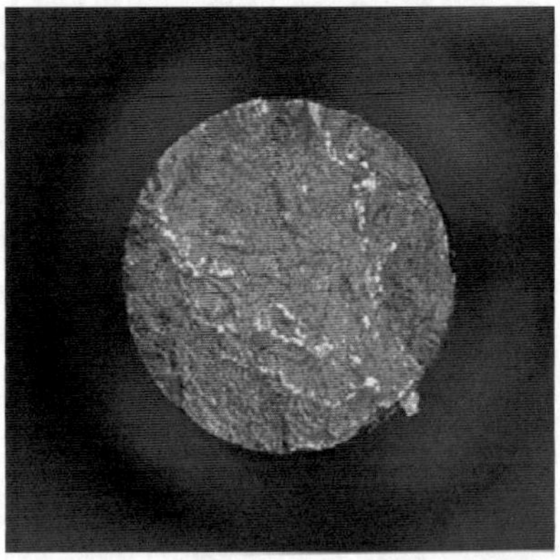

Figure 4: Fractured surface of tensile specimen at 650 °C.

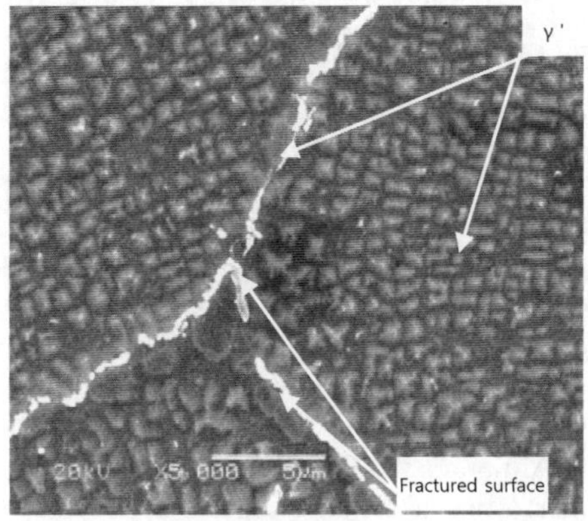

Figure 5: Micrograph image of tensile specimen at 650 °C (X5000 magnification).

3.3. Correlation between yield strength and tensile deformation structures

From the previous study undertaken by Wang et al., it is stated that the finer shape and higher volume of γ', the higher the strength of this superalloy at high temperature. At the high temperature of 650 °C, it was found that the yield strength decreased from RT. At this temperature region, the γ' has shown finer shape with huge volume of γ', and this resulted in high yield strength of the specimen. When the test carried out at 650 °C, the γ' precipitates began to expand their sizes due to exposure at high temperature, and this resulted in reduction of yield strength. This degradation process could also be classified as sphericalisation or coarsening of γ'phase. The tensile properties of GTD-111 DS depend on the condition of γ' during the exposure at high temperature [11, 12]. From the decrement of yield strength less than 10% from RT to 650 °C, it showed that the GTD-111 DS superalloy could maintain its excellent tensile properties (strength) in high temperature environment for a certain period.

4. Conclusions

1. The SEM investigations revealed that the fractured tensile specimens indicated different deformations at RT and 650 ºC. The occurrence of different tensile deformation depends on the test temperature.
2. The decrement of 10% yield strength from RT and 650 °C confirmed that GTD-111 DS alloys has excellent mechanical properties (strength) at high temperature.
3. The effect of temperature on tensile deformation for GTD-111 DS superalloys could be investigated after yielding in the certain temperature range that was determined before carried out tensile test.

References

[1] S.A Sajjadi, S.M Zebarjad, "Effect of temperature on tensile fracture mechanisms of Ni-base superlloy", in International Scientific Journal (2007), Vol.28, Issue 1, pp 34-40.
Arman Dadkhah, Ahmad Keranpur, "On the precipitation hardening of the directionally solidified
[2] GTD-111 Ni base superalloy: Microstructures and mechanical properties", in Materials Science and Engineering (2017), Vol.685, pp.79-86.
[3] Baig Gyu Choi, In Soo Kim, Doo Hyun Kim, Chang Yong, "Temperature dependence of MC decomposition behavior in Ni-base superalloy GTD 111", in Materials Science and Engineering (2007) ,Vol. 478, pp. 329-335.

[4] Z.J. Zhou, L. Wang, D. Wang, L.H. Lou, J. Zhang, "Effect of secondary orientation on room temperature tensile behaviors of Ni-base single crystal superalloys", in Materials Science and Engineering (2006), Vol. 659, pp.130-145.
[5] Seyed Abdolkarim Sajjadi, Said Nategh, Mihaiela Isac, Seyed Motjaba Zebarjad, "Tensile deformation mechanisms at different temperatures in the Ni-base superalloy GTD-111" in Journal of Materials Processing Technology, Vol. 155-156.
[6] Kee Bong Yoon, Tae Gyu Park, and Ashok Saxena, "Elevated Temperature Fatigue Crack Growth Model for DS-GTD-111"
[7] Adam Lipski, Stanisław Mroziński, "The Effects of Temperature On The Strength Properties Of Aluminium Alloy 2024-T3" (2013).
[8] ASTM E21, "Standard Test Methods for Elevated Temperature Tension Tests of Metallic Materials", (2006).
[9] Seyed Abdolkarim Sajjadi, Said Nategh, Roderick I.L. Guthrie, "Study of microstructure and mechanical properties of high performance Ni-base superalloy GTD-111" in Materials Science and Engineering(2002) , Vol. 325, pp. 484-489
[10] S.A. Sajjadi, S. Nategh. "A high temperature deformation mechanism map for the high performance Ni-base superalloy GTD-111", in Materials Science and Engineering, Vol.307, pp. 158-164.
[11] S.A. Sajjadi, S.M Zeberjad, "Study of fracture mechanisms of a Ni-Base superalloy at different Temperature" in Journal of Achievements in Materials and Manufacturing Engineering (2006), Vol.18, pp.1-2.
[12] Leonid B. Getsov, Artem S. Semenov, Elena A. Tikhomirova, Alexander I. Rybnikov, "Thermocyclic- And Static-Failure Criteria For Single-Crystal Superalloys Of Gas-Turbine Blades" in Original scientific article (2014), Vol.48, pp. 255.
[13] A.R. Ibanez, V.S. Srinivasan, Ashok Saxena, "Creep deformation and rupture behaviour of directionally solidified GTD 111 superalloy" in Fatigue & Fracture of Engineering Materials & Structures (2006), Vol.29, Issue 12,
[14] Gas Turbine Blade Materials Handbook Property by EPRI.
[15] Majid RezazadehReyhani, MohammadAlizade, AlirezaFathic, Hiwa Khaledid, "Turbine blade temperature calculation and life estimation –a sensitivity analysis" in Propulsion and Power Research (2013), Vol.2, pp. 148-161.

IOP Conf. Series: Journal of Physics: Conf. Series **914** (2017) 012022 doi:10.1088/1742-6596/914/1/012022

Evaluation on Compression Properties of Different Shape and Perforated rHDPE in Concrete Structures

M Y Yuhazri[1], K M Hafiz[1, 2], Y Z A Myia[2], C P Jia[1], H Sihombing[1], S M Sapuan[3], and N A Badarulzaman[4]

[1]Faculty of Manufacturing Engineering, Universiti Teknikal Malaysia Melaka, Hang Tuah Jaya, 76100 Durian Tunggal, Melaka, Malaysia.
[2]Department of Mechanical Engineering, Politeknik Merlimau, KB. 1031 Pejabat Pos Merlimau, 77300 Merlimau, Melaka, Malaysia
[3]Laboratory of Biocomposite Technology, Institute of Tropical Forestry and Forest Product, Universiti Putra Malaysia, 43400 UPM Serdang, Selangor, Malaysia.
[4]Department of Materials Engineering and Design, Faculty of Mechanical & Manufacturing Engineering, Universiti Tun Hussein Onn Malaysia, 86400 Parit Raja, Batu Pahat, Johor

Email: yuhazri@utem.edu.my

Abstract. The purpose of this study was to develop a concrete structure by incorporating waste HDPE plastic as the main reinforcement material and cement as the matrix via standard casting technique. There are eight different shapes of rHDPE reinforcing structure were used to investigate the compression properties of produced concrete composites. Experimental result shown that the highest shape in compressive strength of rHDPE reinforcing structure were the concrete with the addition of X-perforated beam (18.22 MPa), followed by X-beam (17.7 MPa), square perforated tube (17.54 MPa), round tube (17.42 MPa) and round perforated tube (16.69 MPa). In terms of their compressive behavior, the average concrete containing rHDPE reinforcement was successfully improved by 6 % of the mechanical characteristic compared to control concrete. It is shown that the addition of waste plastic as reinforcement structure can provide better compressive strength based on their shape and pattern respectively.

1. Introduction

Concrete is the most extensive used construction material found worldwide, mainly of its superior properties and economically efficiency compared to other materials. Besides that, these cement-based materials are easy to handle, prepared and formed into various shapes and structural configurations before the curing process. Concrete is very strong in compression but it exhibits very low tensile strength and flexural strength. The main contributor to this low strength of concrete is the quasi-brittle characteristic, associated with their rigid properties, that tends to form micro cracks and propagates when an applied load is subjected [1]. Hence, there is a need to strengthen the concrete by narrowing the defect size or flaws that present inside the concrete so as to create better performance. To improve the tensile and flexural strength of concrete, reinforcing steel is often used.

Apart from the traditional steel reinforcement, there are other types of fibers that can be incorporated into the concrete, including steel fiber, glass fiber, natural fiber as well as synthetic fiber to overcome

IOP Conf. Series: Journal of Physics: Conf. Series **914** (2017) 012022 doi:10.1088/1742-6596/914/1/012022

the strength and toughness challenges found in conventional concrete. The use of fibers to raise the flexural and tensile performance of the concrete matrix, as well as the capability to improve the durability has attract the interest of researchers to developed more new materials to produce fibers. In fact, the adoption of fibers in the concrete matrix can effectively control and prevent crack development, thereby avoid plastic and dry shrinkage crack, maintain concrete integrity, and transform the concrete into much more tougher material with improved crack resistance and ductility. The fibers should possesses high tensile strength and elastic modulus to fulfill as a reinforcement [2].

These days, the utilization of various types of sub-products in cement-based materials has turn into a universal practice in the concrete industry. According to Hannawi *et al.* [3], the exponential growth of plastic waste has come to an alarming rate that it is utmost important to search for alternative route to reduce and dispose these harmful and non-biodegradable plastic waste. Transforming of plastic wastes into building materials and plastic waste valorization was an interesting outlet that provides a major potential market for waste recycling. Ghernouti *et al.* [4] stated that the incorporation of polymeric wastes to concrete conform to another point of view in research interests, by linking the field of concrete technology with environmental technology, thereby enhance better quality of concrete.

High density polyethylene (HDPE) is known to be the third-largest commodity plastic material used worldwide, after polypropylene (PP) and low density polyethylene (LDPE) in terms of volume. HDPE is a crystalline structure and has very little or no branching. Hence, HDPE has strong intermolecular forces and tensile strength. It possess superior characteristic if compared to LDPE, in terms of hardness and temperature resistance [5].

In Malaysia, it is found that the plastic waste is the second highest solid waste after food waste. One of the major applications for HDPE that can found in Malaysia is motor oil container or fuel tank for vehicles. It is found that the large quantity of waste HDPE motor oil container has not yet been fully discovered or recycled for other useful engineering purposes. Therefore, it is utmost important to find an alternative route to dispose and recycled this plastic waste.

There are many researchers have been utilizing the recycling approach to discover more useful substances from the non-biodegradable recycled high density polyethylene (rHDPE) to reduce the environmental impact. The reason why rHDPE is chosen to study scientifically is because of its long life span, lightweight, chemical resistance, and strength [6]. However, there is no scientific research regarding the types of shape, pattern, orientation and dimension for the direct rHDPE that have been used as a reinforcing structure. In this paper, the design structures of the size, thickness and orientation of rHDPE as a constant variable were emphasized. These studies have focused on the effect of different shape and pattern of plastic in the compression strength of the hardened mixtures of concrete. The reports were summarized and classified into many shape of recycle plastic dealing with concrete.

2. Materials and Method

The waste motor oil containers was cleaned and washed with soap so as to remove residue oil. The containers were then cut into sheets when the container was completely dried. Then, the rHDPE sheet was joined by adhesive or hot glued to form different shapes of reinforcing structures. In this research, there were eight designs of rHDPE reinforcing structures produced in order to compare the loading effect in concrete structure. All the shapes of rHDPE reinforcing structure were designed based on the basic shapes for engineering structure. The rHDPE reinforcing structures were manufactured in hollow shapes with constant width of 30 mm for all shapes and the length of the reinforcing structure was 150 mm.

Among the eight designs, there were four designs of rHDPE reinforcing structures with perforated holes on it. The function of the holes was to provide greater interfacial adhesion and as a bridging system between the rHDPE plastic and the cement paste. The diameter of the hole was 6 mm and the distance between two holes was 15 mm. Each surface of the reinforcing structure was designed with 10 holes for a 150 mm long reinforcing structure. Figure 1 showed the shapes and patterns of rHDPE reinforcing structure.

2.1 Making test beams from fresh concrete

Once the workability of the fresh concrete was measured (by slum test), the concrete was cast into predefined molds to prepare test specimens for compression test. The 150 mm size test cubes were made based on BS 1881-108:2013 because the aggregate size used in this study is less than 20 mm. The mold was located on a smooth and firm horizontal surface and filled with concrete in order to eliminate the entrapped air and to generate full compaction of the concrete. Then, the cement paste was filled into the mold in layers about 50 mm deep and every layer was compacted. The minimal number of strokes per layer needed to yield full compaction was 25 strokes for 150 mm cubes. The stroke was circulated evenly over the surface of the concrete and each layer was compacted to its full depth. The excess concrete above the upper edge of the mold was removed and levelled using a float.

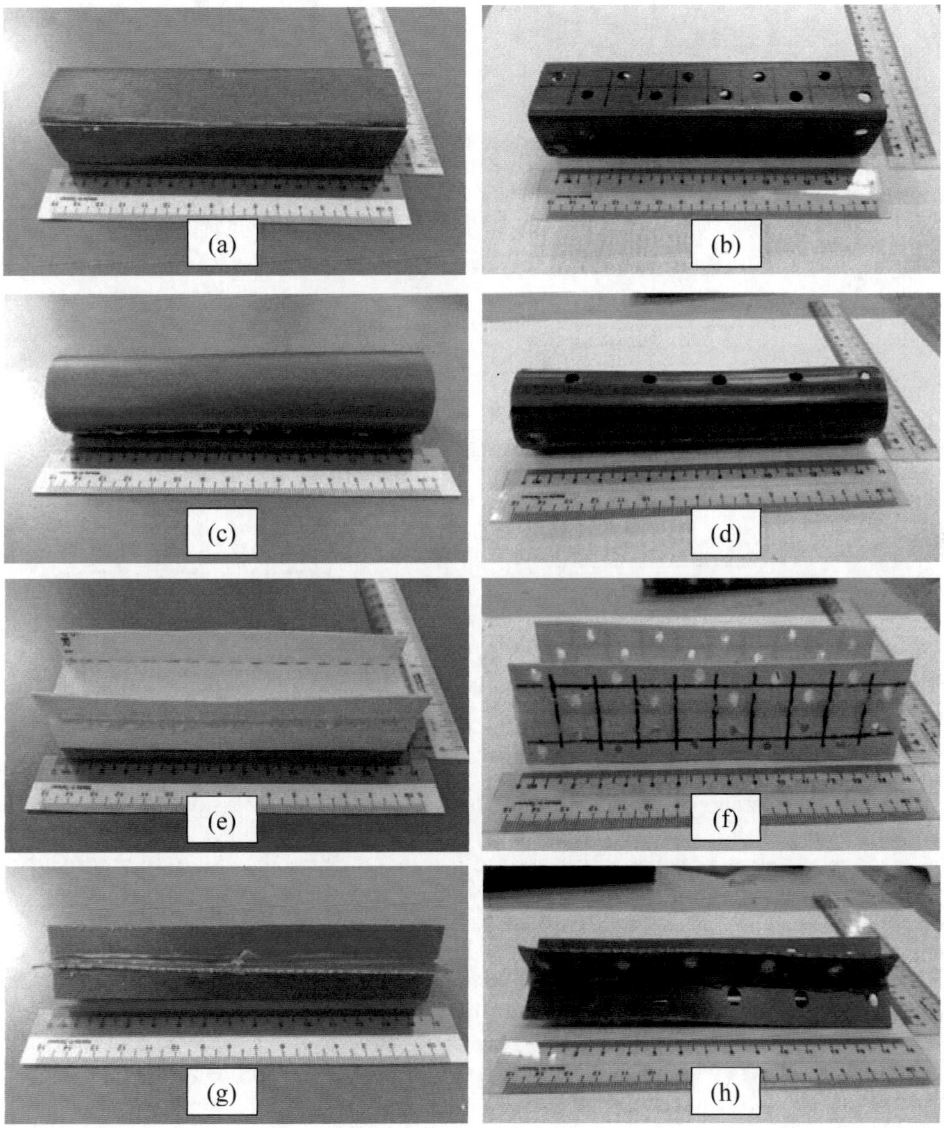

Figure 1. Designs and shapes of rHDPE reinforcing structure: (a) square; (b) square perforated; (c) round; (d) round perforated; (e) I-beam; (f) I-perforated; (g) X-shaped; and (h) X-perforated.

IOP Conf. Series: Journal of Physics: Conf. Series **914** (2017) 012022 doi:10.1088/1742-6596/914/1/012022

2.2 Casting Technique

The test cubes were embedded with a different shape of rHDPE reinforcing structure as shown in Figure 2. The reinforcing structure was placed at the center or the symmetry of the concrete structure. There were 27 test cubes prepared for compression testing. Each form of the rHDPE reinforcement was cast into three test cubes. The specimens were kept in a place that is free from vibration and kept in molds at ambient temperature for about 24 hours to harden before being demoulded and transferred to outside environment for curing for 28 days.

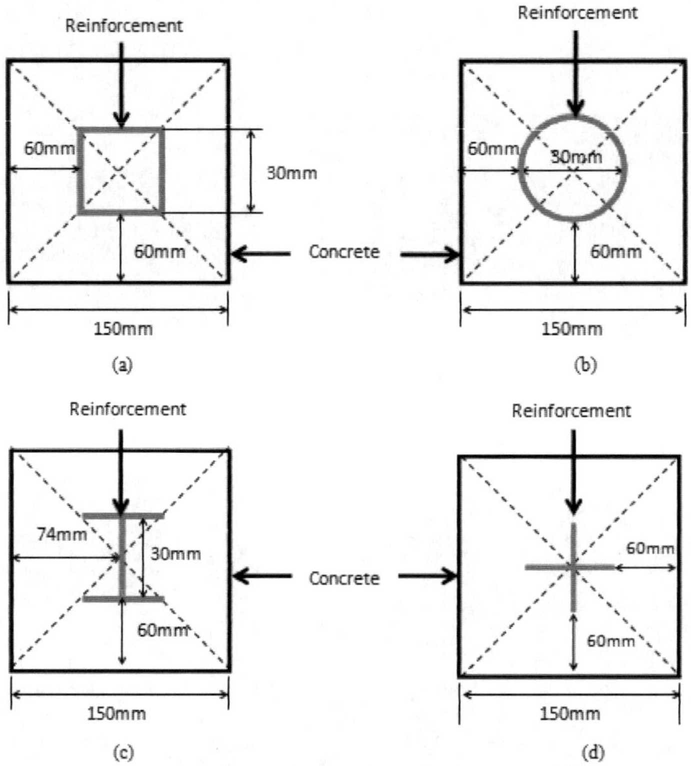

Figure 2. Location of rHDPE reinforcement in fresh concrete: (a) square; (b) round; (c) I-beam; and (d) X-beam.

2.3 Compression Test

The compressive testing performed in this research was using compression machine brand Controls at Politeknik Merlimau Melaka and the test was accordance to the British standard BS 1881-116:2013. Compressive testing is a testing which used to measure a material's ability to withstand compressive forces, where the specimen is squeezed laterally. In this research, the 150 mm x 150 mm x 150 mm test cubes were tested. All cubes were visually inspected. Any cube having broken edge with 20 mm or more were noted. Every test cubes were examined to confirm they are following the perpendicularity requirement based on BS 1881:108:2013. The cross-sectional area of the cube was calculated from the measured dimension. Auxiliary platens were used between the cube and machine platens. The cube specimen was precisely centered on the lower platen. The troweled surface was vertical. Then, the load was applied attentively without shock in such a way so that the stress was raise until no greater load can be sustained. Three concrete specimens for each rHDPE reinforcement design were tested to obtain the average compressive strength. The compressive strength of the cube was calculated by dividing the maximum load by the cross-sectional area. The result should be expressed to the nearest 0.1 MPa.

3. Results and Discussion

Compressive strength is the most important strength for concrete composite. The strength of concrete generally depends on the strength of aggregate, strength of cement, and the bond strength between reinforcement and cement matrix or the interfacial transition zone (ITZ). Albano *et al.* [7] states that the 28-day compressive strength value is approximately equal to 60 days because the concrete attained 75 % to 80 % of its total strength during the first 28 days. According to the ACI 318 Standard (Section 19.2.1.1), the minimum specified compressive strength of a structural lightweight concrete at 28 days should be 2500 psi or 17.24 MPa.

Figure 3 depicts the compressive strength of the fabricated composites incorporated with various shapes of rHDPE reinforcing structure addition. The indicated results are the average of three specimens. The results showed a significant improvement in performance for any substitution pattern of rHDPE reinforcement. The compressive strength of the control concrete is 15.693 MPa whereas the average compressive strength of the reinforced concrete is 16.685 MPa, which is about 6 % increment in strength. However, when we compare the strength of solid structure reinforced concrete (16.464 MPa) and perforated structure reinforced concrete (16.906 MPa) with the control concrete (15.693 MPa), both types of reinforced concrete showed about 5 % and 7% increment in compressive strength. Besides that, the strength of concrete containing perforated rHDPE structure is 3 % higher compared to concrete with solid rHDPE structure addition. Based on the average strength of the reinforced concrete with 16.685 MPa, concrete with the addition of round tube (17.42 MPa), round perforated tube (16.69 MPa), X-beam (17.70 MPa), X-perforated beam (18.22 MPa) and square perforated tube (17.54 MPa) have fulfilled the minimum required compressive strength for reinforced concrete composites.

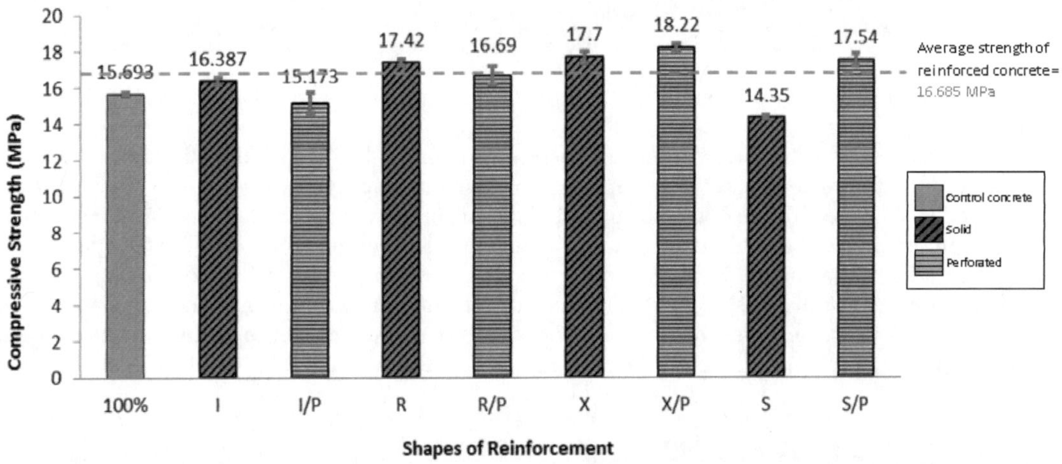

Figure 3. The 28-day compressive strength of different shapes of rHDPE reinforcement addition.

Concrete containing X-shaped perforated reinforcing structure showed the highest compression strength of 18.22 MPa and the lowest strength was observed for concrete containing square reinforcing tube with 14.35 MPa. Based on overall compressive results of reinforced concrete, perforated reinforcing structures promotes better strength than the non-perforated structures in resisting compression. This result proved that the bridge effect does provide better interfacial bonding between the rHDPE plastic and cement matrix. This bridging action tends to transfer the stress from the cement matrix to rHDPE reinforcement. Moreover, the rHDPE reinforcing structure was embedded vertically to direction of maximum applied stress, hence the reinforcement is 360° supported by its entire body from upper to lower part. Therefore, the reinforcement was able to resist the load for a short period before disintegration or split into two fractions. The reduced strength for non-perforated structures reinforced concretes were due to the weak interfacial bond strength resulted from hydrophobic nature of rHDPE plastic surface that impede cement hydration reaction [8]. Thus, the weak interfacial transition zone does not aid in transferring the stress from cement matrix to rHDPE reinforcement. The stress was carried by the cement matrix and if the compressive stress exceeded the ultimate strength of cement matrix, the cement failed.

IOP Conf. Series: Journal of Physics: Conf. Series **914** (2017) 012022 doi:10.1088/1742-6596/914/1/012022

Figure 4 displayed the typical fracture pattern of the normal concrete at front view. The failure mode observed in normal control concrete was meeting the standard crack pattern as stated in BS 1881-116:2003 where all the four exposed faces cracked almost equally after the determination of compressive strength and showed small damage to faces in contact with the platens (upper and bottom of the cube). However, the reinforced concrete did not showed the similar failure mode as observed from the control concrete. The reinforced concrete fractured at the corner or edge of the cube rather than peeling off of the exposed faces.

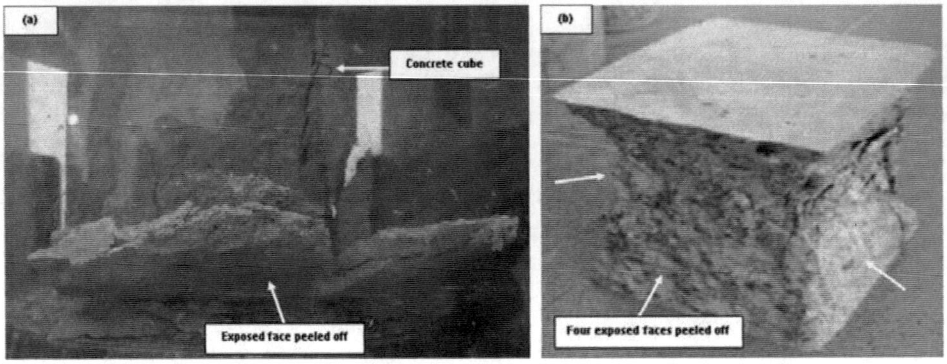

Figure 4. Typical compression failure mode of normal concrete (a) during, (b) after.

Figure 5 shown that the reinforced concrete composites had the similar failure pattern. All the fractured reinforced concretes were split into two halves and the rHDPE reinforcements were remained embedded at one half of the fractured concrete. Besides that, from the fractured surface, there are many pores present in the concrete that may affect the compression strength of the concrete. Apart from that, the fractured surface microstructure showed that the coarse aggregates were broken along the fracture surface. This means that the failure occurred at a direct path through the matrix and aggregates as shown in Figure 6. This scenario was found in concrete where the delayed crack propagation without significant micro-cracks formation and the cracks propagate unhindered and quickly by releasing high amount of energy.

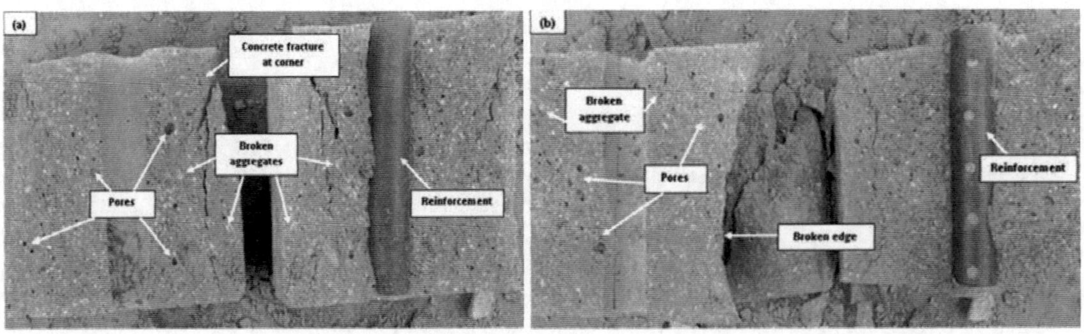

Figure 5. Failure mode of reinforced concrete containing (a) round tube and (b) round perforated tube.

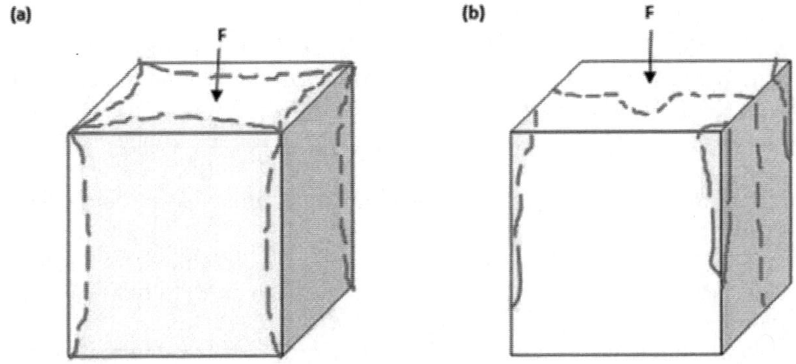

Figure 6. Illustration of fractured pattern of concrete cube (a) control concrete and (b) reinforced concrete.

The failure pattern of the concrete cube without rHDPE structure showed peeling off of the four exposed faces under loading. The fractured concrete cube will displayed curve pattern at the four exposed faces as shown in Figure 6(a). However, for reinforced concrete, the concrete cube fractured at the edges or at the corners of the cube and split into two fractions after the determination of strength. The crack will initiate at the interfacial zone between the plastic surfaces and cement matrix due to the weak interfacial bonding strength as shown in Figure 6(b). The crack line followed the cross-section of the rHDPE reinforcing structure and split into two halves.

4. Conclusions

It was found that the main benefit of adding rHDPE reinforcement into the mix concrete can improve the compressive properties based on their design structure. The following is the main conclusions and significant findings of research.

- The average compressive strength of the rHDPE reinforced concrete is 16.69 MPa whereas the compressive strength of the control concrete is 15.69 MPa. The compressive strength of the reinforced concrete showed 6 % increment compared to control concrete.
- The average compressive strength of the concrete containing perforated rHDPE reinforcing structures is 16.91 MPa while the average compressive strength of the concrete containing solid rHDPE reinforcing structures is 16.46 MPa. The strength of the perforated rHDPE reinforced concrete showed about 2.6 % improvement compared to solid rHDPE reinforced concretes.
- The addition of perforated rHDPE reinforcement and solid rHDPE reinforcement showed 7 % and 5 % increment in compressive strength compared to control concrete.
- X-perforated rHDPE reinforced concrete showed the highest compressive strength of 18.22 MPa among the reinforced concretes, while square tube rHDPE reinforced concrete showed the lowest compressive strength of 14.35 MPa. The X-perforated reinforced concrete showed 21 % increment in strength compared to square tube reinforced concrete.
- Based on the average compressive strength (16.69 MPa) for reinforced concretes, the concrete with the addition of X-perforated beam (18.22 MPa), X-beam (17.7 MPa), square perforated tube (17.54 MPa), round tube (17.42 MPa) and round perforated tube (16.69 MPa) have fulfilled the minimum required compressive strength of reinforced concrete.

Acknowledgments

We are grateful for the MyPhD under MyBrain 15 scholarship from Ministry of Higher Education Malaysia to Author 1and Politeknik Merlimau Melaka. The authors also would like to acknowledge the sponsorship of grant RACE/F3/TK15/FKP/F00253 for this paper and Universiti Teknikal Malaysia Melaka.

References

[1] Silva, R.V., De Brito, J., & Saikia, N. (2013). Influence of curing conditions on the durability-related performance of concrete made with selected plastic waste aggregates. *Cement and*

Concrete Composites, *35*(1), 23-31.

[2] Yin, S., Tuladhar, R., Shi, F., Combe, M., Collister, T., & Sivakugan, N. (2015). Use of macro plastic fibres in concrete: a review. *Construction and Building Materials*, *93*, 180-188.

[3] Hannawi, K., Kamali-Bernard, S., & Prince, W. (2010). Physical and mechanical properties of mortars containing PET and PC waste aggregates. *Waste Management*, *30*(11), 2312-2320.

[4] Ghernouti, Y., Rabehi, B., Safi, B., & Chaid, R. (2011). Use of recycled plastic bag waste in the concrete. *Journal of International Scientific Publications: Materials, Methods and Technologies*, *8*.

[5] Rahim, N.L., Sallehuddin, S., Ibrahim, N.M., Amat, R.C., Jalil, A., & Faizal, M. (2013). Use of Plastic Waste (High Density Polyethylene) in Concrete Mixture as Aggregate Replacement. *Advanced Materials Research*, 701, 265-269.

[6] Pešić, N., Živanović, S., Garcia, R., & Papastergiou, P. (2016). Mechanical properties of concrete reinforced with recycled HDPE plastic fibres. *Construction and Building Materials*, *115*, 362-370.

[7] Albano, C., Camacho, N., Hernandez, M., Matheus, A., & Gutierrez, A. (2009). Influence of content and particle size of waste pet bottles on concrete behavior at different w/c ratios. *Waste Management*, *29*(10), 2707-2716.

[8] Gu, L., & Ozbakkaloglu, T. (2016). Use of recycled plastics in concrete: A critical review. *Waste Management*, *51*, 19-42.

The Effect of Different Shape and Perforated rHDPE in Concrete Structures on Flexural Strength

MY Yuhazri[1], KM Hafiz[1, 2], YZA Myia[2], CP Jia[1], H Sihombing[1], SM Sapuan[3] and NA Badarulzaman[4]

[1]Faculty of Manufacturing Engineering, Universiti Teknikal Malaysia Melaka, Hang Tuah Jaya, 76100 Durian Tunggal, Melaka, Malaysia.
[2]Department of Mechanical Engineering, Politeknik Merlimau, KB. 1031 Pejabat Pos Merlimau, 77300 Merlimau, Melaka, Malaysia
[3]Laboratory of Biocomposite Technology, Institute of Tropical Forestry and Forest Product, Universiti Putra Malaysia, 43400 UPM Serdang, Selangor, Malaysia.
[4]Department of Materials Engineering and Design, Faculty of Mechanical & Manufacturing Engineering, Universiti Tun Hussein Onn Malaysia, 86400 Parit Raja, Batu Pahat, Johor

Corresponding author: yuhazri@utem.edu.my

Abstract. This research was carried out to develop a reinforcing structure from recycled HDPE plastic lubricant containers to be embedded in concrete structure. Different forms and shapes of recycled HDPE plastic are designed as reinforcement incorporate with cement. In this study, the reinforcing structure was prepared by washing, cutting, dimensioning and joining of the waste HDPE containers (direct technique without treatment on plastic surface). Then, the rHDPE reinforced concrete was produced by casting based on standard of procedure in civil engineering technique. Eight different shapes of rHDPE in concrete structure were used to determine the concrete's ability in terms of flexural strength. Embedded round shape in solid and perforated of rHDPE in concrete system drastically improved flexural strength at 17.78 % and 13.79 %. The result would seem that the concrete with reinforcing rHDPE structure exhibits a more gradual or flexible properties than concrete beams without reinforcement that has the properties of fragile.

1. Introduction

According to Apoorv and Patel [1] cement is the most commonly used content in today's globe in all the development performs. Youcef *et al.* [2] also stated that cement materials is trusted in worldwide construction industrial because of their performance in terms of mechanical strength and durability. The addition of polymeric waste to concrete corresponds to a new perspective in research activities, integrating the areas of concrete technology and environmental technology, thus can improve good quality of concrete.

Nowadays, the utilization of various sorts of sub-products in cement-based materials has grown into a universal practice in the concrete industry. According to Hannawi *et al.* [3], the exponential growth of plastic waste has come to an alarming rate that it is utmost important to search for alternative route to reduce and dispose these harmful and non-biodegradable plastic waste. Transforming of plastic wastes into building materials and plastic waste valorization was an interesting outlet that provides a major potential market for waste recycling. Ghernouti *et al.* [4] stated that the incorporation of polymeric wastes to concrete conform to another point of view in research interests, by linking the field of concrete technology with environmental technology, thereby enhance better quality of concrete.

Reis *et al.* [5] emphasize plastic waste is a worldwide concern, solutions are searched to minimize the impacts caused by the use and disposal of these materials because they generate several environmental problems. The production of this waste has increased significantly, making it a challenge for the world's wealthiest countries in environmental, economic and social terms. The environmental impact is reduced, as is the consumption of virgin plastics, when post-consumed polymeric materials are reused. Therefore, with an efficient collection, separation and recycling system, discarded plastics can be recycled into new products with only the addition of energy.

This supported by Mustafa *et al.* [6] and Kumar *et al.* [7] in his recent research, stated that the solutions are possible except for land-filling, recycling of waste plastic tend to produce new elements, such as concrete ingredients and cost savings for raw materials. It seems to be as one of the best solution for losing the plastic waste, due to its economical and environmental advantages. Recently, the manufacturing industry started reusing recycled plastic artifact for the fabrication of many household products like clothes hangers, toys, seeding containers etc. Polyethylene terephthalate (PET), poly vinyl chloride (PVC), high density polyethylene (HDPE), thermosetting plastics, shredded and recycled plastic waste mixtures, expanded polystyrene foams (EPS), glass reinforced plastics (GRP), polycarbonate, polyurethane foams and polypropylene fibers are the plastic wastes that usually used as partial or full substitution in the concrete manufacturing.

In an effort to participate in reducing plastic waste pollution, it is beneficial to introduce and incorporate non-biodegradable construction waste as a part of the resources used in construction. With this in mind, this research presents the development of an effective technique of creating a sustainable concrete material consisting of recycled high density polyethylene plastic. High density polyethylene (HDPE) has been chosen to study scientifically relate to the suitability of the mixture and reinforcement. The life span of the HDPE is one of the factors for selection because of its superior characteristic in terms of rust resistance, temperature resistance, chemical resistance and lightweight characteristics compared with steel reinforcement types [8][9].

Additional of plastic fibers into the concrete improved the flexural tensile strength as proved by many researchers. Based on Ramadevi and Manju [10], the concrete containing PET fibers showed gradual increased in flexural strength as the fiber content rise. The fine natural aggregate was replaced by PET fibers at 0.5 % to 6 % volume of total aggregate in concrete mixture. The flexural strength of concrete showed linear increment for fiber percentage of 0.5 % to 2 %, but the strength became stagnant as the fiber content increased from 4 % to 6 %. Nibudey *et al.* [11] also focused in using post consumed PET bottles to produce plastic fiber of length 25mm with aspect ratio of 35 and 50. The flexural strength of waste PET fiber reinforced concrete was increased to 5.71 MPa (AR35) and 6.00 MPa (AR50) with the addition of 1 % PET fiber in the concrete as compared to control concrete having flexural strength of 4.99 MPa. As the fiber content greater than 1 %, the flexural strength of the reinforced concrete was decreased.

Extensive review shown, it was clearly stated that there were no information available regarding on the design structure of the recycled plastic waste to be embedded in the concrete structure. The summary collected were mainly studied on utilizing waste plastic materials to produce plastic aggregate as fine and coarse aggregate substitution, and plastic fiber to incorporate in concrete structure in order to alter the properties of conventional concrete. Besides, even the results of using recycle plastic improve the flexural strength, there is no scientific research which shape, pattern, orientation and dimension for the plastic that have been used.

In this paper, the studies have focused on the effect of various forms (solid and perforated) and sample of plastic within the flexural strength of the hardened mixtures of concrete. Flexural durability is one of the most important properties of the hardened concrete that represents its load bearing potential. It is also poor in stress thus it is essential to determine the flexing of the concrete beam to avoid the failure. Therefore the flexural tensile strength of concrete should be further analyzed for a wider variety of concrete strength factors considering the lengthy term and confinement conditions of affiliates [12].

2. Materials and Method

The waste motor oil containers was cleaned and washed with soap so as to remove residue oil. The containers were then cut into sheets when the container was completely dried. Then, the rHDPE sheet was joined by adhesive or hot glued to form different shapes of reinforcing structures. In this research, there were eight designs of rHDPE reinforcing structures produced in order to compare the loading effect in concrete structure.

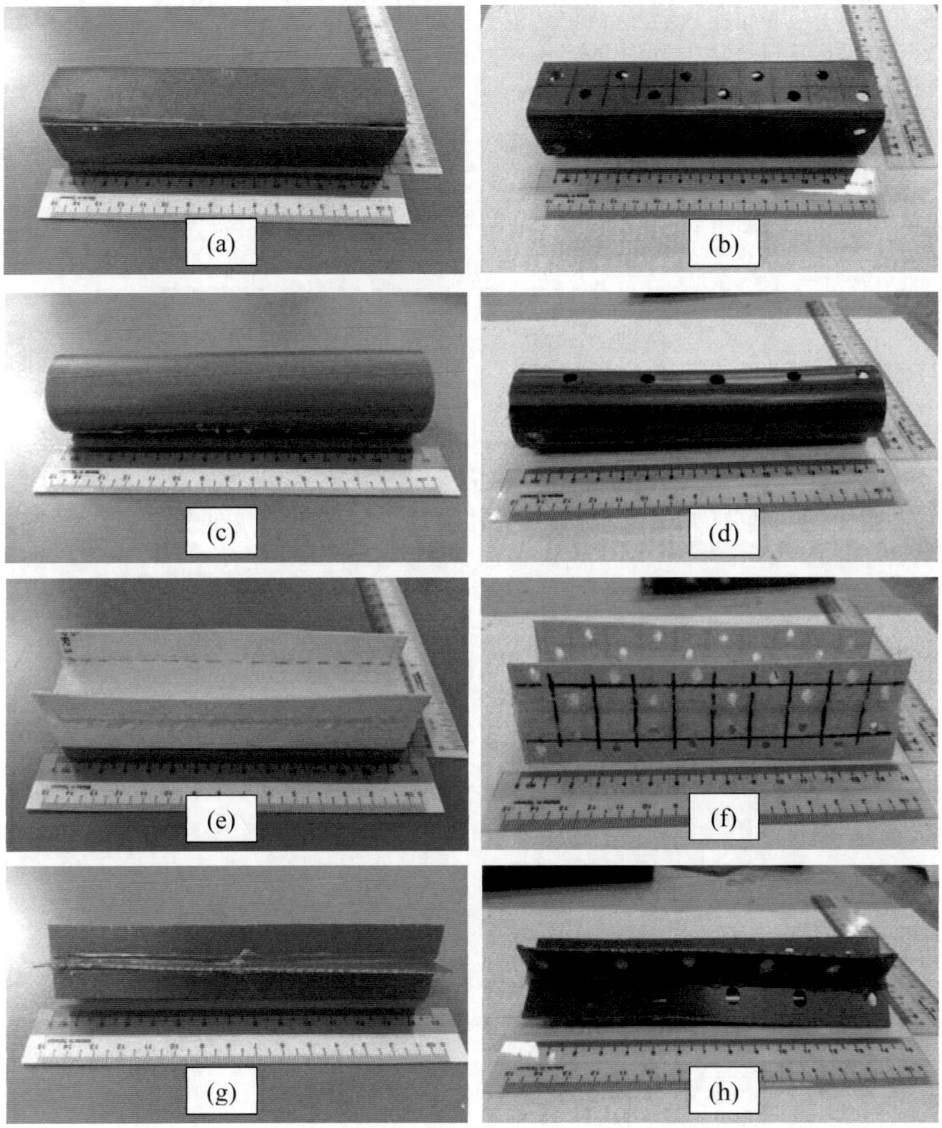

Figure 1. Designs and shapes of rHDPE reinforcing structure: (a) square; (b) square perforated; (c) round; (d) round perforated; (e) I-beam; (f) I-perforated; (g) X-shaped; and (h) X-perforated.

All the shapes of rHDPE reinforcing structure were designed based on the basic shapes for engineering structure. The rHDPE reinforcing structures were manufactured in hollow shapes with constant width of 30 mm for all shapes and the length of the reinforcing structure was 150 mm. Among the eight designs, there were four designs of rHDPE reinforcing structures with perforated holes on it. The function of the holes was to provide greater interfacial adhesion and as a bridging system between the rHDPE plastic and the cement paste. The diameter of the hole was 6 mm and the distance between two holes was 15 mm. Each surface of the reinforcing structure was designed with 10 holes for a 150 mm long reinforcing structure. Figure 1 showed the shapes and patterns of rHDPE reinforcing structure.

2.1 Making test beams from fresh concrete

The test beams dimension were 75 mm x 75 mm x 150 mm long based on BS 1881-109:2013. To prepare test beams, the mold was placed on rigid, smooth horizontal surface and filled with concrete in order to eliminate entrapped air as well as to generate full compaction of concrete. Then, the cement paste was filled into the mold in layers and every layer was compacted by using the compacting bar. Each layer should not be more than 50 mm thick. The strokes were distributed evenly over the surface of the concrete and every layer was compacted to its full depth. Table 1 showed the tolerances for test beam.

Table 1. Test beam tolerances.

Dimensions	± 2 mm (height); ± 1 mm (sides)
Perpendicularity	± 1.5 mm
Parallelism	± 2 mm

2.2 Casting Technique

The test beams were embedded with a different shape of rHDPE reinforcing structure as shown in Figure 2. The reinforcing structure was placed at the center or the symmetry of the concrete structure. There were 27 test beams prepared for mechanical testing. Each form of the rHDPE reinforcement was cast into three test beams. The specimens were kept in a place that is free from vibration and kept in molds at ambient temperature for about 24 hours to harden before being demoulded and transferred to outside environment for curing for 28 days.

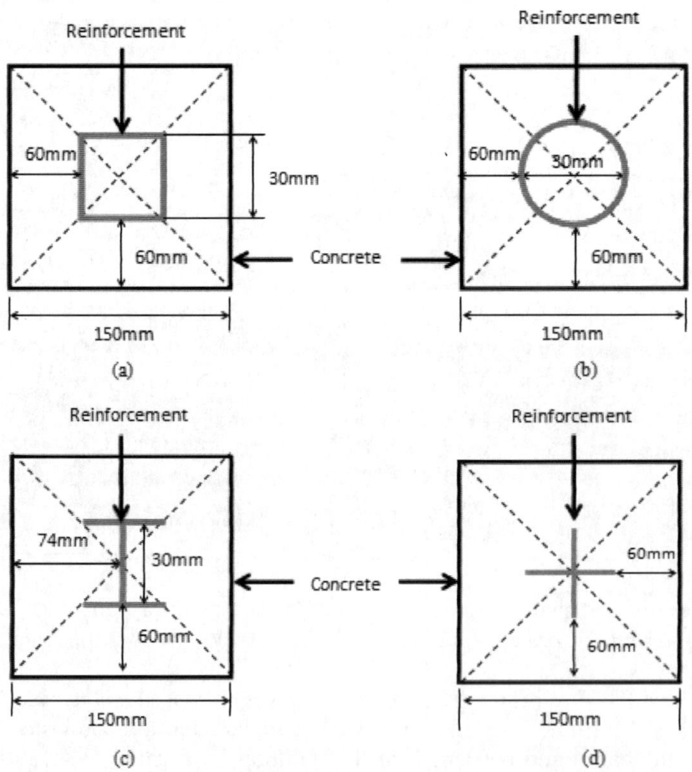

Figure 2. Location of rHDPE reinforcement in fresh concrete: (a) square; (b) round; (c) I-beam; and (d) X-beam.

2.3 Flexural Test

In this research, flexural testing used is a three-point bending method and was performed in accordance to BS 1881-118:2013. The aim of this testing was to measure the flexural strength of the 75 mm x 75 mm x 150 mm specimen dimension. Three beam concrete specimens for each shape rHDPE reinforcing design were tested for flexural strength. All loading and supporting points were uniformly contact with the test beams before load is applied. After engaged with initial load, which does not exceed approximately 20 % of the failure load, test load was applied without shock such that the stress was raise continuously at a selected constant rate of ± 10 % within the range of 0.04 MPa/s to 0.06 MPa/s (450 N/s). The test load was applied until no greater load can be sustained. Once adjusted, the rate of loading was maintained without change until failure occurs.

3. Results and Discussion

The flexural strength is interrelated with the bending ability of the produced concrete composite. From the testing, the flexural strength of the rHDPE reinforced concrete composite can be measured. Figure 3 depicts the flexural strength of the fabricated composites incorporated with various shapes of rHDPE reinforcing structure addition. The indicated results are the average on three specimens. The average strength of the reinforced concrete showed a slight reduction in performance of about 7 % for any substitution pattern of rHDPE reinforcement as compared to control concrete. However, if we compare the flexural strength of concrete containing solid reinforcing structure with control concrete, the solid rHDPE reinforced concrete had better strength than the control concrete. The average strength of solid structure reinforced concrete is 5.192 MPa whereas the strength of control concrete is 5.062 MPa. The average strength of perforated rHDPE reinforced concrete is lower than that of the control concrete and solid rHDPE reinforced concrete for about 16 % and 18 %. Based on the average flexural strength of reinforced concrete (4.734 MPa), concrete with the addition of I-beam (5.834 MPa), round tube (5.962 MPa) and round perforated tube (5.76 MPa) have fulfilled the minimum required flexural strength for reinforced concrete.

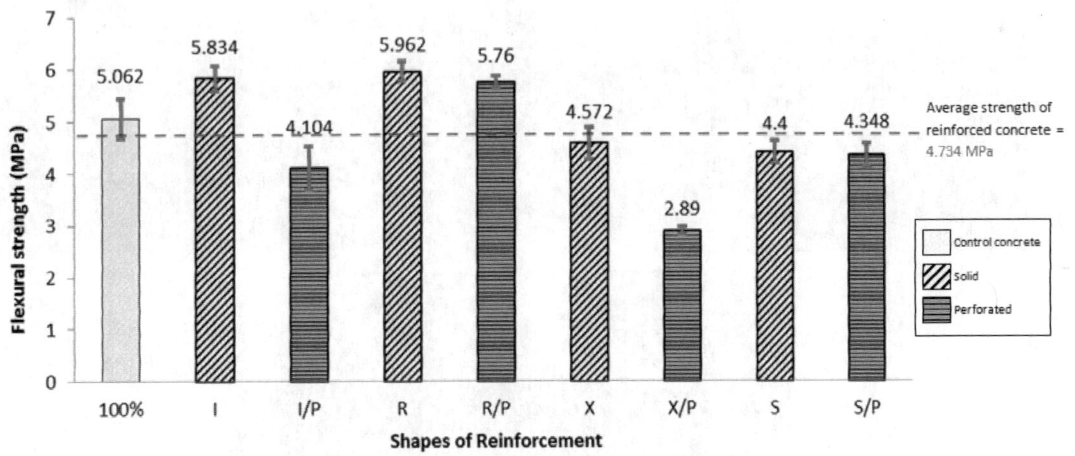

Figure 3. The effect of different shapes of rHDPE reinforcement addition into the 28-days flexural strength of concrete beam.

Figure 3 shown that the flexural strength of the reinforced concretes is improved with the addition of I-beam, round tubing and round perforated rHDPE reinforcing structure with respect to control concrete. The round tube reinforced concrete showed an increment about 18 % in strength, while I-beam and round perforated tube reinforced concrete showed 13 % and 12 % increase in the performance as compared to control concrete. However, concrete containing X-perforated beam, I-perforated beam and square perforated tubing showed 52 %, 31 % and 27 % reduction in flexural strength compared to the round tube reinforced concrete. Therefore, round tube reinforced concrete has the highest compressive strength while X-perforated beam reinforced concrete has the lowest compressive strength.

Based on Figure 4(a), the control concrete is split into two fractions after determination of flexural strength. The controlled concrete displayed brittle failure in the absence of reinforcement. However, the round tube reinforced concrete and I-beam reinforced concrete did not split into two pieces after testing and the observed failure were more of gradual failure or ductile type failure. The elastic nature and non-brittle characteristic of the rHDPE plastic tend to support the entire load by transferring the stress from the cementitious matrix to the reinforcing structure. The rHDPE structure is capable of resisting the load for a short period after failure and acts as a bridge that connects the crack so the concrete beam failed without full disintegration. Thus, the transferred of stress will enhance the tensile strain capacity and increased the flexural strength of the reinforced concrete.

It can be seen from Figure 4 that the crack mode of the three concrete beams is tensile mode of failure. Tensile mode is defined where the crack occurs on the plane with maximum tensile stress. The crack is initiated at the bottom and ended at the center of the upper part of the concrete beam because the bottom part is experiencing tension and concrete is weak in tension compared to compression. This tensile crack mode also applied to all the reinforced concrete regardless of the shapes of rHDPE reinforcing structures. This is because the tensile crack mode is observed for all concrete containing rHDPE reinforcing structure after the flexural test. The only difference after observing all the fractured reinforced concretes are the crack pattern, the crack either initiated from one side at the bottom of the beam and ended at the top as shown in Figure 4(b) or the cracks initiated at both sides at the bottom of the beam and ended at the top of the concrete as shown in Figure 4(c). If the concrete beams showed crack initiated at both sides it means that the stress is distributed evenly around the specimen whereas if the crack initiated at one side it means that the specimen experienced maximum stress at this side.

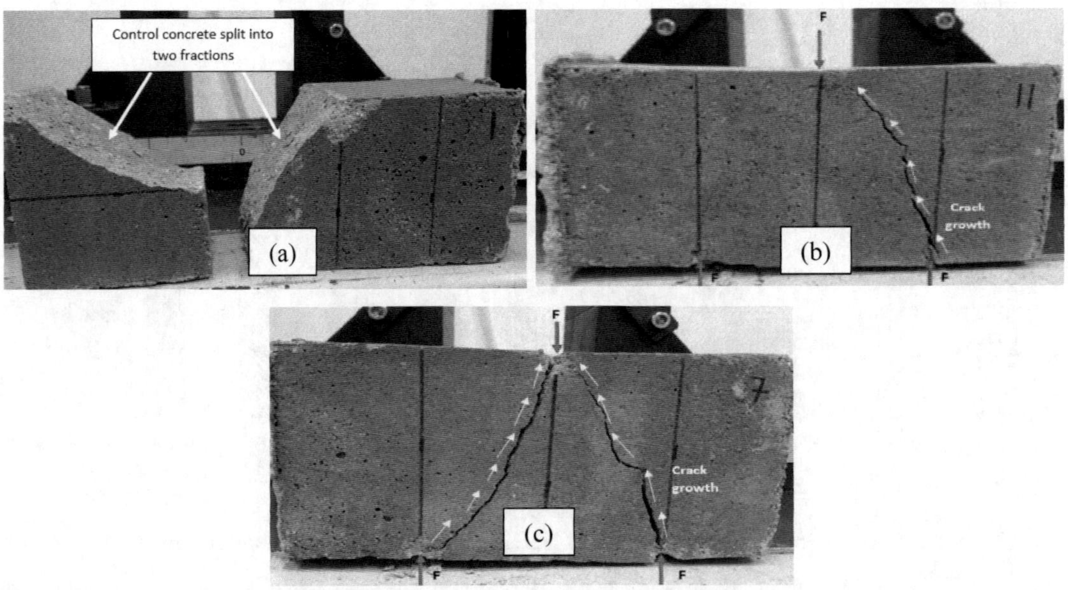

Figure 4. Fracture pattern of concrete beam: (a) Control concrete (b) Round rHDPE reinforced concrete (c) I-beam rHDPE reinforced concrete.

From the testing, both crack patterns can be observed for concrete beams containing similar reinforcing structure because the fractured beams of similar reinforcing structure does not showed only one crack pattern. For example, the first concrete beam containing square reinforcing tube showed crack pattern as Figure 4(b), but the second beam containing similar square reinforcing tube showed crack pattern of Figure 4(c). Besides that, it can be visually observed that the external surface of the produced concrete beams presented cavities and pores, or similar to the shape of honeycomb. The explanation for the reduced split tensile strength due to the internal flaws or microcracks, internal and external pores or cavities, weak interfacial bonding between rHDPE and cement matrix and orientation of the reinforcement applies to the flexural behavior of concrete too.

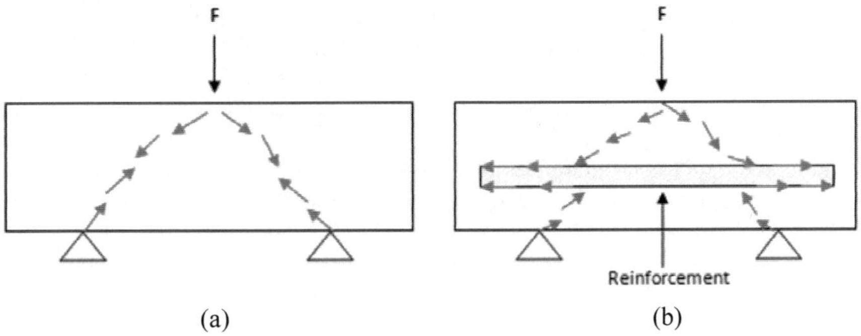

Figure 5. Illustration of stress distribution in concrete beam (a) control concrete and (b) reinforced concrete.

Apart from that, for 3-point flexural test, the stress is applied from the upper part and distributed to the lower part of the concrete beam. During flexural testing, the upper part of the concrete beam will be compressed while the lower part of the concrete beam is in tension. Since the concrete is weak in tension and good in compression, therefore the crack will be initiated from the bottom part of the concrete beam and grow towards the center of the upper part of the beam. For concrete beam without the addition of reinforcement, the concrete showed brittle failure and split into two fractions after the determination of strength. However, the concrete with rHDPE reinforcing structure showed more gradual or ductile type failure. This is because the applied stress is distributed to the rHDPE structure and the plastic structure is capable of resisting the load for a short period before failure. The rHDPE structure will serve as a medium for the transfer of stress between the cement matrices so that the concrete beam does not show brittle failure. Besides, the rHDPE structure will also act as a backbone to the concrete composites as it tends to support the entire load of the concrete and connect the crack surface so the concrete beam does not fracture into two sections. Figure 5 illustrates the stress distribution in control concrete and reinforced concrete.

4. Conclusions

It was found that the main benefit of adding rHDPE reinforcement into the mix concrete can improve the flexural properties based on their design structure. The following is the main conclusions and significant findings of research.

- The average flexural strength of the rHDPE reinforced concrete is 4.73 MPa whereas the flexural strength of the control concrete is 5.06 MPa. The flexural strength of the reinforced concrete showed about 6.5 % reduction in strength compared to control concrete.
- The average flexural strength of concrete containing solid rHDPE reinforcing structures is 5.19 MPa while the average flexural strength of concrete containing perforated rHDPE reinforcing structures is 4.26 MPa. The solid rHDPE reinforced concretes showed 17 % improvement in flexural strength compared to perforated rHDPE reinforced concretes.
- Round tube rHDPE reinforced concrete showed the highest flexural strength among the reinforced concretes that is 5.96 MPa, while the lowest flexural performance is showed by X-perforated beam reinforced concrete, with 2.89 MPa. Round tube reinforced concrete is about 51.5 % better than X-perforated beam reinforced concrete.

- Based on the average flexural strength of 4.73 MPa for reinforced concrete, the concrete with the addition of round tube (5.96 MPa), I-beam (5.83 MPa) and round perforated tube (5.76 MPa) have fulfilled the minimum required flexural strength of reinforced concrete.
- The crack mode of the concrete beams is in tensile mode of failure because the bottom part experiencing maximum tension compared to upper part that is in compression. The crack initiated at the bottom and ended at the center of the upper part of the concrete beam. This tensile crack mode also applied to all the reinforced concrete regardless of the shapes of rHDPE reinforcing structures. The only difference fracture of all reinforced concretes are the crack pattern, either crack initiated from one side at the bottom of the beam and ended at the top side or the cracks are initiated at both sides at the bottom of the beam and ended at the top side of the concrete.
- The crack initiated at both sides of concrete beam means that the stress is distributed evenly around the specimen. It occurs when the embedded reinforcement rHDPE is parallel horizontally to the concrete beam. If the crack initiated at one side, it means that the specimen experienced maximum stress at that side. It happens when the embedded reinforcement rHDPE is misaligned to the concrete beam in horizontal view.

Acknowledgments
We are grateful for the MyPhD under MyBrain 15 scholarship from Ministry of Higher Education Malaysia to Author 1 and Politeknik Merlimau Melaka. The authors also would like to acknowledge the sponsorship of grant RACE/F3/TK15/FKP/F00253 for this paper and Universiti Teknikal Malaysia Melaka.

References

[1] Apoorv, S., and Patel, R. D. (2014). A Comparative Study on Flexural Strength of Concrete Containing Different Mineral Admixtures as Replacement of Cement. *International Journal of Engineering Research & Technology,* 3 (8), pp. 890-893.
[2] Youcef, G., Bahia, R., Brahim, S., and Rabah C. (2014). Use of Recycled Plastic Bag Waste in the Concrete. *Journal of International Scientific Publications: Materials, Methods and Technologies*, 8 (2014), pp. 480-487.
[3] Hannawi, K., Kamali-Bernard, S., & Prince, W. (2010). Physical and mechanical properties of mortars containing PET and PC waste aggregates. *Waste Management*, *30*(11), 2312-2320.
[4] Ghernouti, Y., Rabehi, B., Safi, B., & Chaid, R. (2011). Use of recycled plastic bag waste in the concrete. *Journal of International Scientific Publications: Materials, Methods and Technologies*, *8*.
[5] Reis, J. M. L., Pacheco, L. J., and Costa Mattos, H.S. (2014). Temperature and variable strain rate sensitivity in recycled HDPE. *Polymer Testing*, 39 (2014), pp. 30-35.
[6] Mustafa Al Bakri, A. M., Mohammad Tamizi, S., Rafiza, A. R., and Zarina Y. (2011). Investigation of HDPE Plastic Waste Aggregate on the Properties of Concrete. *Journal of Asian Scientific Research,* 1 (7), pp. 340-345.
[7] Anoop Kumar, M. L., Ramana Reddy, I. V., Sasidhar, C., (2014). Experimental Investigations on the Flexural Strength of PET Reinforced Concrete. *International Journal of Emerging Technology and Advanced Engineering*, 4 (2), pp. 234-240.
[8] Rahim, N.L., Sallehuddin, S., Ibrahim, N.M., Amat, R.C., Jalil, A., & Faizal, M. (2013). Use of Plastic Waste (High Density Polyethylene) in Concrete Mixture as Aggregate Replacement. *Advanced Materials Research*, 701, 265-269.
[9] Pešić, N., Živanović, S., Garcia, R., & Papastergiou, P. (2016). Mechanical properties of concrete reinforced with recycled HDPE plastic fibres. *Construction and Building Materials*, *115*, 362-370.
[10] Ramadevi, K., & Manju, R. (2012). Experimental investigation on the properties of concrete with plastic PET (bottle) fibres as fine aggregates. *International Journal of Emerging Technology and Advanced Engineering*, *2*(6), 42-46.
[11] Nibudey, R.N., Nagarnaik, P.B., Parbat, D.K., & Pande, A.M. (2013). Strength and fracture properties of post consumed waste plastic fiber reinforced concrete. *International Journal of*

Civil, Structural, Environmental and Infrastructure Engineering Research and Development (IJCSEIERD), *3*(2), 9-16.

[12] Ahmed, M., Javed, M., and Abul Hasan, M. (2014). A Study of Factors Affecting the Flexural Tensile Strength of Concrete. *Journal of King Saud University – Engineering Sciences*, 26 (2), pp. 111-116.

Feasibility study of Thermal Electric Generator Configurations as Renewable Energy Sources

Muhammad Akmal Johar[1] , Zulkarnain Yahaya[1] , Omar Mohd Faizan Marwah[1] Wan Akashah Wan Jamaludin[1], Mohamed Najib Ribuan[2]

[1]Faculty of Mechanical & Manufacturing Engineering, Universiti Tun Hussein Onn Malaysia, Johor, Malaysia
[2]Faculty of Electrical & Electronic Engineering, Universiti Tun Hussein Onn Malaysia, Johor.

Email: akmal@uthm.edu.my

Abstract. Thermoelectric Generator is a solid state device that able to convert thermal energy into electrical energy via temperature differences. The technology is based on Seebeck effect that was discovered in year 1821, however till now there is no real application to exploit this capability in mass scale. This research will report the performance analysis of TEG module in controlled environment of lab scale model. National Instrument equipment and Labview software has been choosen and developed to measure the TEG module in various configurations. Based on the experiment result, an additional passive cooling effort has produced a better ΔT by 7°C. The optimal electrical loading of single TEG is recorded at 200Ω. As for circuit connections, series connection has shown superior power output when compared to parallel connection or single TEG. A series connection of two TEGs has produced power output of 416.82μW when compared to other type connections that only produced around 100μW.

1. Introduction

Renewable energy has been a focal point of research in the recent years. Various renewable energy sources ranging from solar [1], wind [2], hydro energy [3] was used to generate electricity in large scale. These energy sources were important due to the advantages towards clean environment and sustainability issues. As the world preparing to reduce dependence on coal based energy generator, various methods of renewable energy have been developed and implemented.

Apart from the renewable energy source mentioned above, there are a lot of research activities in harvesting energy at lower level which is using different physics phenomenon such as piezoelectric[4], vibration [5], acoustic noise, radio frequency (RF) [6] and thermoelectric[7]. Table 1 shows comparison between harvesting technology with respect to power density [8].

Thermoelectric generator or TEG is a device that able to convert thermal energy to electrical energy. It requires temperature differences at the junction in order to create electrical current. High reliability is one of the advantages since there are solid state device with no mechanical moving parts unlike MEMS based energy harvester that may requires additional self-x system to ensure system functionality [9]. TEG can be small in size and easily to be scaled depends on applications. However,

IOP Conf. Series: Journal of Physics: Conf. Series **914** (2017) 012024 doi:10.1088/1742-6596/914/1/012024

there are some drawbacks in this technology such as low in energy conversion efficiency and requires relatively constant heat source. These drawbacks lead to limited application and implementation in mass market.

Table 1. Power generation of several harvesting technology

Harvesting technology	**Power density**
Solar cells (outdoors at noon)	15 mW/cm^2
Piezoelectric (shoe inserts)	330μW/cm^3
Vibration (small microwave oven)	116μW/cm^3
Thermoelectric (10°C gradient)	40μW/cm^3
Acoustic noise (100dB)	960nW/cm^3

In a country like Malaysia, in which received a lot of sunshine throughout the year and thus make source of thermal energy come in abundance. The availability of thermal energy makes a good appeal for implementation of TEG to become one of renewable energy source. This research is focusing on evaluating and analyzing TEG modules in various configurations.

2. Experimental Setup

The evaluation of TEG module is done by using a commercial Peltier module TEC1-12706. It has the size of 40x40x3.5 mm. The device has a total 127 couples of PN junction interconnected as shown in Fig. 1.

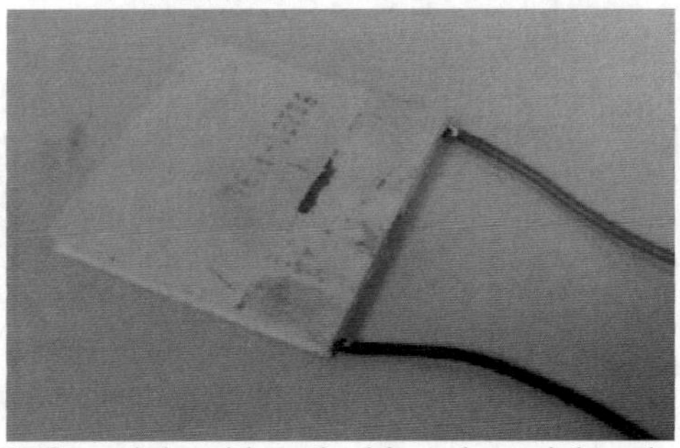

Fig. 1: TEG module with and without heat sink and structure

The experiment setup is divided into two parts. The first part is the heat induced part and the second part is the output measurement part. The first part consists of an AC power supply, two halogen lamps, TEG modules and an aluminum chassis. Each TEGs is exposed to the light produced by halogen lamps with rating of 240V 60Hz to generate heat source on the hot side. A halogen lamp is selected because it is one of the easiest ways to produce controllable heat in laboratory environment. The experiment is done with room temperature is set at 20°C. The TEG module is set at 4cm apart from the halogen lamp.

In this research, the measurement of output produced by TEG is controlled by using NI-CRIO 9014 real-time controller from National Instrument. For temperature measurement, NI-9211 module is used to allow direct connections with thermocouple. This module can handle up to 4-channel of thermocouple measurement simultaneously. In all temperature measurement points, type K thermocouples were used. In performing current measurement, NI-9203 module was used. While in performing voltage measurement, NI-9201 is selected. The overall setup of the experiment is shown in Fig. 2 where as the wiring diagram is shown in Fig. 3.

In order to record the measurement data, LabView software was used to develop a data acquisition programme. In this programme, 2 points of thermocouple measurements and 1 points each for current and voltage measurement was recorded simultaneously. The output measurement is recorded with the interval of 1 second. The measurement data is then processed to calculate the value of ΔT and instantaneous output power. The block diagram of the software is shown in Fig. 4. The red box is referred to the block area where the raw data are acquired from the sensor module. The blue box represents the block area where the data acquired from the sensor are processed to calculate the value of ΔT and output power. The green box area shows the block area for storing measured and calculated value in csv file.

Several sets of experiment were done to evaluate the performance of the TEG. In the first part, the performance of TEG will be compared between modules with and without additional cooling. Passive cooling technique in the form of heat sink will be used in this experiment. The second part of experiment will try to determine the optimum electrical loading value. While in the final part, the performance will be compared between single, series and parallel connections.

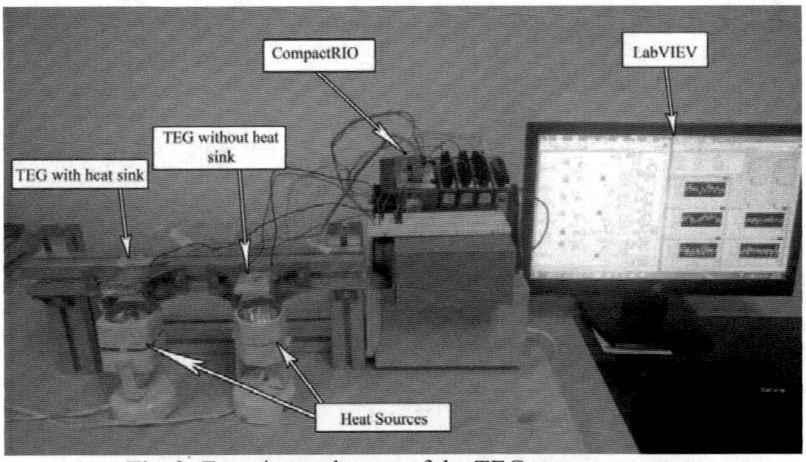

Fig. 2: Experimental setup of the TEG measurement.

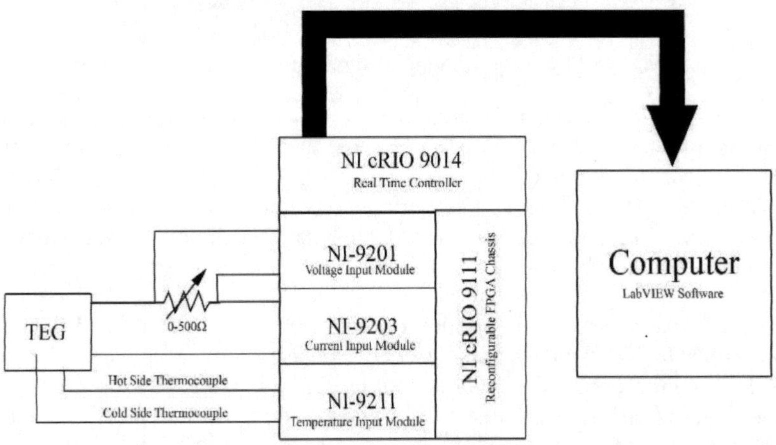

Fig. 3: Wiring layout of the TEG measurement

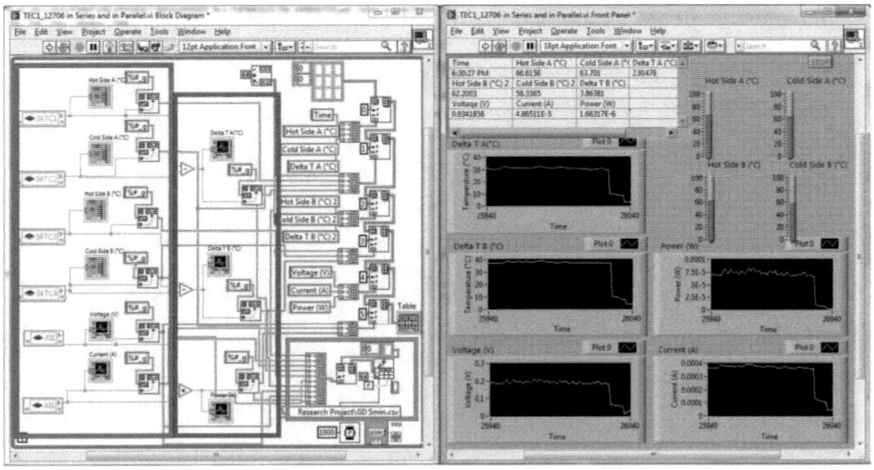

Fig. 4 Block diagram of data acquisition programme used in the experiment

3. Result And Discussion

3.1 Performance Analysis of TEG with and without Heat Sink

In the first set of experiment, the differences in performance of TEG between with and without heat sink were investigated. In this experiment, the load resistance is set to 100Ω. The heat sink which has size of 40x40mm covers the hot plate area. Fig. 5 shows the box plot of data acquired during measurement.

From the results in the **Table *2*:** Box plot parameter of performance analysis of TEG with and without heat sinkTable 2 , the median data of ΔT is 23.45°C for TEG without heat sink and 30.13°C for TEG with heat sink. The temperature difference is larger by 7°C when the TEG is equipped with additional heat sink. The bigger ΔT produced the bigger electricity output. From the experiment the power output produced from TEG without heat sink is 22.7μW while the output of 62.3μW is produced by TEG with heat sink. This is a significant improvement which up to 3 times from without heat sink power output value.

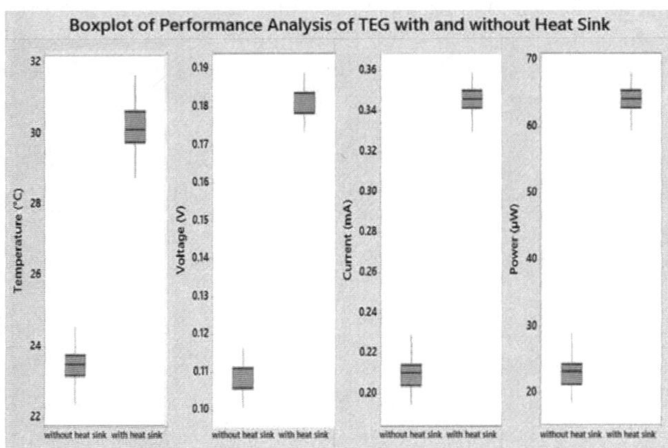

Fig. 5: Box plot of temperature difference of TEG with and without heat sink

Table 2: Box plot parameter of performance analysis of TEG with and without heat sink

Parameter		Lower Quartile, Q_1	Median, Q_2	Upper Quartile, Q_3
ΔT [°C]	Without heat sink	23.15	23.45	23.71
	With heat sink	29.78	30.13	30.63
Voltage [V]	Without heat sink	0.106	0.106	0.111
	With heat sink	0.178	0.178	0.183
Current [mA]	Without heat sink	0.204	0.210	0.214
	With heat sink	0.341	0.346	0.350
Power [μW]	Without heat sink	21.225	22.7	23.875
	With heat sink	60.5	62.3	63.8

3.2 Performance Analysis of TEG with Varied Load Resistance

The next experiment is to define the optimum load resistance range of a single TEG. Using the similar setup, the TEG is equipped with heat sink and the similar heat source. The load resistance is tested ranging from 0.2Ω to 500Ω. The measurement period were taken for 5 minutes which resulted to 300 instantaneous data points for each load resistance.

In this experiment, a median Δ T of 31.5°C has been recorded during all load resistances testing. At 0.2Ω, a median value of instantaneous current of 1.469mA has been recorded. This is the highest value when compared to the other load resistance value. The output instantaneous current showed a reduction as it load resistance increases. At 500Ω, the instantaneous current is recorded at 0.346mA.

The measurements of instantaneous voltage have shown opposite trend. At 0.2Ω, a very low value of 0.003V was recorded. As the load increases, the voltages are also increase respectively. In order to see the optimum instantaneous power, the data need to be calculated using the equation of P=IV. Fig. 6 showed the graph of box plot for current, voltage and power output. Based on this figure, the optimum power for single TEG module can be obtained for the load of 200Ω. This is because at this point, 96.1μW has been produced. As we increased the load resistance, the output power produce has been reduced. Even though at high load resistance, the output voltage were continuously increased, but the output current also decreasing significantly. Thus, the instantaneous power output after 200Ω of load was also reduced.

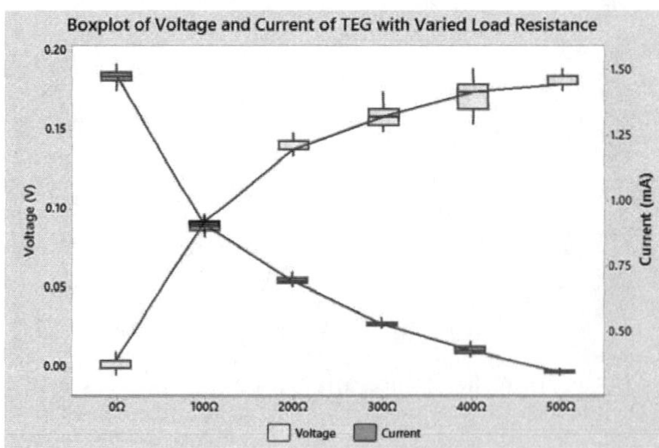

Fig. 6 : Boxplot of voltage and current of TEG with varied load resistance

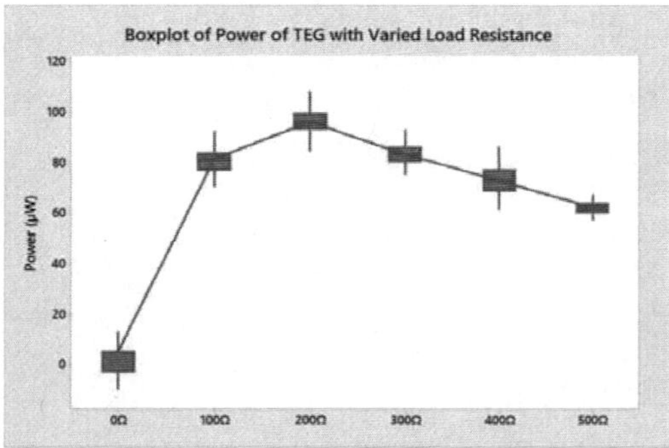

Fig. 7: Boxplot of voltage and current of TEG with varied load resistance

Table 3. Median value of parameter of performance analysis of TEG with varied load resistance

Measurement Parameter	Load Resistance					
	0.2Ω (V)	**100Ω (V)**	**200Ω (V)**	**300Ω (V)**	**400Ω (V)**	**500Ω (V)**
Voltage [V]	0.003	0.091	0.137	0.158	0.173	0.178
Current [mA]	1.469	0.905	0.690	0.527	0.426	0.346
Power [µW]	4.71	81.35	96.1	83.4	72.8	62.3

3.3 Performance Analysis of Multiple TEG with Varied Configuration

In this final part of analysis, the experiment was done to determine the TEG instantaneous performance in various connection configurations. In the previous part, the experiment only used single TEG module. In this part, series and parallel connection are being tested to see the effect towards the TEGs output.

In this experiment, the measurements were done with the median value of ΔT recorded at 33.05°C. The comparisons of performance were compared between single, parallel and series connections. For instantaneous voltage measurement, the parallel connection recorded a median value of 0.122V which is similar with single TEG measurement. The series connection produced 0.250V median value which is double from the single TEG module. These results are in line with the classical electrical law in which the total voltage supply in series is equal to the sum of the voltage across it.

As for output current measurement, the single and parallel connections recorded median values of 0.834mA and 0.824mA respectively. A stark contrast in result can be seen when compared to the median value of series connection which produced 1.671mA which is double the value from other types of connection. The same trend can be determined when the output power is being calculated. The single connection recorded 102.85µW while the parallel connection also recorded about the same value at 101.14µW. The series connection produced significantly higher output which is at 416.26µW.

From these results as shown in Table 4, it is clear that a series connection of TEG module will produce a higher output power generation when compared to parallel connections.

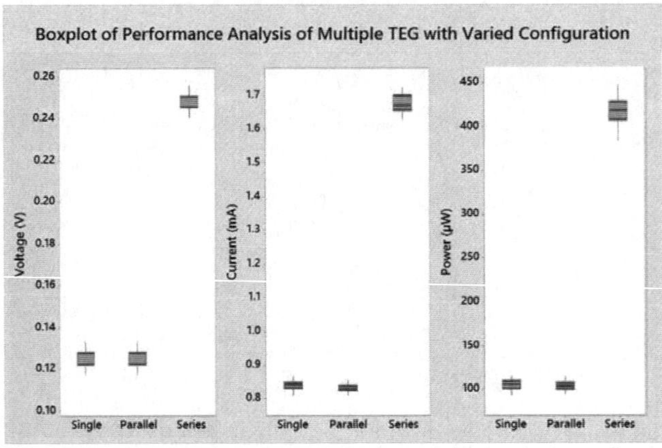

Fig. 8: Boxplot of voltage of multiple TEG with varied configuration

Table 4. Boxplot parameter of performance analysis of multiple TEG with varied configuration

Output Parameter	Type of connections		
	Single	**Parallel**	**Series**
Voltage (V)	0.122	0.122	0.250
Current (mA)	0.834	0.824	1.671
Power (µW)	102.85	101.14	416.26

4. Conclusion

From these experiments, the performance and characteristics of TEG has been confirmed. The TEG will produce a better performance when additional cooling effort is done in the cold part of TEG. In the experiment, passive cooling method has been implemented and a temperature difference of around 7°C has been achieved. This gives a significant improvement in power output produced by TEG. In terms of optimum electrical loading, results shows that a single TEG will produce maximum power output when applied to resistance loading of 200**Ω**. As for circuit configuration, a series connection will give a boost to not only instantaneous voltage but also current. This connection gives the power output of 416.26µW from two TEGs in series where as a single and parallel connections will give a merely 100µW of output power. These results give a better understanding in terms of scalability of TEG in order to build better energy harvester system.

References

1. I. Vidanalage and K. Raahemifar, "Tilt angle optimization for maximum solar power generation of a solar power plant with mirrors," *2016 IEEE Electrical Power and Energy Conference (EPEC)*, Ottawa, ON, 2016, pp. 1-5. doi: 10.1109/EPEC.2016.7771795
2. S. Orrego, K. Shoele, A. Ruas, K. Doran, B. Caggiano, R. Mittal, S. Hoon Kang, Harvesting ambient wind energy with an inverted piezoelectric flag, Applied Energy, Volume 194, 15 May 2017, Pages 212-222, ISSN 0306-2619, doi: 10.1016/j.apenergy.2017.03.016.
3. A. A. Khan and M. R. Khan, "A simple and economical design of micro-hydro power generation system," *2015 Power Generation System and Renewable Energy Technologies (PGSRET)*, Islamabad, 2015, pp. 1-7. doi: 10.1109/PGSRET.2015.7312183
4. V. S. Rana and A. Chauhan, "Energy harvesting from locomotive and coaches by piezoelectric ceramic using MEMS technology," *2015 International Conference on Innovations in Information,*

Embedded and Communication Systems (ICIIECS), Coimbatore, 2015, pp. 1-6. doi: 10.1109/ICIIECS.2015.7193179

5. C. Wei, X. Jing, A comprehensive review on vibration energy harvesting: Modelling and realization, Renewable and Sustainable Energy Reviews, Volume 74, July 2017, Pages 1-18, ISSN 1364-0321, https://doi.org/10.1016/j.rser.2017.01.073.
6. L. Wang *et al.*, "Radio frequency energy harvesting technology," *2016 International SoC Design Conference (ISOCC)*, Jeju, 2016, pp. 219-220. doi: 10.1109/ISOCC.2016.7799866
7. M. Nesarajah and G. Frey, "Thermoelectric power generation: Peltier element versus thermoelectric generator," *IECON 2016 - 42nd Annual Conference of the IEEE Industrial Electronics Society*, Florence, 2016, pp. 4252-4257. doi: 10.1109/IECON.2016.7793029
8. V. Raghunathan, A. Kansal, J. Hsu, J. Friedman and M. Srivastava, "Design considerations for solar energy harvesting wireless embedded systems," *IPSN 2005. Fourth International Symposium on Information Processing in Sensor Networks, 2005.*, 2005, pp. 457-462. doi: 10.1109/IPSN.2005.1440973
9. M. A. Johar, P. Torruella and A. König, "DC MEMS switches with self-x features: Design, simulation and implementation strategies," *2012 10th IEEE International Conference on Semiconductor Electronics (ICSE)*, Kuala Lumpur, 2012, pp. 229-233. doi: 10.1109/SMElec.2012.6417129

IOP Conf. Series: Journal of Physics: Conf. Series **914** (2017) 012025 doi:10.1088/1742-6596/914/1/012025

Optimization of Maghemite (γ-Fe_2O_3) Nano-Powder Mixed micro-EDM of CoCrMo with Multiple Responses Using Gray Relational Analysis (GRA)

Nagwa Mejid Elsiti[1]*, M.Y. Noordin[1], Ani Idris[2], Faraj Saed Majeed[1]

[1]Dept. of Materials, Manufacturing and Industrial Engineering, Faculty of Mechanical Engineering Universiti Teknologi Malaysia, 81310, UTM, Skudai, Malaysia
[2]Department of Bioprocess Engineering, Faculty of Chemical Engineering, c/o Institute of Bioproduct Development, Universiti Teknologi Malaysia, UTM, Skudai 81310, Malaysia

Email: nagwa.mejid@yahoo.com

Abstract. This paper presents an optimization of process parameters of Micro-Electrical Discharge Machining (EDM) process with (γ-Fe_2O_3) nano-powder mixed dielectric using multi- response optimization Grey Relational Analysis (GRA) method instead of single response optimization. These parameters were optimized based on 2-Level factorial design combined with Grey Relational Analysis. The machining parameters such as peak current, gap voltage, and pulse on time were chosen for experimentation. The performance characteristics chosen for this study are material removal rate (MRR), tool wear rate (TWR), Taper and Overcut. Experiments were conducted using electrolyte copper as the tool and CoCrMo as the workpiece. Experimental results have been improved through this approach.

1. Introduction

Electrical Discharge Machining (EDM) is known as a thermal erosion process that electrically produces a spark vaporizing materials that are electrically conductive [1]. This process works through the application of a series of disconnected discharges between a tool, which is generally cathode, and an electrically conductive workpiece, which is normally anode, disjointed using a dielectric liquid medium. For the generation of the spark discharge, an appropriate gap, recognized as spark gap, is upheld between the workpiece and the tool [2]. A spark is created between the closest points between the electrode and the workpiece; then the material is removed by a spark from both the workpiece and the electrode [3]. Powder mixed EDM (PMEDM) is a promising technique which leads to the development of process capabilities and generates near-mirror-like surface finish with reduced surface cracks and homogenization of the white layer [4]. Performance of the PMEDM process depends upon characteristics like powder type, concentration, particle size, electrode area, workpiece constituents and properties [5]. The process parameters play a significant role for material removal. Determination of optimal machining parameters is continuous engineering task whose goals are to reduce the production costs and to achieve the desired product quality [6]. Many researchers studied the optimizations of powder mixed EDM/micro-EDM. Singh and Yeh (2012) found that multiple performance characteristics evaluated for aluminum matrix composite during APM-EDM using grey relational analysis (GRA) result in improvement of material removal rate (MRR), tool wear rate (TWR), and surface roughness (SR) [7]. Response surface methodology has been used with desirability approach for modeling and optimization of the process parameters during EDM of CK-45 die steel using Al2O3 powder suspension into dielectric fluid. The optimal condition of process parameters was found to maximize the MRR [8]. A simplified model based on Taguchi and Utility approach has been used for multi characteristic optimization of the process parameters and to obtain the optimal setting during machining of H-11die steel using Si powder and copper tool [9]. Technique for order of preference by similarity to ideal solution (TOPSIS) was used to optimize multiple responses while machining Al-CuTiB2 to obtain the optimum parameters for machining [10]. Multi-objective optimization of PMEDM has been conducted using Taguchi, GRA, and principal component analysis to control the process parameters [11]. A new approach for the optimization of micro-wire EDM process with multiple performance characteristics based on the statistical-based ANOVA and GRA has been implemented which shows improved machining performance [12]. Multi-objective optimization using Grey Relational Analysis (GRA) was proposed to determine the optimal combination of process parameters [7, 13]. The study by *Tripathy* and *Tripathy* (2016), examined the hybrid multi-response optimization by employing TOPSIS and GRA during the EDM of H-11 die steel using chromium powder-mixed dielectric [14].

Past work reveals that PMEDM involves a huge number of input process variables which influence the quality of the machined component and thus investigating the relative importance of the process variables on the output performance characteristics becomes vital. Though several studies have reported the use of different powders mixed to the dielectric fluid, no study has been done to explore the combined effect of process variables on performance characteristics of the PMμEDM process for CoCrMo. Multi-attribute decision-making techniques like GRA have not yet been implemented to find the optimal setting during PMμEDM of CoCrMo. The present work is a stride in this direction. An effort has been made to find an optimal set of process variables using multi-objective optimization using GRA to get maximum MRR and minimum TWR, Overcut ant Taper using Fe_2O_3 nano-powder mixed to the dielectric fluid.

2.Experimental setup, procedure and equipment

For conducting experiments, the work material CoCrMo is cut into the sample pieces with the dimensions of 50 mm X 90 mm X 1 mm. The electrode copper of diameter 300μm and length 6000μm was used as tool material for machining CoCrMo. The commercially available "EDM 23" oil was selected as the dielectric fluid. All the experiments were conducted on AG40L Sodick die sinking EDM machine. Modified working fluid circulating system has been designed for experimentation. In modified system, a separate tank mounted with pump is installed for better circulation of (γ-Fe_2O_3) nano-powder and surfactant-mixed dielectric fluid. A motorized stirring system is incorporated to avoid settling of powder particles. The chosen process parameters and experimental conditions are presented in the Table 1. The design of experiment (DOE) chosen for this study was a 2-Level factorial design and is presented in Table 2. A digital weighing balance having capacity up to 200 grams with a resolution of 0.1mg was used for weighing the workpieces and electrodes before and after machining. Then the material removal rate (MRR), tool wear rate (TWR), Overcut and Taper are calculated using equations (1-4).

Table 1. Process parameters and experimental EDM conditions

Tool electrode	Copper (300µm diameter and 6mm length)
Workpiece material	Co-Cr-Mo
Dielectric fluid	EDM 23
Nano-powder	γ-Fe_2O_3, size less than 10nm (4 g/l)
Peak current	1.5, 2.25, 3A
Voltage (V)	60, 90, 120V
Pulse on-time (µs)	10, 105, 200µs
Duty factor	95%

$$MRR = \frac{(W_1 - W_2)}{t \times \rho} \times 1000\ (mm^3/min) \quad (1)$$

$$TWR = \frac{(T_1 - T_2)}{t \times \rho} \times 1000\ (mm^3/min) \quad (2)$$

$$OC = \frac{D_w}{2} (mm) \quad (3)$$

$$Taper\ Angle\ (\theta) = tan^{-1} \times \frac{(D_{top} - D_{bottom})}{2h} \quad (4)$$

Where The terms W_1 and W_2 represent the initial and final weights of the workpiece, respectively. The terms T_1 and T_2 denote the weights of the tool before and machining, respectively. The terms ρ denote the density (g/mm^3) of the workpiece whereas t denotes the machining time (min). The terms $D_W = D_1 - D$, where D_1 and D represent the entry hole and electrode (tool) diameters, respectively. The terms θ represent the taper angle; h is machining depth; D_{top} and D_{bottom} are the entry and exit diameters, respectively.

Table 2. Experimental plan and results

Exp no	**Levels of parameters**			**MRR**	**TWR**	**Taper**	**Overcut**
	Current (A)	**Voltage (V)**	**Pulse on (µs)**				
1	1.50	60	10	0.00240	0.00032	0.77	54.02
2	3.00	60	10	0.01730	0.00500	4.63	56.67
3	1.50	120	10	0.01162	0.02915	1.10	61.02
4	3.00	120	10	0.05540	0.03376	5.04	66.44
5	1.50	60	200	0.00063	0.00020	1.77	83.67
6	3.00	60	200	0.01621	0.00248	3.73	103.98
7	1.50	120	200	0.00295	0.00023	0.54	62.00
8	3.00	120	200	0.02946	0.01781	8.86	100.57
9	2.25	90	105	0.00765	0.00997	1.42	65.86
10	2.25	90	105	0.00630	0.00646	0.85	69.68
11	2.25	90	105	0.01804	0.00672	1.83	69.77

3. Analysis method

3.1. Grey rational analysis

Initiator of the Grey system theory (1982) was Deng [15] and it is widely used for measuring the degree of relationship between sequences by Grey relational grade[16]. In GRA, the experimental values of the measured quality characteristics are normalized in a range from zero to one [17]. This is known as grey relational generation. Then the grey relational coefficient (GRC) is calculated. The overall Grey relational grade is then computed by averaging the Grey relational coefficient corresponding to each performance characteristic. As a result, optimal combination of process parameters is evaluated considering the highest Grey relational grade by using the Taguchi method. The overall performance characteristic depends on the computation of the grey relational grade (GRG). Thus, a multiple response process optimization is transformed into a single objective problem [6]. The highest GRG will be evaluated as the optimal parametric combination. Typically the normalization process involves two concepts into the *Taguchi's* technique (nominal the smaller is the better and higher is the better) [18]. The "higher is the better" concept is used for normalizing the MRR by using Eq5, while the lower is the better concept is used for normalizing the variables TWR, Taper, and Overcut by using in Eq 6.

$$x_i^*(k) = \frac{x_i(k) - min\, x_i(k)}{max\, x_i(k) - min\, x_i(k)} \tag{5}$$

$$x_i^*(k) = \frac{max\, x_i(k) - x_i(k)}{max\, x_i(k) - min\, x_i(k)} \tag{6}$$

The terms x_i (k) is the ith series in the original value of k; order $x_i^*(k)$ is the ith series and k order after normalization, $minx_i(k)$ is the minimum value in the ith series, $max\, x_i(k)$ is the maximum value in the ith series. Table 3 shows the normalized results of experimental results obtained for performances.

Table 3. Normalization (data pre-processing) of the experimental results

Exp no	MRR	TWR	Taper	Overcut
1	0.03232	0.99637	0.97236	1.00000
2	0.30436	0.85699	0.50841	0.94696
3	0.20066	0.13744	0.93269	0.85989
4	1.00000	0.00000	0.45914	0.75140
5	0.00000	1.00000	0.85216	0.40653
6	0.28446	0.93207	0.61659	0.00000
7	0.04236	0.99908	1.00000	0.84027
8	0.52638	0.47521	0.00000	0.06826
9	0.12817	0.70886	0.89423	0.76301
10	0.10352	0.81343	0.96274	0.68655
11	0.31788	0.80571	0.84495	0.68475

3.2. Grey rational coefficients (GRC)

Normalization creates a new matrix of difference vectors. From this matrix, a GRC is calculated, expressed as:

$$\vartheta_i(k) = \frac{(\Delta min + \xi \Delta max)}{(\Delta_{0i}(k) + \xi \Delta max)} \tag{7}$$

The term $\vartheta i(k)$ denotes GRC for the k output parameter and $\Delta_{0i}(k) = | x_0*(k) - xi*(k) |$ is the deviation sequence. Lastly, $\Delta min = min| x_0*(k) - x_i*(k)|$ whereas $\Delta max = max |x_0*(k) - x_i*(k)|$ and ζ= weighting coefficient that is 0.5.

3.3. Grey Relational Grades (GRG)

Finally, the GRG is obtained by averaging the GRC corresponding to each performance measures. Thus by applying(Eq 8), all GRGs can be computed

$$(\gamma)i = \frac{1}{n}\sum_{1}^{n} \vartheta_i(k) \tag{8}$$

The term γ denotes the Grey Relational Grade (GRG) while ,*n,* is the number of output parameters. Table 4 presents the grey rational coefficients and grades for each response.

Table 4. GRG for each experimental run

Expt. no	**Grey Relational Coefficient**				**GRG (γ)**	**Rank**
	MRR	**TWR**	**Taper**	**Overcut**		
1	0.34067	0.99278	0.94761	1.00000	0.82027	1
2	0.4181	0.7775	0.50424	0.90409	0.65103	4
3	0.3848	0.3669	0.88136	0.78111	0.60356	9
4	1.0000	0.3333	0.48037	0.66791	0.62040	8
5	0.3333	1.0000	0.77180	0.45726	0.64060	5
6	0.4113	0.8803	0.56599	0.33333	0.54776	10
7	0.3430	0.9981	1.00000	0.75789	0.77477	2
8	0.5135	0.4879	0.33333	0.34922	0.42100	11
9	0.36448	0.63200	0.82540	0.67844	0.62508	7
10	0.3580	0.7282	0.93065	0.61467	0.65790	3
11	0.4229	0.7201	0.76330	0.61331	0.62994	6

4. Results and discussion

Table 5 shows the mean of the GRG for each level of the machining parameters chosen for this study. The orthogonal experiment design separates out the effect of each machining parameter on the GRG at different levels. For example, the mean of GRG for the factor A at level 1 can be calculated by taking the average of the GRG for the experiment no. 1, 3, 5 and 7, respectively (shown in Table 4). Similarly, mean of the GRG for each level of other machining parameters can also be computed. In addition, the total mean of the GRG for the 11 experiments is also calculated and listed in Table 5. The total mean value of the GRG is 0.63566. As stated by Fung, "the grey relational grade represents the level of co-relation between the reference sequence and the comparability sequence" [19]. The greater value of the GRG means that the comparability sequence has a stronger correlation to the reference sequence. Therefore, the optimal level of the machining parameters is the level with the greatest GRG value. The level value marked asterisks (*) in response table, indicates that they results in a better PMμEDM performance. Based on the GRG given in Table 5, the optimal machining performance for

MRR, TWR, Overcut and Taper was obtained for peack current (level 1), voltage (level 1) and pulse current (level 1), pulse on time (level 1). Accordingly, the level constitution of optimal machining parameters are A1, B1 and C1 in the case of multiple performance characteristics optimization for PMμEDM, since higher GRG values yield better quality. The difference between the maximum and the minimum value of the GRG for PMμEDM machining parameters is also calculated and tabulated in Table 5. The tabulated results are follows: 0.1497 for current, 0.0599 for voltage, and 0.0777 for pulse on time, respectively. The most significant factor affecting performance characteristics is determined by comparing these values. This comparison gives the level of significance of the process parameters over the multiple performance characteristics. The most effective controllable factor was the maximum of these values. As per Table 5, the maximum value among the controllable factors is for peak current viz. 0.14975. This higher value indicates that the peak curent has the strongest effect on the multiple performance characteristics among the other machining parameters. The order of importance of the machining parameters to the multiple performance characteristics in the PMμEDM process, in sequence can be ranked as: factor A (peak current), C (pulse on time), B (voltage). Figure 1 shows the main effects plot (response graph) based on GRG where the dash line indicates the value of the total mean of the GRG (viz. 0.63563). Accordingly A1, B1 and C1 are the optimal level of PMμEDM parameters in the case of multiple performance characteristics.

Table 5: Response table for GRG

Machining parameters	Grey Relational Grade			Main Effect Max-Min	Rank
	Level 1	Level 2	Level 3		
Current	0.7098*	0.63764	0.56005	0.14975	1
Voltage	0.66491*	0.63764	0.60493	0.05998	3
Pulse on	0.67381*	0.63764	0.59603	0.07778	2
Total mean value of GRG = 0.63566 * Levels for optimum GRG					

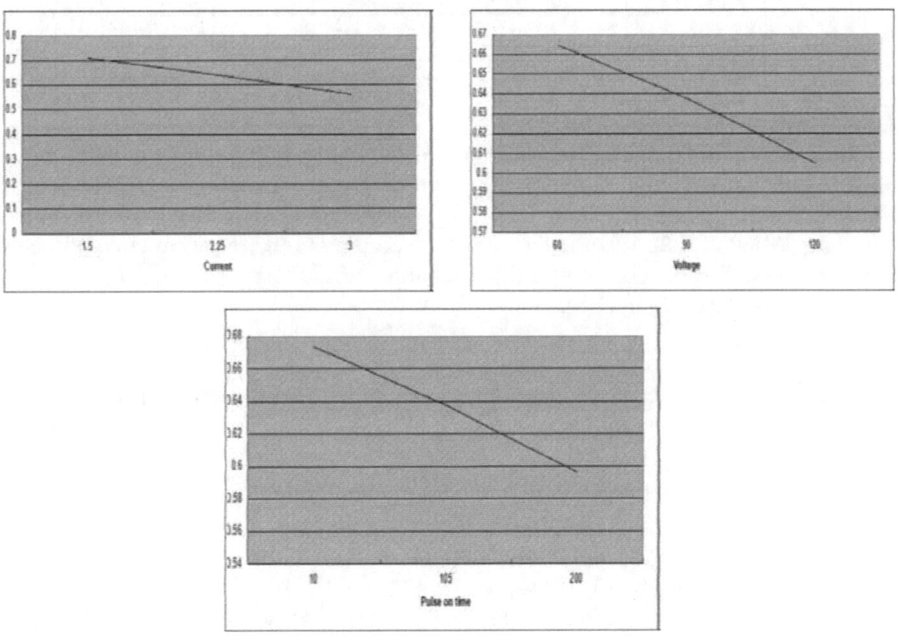

Fig.1. Effects plot for GRG

5. Confirmation tests

Once the optimal level of the process parameters is identified, the final step is to predict and validate the improvement of the performance measures using the optimal level. A good indication of the satisfactory experimental runs is observed by subsequently comparing the results of the confirmation tests with the predicted value. The purpose of the confirmation experiment is to verify the conclusions drawn during the analysis phase. The estimated $\gamma^{\wedge}$ using the optimal levels of the process parameters can be computed by using the following formula:

$$\gamma^{\wedge} = \gamma m + \sum_{1}^{q} (\gamma^{-} i - \gamma m) \tag{9}$$

Where γm is the total mean of the GRG, $\gamma^{\wedge}$ is the mean of the GRG at the optimal level, and q is the number of the process parameters that significantly affects the performance characteristics. The confirmation tests were carried out at the optimum levels predicted by the analysis of the results. From Eq 9, the estimated GRGs using the optimal PMμEDM parameters are computed. Table 6 shows the results of the confirmation tests using the optimal levels of PMμEDM parameters. As noted from Table 6, the MRR is decreased from 0.01162 to 0.0024 $mm^{\wedge 3}$/min, when the tool wear overcut and taper are minimized from 0.02915 to 0.00032 $mm^{\wedge 3}$/min, 61.02 m to 54.02 m and 0.872 to 0.43, respectively. An improvement of 0.20776 is noted in GRG, after validation.

Table 6. Results of performance measures for initial and optimal process parameters

Responses	Initial Conditions	Optimal Factors	
		Prediction	Experiment
Level	A3C3B3	-	A1B1C1
MRR	0.01162	-	0.00240
TWR	0.02915	-	0.00032
T	0.87200	-	0.43000
OC	61.02000	-	54.02000
GRG	0.56193	0.78318	0.76969
Improvements in GRG = 0.20776			

6. Conclusions

The GRA based on the 2-Level Factorial Design was proposed as an approach to investigate the optimization of powder mixed micro-EDM processes parameters for Co-Cr-Mo. The optimal machining parameters were determined by Grey Relational Grade (GRG) for the multi-performance characteristics; MRR, TWR, Taper and Overcut. According to the response table of the average GRG, the optimal process parameters were; 1.5A discharge current, 60V open voltage, 10μs pulse on time with improvement (0.20776) in GRA

Acknowledgement

Sincere regards are to the center for graduate studies of Universiti Teknologi Malaysia for its support during the conduct of this research.

References

1. N. M. Elsiti, M.Y.N., Experimental Investigations into the Effect of Process Parameters and Nano-Powder (Fe2O3) on Material Removal Rate during Micro-EDM of Co-Cr-Mo. Key Engineering Materials, 2017. **740**: p. 125-132.
2. Singh, P.N., K. Raghukandan, and B.C. Pai, Optimization by Grey relational analysis of EDM parameters on machining Al–10%SiCP composites. Journal of Materials Processing Technology, 2004. **155-156**: p. 1658-1661.

3. Jahan, M.P., Y.S. Wong, and M. Rahman, A study on the quality micro-hole machining of tungsten carbide by micro-EDM process using transistor and RC-type pulse generator. Journal of Materials Processing Technology, 2009. **209**(4): p. 1706-1716.
4. Peças, P.H., E., Influence of silicon powder-mixed dielectric on conventional electrical discharge machining. International Journal of Machine Tools and Manufacture, 2003. **43**(14): p. 1465-1471.
5. Peças, P. and E. Henriques, Electrical discharge machining using simple and powder-mixed dielectric: The effect of the electrode area in the surface roughness and topography. Journal of Materials Processing Technology, 2008. **200**(1-3): p. 250-258.
6. Franko Puh, Z.J., Mladen Perinic, Miran Brezocnik, Stipo Buljan, Optimization of machining parameters for turning operation with multiple quality characteristics using Grey relational analysis. Tehnicki vjesnik - Technical Gazette, 2016. **23**(2).
7. Singh, S. and M.F. Yeh, Optimization of Abrasive Powder Mixed EDM of Aluminum Matrix Composites with Multiple Responses Using Gray Relational Analysis. Journal of Materials Engineering and Performance, 2011. **21**(4): p. 481-491.
8. Assarzadeh, S. and M. Ghoreishi, A dual response surface-desirability approach to process modeling and optimization of Al2O3 powder-mixed electrical discharge machining (PMEDM) parameters. The International Journal of Advanced Manufacturing Technology, 2012. **64**(9-12): p. 1459-1477.
9. H K Kansala, S.S., Pradeep Kumarc, Performance parameters optimization (multi-characteristics) of powder mixed electric discharge machining (PMEDM) through Taguchi's method and utility concept. Indian Journal of Engineering & Materials Sciences, 2006. **Vol. 13**: p. 209-216.
10. Senthil, P.V., S.; Singh, A.K, Parametric optimisation of EDM on Al-Cu/TiB2 insitu metal matrix composites using TOPSIS method. International Journal of Machining and Machinability of Materials, 2014. **16**: p. 80-94.
11. Talla, G., et al., Modeling and multi-objective optimization of powder mixed electric discharge machining process of aluminum/alumina metal matrix composite. Engineering Science and Technology, an International Journal, 2015. **18**(3): p. 369-373.
12. Somashekhar, K.P., J. Mathew, and N. Ramachandran, Multi-objective optimization of micro wire electric discharge machining parameters using grey relational analysis with Taguchi method. Proceedings of the Institution of Mechanical Engineers, Part C: Journal of Mechanical Engineering Science, 2011. **225**(7): p. 1742-1753.
13. Talla, G., S. Gangopadhyay, and C.K. Biswas, Multi Response Optimization of Powder Mixed Electric Discharge Machining of Aluminum/Alumina Metal Matrix Composite Using Grey Relation Analysis. Procedia Materials Science, 2014. **5**: p. 1633-1639.
14. Tripathy, S. and D.K. Tripathy, Multi-attribute optimization of machining process parameters in powder mixed electro-discharge machining using TOPSIS and grey relational analysis. Engineering Science and Technology, an International Journal, 2016. **19**(1): p. 62-70.
15. Deng, J.L., Introduction to Grey System Theory. J. Grey Syst, 1989. **1**: p. 1-24.
16. Senthilkumar, N., J. Sudha, and V. Muthukumar, A grey-fuzzy approach for optimizing machining parameters and the approach angle in turning AISI 1045 steel. Advances in Production Engineering & Management, 2015. **10**(4): p. 195-208.
17. Sanjay Kumar Majhi, M.K.P., Hargovind Soni, OPTIMIZATION OF EDM PARAMETERS USING INTEGRATED APPROACH OF RSM, GRA AND ENTROPY METHOD. International Journal of Applied Research in Mechanical Engineering (IJARME), 2013. **3**: p. 1-8.
18. Kumar. S, V. and P. Kumar. M, Optimization of cryogenic cooled EDM process parameters using grey relational analysis. Journal of Mechanical Science and Technology, 2014. **28**(9): p. 3777-3784.

19. Fung, C.-P., Manufacturing process optimization for wear property of fiber-reinforced polybutylene terephthalate composites with grey relational analysis. Wear, 2003. **254**(3-4): p. 298-306.

IOP Conf. Series: Journal of Physics: Conf. Series **914** (2017) 012026 doi:10.1088/1742-6596/914/1/012026

The Effect of Temperature on the Chromizing Process for Ferritic-Martensitic Steel

F F Alia[1], T Kurniawan[1], M H B Ani[2] and A B D Nandiyanto[3]

[1]Structural and Material Degradation Group Faculty of Mechanical Engineering, 26600 Pekan, Pahang Malaysia.
[2]Faculty of Engineering, International Islamic University Malaysia, PO Box 10, 50728 Kuala Lumpur, Malaysia
[3]Department of Chemical, Universitas Pendidikan Indonesia, Jl. Dr. Setiabudhi No. 229, Bandung 40154, Indonesia

Email: farahalia23@gmail.com

Abstract. The formation of protective Cr_2O_3 layer was usually retarded in the high temperature steam oxidation of boiler tube materials. This condition makes the oxidation rate higher than that in dry condition. Therefore in this work, chromizing process is introduced to diffuse chromium on the surface of boiler steel so that it can act as a reservoir for the formation of Cr_2O_3 layer. The chromizing process was conducted on T91 steel by exposing it into alumina crucible. The crucible was exposed at different temperature (600°C-1050°C) under argon environment in the crucible that contains the chromizing mixture powder of masteralloy Cr, activator NH_4Cl and filler Al_2O_3. It was found that Cr diffusion was happened at higher temperature and it formed Cr carbide on the surface. It also clarified that this chromizing process can prevent the retardation of Cr_2O_3 layer.

1. Introduction

The global electricity demand consummation in 2013 to 2040 predicted to be increased by 70% due to the changes of urbanization and economic activity [1, 2]. The increase of electricity leads to the needs of increasing the efficiency at power plant. As stated in Carnot cycle, the easiest way to increase the efficiency is by raising the operating temperature. Currently many developed countries have stated that the operating temperature of future power is targeted to increase above 700°C [3]. Besides, it can increase the efficiency; this improvement will also reduce the greenhouse gasses emission. In the other hand, higher operating temperature may affect the oxidation behaviour of the power plant material, especially the superheater and reheater tubes of boiler, where it exposed to high temperature and high pressure steam [4, 5].

As stated in previous research, the oxidation occurred in steam condition shows worst characteristics morphology compare to the oxidation that occur in dry environment [6-8]. The results show the amount of corrosion product is higher compare to the oxidation in dry condition [9]. The formation of higher corrosion product in steel that exposed to steam condition is due to retardation of protective layer Cr_2O_3 where it cause the formation of non-protective oxide will increase [10].

In order to improve the oxidation resistance, the chromium forming coating or chromizing was conducted to give an idea of the protection for the steel. This idea offers the outer steel is deposited with the chromium and also the chromium diffused into the steel to act as the reservoir to the formation of

Cr_2O_3 [11, 12]. The steel with this method of chemical vapour deposition (CVD) will be embedded in a crucible full of powder mixture, which is composed of masteralloy, inert filler and activator. The crucible then heated to desired coating temperature with the argon gas loaded into the furnace. The masteralloy will react with the activator forming a gaseous transporting element. This will diffuse through the pack to the steel surface where it allows the masteralloy to be deposited to be penetrated into the steel by solid state diffusion [13, 14]. This also depends on the temperature of the surrounding.

The aim of this work is to diffuse the chromium on T91 steel by chromizing process at different temperature. Morphology of the coating will be investigated by FESEM/ EDX and XRD. It is expected that the diffused chromium become the reservoir for the formation of Cr_2O_3 layer during high temperature steam oxidation.

2. Experimental Procedure

2.1. Preparation for chromizing process

T91 ferritic-martensitic steel was used as the sample. Powder mixture of NH_4Cl, Al_2O_3 and Cr, was used as activator, inert filler and master alloy, respectively. First, the sample was cut into the required dimension 10 x 10 x 4mm^3. The material then were prepared and polished the surface finish up to 2000 grit finish using SiC abrasive papers. It then washed with acetone in an ultrasonic bath to remove its surface from rust and grease. After polished and cleaning process, substrate was put inside alumina crucible together with the powder mixture. Chromizing process was conducted inside horizontal tube furnace with different temperature 600°C, 800°C and 1050°C for 2 hours under argon gas environment. FESEM and EDX was used to characterized the cross sectional image of the sample.

The diffusion coating was assembled in powder pack mixture. For chromizing process, there are four components that will chromizing process complete, they are substrate to be coated, master alloy, activators, and filler. In this research, the master alloy used is chromium (Cr) powder, the activators is NH_4Cl powder, and Al_2O_3 powder as filler. Powder pack mixture was homogenized by using a porcelain mortar for one hour. Samples and powder then embedded in the alumina crucible with lid glued by alumina cement. The crucible then was put into a horizontal tube furnace and loaded with argon gas. The exposure time for pack cementation is 2 hour at varies temperature 600°C-1050°C with constant flow rate 150 ml/min. The cross sectional image of the chromized steel was investigated by using FESEM and EDX.

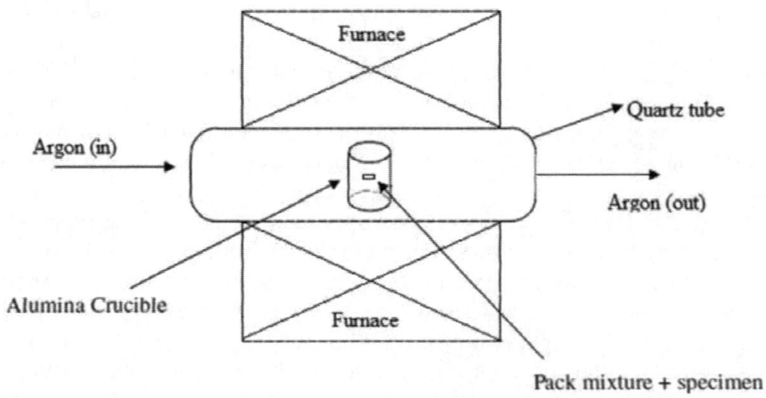

Figure 1. Schematic illustration for chromizing process setup.

2.2. Steam oxidation testing

The schematic illustration for experimental setup which was used in this experiment is illustrated in Figure 2. The temperature for steam oxidation tests was 700°C with constant flow rate at 150 ml/min of high purity argon which is used as a carrier gas. The distilled water was set at a constant value at the mantle heater. The argon gas was carried in the glass frit to make sure a good contact between argon and water. For this experiment, the exposure time for oxidation is exposed to 90h. The samples were weighed before and after oxidation to analyze their weight change.

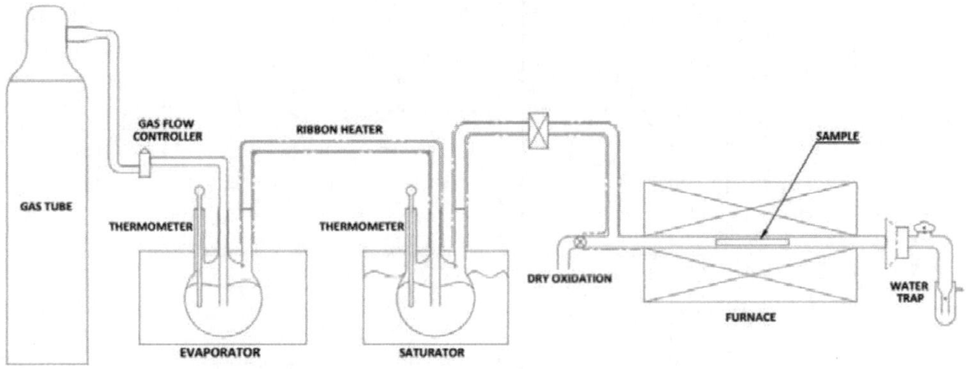

Figure 2. Schematic illustration for steam oxidation experiment setup.

3. Results and Discussions

The formation of Fe-Cr was detected in XRD which also reveals the presence of Cr and Cr_7C_3. Eventhough the morphology in low temperature 600°C do not exhibit chromium, the XRD spectrum shows the differences. A very little amount of Cr peak with only one significant peak Cr reveals on the spectrum image. This can be concluded that the master alloy do react with the activator in the temperature. However, the temperature is not sufficient enough to make the deposition of master alloy and solid state diffusion to occur. In temperature 800°C, the chromium carbide already formed with a few significant peaks of Chromium. Figure 6 shows the XRD spectrum image for all condition; bare T91 sample, low and high temperature.

From the Figure, XRD analysis reveals the chromium carbide start to form at the temperature 800°C. Aside of Cr formation detected on the XRD, Cr_7C_3 also detected on the sample after exposed to lower temperature 800°C. At lower temperature 600°C, there are no phase changes in the steel and remain the same as the bare sample of T91. However, at higher temperature 1050°C the chromium activity is high on the surface where the carbide (Cr_7C_3, Cr_3C_2 and $Cr_{23}C_6$) formation can form at this temperature. The presence of chromium in solid solution cause the large ferrite formed and reduced the amount of carbon in the steel.

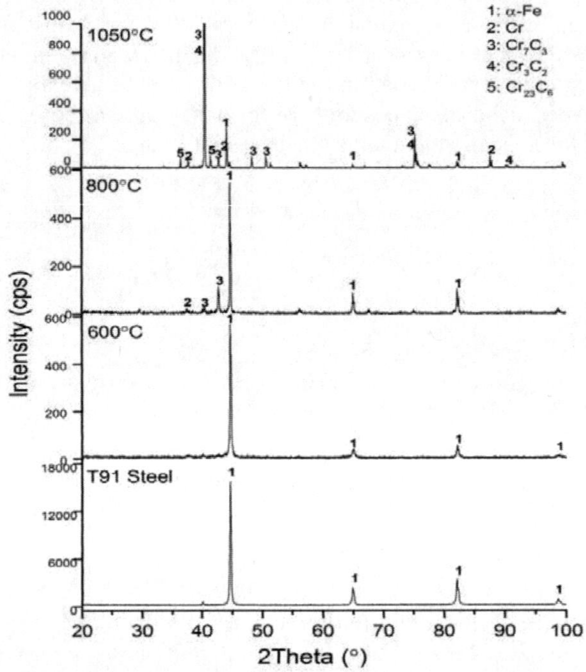

Figure 3. XRD spectrum image for T91 steel and chromizing steel in low and high temperature.

The chromizing process was carried out with varies temperature in order to investigate the effect of deposition behavior with temperature. First, the samples were embedded in furnace with temperature 600°C-1050°C. When exposed to this temperature for 2h, morphology from BSE-FESEM image reveals there are no changes of Cr concentration as shown in Figure 4(a). In the image, the steel become cracks after the exposure shows the temperature did give an effect to the changes of the steel. However, due to the low temperature where the activator cannot react with the masteralloy because of the driving force to diffuse the chromium into the steel is not high enough. Therefore the diffusion does not occur in this temperature..

The temperature of the furnace then was increased to 800°C in order to get the deposition and diffusion of chromium. Figure 4(b) show the steel is different as in Figure 4(a). This is because in this image, we can see here are darker region where the darker region is chromium richer part. Besides, the cracks also do not visible in this condition. In this temperature, the chromium was successfully deposited on top of the steel. However, it is too little and not sufficient enough to be used in industry where the steam environment will be in a very long term.

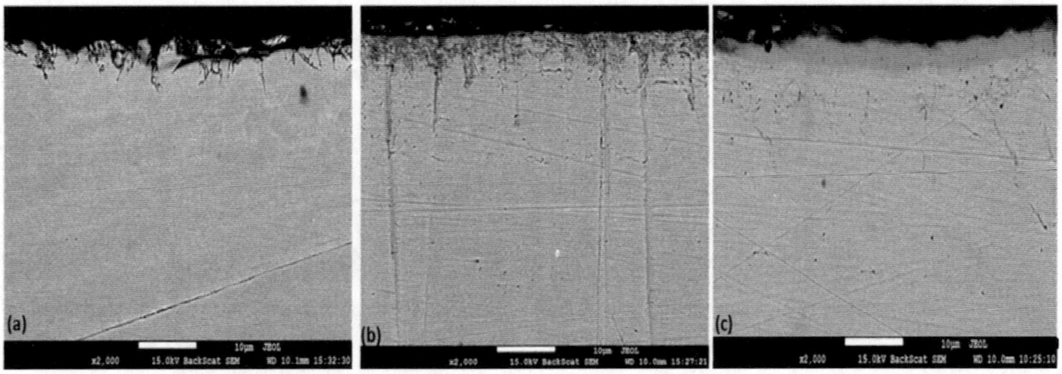

Figure 4. BSE-FESEM image for chromizing temperature at (a) 600°C; (b) 800°C and 1050°C.

The chromizing then was rested at high temperature 1050°C for 2h. By using this temperature, the chromium diffusion leads to the deposition of Cr rich layer on top of the surface where can be seen in Figure 4(c). At this temperature, the formation of Cr rich layer is gives more homogenous and continous layer with average thickness 8.509±1.590μm. In this temperature, we can see the presence of pores on this cross section image between 8 and 15 μm under the layer. The pore were observed all along the steels at around the depth. The pore formation is due to the Kirkendall effect as observed previously in chromising effect [15].

This condition has been proved by the EDX analysis with Cr linescan profile and mapping as shown in Figure 5(a), (b) and (c). The linescan profile which measured the concentration of Cr and comparison with Fe was drawn in the EDX software starting from the surface to the substrate of T91 steel. For sample exposed in low temperature 600°C, the linescan and element mapping profile also show there are no chromium detected either deposited or diffusion in this 600°C temperature.

Microstructure in Figure 5 (b) is too small to measure their thickness in the BSE FESEM image. So, the spot and linescan analysis in EDX has been used to identify the deposition of chromium on top of the steel. The spot is pointed in EDX line scan profile results the surface thickness of the chromium deposition is around 3μm. However, there were no diffusion of chromium occur in this temperature. This can be seen in Figure 5(b). The mapping of element Cr also shows there is chromium deposited on top of the surface.

Figure 5(c) shows the EDX analysis for linescan and mapping profile of Fe and Cr. In the line scan profile, it shows the diffusion of chromium did occur in this condition. By using the spot analysis, the wt% of chromium been investigated along the line to know the depth of diffusion in the steel. From the spot analysis, the diffusion depth in the image was measured by using the image software where it spotted the chromium diffuse until 41.706μm. The diffusion of chromium is very important in order to make sure during the oxidation occurred, the chromium will act as a reservoir to the formation of Cr_2O_3 to prevent the non-protective iron oxide formed on the steel. The elemental mapping from the EDX analysis also shows the formation of chromium is homogenous and continuous along the surface of the sample.

IOP Conf. Series: Journal of Physics: Conf. Series **914** (2017) 012026 doi:10.1088/1742-6596/914/1/012026

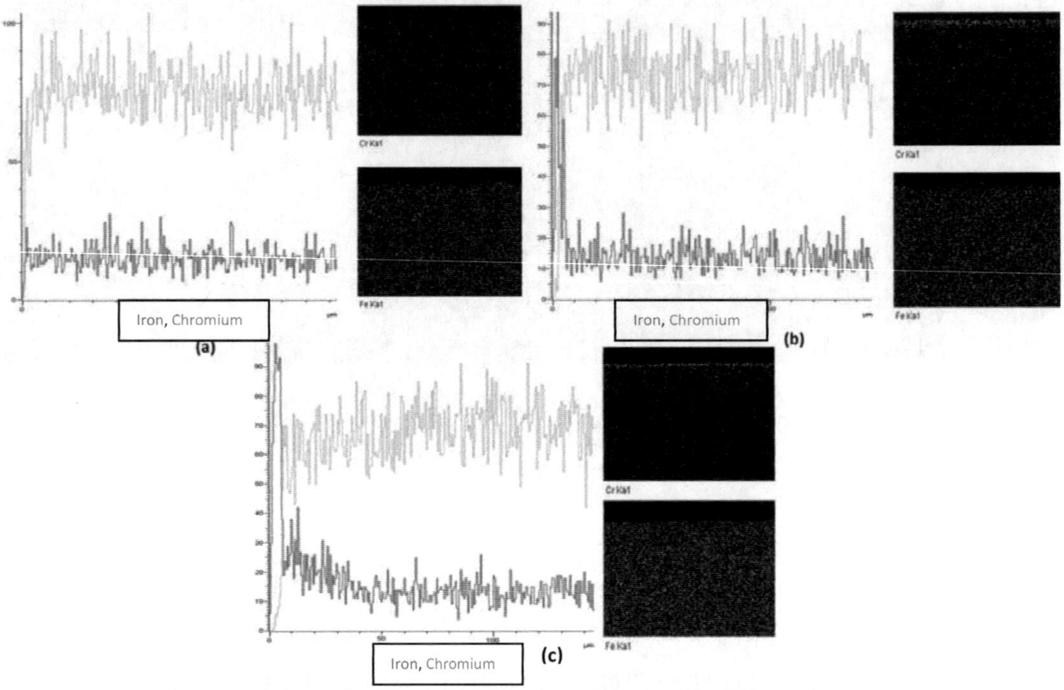

Figure 5. EDX analysis for chromizing steel; (a) 600°C, (b) 800°C and 1050°C

The satisfying characteristics exhibit from the high temperature chromizing lead to the the sample testing in steam oxidation condition. In this condition, the samples was exposed to 90h with constant flow rate in order to investigate their behaviour. The image in Figure 6 shows the chromium layer on the surface become thinner compare to chromized steel before oxidation. The average surface thickness for chromium rich layer after exposed to oxidation is 4.2012±1.129 μm.

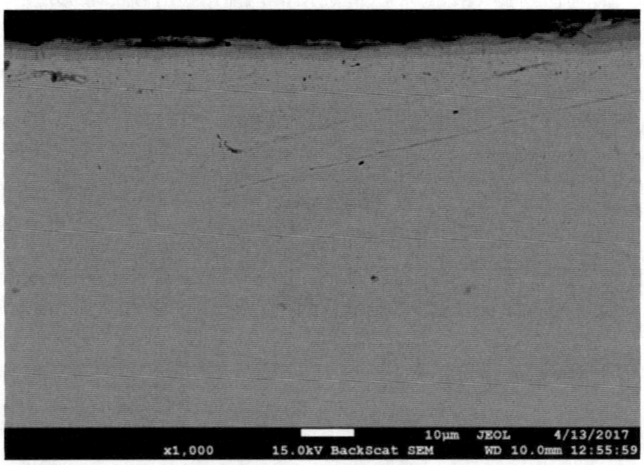

Figure 6. BSE FESEM image for samples exposed to steam oxidation condition.

EDX analysis shows the line and mapping profile for Cr, Fe and O. The linescan profile exhibit the presence of oxygen. However, the oxygen react with chromium to form Cr_2O_3 on the surface. However, the chromium diffusion form the chromizing process is still diffuse in the steel with the increasing diffusion depth with 51.185μm compare to before exposed to oxidation. After exposed in that time, the formation of non-protective oxide Fe_3O_4 and Fe_2O_3 was not detected in the steel. The EDX analysis for samples exposed in steam condition for 90h at high temperature 1050°C can be seen in Figure 7.

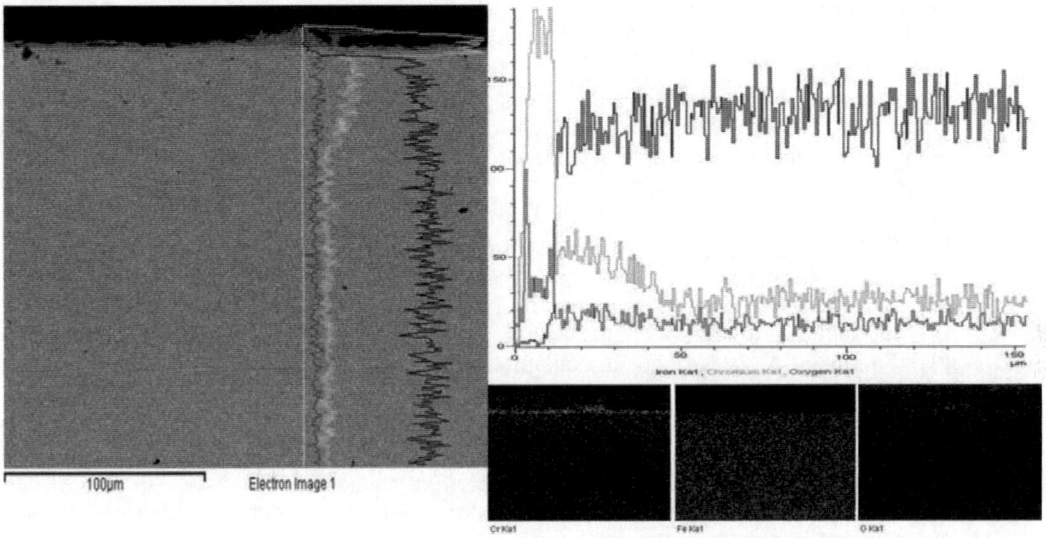

Figure 7. EDX analysis for sample exposed in high temperature 1050°C.

In order to identify the formation of chromium and chromia after doing chromizing and steam oxidation, the XRD analysis was carried out to analyze the spectrum, as shown in Figure 8. The formation of Cr_2O_3 can be detected in the XRD. Beside of chromium carbide, the Fe-Cr spinel also formed in the steel when exposed to steam oxidation. Both in chromizing process and corrosion testing the chromium carbide(Cr_3C_2 and Cr_7C_3) coming from the Cr rich layer that react with steel was present during the exposure of chromizing and steam oxidation.

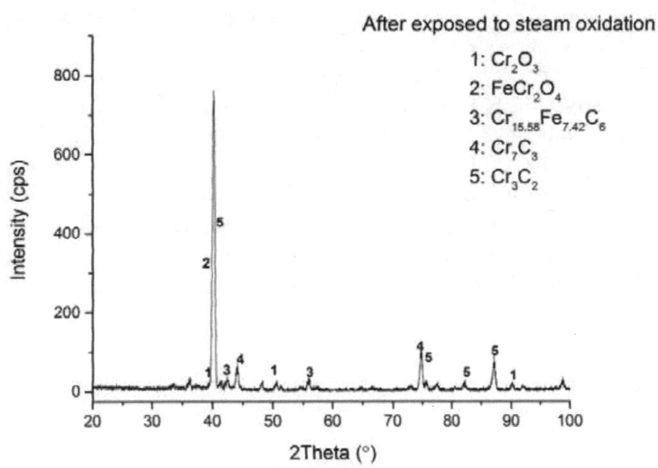

Figure 8. XRD spectrum image for T91 sample after exposed to steam oxidation.

4. Conclusions

The chromium was succesfully diffused at temperature above 800ºC. At higher temperature 1050ºC, carbon from the steel also diffuse to the surface and form chromium carbide. The difusion is essential to become reservoir for the formation of Cr_2O_3 during high temperature steam oxidation as shown by the oxidation result at 700ºC.

Acknowledgements

This facilities and resources for this research are fully supported by Fundamental Research Scheme Grant from Ministry of Higher Education Malaysia, under grant No. RDU140110 and UMP research grant for Postgraduate Research Scheme under grant No. GRS160310.

References

[1] Parker JD. *International perspectives on plant life assessment.* Materials at high temperatures. 2002;**19**(2):47-57.
[2] Bromhead A, Alessandro Blasi, Ikeyama S. Southeast Asia Energy Outlook 2015. France: International Energy Agency, 2015.
[3] Tominaga A, Wakazono O, Nakajima M, Fujikawa T, Sato S, Sugitani T. *Mitsubishi heavy industry* 2003.
[4] Viswanathan R, Romanosky R, Rao U, Purget R, Johnson H, editors. Boiler materials for USC plant. 17 th annual conference on fossil energy materials; 2003.
[5] Kurniawan T, Fauzi FAB, Asmara YP. *High-temperature Oxidation of Fe-Cr Steels in Steam Condition–A Review.* Indonesian Journal of Science and Technology. 2016;**1**(1):107-14.
[6] Zurek J, Quadakkers WJ. *Effect of Surface Condition on Steam Oxidation of Martensitic Steels and Nickel-Based Alloys.* Corrosion-Us. 2015;**71**(11):1342-59.
[7] Fujii C, Meussner R. *The Mechanism of the High-Temperature Oxidation of Iron-Chromium Alloys in Water Vapor.* Journal of the Electrochemical Society. 1964;**111**(11):1215-21.
[8] E. Essuman GHM, J. Zurek , M. Hansel, W. J. Quadakkers. *The Effect of Water Vapor on Selective Oxidation of Fe–Cr Alloys.* Oxidation of Metals. 2008;**69**(3-4):143-62.
[9] Fauzi F, Kurniawan T, Asmara Y, Ani M, Aziz M. *Oxidation behavior of pack-cemented P91 steel in steam condition.* Oxidation behavior of pack-cemented P91 steel in steam condition. 2016.
[10] Lepingle V, Louis G, Petelot D, Lefebvre B, Vaillant J, editors. High temperature corrosion behaviour of some boiler steels in pure water vapour. Materials science forum; 2001: Trans Tech Publ.
[11] Bose S. *High Temperature Coatings.* 2007;**1**:73-154.
[12] Fauzi F, Kurniawan T, Salwani M, Bin Y, Harun W, editors. Chromium Enrichment on P11 Ferritic Steel by Pack Cementation. MATEC Web of Conferences; 2016: EDP Sciences.
[13] Geib FD, Rapp RA. *Simultaneous chromizing—Aluminizing coating of low-alloy steels by a halide-activated, pack-cementation process.* Oxidation of Metals. 1993;**40**(3-4):213-28.
[14] Kung S, Rapp RA. *Kinetic study of aluminization of iron by using the pack cementation technique.* Journal of the Electrochemical Society. 1988;**135**(3):731-41.
[15] Fan HJ, Gösele U, Zacharias M. *Formation of nanotubes and hollow nanoparticles based on Kirkendall and diffusion processes: a review.* small. 2007;**3**(10):1660-71.

Effect of lithium nitrate and calcium nitrate composition on the thermal properties of quaternary molten salts mixture for heat transfer application

Nuratikah Nadhirah Binti Ahmad, Nursyafiqah Binti Yunos, Wan Nur Azrina Binti Wan Muhammad, Md Nor Anuar Bin Mohamad,, Farazila Binti Yusof

Universiti Tun Hussein Onn Malaysia, 86400 Parit Raja, Batu Pahat, Johor, Malaysia
Universiti Malaya, 50603 Kuala Lumpur, Malaysia.

Email: wazrina@uthm.edu.my

Abstract. Mixed molten salt is considered as a promising medium for both heat transfer and energy storage in thermal power because of its many advantages such as low the melting point, large heat capacity, good thermal stability and low cost. In order to determine the thermal properties of the molten salt, the nitrate quaternary mixture of the molten salt are prepared based on the different composition of the lithium nitrate and calcium nitrate. The other salts in the mixture are potassium nitrate and sodium nitrate. Mixture of molten salts were heated in furnace at 150°C for 4 hours and increased the temperature to 400°C for 8 hours for homogenize the salt mixture. Then, decreased the temperature to 115°C for an hour. The melting point and thermal stability of the mixture was determined by using thermogravimetric analysis [TGA] while the heat capacity was determined by using differential scanning calorimetry [DSC]. The lowest melting point using different composition of lithium nitrate, 10wt%$NaNO_3$, 4Owt%KNO_3, 20wt%$Ca[NO_3]_2$, 30wt%$LiNO_3$ which is 97.1°C while the heat capacity is 6.33 J/g °C. For the experiment using various composition of calcium nitrate, the lowest melting point is from 14wt% $NaNO_3$ + 48wt% KNO_3 + 13wt% $LiNO_3$ + 25wt% $Ca[NO3]_2$ which is 111.7°C and the heat capacity is 2.06 J/g °C. From the result, the addition of lithium nitrate in the quaternary molten salts give more effect to the reduction of melting point value but higher heat capacity.

1. Introduction

Molten salt is a means of storing heat at a high temperature. Molten salt is constituting of potassium nitrate, calcium nitrate, sodium nitrate, lithium nitrate. It has the property to absorb heat and to store the heat energy that is released to the water to transfer energy when needed. Molten salt that have high thermal energy storage capacity will increase the efficiency and enhance the performance for thermal energy storage. Thermal energy storage capacity of a salt mixture is dependent on melting point, density and specific heat capacity. Lower the melting point, more efficiency the molten salt. Then for thermal stability of the salt mixtures plays an important role in determining its utility as a potential thermal energy storage media.

Nitrate mixtures and mixtures of nitrate and nitrite are being used in the applications because of low melting point, low unit cost, high heat capacity and moderate density which imply excellent energy storage capability. Binary mixtures of alkali molten nitrates or nitrites present phase diagrams with a simple eutectic point. By adding one or more components, it is expected that the resulting mixture will have a lower melting point compared to the initial eutectic binary mixture. In this direction, the design of multicomponent mixtures by the addition of alkali/alkaline earth nitrates will make more working temperature range of thermal energy storage. The additives that present greater potential to be used for solar energy storage are $Ca[NO_3]_2$ and $LiNO_3$ because these salts decrease the melting point and improve the thermal stability, respectively. Other researcher proposed the study of molten salts of more than three components approach to achieve an effect between the properties calcium nitrate, $Ca[NO_3]_2$ and lithium nitrate, $LiNO_3$ of additives and the components of the solar salt. The most notable property of the molten salts using calcium and lithium nitrate is its energy density which is based on the heat capacity, which represents an important improvement over the solar salt currently used. Based on this parameter, a smaller amount of salt could be used to store the same amount of energy.

Most of the existing commercial solar thermal power plants currently use synthetic oil as the heat transport fluid in the collector field. It offers a very low freezing point less than 0°C and tolerable temperature limit 393°C among those available heat transfer fluids. The removal of the oil to nitrate salt, heat exchangers also reduces the unit storage system costs. Such as direct system would assistance impressively from a reformed parabolic through system such as the Super trough describe by Kolb and Diver [1]. The possessions of molten nitrate salt mixtures as heat transfer fluids for solar thermal systems are estimated, at the same time as the focus is placed on the melting temperature of salt mixtures with varieties compositions. In addition, enhance molten salt composition will beneficial for a wide range of industrial development heat applications.

Binary mixture and ternary mixture was used as heat transfer fluid in several molten salt but the most efficient mixture is from quaternary mixture. developing quaternary mixture will improve the performance of salt drastically. Researcher found that KNO_3-$NaNO_3$ with other several alkali nitrate has quite low melting point. So they proposed new formulation of quaternary mixture with composition 10 wt% $LiNO_3$ + 20 wt% $NaNO_3$ + 60 wt% KNO3 + 10 wt% Ca[NO3]2. The composition of this system having melting point below or closely 95°C and high thermal stability up to 500°C.

Research from [2] state that heat transfer fluid made of mixture of four inorganic nitrate salts including 9-18 wt% $NaNO_3$, 40-52 wt% KNO3, 13-21 wt% $LiNO_3$ and 20-27 wt% $Ca[NO_3]_2$ can have liquidus temperature less than 100°C. Based on [3], the research revealed that ternary mixture of molten salt which 63.7 wt% KNO_3, 27.3 wt% Ca[NO3]2, 9.0 wt% $LiNO_3$ having 76°C as melting point temperature. In their paper, the melting point, heat capacity and chemical stability and viscosity of molten salt mixtures were measure through experiments. Besides that, heat capacity curve shows at about 150°C. This means the composition cannot form the eutectic at about 80°C. It may because of operational of a freezing-up if the temperature drops unexpectedly. According to the [4], 16 kinds of mixed nitrate were prepared based on different mass proportions of KNO_3-$LiNO_3$-$NaNO_3$-$Ca[NO_3]_2.4H_2O$. The melting point of the samples were obtained by analysing the heating DSC curve. Melting point Nos. 12 until 16 are about 80°C.

Quaternary mixtures are mixtures consisting of four nitrate salt mixture used as liquid energy storage. In previous study state that quaternary mixtures are more effective compared to binary and ternary mixtures because of it low melting temperature condition. Thus, this study contributes to examines the design quaternary mixture of molten nitrate innovative, with the goal to improve solar salt is used at present as heat transfer fluid in solar energy. The quaternary salt involved which contains different weight percentages of $NaNO_3$, KNO_3, $LiNO_3$, and $Ca[NO_3]_2$, exhibits better physical and chemical properties than the binary solar salt. In several light studies [5],where involved the $LiNO_3$ and $Ca[NO_3]$. From the other researcher [6] is known to disclose anhydrous compositions mixture belonging to the quaternary $LiNO_3$-$NaNO_3$-KNO_3-$Ca[NO_3]_2$ system, said this compositions having a melting temperature below or closely 95 °C and a high thermal stability up to the temperature of 500 °C. Molten salt mixture which containing calcium nitrate and lithium nitrate lowers the melting point of a salt mixture based on sodium nitrate and potassium nitrate. The transition observed in the thermal study the KNO_3 content temperature above 300°C did not properly melt with the other components and the signal coincides with values reported in the literature and experimental values obtained for the melting of potassium nitrate [7]. Therefore, in this studied and proposed new heat transfer fluids [HTFs] with various additions of $Ca[NO_3]_2$, with the objective of replacing the binary salt that is currently used. The heat capacity will increase slightly if the mixtures is added by the calcium nitrate and lithium nitrate. The calcium nitrate heat capacity an approximate value is 120-140 J/mol/K by previous researcher [8].

In general, the development of the molten salt preparation is important in order to find the molten salt mixture with low melting point and wide usable temperature. As the mentioned above is greatly effort of development of molten salt as heat transfer fluid and storage technology. Therefore, this study will determine the effect of lithium nitrate and calcium nitrate in nitrate quaternary molten salt in order to develop low melting point and high heat capacity of molten salts as heat transfer fluid in heat recovery system.

2. Methodology

2.1. Preparation of mixture

The primary salt that used in this research are sodium nitrate $NaNO_3$, potassium nitrate KNO_3, calcium nitrate tetrahydrated $Ca[NO_3]_2.4H_2O$ and lithium nitrate $LiNO_3$. For those salt, it is found in the form of anhydrous except for calcium nitrate which is tetrahydrate and not dehydrated before melted. The percentage of weight for each component of salt in the quaternary mixture were developed. The salt is placed into the container and weighed according to its position. The mass of all the salt mixture is the same which is 50 grams. Table 2.1 shows the mass of the component is according to their percentage.

After weighed each component of the salt, the four salt were mixed all together. Then the sample of mixture was heated in a furnace at 150°C for 4 hours for the drying process. The process will continue by increased the temperature to 400°C for 8 hours for homogenization process to ensure homogenous between each of salt. Then the sample will be take out from the furnace and cooled the sample at the surrounding temperature.

Table 2.1 Mass of the component according to percentage with addition of calcium nitrate.

Sample	Mixture	Composition (wt%)	Weight (g)
Sample 1	$NaNO_3$+KNO_3+$LiNO_3$+$Ca(NO3)_2$	20 + 48 + 22 + 10	10 + 24 + 11 + 5
Sample 2	$NaNO_3$+KNO_3+$LiNO_3$+$Ca(NO3)_2$	22 + 40 + 23 + 15	11 + 20 + 11.5 + 7.5
Sample 3	$NaNO_3$+KNO_3+$LiNO_3$+$Ca(NO3)_2$	10 + 50 + 20 + 20	5 + 25 + 10 + 10
Sample 4	$NaNO_3$+KNO_3+$LiNO_3$+$Ca(NO3)_2$	14 + 48 + 13 + 25	7 + 24 + 6.5 + 12.5

Table 2.2 Mass of the component according to percentage with addition of lithium nitrate.

Sample	Mixture	Composition (wt%)	Weight (g)
Sample A	$NaNO_3$+KNO_3+$LiNO_3$+$Ca(NO3)_2$	10 + 60 + 10 + 20	5 + 30 + 5 + 10
Sample B	$NaNO_3$+KNO_3+$LiNO_3$+$Ca(NO3)_2$	10 + 50 + 20 + 20	5 + 25 + 10 + 10
Sample C	$NaNO_3$+KNO_3+$LiNO_3$+$Ca(NO3)_2$	10 + 40 + 30 + 20	5 + 20 + 15 + 10
Sample D	$NaNO_3$+KNO_3+$LiNO_3$+$Ca(NO3)_2$	10 + 30 + 40 + 20	5+ 15 + 20 + 10

2.2. Characterization of molten salt

The characteristic of all sample of mixture will be measured and analysed. Thermodynamic properties were measured including melting point, heat capacity and thermal stability. For this research, Differential Scanning Calorimetry [DSC] and Thermogravimetric Analysis [TGA] is used. Differential

Scanning Calorimetry [DSC] have been used to study or test the heat capacity of the sample of mixture while Thermogravimetric analysis have been used to determine the melting point and thermal stability of the molten salt.

3. Result and discussion

3.1. Discussion of result to the effect of lithium nitrate

Melting point is one of the characteristic of the molten salt as the heat transfer fluid and one of the properties that need to determine in this project. Melting point is when the salt mixture starts to melt and liquid is form. Figure 1 shows the melting point for all samples quaternary mixture various of Lithium Nitrate.

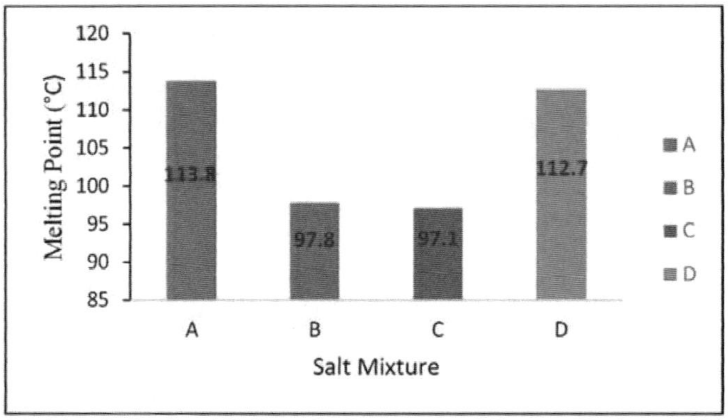

Figure 1. Melting Point for All Sample Quaternary Mixture Various of Lithium Nitrate

Many researchers found that molten salt that contain four inorganic nitrate salt having low melting point. This statement was confirmed by [2] that stating that molten salt with four different composition of nitrate salt can reach temperature below than 100°C and their paper also state that by addition or replace of LiNO3 is suitable and can lower the melting point. Besides that, other researchers found that this quaternary molten salt has a low melting point and can be approaching 95°C. However, the melting temperature of quaternary mixture for sample A which is 10wt%NaNO3, 6Owt%KNO3, 20wt%Ca[NO3]2, 10wt%LiNO3 is 113.8°C.

Composition of molten salt mixture for sample A and B are the same with composition of molten salt that was done by [4]. The melting point for sample 2 which is 10wt%$NaNO_3$, 5Owt%KNO_3, 20wt%$Ca[NO_3]_2$, 20wt%$LiNO_3$ obtained from this mixture is 97.4°C. For sample C and D, the development composition of the mixture is to make a variety of composition and to see the behaviour of the molten salt if we are adding more composition of lithium nitrate to the mixture. The mixture for sample C and D is 10wt%$NaNO_3$, 5Owt%KNO_3, 20wt%$Ca[NO_3]_2$, 20wt%$LiNO_3$ and 10wt%$NaNO_3$, 5Owt%KNO_3, 20wt%$Ca[NO_3]_2$, 20wt%$LiNO_3$ respectively. The melting point obtained from the result is 97.1°C and 112.7°C respectively. Sample C is lowest melting point compared to others mixture.

According to [4], their paper reported that the melting point of this both composition of molten salt is 80.4°C and 82.2°C respectively. Lithium nitrate which contain in the mixture is actually help the mixture to force melting point based sodium, potassium and calcium nitrate to the lower point. The result obtained is differ from the researchers maybe because the percentage usage of potassium nitrate is too high compared to others three salt. The melting temperature of potassium nitrate also high. According to [9] state that the potassium nitrate content did not melt completely with other salts at temperature above 300°C.

For the heat capacity, heat capacity is the amount of heat energy a unit of matter can hold. Definition also same with the amount of heat required to increase temperature of material. All material shows this increase in heat capacity with temperature. It is measured in unit J/mol or J/g.

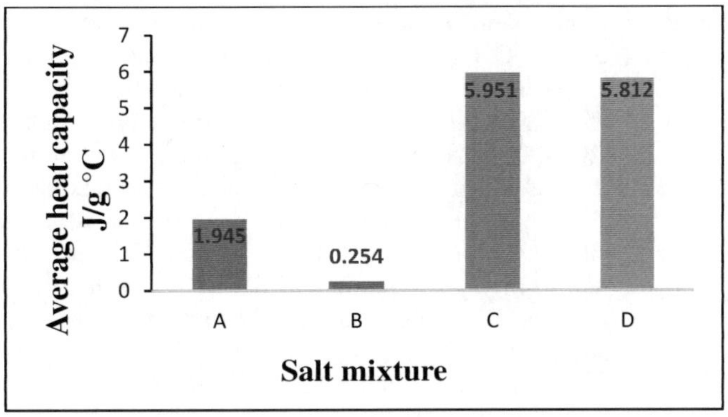

Figure 2. Heat Capacity Value of Liquid Phase in J/g °C

Figure 2 shows the value of heat capacity for the quaternary mixture of molten salts. From the result above, it shows that heat capacity for the sample A and B are lowest compared to the sample C and D. The recent research the addition of lithium nitrate can lower the melting point and provide high heat capacity. So it can be concluded that sample C and D can be considered as best composition of the quaternary molten salt.

For thermal stability, thermal stability is another important characteristic of quaternary mixture molten salt in order to look its behaviour and to determine best composition of quaternary mixture molten salt as heat transfer fluid. Generally, the reaction occur in the process is the endothermic reaction. This is due to break down the bonds that holding the reactant and the energy will obtain as the new bond are formed. In the endothermic reaction, the energy that we obtained is less than the energy we put on it.

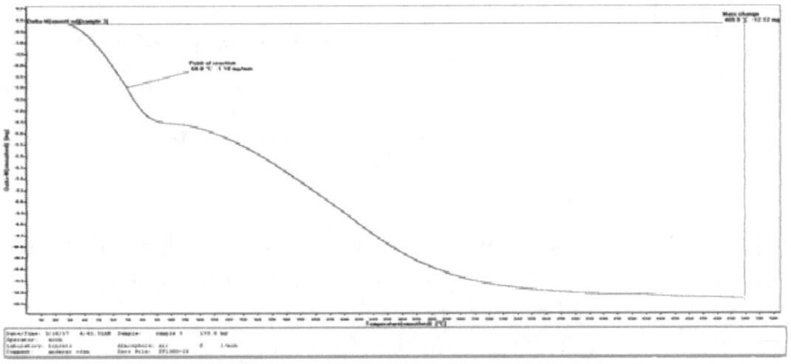

Figure 3. TG curve for sample C

From the TG curve (Figure 3), its shows the number of mass loss difference is not much. According to the [10], in their paper research stated that the mass losses with gas evolution of alkali metal nitrate salts may occur due to three reason that is nitrite formation in the melt and oxygen release , alkali metal oxide formation in the melt and nitrogen/nitrogen oxide release and vaporization of the nitrate salts. In their research also gives information about the aspects that affect the decomposition temperature. For examples is the method when conducted the experimental which is the value of heating rate is important parameter to be considered. Other than that is crucible material and atmosphere. Most measurement of thermal stability of the molten salt, the mass losses occur at temperature range 500 °C up to 600°C.

3.2. Discussion result to the effect of calcium nitrate
The transition in each composition between solid to liquid is very high for a small sample of a substance and it can be measured to 0.1°C. Quaternary system gives a much lower melting temperature than binary and ternary systems. In recent studies have proved that mixture of $NaNO_3$-KNO_3 shows high melting point in the mixture. This is because, there is no low melting point in this system that have been reported by Berg & Kerridge [2004] state the minimum melting point stood close to 220°C for this salt mixture [11]. Figure 4 shows the melting point for all sample quaternary mixture several of calcium nitrate. The composition for sample 1 shows the higher melting point compared to sample 2, sample 3 and sample 4. The result for sample 1 which the composition is 20wt% $NaNO_3$ + 48wt% KNO_3 + 22wt% $LiNO_3$ + 10wt% $Ca[NO3]_2$ shows the higher temperature of 122.6°C in the mixture. It is because this composition having a large amount of potassium nitrate and sodium nitrate. In recent research state that $NaNO_3$ and KNO_3 with higher range wt% will increase the melting point of the system [12]. The addition of $Ca[NO_3]_2$ and $LiNO_3$ in specific range of 13-22 wt% $LiNO_3$ and 10-27 wt% $Ca[NO_3]_2$ was obtained in this studies resulted the greater potential to improving the lower melting point which both salt mixture can be depressing melting point for the molten salt system. It have been proved that $NaNO_3$-KNO_3-$LiNO_3$-$Ca[NO_3]_2$ system which known as quaternary composition having a melting temperature closely to 130°C [12]. In this project, the lowest melting point are found in the sample 4 which 14wt% $NaNO_3$ + 48wt% KNO_3 + 13wt% $LiNO_3$ + 25wt% $Ca[NO_3]_2$ with value of melting point is 111.8°C.

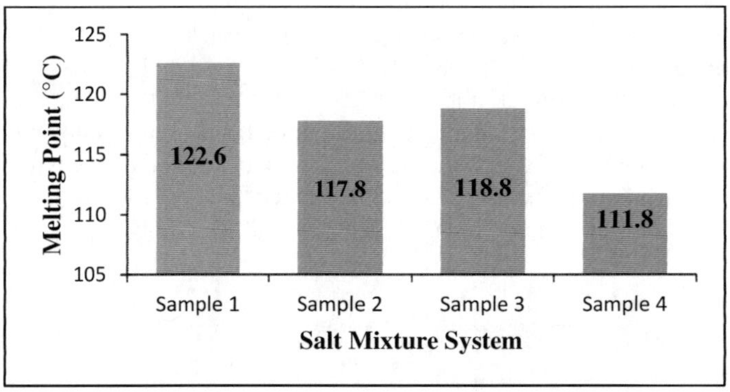

Figure 4. Melting Point for All Sample Quaternary Mixture Various composition of Calcium Nitrate

For thermal stability and heat capacity determination, the thermal stability result was obtained in this study by using TG curve of quaternary mixture system. This curve also can determine the decomposition temperature of molten salt mixtures in the system which is known as temperature at which the substance is chemically decomposed. The reaction generally endothermic system as heat is required to break the chemical bonds in the compound undergoing decomposition.

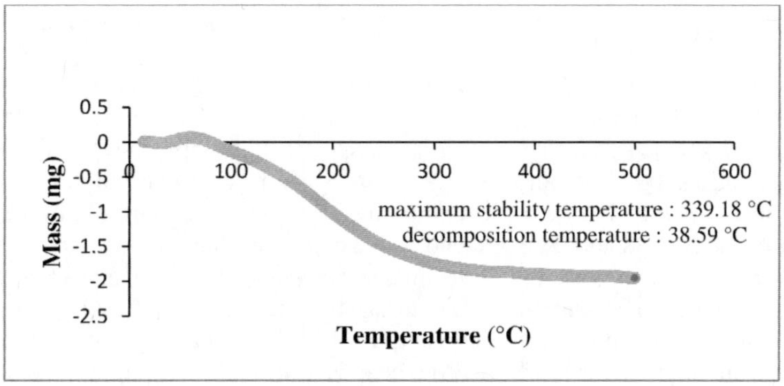

Figure 5. TG curve for salt mixture sample 3: 10wt% $NaNO_3$ + 50wt% KNO_3 + 20wt% $LiNO_3$ + 20wt% $Ca[NO_3]_2$

Decomposition temperature is an important parameter for setting the maximum usable temperature which can be obtained from TG curve. This comparison of the sample mixture shows sample 1 with the rapid weight loss begin at 300.1 °C with the composition temperature for this sample is 58.5 °C. For sample 2 shows the maximum stability temperature at 322.8 °C with it decomposition temperature starts at 42.71 °C. Figure 5 shows TG curve for sample 3 of quaternary system. The maximum stability temperature with the rapid weight loss begin at 339.18 °C followed by decomposition temperature at 38.59 °C. Maximum stability temperature in sample 4 shows the rapid weight loss begin at 348.93 °C and the decomposition temperature started at 36.55 °C. For quaternary curve illustrated the small curved at initial temperature where it is referred as the transition and transformations that occur which can be seen that was caused by the loss of water at initial temperature.

This study proves the elimination of water will undergo by the stability of the sample mixture will produced maximum temperature stability in the systems. In the previous studies state that the thermal stability of the sample was defined as the temperature at which the sample had loss up to 1.2% of it mass due to strongly hygroscopic property and interior moister and impurities contained in this kind of molten salt [4]. Hence, there is relatively large mass loss was observed in the cycle that is mainly contribute the presence of impurities and also moisture which is arise from the atmosphere in the process of the sample. The moisture and impurities mostly were removed during the cycle and the mass loss is attributable to the decomposition of the salt component [13]. The curve seen to reach the equilibrium state at certain amount of time that is closely to 500 °C which undergoes in this quaternary salt system. The thermal stability can be improved with additive of $Ca[NO_3]_2$ in nitrate system and it is suitable in improving the range of thermal stability as state by the recent research [2].

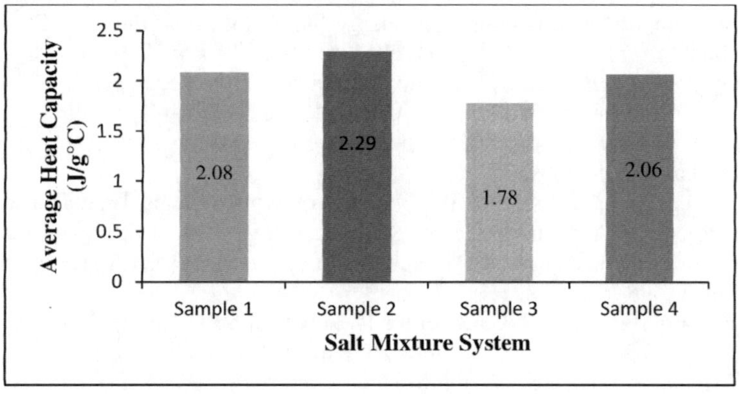

Figure 6. Heat Capacity Value of Quaternary Mixture various composition of Calcium Nitrate

For heat capacity, value of heat capacity was determined by using the differential scanning calorimetry. The result of four samples mixture shown in the bar graph (Figure 6). This evaluation curved in DSC graph is to determine an average of specific heat capacities for temperatures between the solid-liquid transitions for four salt mixtures in this system. From all the system sample 3 shows the lowest heat capacity which is 1.78 J/g°C compare to the other three salt mixture. Sample 1 and sample 4 shows heat capacity 2.08 and 2.06 respectively which is not much different between them. In recent studies the higher heat capacity value of the quaternary salt system indicates greater potential of energy storage capacity for the solar thermal energy storage applications [13]. In solid state, mobility is low due to the rigid arrangement and the specific heat is mostly determined by vibration effect while in liquid state, the huge amount increase of specific heat is attributed the higher mobility which result of improvement of fluidity.

The statement was approved that in addition of $LiNO_3$ and $Ca[NO_3]_2$ into formula mixtures are able to reduce the melting point, improve the thermal stability and heat capacity of mixtures which can makes the resulting high heat capacity where it can see in this quaternary system [14]. It is important feature in molten salt system as heat transfer fluid that specific heat must be able to storing the greatest possible amount of energy as the lowest possible temperature in the salt system.

4. Conclusion

For the conclusion of this project, aim of this study were achieved which is to determine the best molten salt composition with various of lithium nitrate and calcium nitrate in order to develop lower the melting point, high heat capacity and good thermal stability molten salts as heat transfer fluid in heat recovery system. It is important to highlight that lithium nitrate and calcium nitrate containing with other two nitrate salt which is potassium and sodium can give best performance as thermal storage and heat transfer fluid because of its lower melting point. Lower melting point can be obtaining based on the composition of each salt to become mixture. Each mixture has their own physical and chemical properties that to be discussed.

References

1. Bradshaw RW, Vegas L, Mixtures NS, Bradshaw RW, Affect EF. [12] United States Patent Temperature [° C]. 2009;1[12].
2. Bradshaw RW, Siegel NP. Molten Nitrate Salt Development For Thermal Energy Storage In Parabolic Trough Solar Power Systems. Proc 2nd Int Conf Energy Sustain. 2008;631–7.
3. Zhao CY, Wu ZG. Thermal property characterization of a low melting-temperature ternary nitrate salt mixture for thermal energy storage systems. Sol Energy Mater Sol Cells. 2011;95[12]:3341–6.
4. Ren N, Wu YT, Ma CF, Sang LX. Preparation and thermal properties of quaternary mixed nitrate with low melting point. Sol Energy Mater Sol Cells [Internet]. 2014;127:6–13. Available from: http://dx.doi.org/10.1016/j.solmat.2014.03.056
5. R. G. Wang TM. Novel low melting point quaternary eutectic system for solar thermal energy storage. Appl Energy. 2013;
6. Siegel B and. Development of molten nitrate salt mixtures fo concentrating solar power system. procSolarPACES. 2009;
7. F. Fernández, A.G. Lasanta MIP. Synthesis of Molten Salt as Heat Transfer Fluid For Waste Heat Recovery. Molten salt Corros Stain steels low-Cr steel CSP plants. 2012;
8. Cordaro JG, Kruizenga AM, Altmaier R, Sampson M, Nissen A. Thermodynamic properties of molten nitrate salts. Vol. 1. 2017. p. 8.
9. Fernández AG, Ushak S, Galleguillos H, Pérez FJ. Solar Energy Materials & Solar Cells Thermal characterisation of an innovative quaternary molten nitrate mixture for energy storage in CSP plants. Sol Energy Mater Sol Cells [Internet]. 2015;132:172–7. Available from:

http://dx.doi.org/10.1016/j.solmat.2014.08.020
10. Bauer T, Breidenbach N, Eck M. Overview of Molten Salt Storage Systems and Material Development for Solar Thermal Power Plants. Wref. 2012;1–8.
11. D. Berg RW and K. The NaNO3/KNO3 system: the position of the solidus and subsolidus. Dalt Trans. 2004;2224–2229.
12. Bradshaw RW, Brosseau DA. Improved Molten Salt Formulation For Heat Transfer Fluids in Parabolic Trough Solar Power System. In: SolarPACES 2008, Las Vegas, NV, Mar 4-7, 2008. 2008. p. 1–9.
13. Wang, T., Mantha D and RG. Novel low melting point quaternary eutectic system for solar thermal energy storage. Appl Energy. 2013;
14. Fernández AG, Galleguillos H, Fuentealba E, Pérez FJ. Thermal characterization of HITEC molten salt for energy storage in solar linear concentrated technology. J Therm Anal Calorim. 2015;122[1]:3–9.

IOP Conf. Series: Journal of Physics: Conf. Series **914** (2017) 012028 doi:10.1088/1742-6596/914/1/012028

Iron deficiency anaemia: with the conclusion of a need for iron reader

Wai Feng Lim[1,2], Boon Kar Yap[1], Mei I Lai[2,3], Noorazrina Talik[1], Ammar Ahmed Nasser[1], Ahmed Mubarak Ahmed Al-Haiqi[1] and Prajindra Sankar Krishnan[1]

[1]Department of Electronics and Communication Engineering, College of Engineering, Universiti Tenaga Nasional, 43000 Kajang, Selangor, Malaysia
[2]Department of Pathology, Faculty of Medicine and Health Sciences, Universiti Putra Malaysia, 43400 UPM Serdang, Selangor, Malaysia
[3]Genetics and Regenerative Medicine Research Centre, Faculty of Medicine and Health Sciences, Universiti Putra Malaysia, 43400 UPM Serdang, Selangor, Malaysia

Abstract. In our bloodstream, there are plenty of red blood cells (RBC), which function as an important oxygen carrier in our bodies. Each RBC consists of millions of haemoglobin (Hb), which is made up from globin and iron. If any deficiency/malfunction of any globin, it will lead to anaemia as indicated in low Hb level while iron deficiency anaemia (IDA) is anaemic due to the lacking of iron as indicated in low Hb and ferritin levels. IDA affects almost two billion people globally while anaemia without iron deficiency, such as thalassaemia, affects almost 4.5% in Malaysian population. These anaemic conditions have similar clinical symptoms like fatigue, dizziness, in which disturb their cognitive development and productivity in workplace. In areas without proper medical access, many anaemic individuals were misdiagnosed and treated with iron tablets because they were thought to have iron deficiency anaemia due to low Hb content. But, excess iron is toxic to the body. Misdiagnosis can be avoided by iron status assessment. We hereby review the currently available iron status parameters in laboratory and field study with the conclusion of demonstrating the importance of a need for iron reader, in the effort to reduce the prevalence of IDA globally.

1. Introduction

Iron deficiency anaemia (IDA) is a common nutritional deficiency as well as a major cause of anaemia globally that affects approximately two billion people globally in many developing countries [1-2]. Dietary iron deficiency, iron malabsorption, chronic blood loss and certain infectious disease are the causes of IDA [3]. Among these two billion anaemic people, around 50% of them are anaemic with iron deficiency (Figure 1) [2, 4].

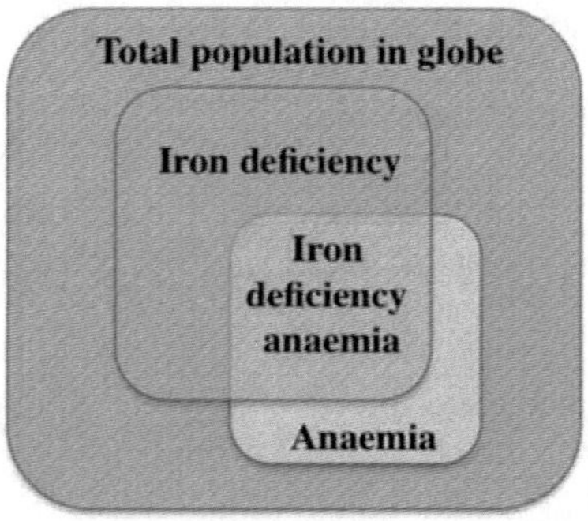

Fig. 1. Prevalence of iron deficiency anaemia in global view.

Moreover, the clinical symptoms of IDA and other haemoglobinopathies, such as thalassaemia are very similar in terms of their clinical appearance and diagnosis. Thalassaemia is characterised by either a reduced or absent production of haemoglobins but a normal or increased iron absorption that affects almost 4.5% of Malaysian population [5]. In areas without proper medical access, many haemoglobinopathic-affected individuals were misdiagnosed and treated with iron tablet due to low haemoglobin (Hb) content. Wrong diagnosis in addition with iron supplements will further worsen the disease leading to cardiac failure and kidney dysfunction. Therefore, one of the ways to differentiate IDA and thalassaeamia is by quantifying the iron level in the body.

Bone marrow iron is a gold standard to estimate body iron stores accurately [6]. Due to its invasive technique in bone marrow aspiration, several other indirect tests such as red cell and biochemical parameters are generally used to assess iron status [7]. In most cases, combining iron status parameters is necessary to make a definitive diagnosis to avoid the diagnostic error of individual biochemical test, which is attributed by other clinical factors. It is of great importance to summarize the currently available laboratory parameters followed by reviewing the potential need of an iron reader for IDA screening, especially in remote and rural areas.

2 Red cell parameters

Red cell parameters can be analysed from automated blood cell analyser for red cell and other blood cell counting as well as haemoglobin (Hb) content [7]. The red cell patterns of iron deficiency anaemia (IDA) are reviewed in the following section.

2.1 Haemoglobin, mean cell volume, mean cell haemoglobin, haematocrit and red cell distribution width

A standard automated blood cell analyser measure haemoglobin (Hb), mean cell volume (MCV), mean cell haemoglobin (MCH), haematocrit (HCT) and red cell distribution width (RDW). Hb is a direct indicator in diagnosing whether an individual suffers from anaemia when Hb level is below the reference range, regardless anaemia with or without iron deficiency. Iron deficiency anaemia (IDA) individuals have anaemic red cells with reduced cell size as evaluated by MCV and reduced Hb content as evaluated by MCH, respectively. A decrease in Hb, HCT and red cell levels as well as an increase in red blood cell distribution width (RDW) – the red cell size variation, can also be seen in IDA individuals.

In Aslan study focusing on RDW level, IDA group (18.00±1.94%) showed significant differences from normal (12.49±0.63%) as well as β-thalassaemia trait (14.88±1.77%) individuals [8]. However, RDW demonstrated only 53.4% specificity in detecting IDA among microcytic hypochromic anaemic children [9]. To diagnose iron deficiency during pregnancy, RDW demonstrated the highest sensitivity (82.3%) and specificity (97.4%) as compared to Hb level, MCV, MCH, mean cell haemoglobin concentration (MCHC) and peripheral blood film [10]. According to a recent study, MCH and MCHC are only moderately accurate to detect empty iron stores, implying that individuals with normal MCH and MCHC might have a risk of empty iron stores [11].

2.2 Zinc protoporphyrin

In iron deficient condition, the iron's binding site is being replaced by zinc, thereby forming zinc protoporphyrin (ZPP). ZPP represents an inefficient red cell synthesis in a condition where a lack of iron in the body.

A study illustrated that both MCH and ZPP parameters could be used to discriminate between iron deficiency and beta-thalassaemia carriers [12]. Other study displayed that combined both MCV and ZPP could differentiate patients with iron deficiency and thalassaemia effectively, up to 95% [13]. A positive response of iron therapy in paediatric population showed a reduction of ZPP [14]. However, ZPP could not distinguish those with and without iron deficiency among pregnant women from Kenya [15]. This implies that the diagnostic potential of ZPP is limited to only certain groups of people.

In short, red cell parameters are mainly detecting the severe stages of iron deficiency, whereby functional iron deficiency has not been addressed unless with combined analysis with biochemical parameters.

3. Biochemical parameters

Biochemical parameters are used to assess iron status using immunoturbidimetric assay by biochemistry autoanalyser and/or enzyme-linked immunosorbent assay (ELISA). Biochemical parameters can be divided into two groups: transport iron and storage iron.

3.1 Transport iron: plasma iron, total iron binding capacity, transferrin saturation, soluble transferrin receptor

Both transferrin and transferrin receptor are transport proteins. Plasma iron (PI) is determined by iron bound to transferrin. The capacity of transferrin to bind iron is indicated as total iron binding capacity (TIBC). A derivative parameter from PI and TIBC known as transferrin saturation (%TSAT), which is calculated from the ratio of [PI/TIBC]. More transferrin receptors are being expressed in iron deficient cells, indicating by the presence of soluble transferrin receptor (sTfR) in plasma. This suggests a balance between cellular iron demand and iron supply.

sTfR is the only biochemical marker to detect mild tissue iron deficiency, also known as iron deficient erythropoiesis (IDE) [16]. A study demonstrated sTfR is the only reliable marker to diagnose advanced IDE individuals with ZPP level >70 μmol/mol heme [17]. Combined sTfR and ZPP analysis could better illustrate the status of erythropoietic activity.

3.2 Storage iron: plasma ferritin

Plasma ferritin is the important storage protein. It is the widely accepted parameter that reflects the size of the iron stores similar to bone marrow iron [18]. However, ferritin is an acute phase protein that can be confounded by inflammation and infection. This suggests that in the presence of inflammation and infection, ferritin level might not reflect the iron status accurately. Therefore, ferritin level should be evaluated together with inflammatory marker, such as C-reactive protein (CRP) [19].

To better demonstrate body iron status in both storage iron and functional iron, the ratio of transferrin receptor to ferritin (sTfR-F index) is a widely applied parameter together with at least one acute phase protein in order to provide a more reliable diagnosis than a single test [19-20].

4. Classification of iron status

Iron deficiency can be divided into three major stages, i.e. iron depletion (IDP), iron deficiency erythropoiesis (IDE) and iron deficiency anaemia (IDA). Those stages are classified with respect to the iron balance, which is regulated by our bodies. There are functional iron (haemoglobin, myoglobin, cytochromes, enzymes), transport iron (transferrin) and storage iron (macrphage, ferritin, haemosiderin).

4.1 Iron depletion, iron deficiency erythropoiesis and iron deficiency anaemia

Iron depletion (IDP) is the onset of iron deficiency with a reduced in storage iron but still with normal iron for erythropoiesis in functional compartment, as examined in low plasma ferritin (PF).

As the iron deficiency progresses, there is insufficient iron delivery for erythropoiesis, as examined in reduced plasma iron (PI), increased total iron binding capacity (TIBC), reduced percentage of transferrin saturation (%TSAT), increased plasma transferrin receptor (TfR) and increased red blood cell zinc protoporphyrin (ZPP). This iron-deficient condition termed as iron deficiency erythropoiesis (IDE).

Lastly, long-term iron deficiency can be very severe that causes IDA, an iron-deficient condition to a certain extent that causes anaemic condition. There is a reduction in functional iron leading to inadequate haemoglobin synthesis, as indicated in mean corpuscular volume (MCV) and mean corpuscular haemoglobin (MCH) [19, 21]. Figure 2 demonstrate the iron status classification with respect to the iron distribution compartment [22].

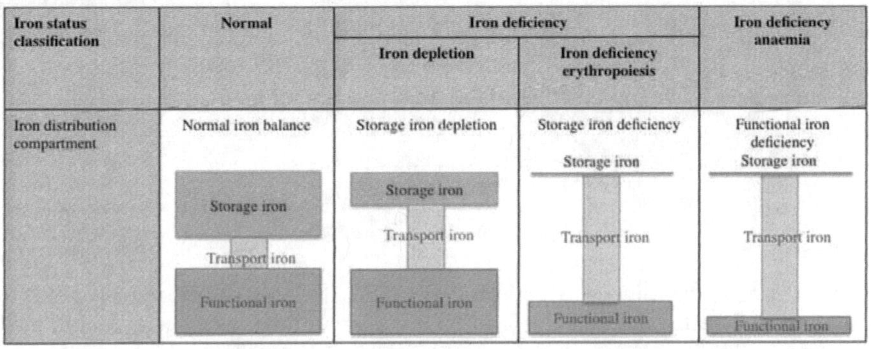

Fig. 2. Classification of iron deficiency severity with respect to the iron distribution compartment.

5. A need for iron reader

To date, global estimates of iron deficiency is only relied on the global estimates of anaemia based on haemoglobin (Hb) level. Hb level has often been applied as a proxy marker for iron deficiency [2, 23-24]. In this section, we emphasize the need for an iron reader by reviewing the current medical facilities among countries with high income inequality and the existing instruments in the market.

5.1 Current medical facilities

Although middle- and high-income countries having public and private healthcare center, it is always full of people, crowded which results in long waiting time. Furthermore, they might need to have a long walk to hospital due to limited parking spaces. In poor countries, malnutrition might lead to anaemia and there is almost 72% children in India are anaemic. Although there are mobile clinics to support them, some of them have to walk from as early as 5 a.m. to reach the place to get treatment and this mobile clinic limited to only once a week due to lack of medical personnels. Again, results take time to release as it involves the sample transportation to the laboratory (Figure 3).

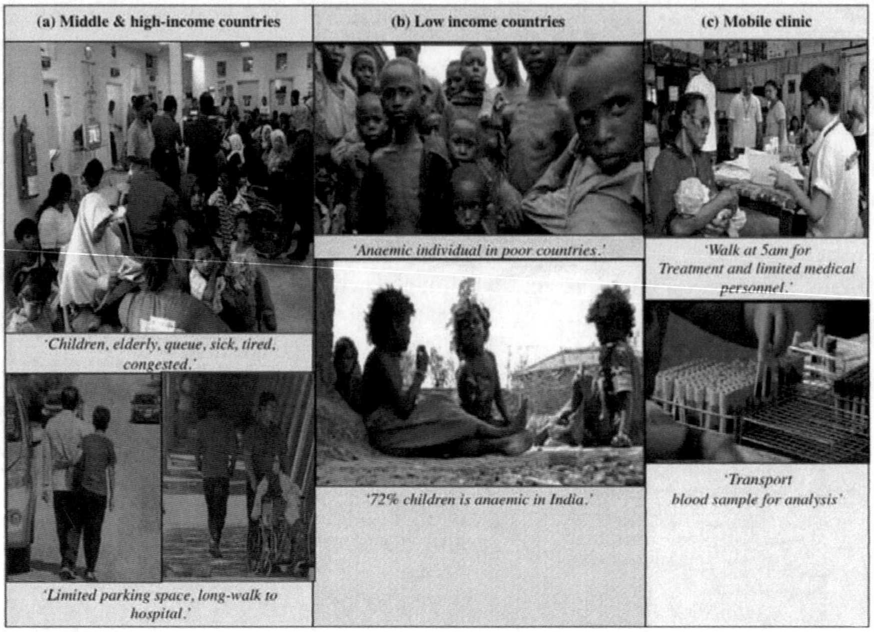

Fig. 3. Issue we want to solve in low-, middle- and high-income countries: a need of an iron reader.

Issue we want to solve aligned with what World health organization stress: 'It is time to act to eliminate iron deficiency anaemia' [25]. In order to target low-, middle- and high-income countries, there is an urgent need to devise a portable, low-cost and user-friendly iron reader to not only save time, money and resources, but also save more lives eventually.

5.2 Current existing instruments

To diagnose iron deficiency anaemia (IDA), the conventional methods are based on current existing instruments, i.e. red cell analyser and biochemistry autoanalyser to ensure definitive diagnosis. Both of these analysers are big and bulky and cost up to RM75.000 (DXH500, Beckman Coulter) and RM200.000 (ILAB Aries, Instrumentation Laboratory). Although these instruments enable definitive diagnosis, they are inconvenient for field study as a trained personnel is needed to operate and analyse the results.

Single-parameter device such as haematofluorometer and hemocue are used to measure haemoglobin (Hb) and red blood cell protoporphyrin content, respectively. These portable devices are relatively low-cost as compared to the previously discussed analysers. Furthermore, these devices require only minimal sample processing and immediate result will be possible. Table 1 summarises the instruments/devices available in the laboratory and for field study.

Table 1. The instruments/devices available to diagnose iron deficiency anaemia.

Instrument	Multi-parameters laboratory instruments		Single-parameter devices	
Name	Red cell analyser	Biochemistry autoanalyser	Haematofluorometer	Hemocue
Brand	Beckman Coulter	Instrumentation Laboratory	Aviv Biomedical	Haemocue®
Cost	~RM 75,000	~RM 200,000	Not available	~RM 1,000
Pros	Multi-parameters for definitive diagnosis	Multi-parameters for definitive diagnosis	Small and portable Convenient and affordable for rural area	Small and portable Convenient and affordable for rural area
Cons	Big, bulky, high-cost and inconvenient	Big, bulky, high-cost and inconvenient	Moderate in size	Detect only end-stage of iron deficiency anaemia and other anaemia without iron deficiency
Image				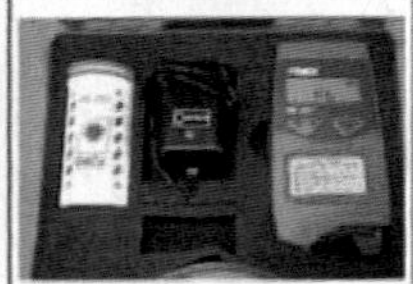

6. Conclusion

Thus far, there is no portable iron reader in the market especially for iron parameter. There is an urgent need for a portable iron reader in an affordable format to reduce/eliminate the iron deficiency anaemia (IDA), especially in developing countries and rural areas. Currently, our group are devising a novel, low-cost and portable iron reader that is able to commercialise globally. Eventually in years to come, the prevalence of anaemia/IDA could be reduced, especially in rural areas.

References

1. A.V. Hoffbrand, P.A.H. Moss, J.E. Pettit. *Blackwell Publishing* (2006)
2. WHO. Retrieved 30 May 2017 from http://apps.who.int/iris/bitstream/10665/66914/1/WHO_NHD_01.3.pdf?ua=1 (2001)
3. J.L. Miller. Cold Spring Harb. Perspect. Med. **3** (2013)
4. WHO. Retrieved 30 May 2017 from http://apps.who.int/iris/bitstream/10665/85839/3/WHO_NMH_NHD_MNM_11.1_eng.pdf?ua=1 (2011)
5. S. Jameela, S.O. Sabirah, J. Babam, C.L. Phan, P. Visalachy, K.M. Chang, M.A. Salwana, A. Zuraidah, Y. Subramanian, A. Rahimah. Med. J. Malaysia. **66**, 522-524 (2011)
6. R.D. Sundberg, H. Broman. Blood. **10**, 160-166 (1955).
7. B.J. Bain, I. Bates, M.A. Laffan, S.M. Lewis. *Churchill Livingstone* (2007)
8. D. Aslan, F. Gümrük, A. Gürgey, Ç. Altay. Am. J. Hematol. **69**, 31-33 (2002)
9. R. Aulakh, I. Sohi, T. Singh, N. Kakkar. Indian J. Pediatr. **76**, 265-267 (2009)
10. G.S. Sultana, S.A. Haque, T. Sultana, Q. Rahman, A.N.N. Ahmed. Bangladesh Med. Res. Counc. Bull. **37**, 102-105 (2011)
11. A.E. Åsberg, G. Mikkelsen, M.W. Aune, A. Åsberg. Int. J. Lab. Hematol. **36**, 98-104 (2013)
12. E. George, M.L. Ng, J.A.M.A. Tan. Malays. J. Med. Health Sci. **4**, 51-55 (2008)
13. E.J. Harthoorn-Lasthuizen, J. Lindemans, M.M.A.C. Langenhuijsen. Eur. J. Haematol. **60**, 245-251 (2009)
14. H. Magge, P. Sprinz, W.G. Adams, M-L. Drainoni, A. Meyers. JAMA Pediatr. **167**, 361-367 (2013)
15. M.N. Mwangi, S. Maskey, P.EA. Andang'o, N.K. Shinali, J.M. Roth, L. Trijsburg, A.M. Mwangi, H. Zuilhof, B. van Lagen, H.FJ. Savelkoul, A.Y. Demir, H. Verhoef. BMC Med. **12** (2014)
16. B.S. Skikne, C.H. Flower, J.D. Cook. Blood. **75**, 1870-1876 (1990)

17. G. Metzgeroth, V. Adelberger, A. Dorn-Beineke, C. Kuhn, M. Schatz, O. Maywald, T. Bertsch, H. Wisser, R. Hehlmann, J. Hastka. J. Eur. J. Haematol. **75**, 309-317 (2005)
18. D. Haskins, A.R. Stevens, S. Finch, C.A. Finch. J. Clin. Invest. **31**, 543-547 (1952)
19. WHO. Retrieved 30 May 2017 from http://www.who.int/nutrition/publications/micronutrients/anaemia_iron_deficiency/9789241596107.pdf (2007)
20. J.D. Cook, C.H. Flowers, B.S. Skikne, Blood. **101**, 3359-3363 (2003)
21. S. Lynch. Retrieved 30 May 2017 from http://www.who.int/nutrition/publications/micronutrients/background_paper3_report_assessment_vitAandIron_status.pdf (2012)
22. V. Herbert, S. Shaw, E. Jayatilleke, T. Stopler-kasdan. Stem Cells. **12**, 289-303 (1994)
23. WHO. Retrieved 30 May 2017 from http://apps.who.int/iris/bitstream/10665/177094/1/9789241564960_eng.pdf (2015)
24. S. Pasricha. Blood. **123**, 611-612 (2014)
25. WHO. Retrieved 30 May 2017 from http://www.who.int/nutrition/topics/ida/en/

IOP Conf. Series: Journal of Physics: Conf. Series **914** (2017) 012029 doi:10.1088/1742-6596/914/1/012029

Synthesis of carbon nanostructures from high density polyethylene (HDPE) and polyethylene terephthalate (PET) waste by chemical vapour deposition

M N M Hatta, M S Hashim, R Hussin, S Aida, Z Kamdi, AR Ainuddin and MZ Yunos

Nano and Surface Modification Research Group, Faculty of Mechanical and Manufacturing Engineering, Universiti Tun Hussein Onn Malaysia, Parit Raja 86400 Batu Pahat Johor.
Faculty of Engineering Technology, Universiti Tun Hussein Onn Malaysia, Parit Raja 86400 Batu Pahat, Johor

Email: mnasrul@uthm.edu.my

Abstract. In this study, carbon nanostructures were synthesized from High Density Polyethylene (HDPE) and Polyethylene terephthalate (PET) waste by single-stage chemical vapour deposition (CVD) method. In CVD, iron was used as catalyst and pyrolitic of carbon source was conducted at temperature 700, 800 and 900ºC for 30 minutes. Argon gas was used as carrier gas with flow at 90 sccm. The synthesized carbon nanostructures were characterized by FESEM, EDS and calculation of carbon yield (%). FESEM micrograph shows that the carbon nanostructures were only grown as nanofilament when synthesized from PET waste. The synthesization of carbon nanostructure at 700ºC was produced smooth and the smallest diameter nanofilament compared to others. The carbon yield of synthesized carbon nanostructures from PET was lower from HDPE. Furthermore, the carbon yield is recorded to increase with increasing of reaction temperature for all samples. Elemental study by EDS analysis were carried out and the formation of carbon nanostructures was confirmed after CVD process. Utilization of polymer waste to produce carbon nanostructures is beneficial to ensure that the carbon nanotechnology will be sustained in future.

1. Introduction

The first discovery of carbon nanotubes (CNTs) was discovered by Sumio Iijima in 1991 at NEC laboratories Japan [1]. This discovery has attracted more research in related applications such as in mechanical and electrical field. The structure of nanocarbon is main criteria and has been explored to be used in electronic materials, optoelectronic, sensor application and optimization of polymer composite [2]. Generally the structure of carbon nanotubes can be divided into two; single walled carbon nanotubes (SWCNTs) which describe as a single sheet of graphene with diameter of 1nm and multi-walled carbon nanotubes (MWCNTs), which contains of layers of single sheet graphene. The MWCNTs have a diameter in the range of 2 to 100 nm and a length in micrometers [3].

IOP Conf. Series: Journal of Physics: Conf. Series **914** (2017) 012029 doi:10.1088/1742-6596/914/1/012029

There are currently three main synthesis methods used to produce graphitic nanofibres GNFs and/or carbon nanotubes CNTs. The methods are electric arc discharge (EAD) [4], laser ablation technique (LA) [5] and chemical vapour deposition technique (CVD) [6]. Among these techniques, the CVD is a common method used to produce GNFs and CNTs because of providing a good carbon yields and low amount of unwanted soot [7]. In CVD method, the carbon nanotubes are generated from carbon source whereas the main carbon sources are from petrochemical based such as styrene and methane [6,8]. In future, the petroleum source will cease and implicate to nanotechnology as well. Therefore to sustain this technology, it is needed to find other source materials to replace current petrochemical based source.

Plastics are among the materials that are used widely in industry such as in packaging, textiles, medicine and others. The difficulty to recycle this material and bad waste management resulting plastic is one of the largest contributions in solid waste disposal sector in Malaysia. The plastic will take longer time to decompose. Thus, to utilize this material will be helpful to reduce this problem. The main source of plastic production is petroleum, known as petrochemical based material. Since plastic material is petrochemical based, it is believed that this material has potential to be a carbon source for development of carbon nanostructure materials [9,10].

Thus, this study was carried out to explore the capability of polymer waste to be used in production of carbon nanostructures. Carbon nanostructures were synthesized by CVD method with two major polymer wastes; HDPE and PET were selected to replace current carbon source. The effect of different polymer waste and CVD temperature on synthesized carbon nanostructures was characterized.

2. Experimental

2.1. Sample Preparation

The chemical vapour deposition (CVD) system was set up by using tube furnace (Elite furnace model TSH/16/50/180) as schematic shown in figure 1. The synthesis process of carbon nanostructures consist of two step, catalyst preparation and CNT deposition. For catalyst preparation, the 5g of Iron (III) Nitrate Nanohydrate dissolved in 100ml of distilled water and mixed with 2 mol of ammonia solution. Then, the sediment form in solution was dried and calcined at temperature of 400ºC for 4 hours to form of Iron hydroxide (FeOH) catalyst precursor.

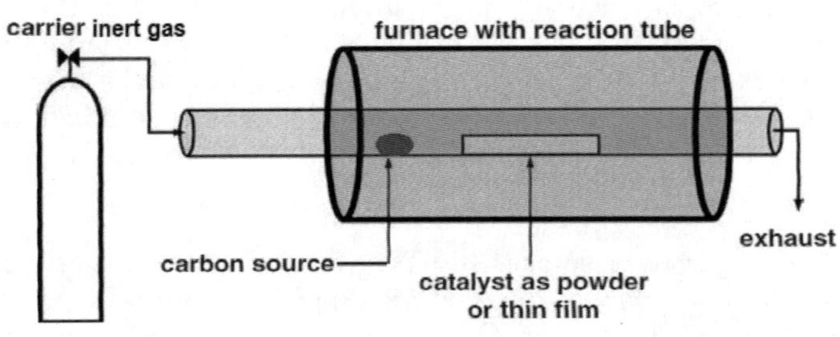

Figure 1: Schematic of single-stage chemical vapour deposition (CVD) system.

The CVD experiment was carried out by flowing 90 sccm argon gas as carrier gas into the CVD reactor for 20 minutes to flush out the oxygen. Prior, 5g of shredded polyethylene terephthalate (PET)/ High-density polyethylene (HDPE) as carbon source and 10mg of iron catalyst were placed inside the quartz tube as shown in figure 1. Then the furnace was heating up to required temperature (700, 800 and 900°C) with heating rate of 15°C/min. The reaction was kept for 30 minutes.

2.2. *Characterization*

The morphology and structure of synthesized carbon nanostructure was characterized by using Field Emission Scanning Electron Microscope (FESEM) model JEOL-JGM 7600F. The sample was not coated with metal prior characterization because of carbon nanostructure was conductive material. Energy Dispersive X-ray spectroscopy (EDS) attached with FESEM was used to determine the element present in catalyst and after CVD reaction. Carbon yields were calculated based on equation 1 in order to measure the amount of carbon has been deposited on catalyst.

$$\text{Carbon yields (\%)} = (\text{Weight after CVD - Weight of catalyst})/\text{Weight of catalyst} \times 100 \quad [1]$$

3. Results and Discussion

The formation of carbon nanostructures on catalyst were observed by FESEM analysis. Figure 2 shows the samples synthesized from HDPE waste at 700, 800 and 900°C, while figure 3 shows the samples synthesized from PET waste at 700, 800 and 900°C respectively. It is observed that the deposited carbon nanostructures on catalyst were not in filament form as expected. The carbon tends to agglomerate and form lookalike carbon ball with variant shape on the catalyst. At higher temperature 900°C, the carbon started to deposit in filament form. However the diameter of filaments was too big. The diameter was measured around more than 100 nm.

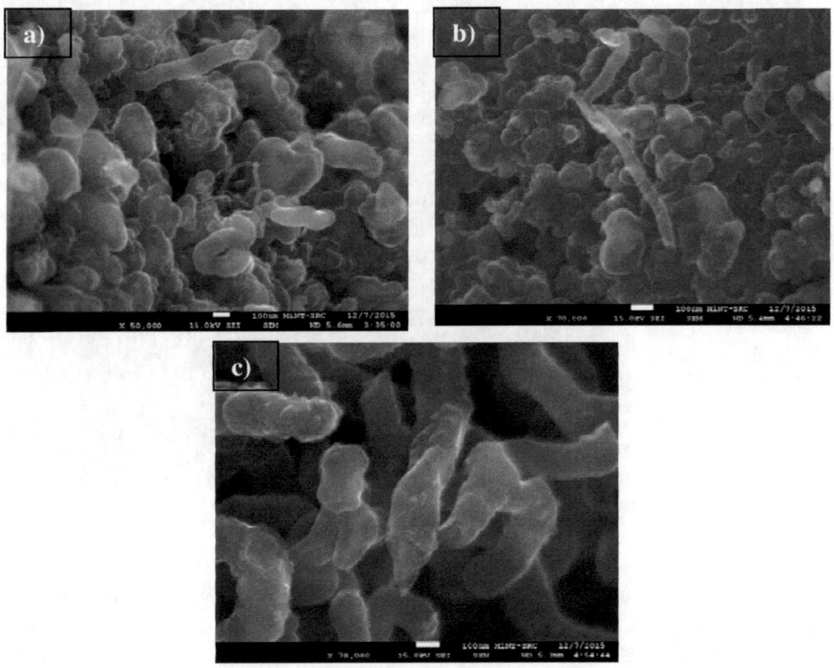

Figure 2: SEM micrograph of carbon structures synthesized from HDPE at different CVD temperature a) 700°C, b) 800°C and c) 900°C.

For the sample synthesized from PET, it is observed in figure 3 that the carbon nanostructures were grown as nanofilament at all temperatures. At 700°C, the nanofilaments were grown with smooth surface and morphology. The diameter was measured around 20 to 30 nm. When the temperature was increased to 800°C, the carbon was still grown as nanofilament. However the diameter of carbon nanofilament was bigger compared to the sample at 700 °C. The diameter of nanofilament was around 80 to 100nm. Meanwhile, when observed the sample synthesized at 900°C, it can be seen that the carbon nanofilaments were grown with rougher and crook morphology. The diameter of the samples was slightly lower than sample synthesized at 800°C. The measured diameter was around 70 to 80 nm respectively. The formation of carbon nanofilament on the catalyst is involved with diffusion process of carbon in pure metal catalyst and grows to form filament wall at the edge of catalyst [6,11-12]. However, the prior catalyst placed in tube furnace was iron hydroxide (FeOH). This iron hydroxide is needed to reduce to pure catalyst by react with hydrogen [14]. No hydrogen gas was supplied means the hydrogen is mainly obtained from decomposition of PET and HDPE. Based on chemical structure PET ($(C_{10}H_8O_4)n$) and HDPE ($(C_2H_4)n$), it is believed that the hydrogen supplied by PET is more than HDPE. The catalyst in PET more active compared to the catalyst in HDPE. Thus carbon from PET tends to deposit on catalyst by grow as nanofilament form.

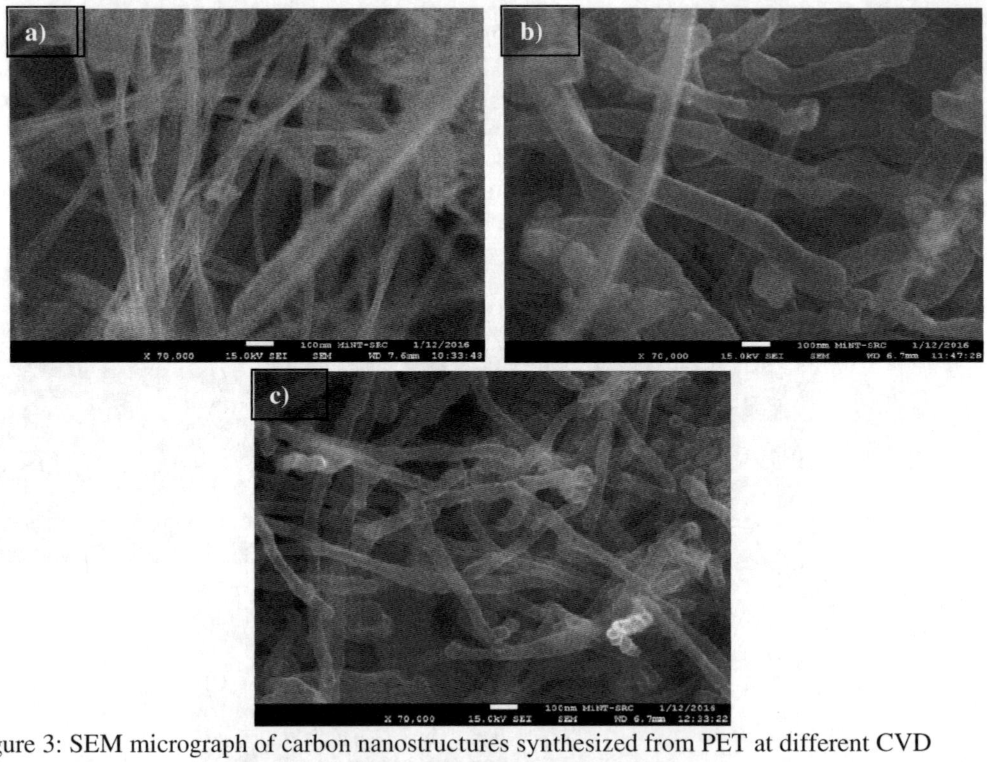

Figure 3: SEM micrograph of carbon nanostructures synthesized from PET at different CVD temperature a) 700°C, b) 800°C and c) 900°C.

The experiment was further carried out to obtain the carbon yield corresponded to the amount of carbon structures deposited on the catalyst. It is observed in figure 4 that the carbon yield of carbon nanostructures synthesized from PET were increased with increasing of reaction temperature from 700 to 900°C. It is agree with SEM micrograph whereas the diameter of carbon nanofilaments at 800°C and 900°C are higher than 700°C indicating more carbons were deposited to form a thicker filament wall. Furthermore, it is also can be seen in the figure that the carbon yields of samples synthesized from HDPE were drastically higher compared to the samples synthesized from PET. It can be noticed from the SEM micrograph where the carbons were deposited densely on the catalyst and cover up entirely all the catalyst surface. Therefore the amount of carbon deposited from HDPE is as expected excessively higher compare than PET.

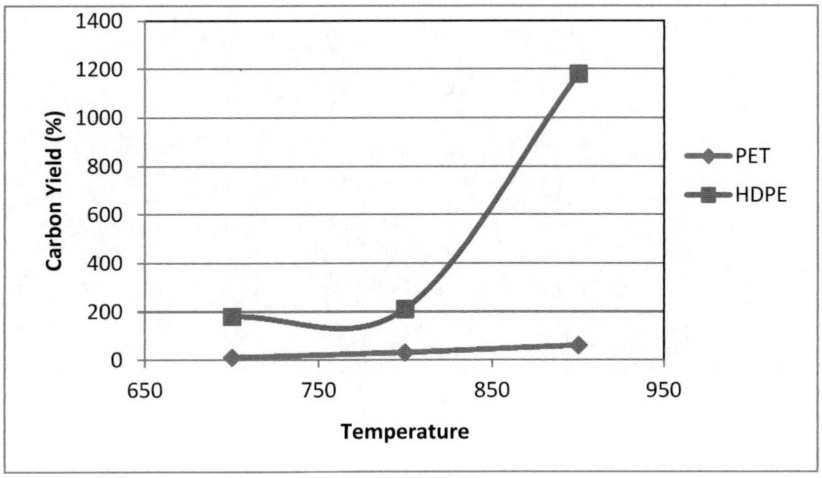

Figure 4: Carbon yield of carbon structures synthesized from PET and HDPE waste at different CVD temperature.

Beside the morphological observation, Energy Dispersive X-ray Spectroscopy (EDS) analysis also was carried out on the carbon nanostructure synthesized from PET at 700°C to trace the existence of carbon element after CVD process. Table 1 shows the EDS results of catalyst. There were 2 main elements were studied; iron and oxygen. The average value of Fe is 76.07%, while the average for Oxides (O) is 23.93%. The existence of oxygen in the catalyst indicated that the catalyst is exists as metal hydroxide (iron catalyst). This metal hydroxide name as catalyst precursor. No carbon element was observed at this stage. However after the CVD process as shown in figure 5, the carbon element is detected in EDS analysis indicating that carbon was synthesized after went through the CVD reaction. The element of iron is still detected representative that the formation of carbon nanostructure was occurred at the catalyst surface. In addition, no oxygen element is detected after CVD process. This is shows that catalyst precursor was converted to pure catalyst before the carbon can diffuse and deposit on top of the catalyst.

Table 1: EDS analysis of iron precursor

No of testing	Percentage Iron (Fe)%	Percentage Oxide (O)%
1	73.69	26.31
2	79.04	20.96
Average	76.07	23.93

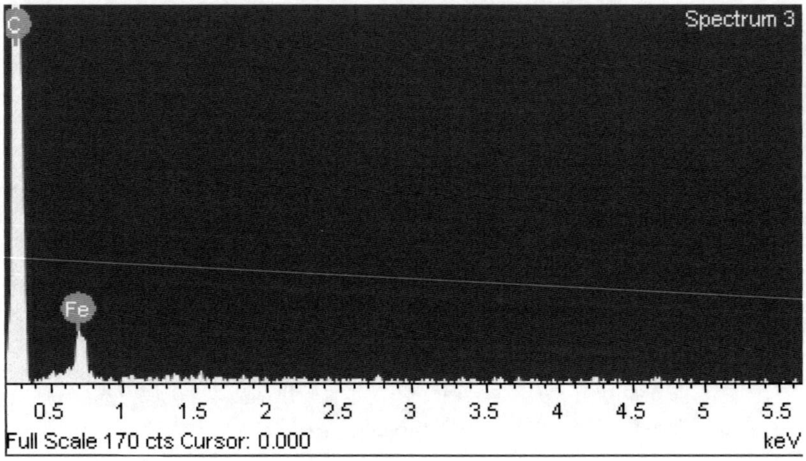

Figure 5: EDS analysis of carbon nanostructures synthesized from PET at 700ºC.

4. Conclussion

Carbon structures in nano and micron size were successful produced from HDPE and PET waste by CVD method. The different type of polymer waste was produced different structures whereas the PET tends to produce carbon nanofilaments while HDPE source tends to produce agglomerate carbon ball. The formation of carbon structures also were influenced by temperature reaction. For PET, carbon nanofilament with higher diameter was produced at higher reaction temperature of 900ºC. The carbon yield of carbon nanostructure synthesized from PET was lower than HDPE due to different morphology and formed structures. Finally, EDS analysis was carried out in order to confirm the formation of carbon nanostructures after CVD reaction.

Acknowledgement

I acknowledge Universiti Tun Hussein Onn Malaysia (UTHM) for the short-term grant of this related project.

References

[1] Iijima, S. 1991 *Nature*. **354**(6348) L56
[2] Musatov A, Kiselev NA, Zakharov DN, Kukovitskii EF, Zhbanov AI, Izrael'yants KA and Chirkova EG, 2001 *Applied Surface Science*. **183**(1-2) L111
[3] Zhao X, Ohkohchi M, Wang M, Iijima S, 1997 *Carbon*. **35**(6) L775
[4] Shi Z, Lian Y, HuiLiao F, Zhou X, Gu Z, Zhang Y, Iijima S, Li H, ToYue K, Zhang S, 2000 *Journal of Physics and Chemistry of Solids*. **61**(7) L1031
[5] Iijima S and Ichihashi T, 1993 *Nature*. **363**(6430) L603
[6] Hatta MNM, Xu F. Xia Y and Zhu Y, 2014 *Applied Mechanics and Materials*. **465** L927
[7] Lee CJ, Lyu SC, Cho YR, Lee JH and Cho KI, 2001 *Chemical Physics Letters*. **343**(1-2) L33
[8] Li WZ, Wen JG and Ren ZF, 2002 *Applied Physics A: Materials Science & Processing*. **74**(3) L397
[9] Harding KG, Dennis JS, Von Blottnitz H and Harrison STL, 2007 *Biotechnology* **130**(1) L57
[10] Mishra N, Das G, Ansaldo A, Genovese A, Malerba M, Povia M and Sharon M, 2012. *Journal of Analytical and Applied Pyrolysis*. **94** L91
[11] Chen J, Xiong X , and Xiao P, 2009 *Materials Chemistry and Physics*. **116**(1) L57
[12] Hofmann S, Csanyi G, Ferrari AC, Payne MC and Robertson J, 2005 *Physical Review Letters*. **95**(3) L101.

The effect of alkali treatment under various conditions on physical properties of kenaf fiber

Mohd Yussni Hashim, Azriszul Mohd Amin, Omar Mohd Faizan Marwah, Mohd Hilmi Othman, Mohd Radzi Mohamed Yunus, Ng Chuan Huat

Advanced Forming Research Group, Universiti Tun Hussein Onn Malaysia, 86400 Parit Raja, Batu Pahat, Johor, Malaysia

E-mail: yusni@uthm.edu.my

Abstract. Alkali treatment was known as one commonly used treatment for natural fiber and obtained a good fiber matrix adhesion, thermal and mechanical properties of composite. This paper reports the effect of alkali treatment conditions on the physical properties of kenaf fiber such as density, weight loss, diameter, cross sectional and fiber morphology. In this study, kenaf fiber were treated under various conditions alkali concentration at 2, 6 and 10 (w/v%), immersion duration at 30, 240 and 480 minute and immersion temperature at 27, 60 and 100°C. The effects of alkali treatment on kenaf fiber physical properties under various conditions were explored. Alkali treatment conditions setting have an impact on kenaf bast fiber physical characteristics. In kenaf fiber density, it was found that a small increase of kenaf density value after alkali treatment compared to untreated kenaf fiber. When each of alkali treatment parameters increased, kenaf fiber density also increased slightly. Kenaf fiber weight loss also exhibit a similar pattern of small increment with the increase of treatment parameters. Additionally, kenaf fiber diameter and cross sectional area show a decline pattern after treated with alkali. Furthermore, this decline pattern was corresponding with the increasing level of alkali treatment parameters. The maximum of kenaf diameter reduction was recorded at high level of each alkali treatment parameters.

1. Introduction

The growing interest in utilization of plant fiber as environment friendly reinforcement material in polymer composite industry is mainly due to their renewable origin, low-cost and abundantly available, relative high specific strength and modulus, light weight, low density, less abrasiveness, desirable fiber aspect ratio, minimal health hazards and also good thermal, electrical and acoustic insulating character. These advantageous make plant fiber as potential replacement for synthetic fiber in composite industries [1-3]. A renaissance in the use of natural fibers as reinforcements in technical applications is taking place mainly in the automobile [4-5], packaging [6], insulation [7] and construction industries [8].

However, there still several drawbacks that limit the potential of plant fiber to be used as alternative to synthetic fibers. Most highlighted problem in dealing with natural fiber was it hydrophilic nature, which leads to an adhesion problem with hydrophobic nature of the polymer matrix [9]. Hydrophilic character of natural fiber is incompatible with hydrophobic polymer matrix and has a tendency to form aggregates. Furthermore, it exhibit poor resistant to moisture, which lead to high water absorption, subsequently resulting in poor mechanical properties and dimensional stability of the natural fiber reinforced composites [10-11]. Therefore, chemical modifications either

IOP Conf. Series: Journal of Physics: Conf. Series **914** (2017) 012030 doi:10.1088/1742-6596/914/1/012030

on natural fiber, polymer matrix or both materials are an alternative solution to overcome these challenges [12-16]. One of the common fiber treatment that extensively used by the number of researcher is alkali treatment process [17-20]. Alkali treatment was selected as a fiber treatment process in this work due to its potential to provide a good fiber matrix adhesion, thermal and mechanical properties of composites .

There are three variables involved during conducting alkali treatment process which are alkali solution concentration, immersion duration and immersion temperature. This treatment variable was set differently according to the types of fiber by the researcher. An attempt to investigate the effect of immersion time during alkali treatment at specific alkali concentration and immersion temperature on fiber was conducted by Symington *et al.*[21] and Obi [22]. For example, Symington *et al.*[21] set the immersion duration at 10, 20 and 30 minute where the alkali solution concentration was fixed as constant at 3% over room temperature condition. The tensile strength of all studied fibers was found to give a decrease pattern when the immersion time increased. On the other hand, Obi [22] used immersion duration at 1,4,8 and 8 hour with 5% alkali concentration at 30°C and found that 8 hour give a best results in improving Borassus fiber tensile properties. Another work also has been done to investigate the effect of various alkali concentrations at fixed immersion time and temperature. Mwaikambo [23] studied the tensile properties of jute fibers after treated with alkali treatment at different range of concentration. He found that jute fiber tensile properties optimized at 0.24% weight by weight (w/w) alkali concentration. Similar type of experiment was reported by Modibbo *et al.* [24], where alkali treatment at different range of alkali solution concentration at fixed immersion duration and temperature had been conducted to observed the bast fiber physical properties changes. Cho [20] determined pre treatment with 10% weight (wt) alkali solution give positive role in the formation of jute based carbon fiber. Boopathi *et al.* [25] investigated the physical, chemical and mechanical properties of alkali treated Borassus fruit fiber. The fiber was treated with 5%, 10% and 15% alkali about 30 minute at room temperature and the effect of alkali treatment on the fiber properties were explored.

Although most of the researcher focused on specific variables at one time during conducting alkali treatment, there are some works that also focus on investigating the relation between alkali treatment variables. Wang *et al.* [26] investigated the effect of chemical treatment process of jute fiber. The examined treatment process was sodium hydroxide concentration, treatment time, temperature, sodium silicate concentration, fiber to liquor ratio, penetrating agent concentration and degumming agent concentration. An orthogonal experiment design was used to optimize the treatment process. It was found that sodium hydroxide concentration, sodium silicate concentration and treatment time were the three most important variables for degumming process. Edderozey [27] studied the morphology and tensile strength of kenaf fiber that modified at different alkali solution concentration and temperature. Alkali concentration used was 3%, 6% and 9% for 3 hour at room temperature. Another treatment was immersion of kenaf fiber in 6% alkali concentration at 95°C. The result shows tensile strength of kenaf fiber that treated with 6% alkali solution in water bath at 95°C was better compared to other treatment condition at room temperature. This studied result has good agreement with Saha *et al.* [28] work which applied alkali treatment at ambient and elevated temperatures and high pressure steaming conditions. The results indicated that jute fiber uniaxial tensile strength increased up to 65% for alkali steam treatment.

Studied regarding tensile properties of kenaf fiber due to various condition of alkali treatment was reported by Mahjoub *et al.* [29]. They employed different alkali concentration and different immersion time. Untreated kenaf fiber was used as control unit. As a result, they found alkaline with 10% and 15% concentration solution damaged the texture of fiber as the treated fiber was more twisted, much finer and so brittle than the untreated fiber. Three methods which are system compliance method, regression method and averaging the data method were performed in their analysis work. Unfortunately, variables interaction or main effect analysis was not discussed in their work.

Natural fiber characters that commonly analyzed by researcher are fiber density, fiber diameter, fiber cross-section area and fiber surface morphology. According to Mwaikambo[30], there are two types of density, which are absolute density and bulk density. The absolute density excludes all the pores and lumen and therefore a measure of the solid matter of the fibers. The apparent density on the other hand includes all the solid material and the pores of the fibers. Fiber apparent density is always less than the absolute density due to the buoyancy effect caused by the trapped air. The absolute density of most natural fiber is between 1.4 – 1.5 g/cm^3. The common method of determining fiber apparent density is using Archimedes principle. This method involves the immersion of a known weight of fiber into a solvent of lower density than the fiber like water or benzene. The volumetric displacement was observed and the weight to volume ratio yielded the density value [25, 31]. For standard method of determining fiber density, a Technical Association of Pulp and Paper Industry (TAPPI) was used by Modibbo *et al.* [24] while American Society for Testing and Materials (ASTM) D 1895 was used by Jawaid *et al.* [32].

One of the challenges in determining natural fiber diameter is the irregularity within or between the single fibers. Mwaikambo *et al.* [33] calculated mean, standard deviation and standard error of hemp fiber diameter using image analysis captured by scanning electron microscope (SEM). The fiber diameter was determined along the width of the fiber bundles. Similar measuring method was used by Khan *et al.* [34] and Andersons *et al.* [35] to determine fiber diameter. However, in Anderson *et al.* [35] work, a fiber cross section was considered as circular. Another studied was conducted by Kabir [36] which used two different approach in calculating the hemp fiber diameter. The first method used was averaged fiber widths at different locations along the fiber's length. For this, fibers were placed horizontally under the microscope and attached with glue tag to avoid movement. The second method was measured the diameters around the breakage of the tested fibers. For this case, tested fibers were perpendicularly attached with glue tag and the equivalent diameters were obtained through image analysis.

For simplicity purposed, a typical circular fiber area assumption was widely applied during fiber diameter or cross-section measurement. Thus, it introduces a considerable error into the calculated mechanical properties. To minimize this variation, Aslan [37], studied two method of measuring cross sectional area of the flax fiber. First method was measured a fiber width along a fiber with assuming the fiber have a circular cross section area. Second method was true fiber area measurement by embedded the fiber in epoxy resin and conduct a measurement on a series of cross sectional image. The difference between the mean values suggests that the circular fiber area measured by the first method is overestimating the true fiber area by 39%. Nitta *et al.* [38] also conducted similar work which investigated the microstructural changes of kenaf fibers after highly concentrated alkali treatment. However, the approximated shapes of the cross-section was not only a circle but including an, ellipse, hexagon, octagon, dodecagon, and icositetragon. Micrograph observation showed each cell in the fiber cross-section changed from a polygon shape to an elliptical, accompanied with many gaps between cells. Their results indicated the tensile strength was greatly improved. Particularly, the alkali treated fiber on the condition of 15 wt % of alkali concentration for one hour is comparable to untreated fiber.

Based from above literature, it shows a number of works that used alkali treatment as chemical modification to improve natural fiber surface and it compatibility with matrix to enhance composite properties. However, there are limited work reported regarding attempts to determine the main effect and the interaction between these three variables on it effect to natural fiber properties enhancement. Therefore, it is necessary to fill this gap by conducting an appropriate experiment design. This experimental work potentially provides significant knowledge on alkali treatment variables relation and it optimum treatment condition if performance of the natural fiber in composite material form is to be maximized. Thus, this study aim to determine the effect of alkali treatment under various conditions on the physical properties of kenaf fiber such as density, weight loss, diameter, cross sectional and fiber morphology.

IOP Conf. Series: Journal of Physics: Conf. Series **914** (2017) 012030 doi:10.1088/1742-6596/914/1/012030

2. Experimental Work

2.1 Kenaf Fiber & Alkali Treatment

Bast type kenaf fibers was supplied from Kenaf Natural Fiber Industries (Malaysia) Sdn.Bhd. This fiber was subjected to water retting process before supplied to the laboratory. Sodium hydroxide (NaOH) used in this studied was supply by BDH Prolabo (UK) in 1 kg container at small pallet shape. Alkali solution concentration was prepared using weight per volume percentage (w/v %). Alkali treatment setting is according to three factors at three levels full factorial design matrix as shown in Table 1. The setting was set manually at laboratory dyeing and the mechanism of temperature control in this dyeing machine is using water. Initially, tab water was poured into water bath container until specific marking line. Temperature and time can be control using a knob at control panel. When machine is switch on, heating coils inside water bath will generate heat so that the water temperature will increase and maintain at selected temperature level. Portable infrared thermometer device was used to confirm the solution temperature. After temperature was set correctly, kenaf fiber was inserted into stainless steel container and this container was placed into dyeing machine until it specific immersion time. Once treatment process is finished, kenaf fiber was taken out from container and washed in running tab water. A drop of acetic acid was dropped into a beaker that contained washed kenaf fiber to remove any excessive sodium hydroxide. This washing process was continued until pH level in distilled water inside a beaker show 7.0 ± 0.5 values. A digital pH tester was used to check the rinse solution pH value to ensure the alkali treatment process is complete. Alkali treated kenaf fiber were conditioned at 100°C ~105°C for one hour in high temperature steamer for fiber drying process.

Table 1: Independent variables matrix of alkali treatment conditions for kenaf fiber physical properties evaluation

Independent variables	Actual variables value		
	Low	**Center**	**High**
X_1: Alkali concentration (w/v%)	2	6	10
X_2: Immersion time (min)	30	240	480
X_3: Immersion temperature (±1°C)	27	60	100

2.2 Fiber density

Alkali treated and untreated kenaf fiber was select randomly and cut into a nominal length of 20 mm. Then, this fiber was weighted using Mettler Toledo analytical balance (Model XS 64) to measure it weight in air and in distilled water. The fiber nominal weight was 6.0 ± 1.0 mg. Fiber density value was determined using the principle of Archimedes by calculating the different of fiber weight in air and in distilled water according to ASTM D3800-99 (2010)[39]. The fiber density measurement was replicated three times at each treatment conditions.

2.3 Fiber diameter

Kenaf fiber was select and separated from bast fiber to bundle fiber and finally to single fiber using Waltex magnifying glass as shown in Figure 1. A kenaf fiber diameter was measured at 4.5 mm interval along its length using Leica Stereo Microscopic Video Analyzer 2000 (Model 250D). The mean diameter of single kenaf fiber was measured from the average of ten fibers, in which the diameter for every fiber is the average value of the diameters measured ten times at 4.5 mm interval along the fiber length. The measurement was replicated three times at each treatment conditions.

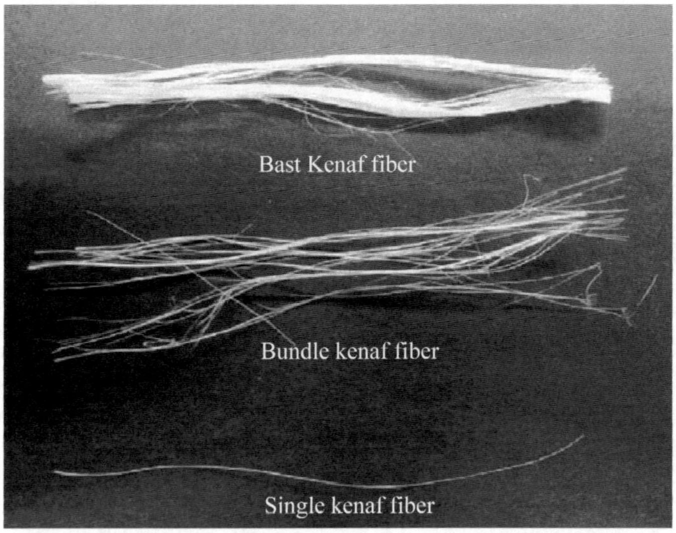

Figure 1: Selection of kenaf fiber from bast fiber to single fiber

2.4 Fiber cross-sectional area
For determining fiber cross sectional area, kenaf fiber with no obvious defects was randomly selected as test sample under Waltex magnifying glass observation. A test specimen mounting tab according to ASTM C1557-03 [40] was prepared for each alkali treatment conditions. Kenaf fiber was assumed to have an elliptical shape with a major,*a* and a minor,*b* diameter. A kenaf fiber cross sectional area was measured at five locations in 6 mm interval of fiber length along two orthogonal directions using Leica Stereo Microscopic Video Analyzer 2000 (Model 250D). The first diameter was measured and followed by next diameter measured at about 90° from the earlier measurement. This measurement was repeated three times to get the average value and it has been replicated three times for full factorial analysis. The fiber cross sectional area, A is then calculated using Equation 1.

$$A = \pi \frac{ab}{4} \qquad (1)$$

2.5 Fiber morphology
The surface morphology of kenaf fiber at various alkali treatment conditions was examined using a scanning electron microscope (SEM). Kenaf fiber was randomly cut into appropriate length and placed into SEM specimen stub. Then this specimen stub was place inside a coating machine to be sputter coated with gold layer to make the surface of the samples conductive prior to SEM observations.

3. Results and Discussion

3.1. Kenaf fiber density and weight loss
In this work, apparent density approach that included all solid material and the pores of kenaf fiber was used to investigate the impact of alkali treatment condition on kenaf fiber density changes. Untreated kenaf fiber density was used as control unit and it measured value was 1.254 (g/cm^3). The amount of kenaf fiber weight reduction and kenaf fiber density at different alkali treatment conditions was presented in Table 2. The result shows that there was a slight increment of kenaf fiber density value after alkali treatment process conducted. Measured responses data in Table 2 was transform into chart form as shown in Figure 2 and Figure 3 to clarify the effect of alkali treatment conditions pattern

on kenaf fiber weight loss and density. Figure 2 shows the bar chart of kenaf fiber weight lost after alkali treatment at various conditions setting. Its indicated nearly similar range of changed (18% ~21%) at constant room temperature with various alkali concentration and immersion time. However, at 2 w/v% alkali concentration, the weight loss was 13% recorded. On the other hand, at immersion temperature 60°C and 100°C, kenaf weight loss shows an increment pattern with the increment of alkali concentration and immersion duration value.

The line graph in Figure 3 shows the kenaf fiber density mean value at different alkali concentration. It depict that kenaf fiber density value became higher when it was treated with higher alkali concentration (10w/v%) with higher immersion temperature and longer immersion time. Furthermore, kenaf fiber density at lower alkali concentration (2w/v%) exhibit nearly constant changes at room temperature and 60°C immersion temperature. However, at 100°C and longer immersion time, the kenaf fiber density value was increased slightly. The gradual increase in weight loss after alkali treatment with increasing alkali concentration and immersion time was reported by Roy *et al.* [41]. In their elemental analysis of untreated and alkali treated jute fiber, they suggested that the weight loss was due to leaching of lignin and/or hemicellulose. This gradual removal of hemicellulose and lignin increased the weight loss. The alkali treatment breaks hydrogen bonds between the hydroxyl groups (–OH) of the cellulose, hemicellulose and lignin and leads to defibrillation, the breakdown of the fiber bundle into smaller fibers. This change increases the effective surface area [42].

Table 2: Amount of kenaf fiber weight reduction and fiber density measured after various alkali treatment conditions

Responses to the kenaf fiber properties	**Time (min)**	**Temperature level (°C)**								
		Room temperature (±27)			**60**			**100**		
		Alkali concentration (w/v %)								
		2	**6**	**10**	**2**	**6**	**10**	**2**	**6**	**10**
Fiber weight after treatment (g)		0.82	0.81	0.81	0.84	0.85	0.83	0.74	0.71	0.67
Weight loss %	30	18	19	19	16	15	17	26	29	33
Fiber density (g/cm^3)		1.262	1.270	1.276	1.268	1.269	1.280	1.272	1.273	1.281
Fiber weight after treatment (g)		0.87	0.80	0.79	0.84	0.80	0.77	0.75	0.67	0.64
Weight loss %	240	13	20	21	16	20	23	25	33	36
Fiber density (g/cm^3)		1.269	1.267	1.282	1.271	1.275	1.285	1.279	1.283	1.285
Fiber weight after treatment (g)		0.81	0.79	0.79	0.81	0.75	0.71	0.60	0.62	0.60
Weight loss %	480	19	21	21	19	25	29	40	38	40
Fiber density (g/cm^3)		1.269	1.273	1.282	1.269	1.280	1.287	1.279	1.283	1.288

IOP Conf. Series: Journal of Physics: Conf. Series **914** (2017) 012030 doi:10.1088/1742-6596/914/1/012030

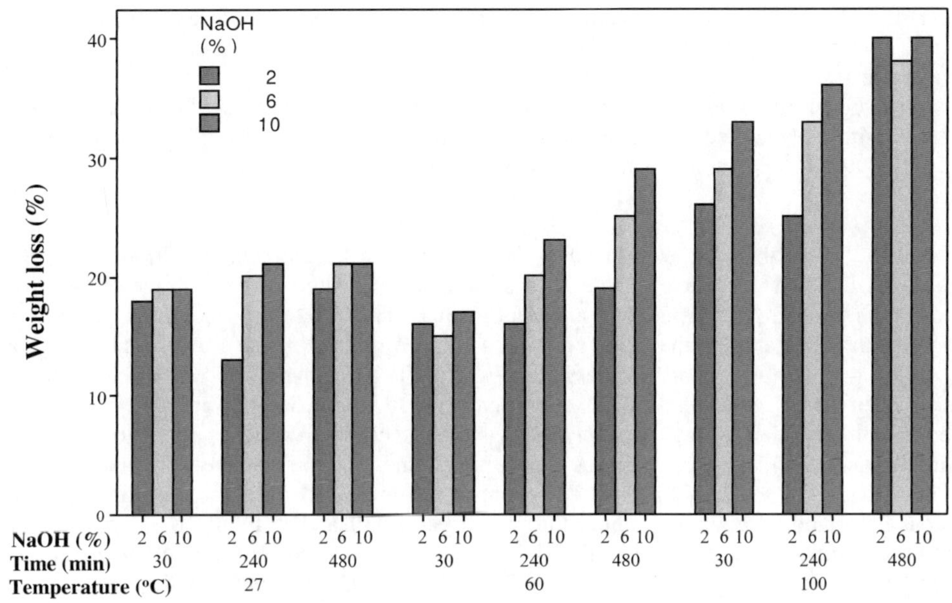

Figure 2: Kenaf fiber weight loss after treated at various alkali treatment conditions

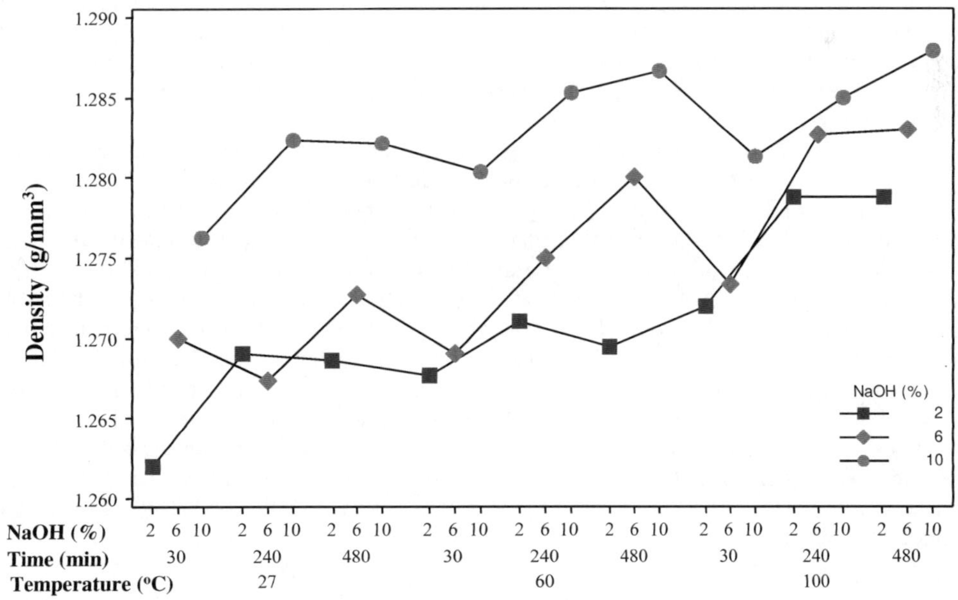

Figure 3: Kenaf fiber density changing pattern after treated at various alkali treatment conditions

Furthermore, the density of kenaf fiber at various alkali treatment conditions shows a gradual increment pattern at small amount with the increment of all treatment parameters value about 2.32%. On top of that, alkali treated kenaf fiber density for all conditions show a higher density value when compared to untreated fiber at about 0.6% to 2.7% increment. This similar pattern was reported by Aziz *et al.* [31] for alkali treated kenaf and hemp fiber. This positive change or increment in fiber densities normally signifies cell wall densification. However, a negative change would signify cell wall damage leading to de-polymerization of the cellulose molecule. As reported by Modibbo *et al.* [24], when higher alkali concentration was used during treatment process, the density of kenaf fiber decreased with the increment of alkali concentration. At their work, the alkali concentration ranges was 10-25% at 5°C for 20 minutes immersion duration.

3.2. Kenaf fiber diameter

The result of kenaf fiber diameter changes after various alkali treatment conditions was presented in Table 3. At the first column of Table 3, it shows the measured responses, which are kenaf fiber diameter and standard deviation measurement values. The rest of the columns present the alkali treatment conditions and the measured responses value. Untreated kenaf fiber diameter average value was 104.69 μm with 15.5 standard deviation value. The results indicated that kenaf fiber diameter reduced after treated with alkali at all conditions. At 30 minutes constant immersion duration, the percentage of kenaf diameter decrement was 27.4% when others treatment parameter increased.

Similarly, at 240 and 480 minutes immersion time, the percentage of kenaf fiber diameter decrement was 33.33% and 40.12%. This percentage increased with the increment of immersion duration. For standard deviation value, the range was small and it about 10 ~ 20. The kenaf fiber diameter was little bit larger compared to kenaf diameter reported by Mahjoub *et al.* in their recent studied [29] which was around 55 ~ 73 μm with 10 ~ 17 standard deviation value. Figure 4 shows the decrement pattern of kenaf fiber diameter after treated at various alkali treatment conditions. From the line graph, it was clearly understood that at constant immersion temperature and immersion time, an increment of alkali concentration reduced the diameter of kenaf fiber. Furthermore, the increment of all parameters setting value also reduced the diameter of kenaf fiber. This finding could be explained by fiber degradation and fiber delignification phenomenon that occurred in kenaf fiber as alkali treatment variables increased. As the alkali concentration, immersion time and immersion temperature increased, kenaf fiber became finer due to the leaching of non-cellulosic materials like hemicellulose, lignin and pectin [41].

Other researchers also reported the decrement pattern of fiber diameter after treated with alkali solution. Gomes et. al [43] emphasized that the decreased in fiber diameter with the increment of alkali concentration was caused by change of morphology of the treated fiber. From scanning electron microscope observation, they revealed that the fiber was not truly monofilament. It was a bundle of monofilament bonded and covered by lignin. Alkali treatment was inferred to provoke removal of a great amount of lignin. The removal of lignin reduced the fiber diameter. This statement was supported with the results from Oby [22] work, where their found that average diameter of borassus fine fiber also decreased with the increment of alkali treatment period. The alkali treatment removes certain hemicellulose and wax covering the external surface of the fiber and therefore it increases the aspect ratio (length/diameter) values. In addition, Saha *et al.* [28] also found that 30 minutes alkali treatment with 1% alkali solution at 90°C was contributed to fiber diameter reduction at 32%. The degradation of pectin together with lignin removed from the fiber cell wall that causing separation of the ultimate fiber which lead to the fiber diameter decreased was about between 0 to 0.03% alkali concentrations [23].

Table 3: Kenaf fiber diameter changes after various alkali treatment conditions

Kenaf fiber diameter changes	Time (min)	Temperature Level (°C)								
		Room temperature (±27)			60			100		
		Alkali Concentration (w/v %)								
		2	6	10	2	6	10	2	6	10
Average (µm)	30	96.46	95.16	91.46	85.12	83.08	82.12	73.34	72.27	70.03
STD		15.98	16.17	18.29	18.94	10.48	11.55	19.34	19.54	15.96
Average (µm)	240	95.25	93.03	91.48	85.49	80.05	80.79	70.29	68.86	63.50
STD		14.22	15.42	16.01	19.64	19.57	17.47	15.78	15.71	15.13
Average (µm)	480	92.17	91.85	89.05	78.66	75.19	72.89	61.19	59.54	55.19
STD		18.03	16.98	16.20	19.47	19.72	11.88	18.07	18.05	17.27

STD = Standard deviation

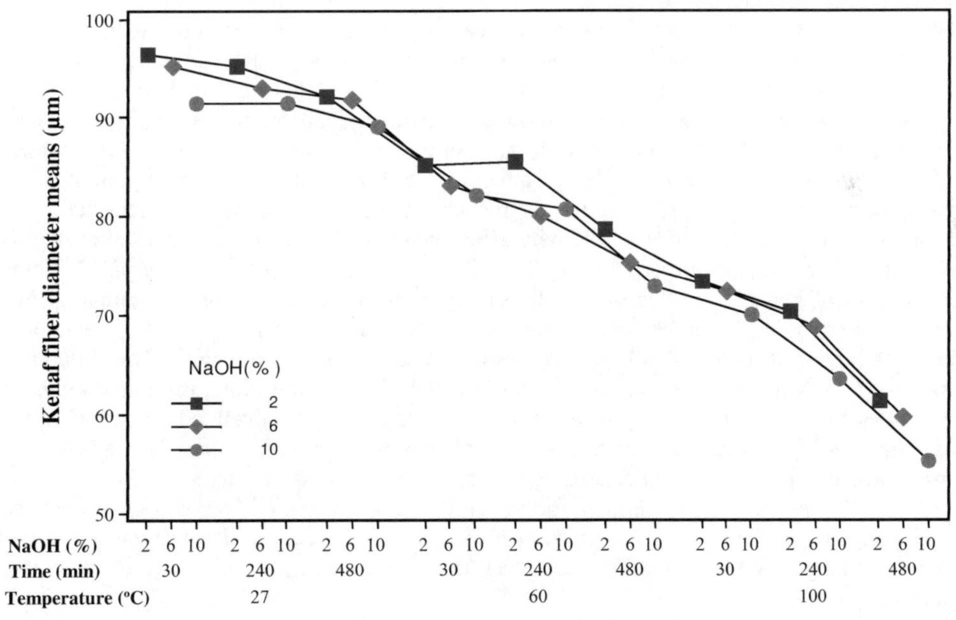

Figure 4: Kenaf fiber diameter (µm) changing pattern after treated at various alkali treatment conditions

3.3. Kenaf fiber cross-sectional area

The result of kenaf fiber cross sectional area changes after various alkali treatment conditions was presented in Table 4. The first column of Table 4 shows the measured responses, which are kenaf fiber diameter and standard deviation measurement values. The rest of the columns present the alkali treatment conditions and the measured responses value. Untreated kenaf fiber cross sectional area.

Table 4: Kenaf fiber cross sectional changes after various alkali treatment conditions

Kenaf fiber cross sectional changes (μm^2)	**Time (min)**	**Temperature Level (°C)**								
		Room temperature (±27)			**60**			**100**		
		Alkali Concentration (w/v %)								
		2	6	10	2	6	10	2	6	10
Average	30	6034	5966	5849	5832	5794	5654	5612	5417	5351
STD		131	114	63	92	136	106	134	75	113
Average	240	5963	5870	5817	5806	5713	5637	5445	5233	5151
STD		84	92	111	111	129	93	73	102	64
Average	480	5865	5877	5775	5797	5718	5595	4841	4633	4604
STD		45	108	68	87	118	94	73	72	56

STD = Standard deviation value was 6472.23 μm^2 with 138.15 standard deviation value. The results indicated that kenaf fiber cross sectional area reduced after treated with alkali at all conditions. At 30 minutes constant immersion duration, the percentage of kenaf cross sectional area decrement was 11.32% when others treatment parameter increased. Similarly, at 240 and 480 minutes immersion time, the percentage of kenaf fiber cross sectional area decrement was 13.62% and 21.5%. This percentage increased with the increment of immersion duration. Figure 5 shows the decrement pattern of kenaf fiber cross sectional area after treated at various alkali treatment conditions. From the line graph, it shows a slight decrement pattern at constant immersion temperature, immersion time and alkali concentration. However, at 100°C immersion setting, a rapid decrement pattern of kenaf fiber cross section area was recorded when immersion time and alkali concentration increased.

Similar to the reduction of fiber diameter, the reduction of fiber cross-sectional area possibly was due to the changed of fiber morphology after alkali treatment process. Nitta *et al.* [38] reported that cross-section of kenaf fiber was drastically changed by the alkali treatment compared to untreated fiber. The increments of immersion duration give a fluctuated pattern where the cross-sectional area value was higher at higher alkali concentration and longer immersion time. They found that cross-sectional of alkali treated fiber was less then untreated fiber. Furthermore, each cell in the fiber cross-section was changed from a polygon shape to an elliptical one, accompanied with many gaps between cells. Thus, during measuring fiber cross-sectional area, the assumption that fiber cross-sectional is an ellipse shape will produce more accurate results compare to round shape assumption.

On the other hand, Kabir *et al.* [36] found that the cross section of fiber was not show a continuous decrease pattern as the alkali concentration increased. Their studied also present a contradict result where the untreated fiber cross section was less than alkali treated cross section at particular concentration.

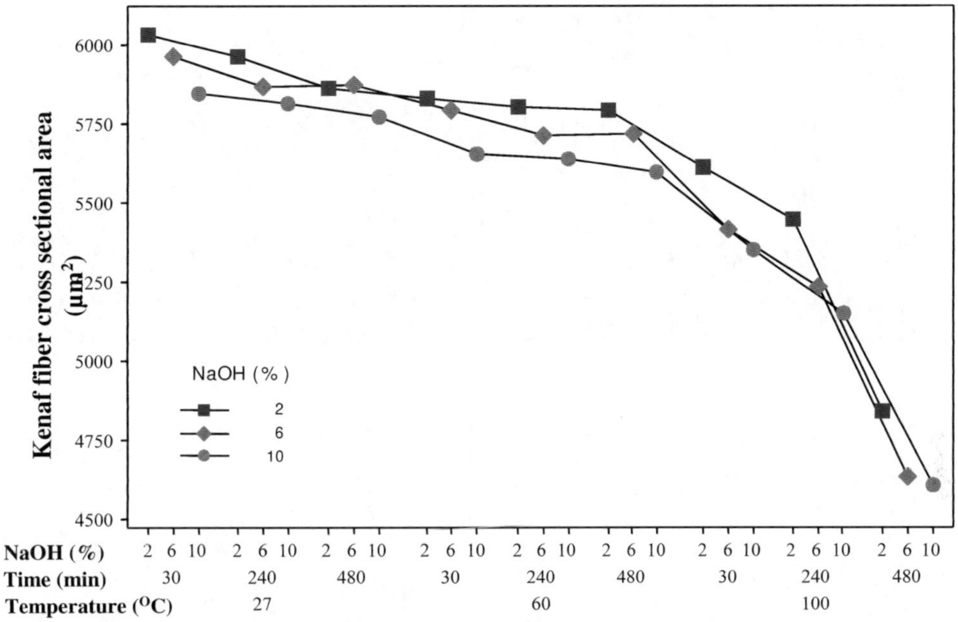

Figure 5: Kenaf fiber cross sectional area (μm^2) changing pattern after treated at various alkali treatment conditions

3.4. Kenaf fiber morphology

Figure 6 (a) – (f) shows the scanning electron microscope image of untreated kenaf fiber and alkali treated kenaf fiber at different conditions. From Figure 6 (a) it appears that the surface of untreated kenaf fiber is not smooth but rather has longitudinal ridges, which could possibly enhance the mechanical interlock with the polymer matrix. It is observed that the impurities such as hemicelluloses, lignin, and waxes substances were presented in untreated kenaf fiber. The white spots on the surface of the fiber may indicate impurities or localized damage of the fibers due to the handling process. Similar conditions was reported by Lai *et al.* [44]. From Figure 6 (b), it shows the considerable deposits of natural compounds such as hemicellulose, wax and lignin which are still present on the fiber surface after treatment with a 2% solution of alkali for 30 minute at room temperature, RT (± 27°C). This indicates that at lower setting alkali treatment conditions, the alkali treatment was not significantly effective to clean the fiber surface.

In Figure 6 (c) at lower alkali concentration at room temperature over 240 minute immersion time, the impurities is still can be detected at the fiber surface. Furthermore, delignification effect also can be observed in fiber surface image. The cleanness and roughness in fiber surface due to fiber delignification process was observed to increase by increasing the alkali concentration, immersion time and immersion temperature as presented in Figure 6 (d), (e) and (f). This observation had good agreement with recent research work reported by Azelee *et al.* , which found that temperature increase in pretreatments for delignification has increase the percentage of lignin removal. Increasing temperature from 40°C to 100°C significantly increased the percentage of lignin removal to more than 70% in kenaf fiber [45]. Moreover, thermo gravimetric analysis (TGA) indicates that the kenaf fibers were thermally stable below 218 and that, as such, the fibers could be effectively used as reinforcement when the moulding temperature was set under this temperature [46].

IOP Conf. Series: Journal of Physics: Conf. Series **914** (2017) 012030 doi:10.1088/1742-6596/914/1/012030

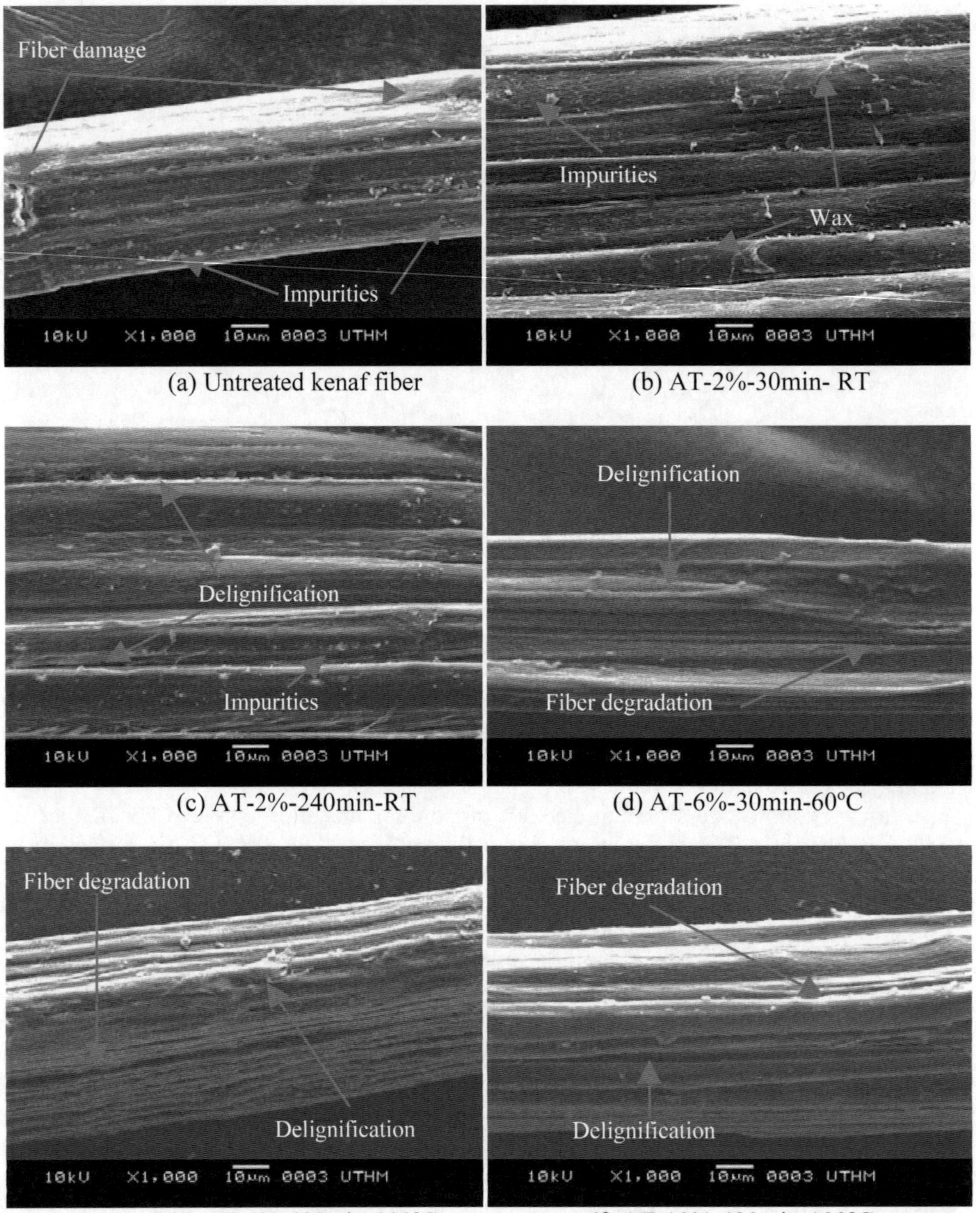

(a) Untreated kenaf fiber (b) AT-2%-30min- RT

(c) AT-2%-240min-RT (d) AT-6%-30min-60°C

(e) AT-6%-480min-100°C (f) AT-10%-480min-100°C

Figure 6: (a) – (f) Scanning electron microscope image of untreated kenaf fiber and alkali treated (AT) kenaf fiber at different conditions

4. Conclusion

The effects of alkali treatment under various conditions setting on physical properties of kenaf fibers were investigated. For kenaf fiber density, it was found that a small increase of kenaf density value after alkali treatment compared to untreated kenaf fiber. When each of alkali treatment parameters increased, kenaf fiber density also increased slightly. Kenaf fiber weight loss also exhibit a similar pattern of small increment with the increase of treatment parameters. Additionally, kenaf fiber diameter and cross sectional area show a decline pattern after treated with alkali. Furthermore, this decline pattern was corresponding with the increasing level of alkali treatment parameters. The maximum of kenaf diameter reduction was recorded at high level of each alkali treatment parameters.

5. Acknowledgements

Authors wish to thank University Tun Hussein Onn Malaysia (UTHM) and Ministry of Higher Education, Malaysia for financial support and assistance in order to complete this study.

6. References

[1] C. Asasutjarit, *et al.*, "Materials and mechanical properties of pretreated coir-based green composites," *Composites Part B: Engineering,* vol. 40, pp. 633-637, 2009.
[2] K. Murali Mohan Rao, *et al.*, "Fabrication and testing of natural fibre composites: Vakka, sisal, bamboo and banana," *Materials & Design,* vol. 31, pp. 508-513, 2010.
[3] S. M. Sapuan, *et al.*, "Mechanical properties of woven banana fibre reinforced epoxy composites," *Materials & Design,* vol. 27, pp. 689-693, 2006.
[4] G. Koronis, *et al.*, "Green composites: A review of adequate materials for automotive applications," *Composites Part B: Engineering,* vol. 44, pp. 120-127, 2013.
[5] C. Borsellino and G. D. Bella, "Paper-reinforced biomimetic cellular structures for automotive applications," *Materials & Design,* vol. 30, pp. 4054-4059, 2009.
[6] G. Davis and J. H. Song, "Biodegradable packaging based on raw materials from crops and their impact on waste management," *Industrial Crops and Products,* vol. 23, pp. 147-161, 2006.
[7] H.-R. Kymäläinen and A.-M. Sjöberg, "Flax and hemp fibres as raw materials for thermal insulations," *Building and Environment,* vol. 43, pp. 1261-1269, 2008.
[8] A. Elsaid, *et al.*, "Mechanical properties of kenaf fiber reinforced concrete," *Construction and Building Materials,* vol. 25, pp. 1991-2001, 2011.
[9] N. Sgriccia, *et al.*, "Characterization of natural fiber surfaces and natural fiber composites," *Composites Part A: Applied Science and Manufacturing,* vol. 39, pp. 1632-1637, 2008.
[10] M. J. John, *et al.*, "Effect of chemical modification on properties of hybrid fiber biocomposites," *Composites Part A: Applied Science and Manufacturing,* vol. 39, pp. 352-363, 2008.
[11] P. A. Sreekumar, *et al.*, "Effect of fiber surface modification on the mechanical and water absorption characteristics of sisal/polyester composites fabricated by resin transfer molding," *Composites Part A: Applied Science and Manufacturing,* vol. 40, pp. 1777-1784, 2009.
[12] X. Li, *et al.*, "Chemical Treatments of Natural Fiber for Use in Natural Fiber-Reinforced Composites: A Review," *Journal of Polymers and the Environment,* vol. 15, pp. 25-33, 2007.
[13] V. Vilay, *et al.*, "Effect of fiber surface treatment and fiber loading on the properties of bagasse fiber-reinforced unsaturated polyester composites," *Composites Science and Technology,* vol. 68, pp. 631-638, 2008.
[14] S. Kalia, *et al.*, "Pretreatments of natural fibers and their application as reinforcing material in polymer composites—A review," *Polymer Engineering & Science,* vol. 49, pp. 1253-1272, 2009.
[15] Y. Xie, *et al.*, "Silane coupling agents used for natural fiber/polymer composites: A review," *Composites Part A: Applied Science and Manufacturing,* vol. 41, pp. 806-819, 2010.

[16] I. M. De Rosa, *et al.*, "Effect of chemical treatments on the mechanical and thermal behaviour of okra (Abelmoschus esculentus) fibres," *Composites Science and Technology,* vol. 71, pp. 246-254, 2011.

[17] D. Bachtiar, *et al.*, "The effect of alkaline treatment on tensile properties of sugar palm fibre reinforced epoxy composites," *Materials & Design,* vol. 29, pp. 1285-1290, 2008.

[18] M. S. Islam, *et al.*, "Influence of alkali treatment on the interfacial and physico-mechanical properties of industrial hemp fibre reinforced polylactic acid composites," *Composites Part A: Applied Science and Manufacturing,* vol. 41, pp. 596-603, 2010.

[19] H. M. Akil, *et al.*, "Kenaf Fiber Reinforced Composites: A Review," *Materials & Design,* vol. Volume 32, September 2011, pp. Pages 4107-4121 2011.

[20] D. Cho, *et al.*, "Effect of alkali pre-treatment of jute on the formation of jute-based carbon fibers," *Materials Letters,* vol. 65, pp. 1492-1494, 2011.

[21] M. C. Symington, *et al.*, "Tensile Testing of Cellulose Based Natural Fibers for Structural Composite Applications," *Journal of Composite Materials,* vol. 43, pp. 1083-1108, May 1, 2009 2009.

[22] K. Obi Reddy, *et al.*, "Tensile and structural characterization of alkali treated Borassus fruit fine fibers," *Composites Part B: Engineering,* vol. 44, pp. 433-438, 2013.

[23] L. Y. Mwaikambo, "Tensile properties of alkalised jute fibres," *Bioresources,* vol. 4, pp. 566-588, 2009.

[24] U. Modibbo, *et al.*, "The effect of mercerization media on the physical properties of local plant bast fibres," *International Journal of Physical Sciences,* vol. 4, pp. 698-704, 2009.

[25] L. Boopathi, *et al.*, "Investigation of physical, chemical and mechanical properties of raw and alkali treated Borassus fruit fiber," *Composites Part B: Engineering,* vol. 43, pp. 3044-3052, 2012.

[26] W.-m. Wang, *et al.*, "Study on the chemical modification process of jute fiber," *Journal of Engineered Fibers and Fabrics,* vol. 3, pp. 1-11, 2008.

[27] A. M. M. Edeerozey, *et al.*, "Chemical modification of kenaf fibers," *Materials Letters,* vol. 61, pp. 2023-2025, 2007.

[28] P. Saha, *et al.*, "Enhancement of tensile strength of lignocellulosic jute fibers by alkali-steam treatment," *Bioresource Technology,* vol. 101, pp. 3182-3187, 2010.

[29] R. Mahjoub, *et al.*, "Tensile properties of kenaf fiber due to various conditions of chemical fiber surface modifications," *Construction and Building Materials,* vol. 55, pp. 103-113, 2014.

[30] L. Y. Mwaikambo and M. P. Ansell, "The determination of porosity and cellulose content of plant fibers by density methods," *Journal of Materials Science Letters,* vol. 20, pp. 2095-2096, 2001.

[31] S. H. Aziz and M. P. Ansell, "The effect of alkalization and fibre alignment on the mechanical and thermal properties of kenaf and hemp bast fibre composites: Part 1 - polyester resin matrix," *Composites Science and Technology,* vol. 64, pp. 1219-1230, 2004.

[32] M. Jawaid, *et al.*, "Hybrid Composites Made from Oil Palm Empty Fruit Bunches/Jute Fibres: Water Absorption, Thickness Swelling and Density Behaviours," *Journal of Polymers and the Environment,* pp. 1-4, 2010.

[33] L. Y. Mwaikambo and M. Ansell, "Mechanical properties of alkali treated plant fibres and their potential as reinforcement materials. I. hemp fibres," *Journal of Materials Science,* vol. 41, pp. 2483-2496, 2006.

[34] M. M. Rahman Khan, *et al.*, "Fineness and tensile properties of hemp (Cannabis sativa L.) fibres," *Biosystems Engineering,* vol. 108, pp. 9-17, 2011.

[35] J. Andersons, *et al.*, "Modeling strength scatter of elementary flax fibers: The effect of mechanical damage and geometrical characteristics," *Composites Part A: Applied Science and Manufacturing,* vol. 42, pp. 543-549, 2011.

[36] M. M. Kabir, *et al.*, "Tensile properties of chemically treated hemp fibres as reinforcement for composites," *Composites Part B: Engineering,* vol. 53, pp. 362-368, 2013.

[37] M. Aslan, *et al.*, "Strength variability of single flax fibres," *Journal of Materials Science,* vol. 46, pp. 6344-6354, 2011.

[38] Y. Nitta, *et al.*, "Cross-sectional area evaluation and tensile properties of alkali-treated kenaf fibres," *Composites Part A: Applied Science and Manufacturing,* vol. 49, pp. 132-138, 2013.

[39] A. S. f. T. a. M. A. International, "Standard Test Method for Density of High-Modulus Fibers," ed. Pennsylvania, D3800 – 99 (Reapproved 2010).

[40] A. International, "Standard Test Method for Tensile Strength and Young's Modulus of Fibers," ed. Pennsylvania, US, 2008.

[41] A. Roy, *et al.*, "Improvement in mechanical properties of jute fibres through mild alkali treatment as demonstrated by utilisation of the Weibull distribution model," *Bioresource Technology,* vol. 107, pp. 222-228, 2012.

[42] L. Y. Mwaikambo and M. P. Ansell, "Chemical modification of hemp, sisal, jute, and kapok fibers by alkalization," *Journal of Applied Polymer Science,* vol. 84, pp. 2222-2234, 2002.

[43] A. Gomes, *et al.*, "Effects of alkali treatment to reinforcement on tensile properties of curaua fiber green composites," *JSME International Journal Series A,* vol. 47, pp. 541-546, 2004.

[44] W. L. Lai, *et al.*, "The Properties of Woven Kenaf and Betel Palm Reinforced Unsaturated Polyester Composites," *Polymer-Plastics Technology and Engineering,* vol. 47, pp. 1193 - 1199, 2008.

[45] N. I. Wan Azelee, *et al.*, "Efficient removal of lignin with the maintenance of hemicellulose from kenaf by two-stage pretreatment process," *Carbohydrate Polymers,* vol. 99, pp. 447-453, 2014.

[46] Y. Cao, *et al.*, "Effects of heat and alkali treatments on mechanical properties of kenaf fibers," in *Proceedings of 16th International Conference on Composite Materials, Kyoto Japan*, 2007, pp. 1-4.

Mechanical Behaviour Investigation Of Aluminium Alloy Tailor Welded Blank Developed By Using Friction Stir Welding Technique

Agus Dwi Anggono, Bibit Sugito, Agus Hariyanto, Subroto and Sarjito

Department of Mechanical Engineering, Universitas Muhammadiyah Surakarta, Jl. Ahmad Yani, PO. BOX I Pabelan, Surakarta 57162 Indonesia.

Email: agus.d.anggono@ums.ac.id

Abstract. The objective on the research was to investigate the mechanical properties and microstructure of tailor welded blank (TWB) made from AA6061-T6 and AA1100 using friction stir welding (FSW) process. Due to the dissimilar mechanical properties of the two aluminium alloys, microhardness test was conducted to measure the hardness distribution across the weld nugget. The mixing of two distinct materials was influenced by tool rotation speed. Therefore, microstructure analysis was carried out to investigate the grain size and shape. The grain size of AA6061-T6 has increased in the heat affected zone (HAZ) while for AA1100 has decreased. In the weld nugget, it has found a hook defects in the dissimilar aluminium joining. By using monotonic tensile load, the different weld line direction was observed with the expansion in tool rotation. The joints failure were consistently on the area of AA1100 series. Furthermore, two specimens were investigated, one through the dissimilar aluminium and the other through similiar material. Inspection of the weld nugget hardness was shown that nonhomogen material intermixing during the stiring process as confirmed by microhardness measurement.

1 Introduction

Tailor welded blank (TWB) has being used in automotive industries to develop lighter car body. It can decrease fuel consumtion, have local strength and reduce production cost. By weight reducing, it can make 0.43 km/litre every 68 kg reduction [1]. It was created from similiar or different properties and thickness of sheet metal. TWBs are used to reinforce the car body in specific area where a stiffness and high strength is needed such as pillars, doors and side frames [1], [2]. The seam of welding could be non-linear or linear. It was called engineered blanks for non-linear welding. The joining of TWB can be done by laser welding, induction welding, electron beam and friction stir welding (FSW). The analysis of TWB can be conducted by using numerical analysis and finite element method. The results have a good agreement with the experimental results [3].

To reduce car body weight, the use of low density material can be applied as well such as aluminium alloys (AA). But the material has low formability and hard to weld by using fusion welding. Joining different aluminium alloy cannot be conducted by using fusion welding due to the formation of intermetallic compounds (IMCs) that cause massive brittle. Voids and cracks usually appear in weld line as micro defects. Therefore, friction stir welding has good technique to joint aluminium alloy and other metal with solidification problem during welding process. The process can avoid material distortion and residual stress because of the low temperature during FSW compared with the conventional arc welding. Recently, several hard to weld material have been connected successfully by using FSW method such as Al-Al [4], Al-Cu [5], Al-Mg [6], Al-Fe [5], Al-Ti [7], and Mg-Zn. It can be obtained due to the process of welding was in low temperature to minimize the formation of IMCs.

In the FSW process, many parameters were influenced weld performance such as position of specimen, tool rotation, travel speed, tool tilt angle and tool shape. However, FSW performance related to the welding strength still not satisfying. Many efforts have been conducting to meet the industrial need of welding strength, to improve tensile, bending and fatigue, low shrinkage, low distortion, robust technique and quality repeatable. Based on the literature review above, it can be known that the welding parameters needs to be optimized to get a good performance of FSW.

The research objectives to investigate the influences of tool rotation and welding speed on the mechanical properties and microstructure of the FSW joint in the aluminium sheet of AA6061-T6 and AA1100.

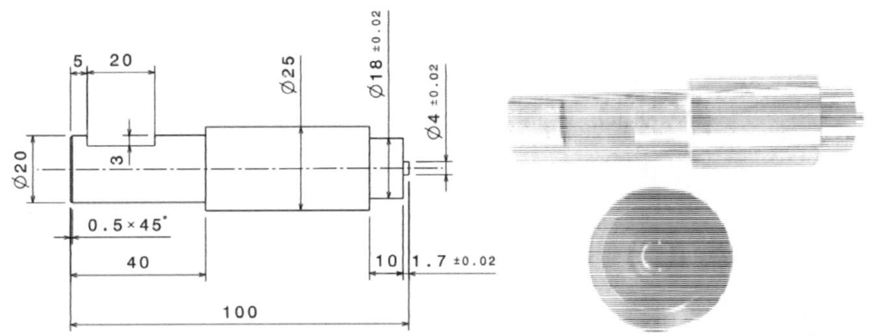

Fig 1. Tool joint design.

2 Materials and Method

In this research, thickness of 2 mm aluminium alloys sheet were selected as a specimens to develop tailor welded blank by using FSW method. Aluminium alloys are widely use in many structures such as building, automotive and aircrafts. That have both of strength and lightweight. Another advantages of AA materials are resistant to corrosion, stress, cracking and toughness [8].

2.1 Materials and tools

There are two types of 2 mm aluminium sheet in this study i.e AA 1100 and AA 6061-T6. Aluminium sheet of 2 mm thick rolled were used as a specimens of FSW. AA1100 is an aluminium family of 1000 or 1xxx series of aluminiun based alloy with 99.0% aluminium. AA6061-T6 is an aluminium alloy which have both strength and toughness. The chemical composition of materials can be seen in Table 1. In the FSW process, the tools used in the research were conventional milling with adjustable spindle head, shearing machine, dial indicator, centrofix, caliper and hand grinding. The tool joint for the FSW process was made of Bohler K100 steel without any treatment. Pin tool profile was cylindrical type with CNC lathe finishing. Dimensions of the FSW tool is shown in Figure 1.

The milling machine specifications are able to move in X, Y and Z direction, Richon X8140A type and the power consumption of 4.5 kVA. The shearing machine was used Ciang Hai Machine Tool brand, QH-II-6x2500 type, cutting thickness of 6mm, cutting width of 2500 mm maximun and the power supply 380V 50Hz. Tensile test was conducted by using universal testing machine with the brand of Instron, 3367 model and the maximum load of 3500 kgf. Micro hardness test was carried out by using Highwood TTS Unlimited Inc, with HWMMT-X7 type.

2.2 FSW parameters and method

The FSW parameters used to produce the joints are shown in Table 2, such as tool speed, welding speed, tilt angle of tool and depth plunge. The ilustration of tool position and the specimen can be seen in Figure 2.

Table 1. Chemical composition of the specemens

No.	Element	Percentage (%)	
		Material 1	Material 2
1	Aluminium (Al)	99.28	96.27
2	Zinc (Zn)	0.036	0.25
3	Ferit (Fe)	0.289	<=0.70
4	Copper (Cu)	0.170	0.35
5	Silicon (Si)	0.134	0.6
6	Manganese (Mn)	<0.0200	0.15
7	Magnesium (Mg)	<0.0500	0.85
8	Chrom (Cr)	<0.0150	0.35
9	Nikel (Ni)	0.0249	0.036
10	Tin (Sn)	<0.0500	<0.04
11	Titanium (Ti)	0.0194	0.15
12	Lead (Pb)	<0.0300	<0.02
13	Calsium (Ca)	0.0027	0.0018
14	Vanadium (V)	0.012	0.013
15	Zirconium (Zr)	0.0163	0.015

The prelimininary investigation will be conducted by visualize the weld results to check the influence of FSW parameters. It will be about the shape, welding contour and defects which can be seen normaly. The welding process will be said successfully if the welding line was smooth and without any hole defects. The specimen is then cut to the standard of ASTM E8, E3 and E384 as seen in Figure 3.

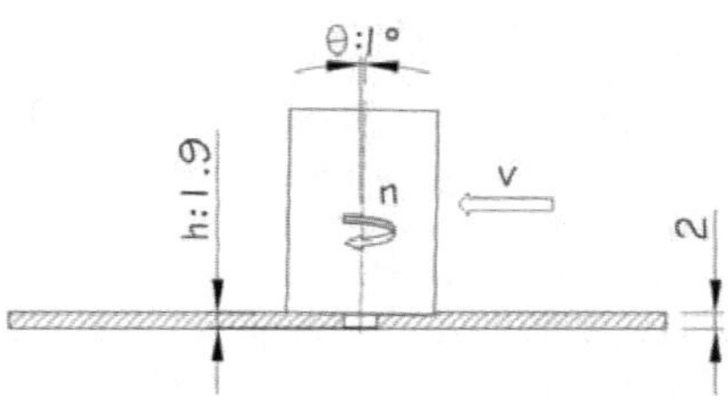

Fig 2. Tool tilt angle and position.

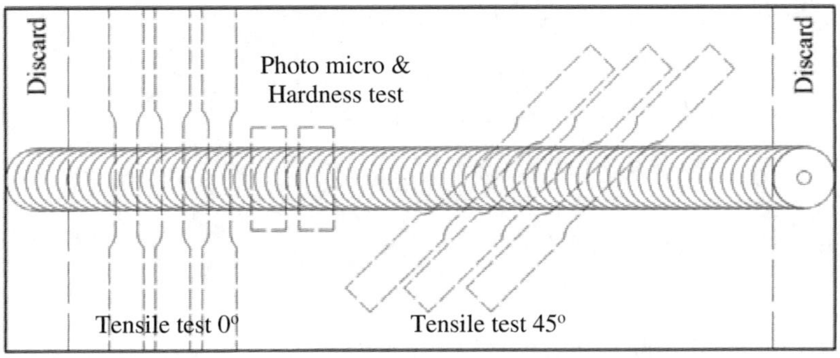

Fig 3. Draft of specimens cutting for testing.

Table 2. FSW parameters

Materials	Tool speed (rpm)	Welding speed (mm/mnt)	Tilt angle (deg)	Depth plunge (mm)
1100 x 1100	1250 1600	10 12.5	1	1.9
6061-T6 x 6061-T6	1250	12.5	1	1.9
1100 x 6061-T6	1250	12.5	1	1.9

3 Results and Discussion

Figure 4(a) was shown the joining results of the same AA1100 series with the parameters of 1250 rpm and 12.5 mm/minutes. The FSW result have many defects such as holes, cracks and rough surface. Unstable welding process and not enough temperature during FSW process will cause many defects [9], [10]. To improve weld quality, another welding parameters were applied i.e 1600 rpm of tool rotation and 10 mm/minutes feedrate. Employing the new parameters has improved the joining quality as seen in thr Figure 4(b). Although the welding line surface was seen rough enough.

The joining of same material of Aluminium Alloy 6061-T6 was carried out by using 1250 rpm of tool rotation and welding speed of 12.5 mm/minutes. The result of the FSW process can be seen in Figure 5(a). According to the visualization of the weld line, it was generated smooth surface along the welding line.

The joining results of the same AA 6061-T6 can be seen in the Figure 5(a). The parameters of FSW process were 1250 rpm tool rotation and 12.5 mm/minutes welding speed. The visualization of the specimen is then investigated in the naked eyes. It was clearly seen that the weld line surface was smooth. The hole defects were not appeared on the weld surface. Therefore, the FSW process was run stable and succesfully process.

The different aluminium alloy of AA1100 and AA6061-T6 was joined in the same parameters of tool rotation and feeding by using FSW method. According to the investigation of the welding speciment, it was looked in good condition such as smooth welding surface and no hole detected along the welding surface. Nevertheless, on the weld surface was appeared the smooth grain. It can be caused by the difference properties of material. During the stirring process, the mixing of material grain has happened in non homogeneous condition.

IOP Conf. Series: Journal of Physics: Conf. Series **914** (2017) 012031 doi:10.1088/1742-6596/914/1/012031

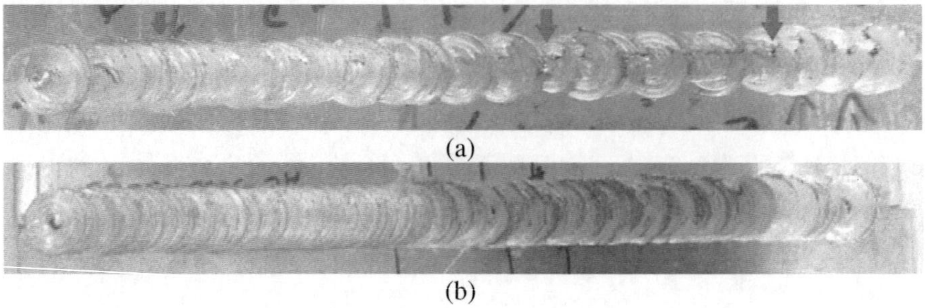

Fig. 4. FSW result of 1100x1100, (a) 1250 rpm and 12.5 mm/minutes feed rate; (b) 1600 rpm and 10 mm/minutes feet rate.

Fig.5. FSW result of 1250 rpm and 12.5 mm/minutes feed rate; (a) 6061-T6x6061-T6; (b) 1100x6061-T6

To see the grain size and shape, the photo-micro is then conducted. The first investigation is done for the base metal. Base metal is the area that it was not influenced by the temperature field and stiring during the FSW process. It can be seen clearly in Figure 6 that the grain shape was ellipse. The shape was developed due to the material forming process and grain size will be smaller then before. Rolling process usually used to form metal plate with the specific thickness. The ellipse thickness of grain was influenced by the ratio of plate thickness reduction.

IOP Conf. Series: Journal of Physics: Conf. Series **914** (2017) 012031 doi:10.1088/1742-6596/914/1/012031

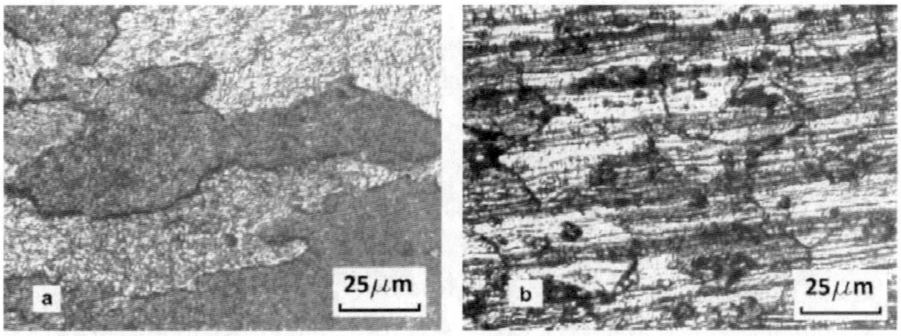

Fig. 6. Photo micro of base metal, (a) AA 1100; (b) AA 6061-T6

Heat affected zone (HAZ) is a area that influenced by thermal cycle without any plastic deformation in the welding process [11]. Therefore, in the HAZ are, the grain size will change which the size transformation is depend on the material characteristic, themperature, time of heating, and the cooling rate. As seen in Figure 7, the grain size of AA 1100 was transformed into smaller than the base metal. It could be caused by the heat generation, stiring and pressure during the process of FSW. On the contrary, the bigger grain size was delivered by AA 6061-T6 compared to the base metal. Phenomenon of grain size transformation will influence the strength of material joint. The higher grain size will decrease its hardness and tensile strength.

Figure 8 shows the micro photo of weld nugget area. That is a area which was influenced by temperature field generated from friction of stirring process during FSW. It has plastic deformation as well. Hence, the stiring process has delivered recrystallization which has produced smaller grain size in the stiring area. More crystal boundary has significant mean of smaller grain size which has higher barrier energy to the dislocation movement. It was denoted the stronger material [6], [12].

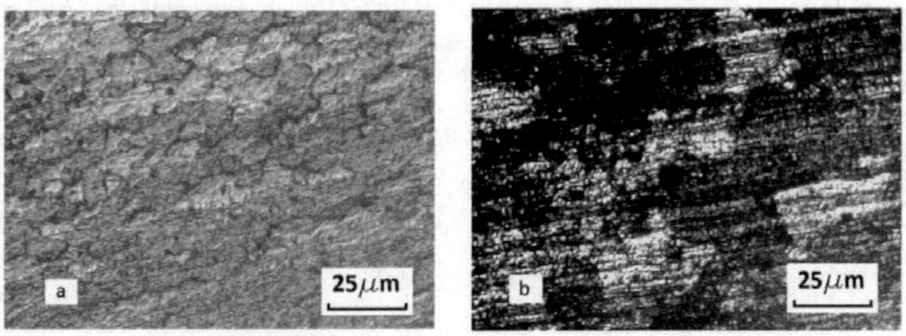

Fig. 7. Photo micro of HAZ, (a) AA 1100; (b) AA 6061-T6

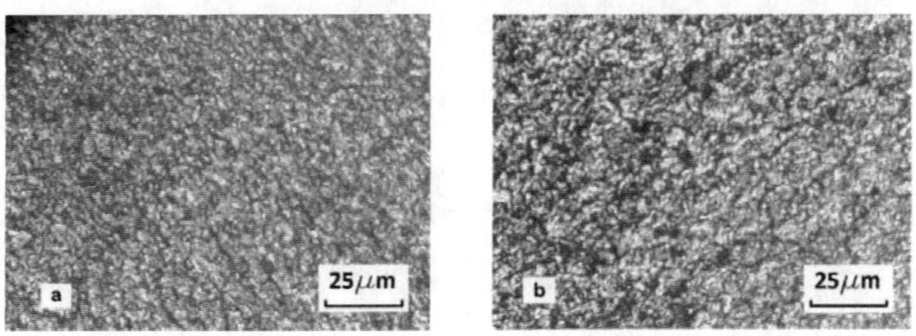

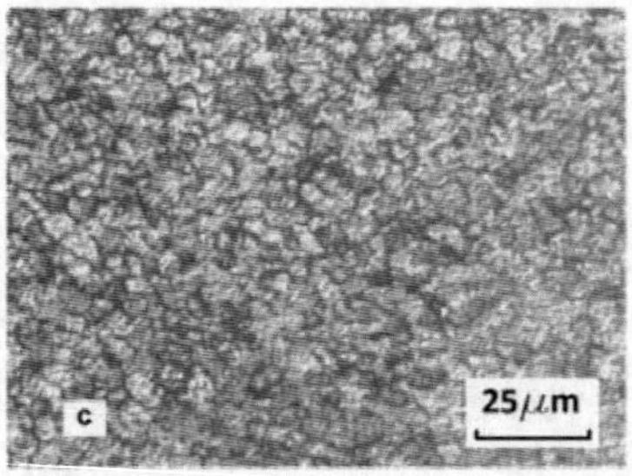

Fig. 8. Photo micro of weld nugget, (a) 1100x1100; (b) 6061-T6x6061-T6; (c) 1100x6061-T6

In the tensile test of 0° weld angle, it was broken in the weld nugget area for both specimens 1100x1100 and 6061-T6x6061-T6. The broken section has perpendicular to the load direction. It has indicated appearence defects in the weld nugget that could decrease the joint strength. On the other hand, the FSW of 1100x6061-T6 has broken on the HAZ area of AA1100. It could be caused by the stiring process of two different materials increasing the weld strength. So, the broken area was not in the weld nugget but in the low strength materials i.e AA 1100.

Almost all specimens of FSW 45° weld angle have broken in the HAZ area during tensile test. Only one specimen of 1100x1100 has broken on the weld nugget due to many defects and voids. It can be seen that the weld angle of 45° was able to employ more tensile loading compare with the weld angle of 0° as shown in Figure 8.

Hardness test of the specimen 1100x1100 has increased on the weld zone compared with base metal. As known that Aluminium 1100 is a non heat-treatable material. Increasing of hardness could be caused by stiring process and high pressure from the tool. The highest hardness was 78.1 HV on point 3 of the weld nugget as seen in Figure 9. From the hardness investigation of FSW AA 6061-T6, it was decreased in the weld zone compared with the base AA 6061-T6. The characteristic of AA 6061-T6 is a heat-treatable material. During the process of FSW, the material was suffered of heat due to the tool friction. Therefore, it could be overaging in the HAZ area. So, the hardness was decreased as 47.7 HV on point 1 (Figure 9). The fluctuative value of hardness was delivered by the specimen of 1100x6061-T6. The hardness was shown increase compared with the 1100 but it was decreased when compared with the 6061-T6.

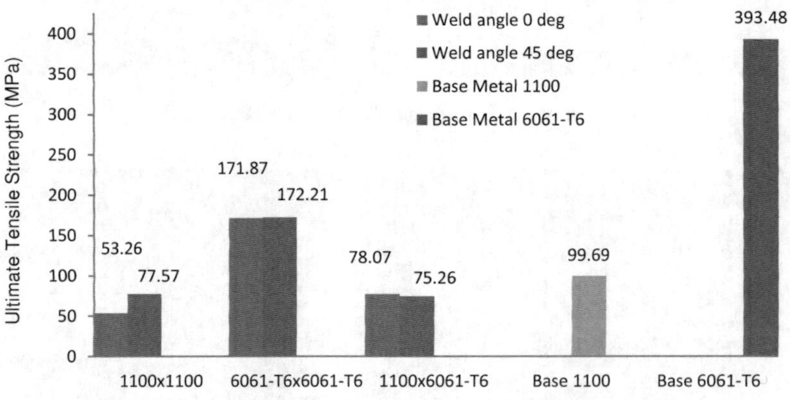

Fig. 9. Comparison of ultimate tensile strength.

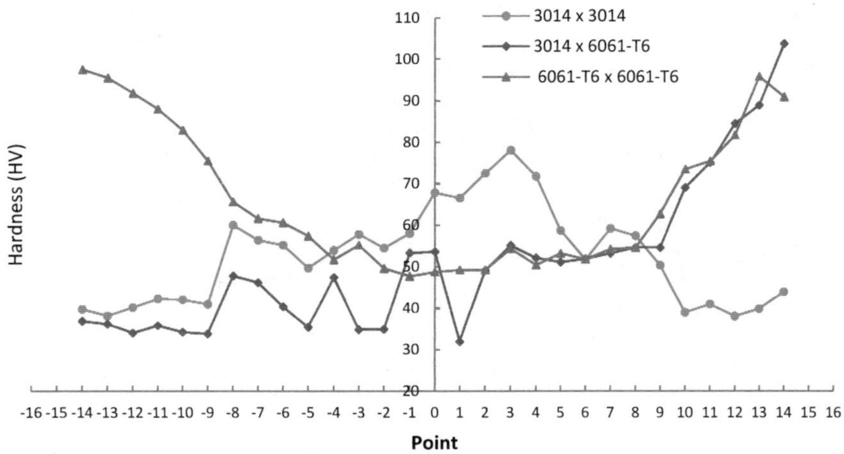

Fig. 10. Comparison of hardness

4 Conclusions

The tensile test was shown the highest value of 78.07 MPa delivered by joinning material of 1100x6061-T6 with the welding direction of 0°. For the weld direction of 45°, it was given 75.26 MPa. Whereas, the strain value was delivered 10.99% and 13.8% for the weld direction of 0° and 45° respectively. Therefore, in this research, the highest tensile strength of the FSW was given by specimen of 1100x6061-T6.

The hardness test of the weld zone of 1100x6061-T6 specimen was dedected fluctuative. The highest value of hardness test was 55.1 HVN and the lowest was 31.9 HVN. It could be caused by many factors such as homogenity and the deffects or void.

In the investigation of micro structure, it can be said that it was grain size transformed happen during FSW process. In the HAZ area of 1100 material, it was smaller grain size compared to the base metal. The bigger grain size was indicated in the material of 6061-T6. While on the weld nugget, it was grain refinement in materials both 1100 and 6061-T6. It can be influenced by temperature and stiring during the welding process.

Acknowledgement

The authors would like to acknowledge Lembaga Penelitian dan Pengabdian Masyarakat (LPPM) Universitas Muhammadiyah Surakarta and Direktur Jenderal Penguatan Riset dan Pengembangan, Ministry of Research Technology and Higher Education, Republic of Indonesiafor the funding support of the research project [Project No: 211.63/A.3-III/LPPM/V/2017 and No: 211.19/A.3-III/LPPM/V/2017].

References

[1] J. Hirsch, "Recent development in aluminium for automotive applications," *Trans. Nonferrous Met. Soc. China (English Ed.*, vol. 24, no. 7, pp. 1995–2002, 2014.

[2] M. Merklein, M. Johannes, M. Lechner, and A. Kuppert, "A review on tailored blanks - Production, applications and evaluation," *J. Mater. Process. Technol.*, vol. 214, no. 2, pp. 151–164, 2014.

[3] A. D. Anggono and T. W. B. Riyadi, "Finite Element Simulation of the Drawability of Tailor-

Welded Blank," in *Advances in Mechanical, Materials and Manufacturing Engineering*, 2014, vol. 660, pp. 3–7.

[4] R. I. Rodriguez, J. B. Jordon, P. G. Allison, T. Rushing, and L. Garcia, "Materials & Design Microstructure and mechanical properties of dissimilar friction stir welding of 6061-to-7050 aluminum alloys," *J. Mater.*, vol. 83, pp. 60–65, 2015.

[5] P. Kannan, K. Balamurugan, and K. Thirunavukkarasu, "An experimental study on the effect of silver interlayer on dissimilar friction welds 6061-T6 aluminum MMC and AISI 304 stainless steel," vol. 21, no. December, pp. 635–646, 2014.

[6] P. Pourahmad and M. Abbasi, "Materials flow and phase transformation in friction stir welding of Al 6013/Mg," *Trans. Nonferrous Met. Soc. China (English Ed.*, vol. 23, no. 5, pp. 1253–1261, 2013.

[7] B. Li, Z. Zhang, Y. Shen, W. Hu, and L. Luo, "Dissimilar friction stir welding of Ti-6Al-4V alloy and aluminum alloy employing a modified butt joint configuration: Influences of process variables on the weld interfaces and tensile properties," *Mater. Des.*, vol. 53, pp. 838–848, 2014.

[8] J. R. Hirsch, "Recent development in aluminium for automotive applications Recent development in aluminium for automotive applications," *Trans. Nonferrous Met. Soc. China*, vol. 6326, no. June, pp. 1995–2002, 2014.

[9] S. Y. Tarasov, V. E. Rubtsov, and E. A. Kolubaev, "A proposed diffusion-controlled wear mechanism of alloy steel friction stir welding (FSW) tools used on an aluminum alloy," *Wear*, vol. 318, no. 1–2, pp. 130–134, 2014.

[10] B. Fu, G. Qin, F. Li, X. Meng, J. Zhang, and C. Wu, "Friction stir welding process of dissimilar metals of 6061-T6 aluminum alloy to AZ31B magnesium alloy," *J. Mater. Process. Technol.*, vol. 218, pp. 38–47, 2015.

[11] N. D. Ghetiya and K. M. Patel, "Prediction of tensile strength and microstructure characterization of immersed friction stir welding of aluminium alloy AA2014-T4," vol. 22, no. April, pp. 133–140, 2015.

[12] W. Y. Li, T. Fu, L. Hutsch, J. Hilgert, F. F. Wang, J. F. dos Santos, and N. Huber, "Effects of tool rotational and welding speed on microstructure and mechanical properties of bobbin-tool friction-stir welded Mg AZ31," *Mater. Des.*, vol. 64, pp. 714–720, 2014.

Review on recent Developments on Fabrication Techniques of Distributed Feedback (DFB) Based Organic Lasers

[1**]**Noor Azrina Talik** [2*]**Yap Boon Kar**, [3] **Siti Noradhlia Mohamad Tukijan**, [4]**Chuan Ling Wong**

[1,2,3,4]Department of Electronic and Communications, College of Engineering, Universiti Tenaga Nasional, Jalan IKRAM-UNITEN, 43000 Kajang, Selangor, Malaysia

*kbyap@uniten.edu.my / **azrina@uniten.edu.my

Abstract. To date, the state of art organic semiconductor distributed feedback (DFB) lasers gains tremendous interest in the organic device industry. This paper presents a short reviews on the fabrication techniques of DFB based laser by focusing on the fabrication method of DFB corrugated structure and the deposition of organic gain on the nano-patterned DFB resonator. The fabrication techniques such as Laser Direct Writing (LDW), ultrafast photo excitation dynamics, Laser Interference Lithography (LIL) and Nanoimprint Lithography (NIL) for DFB patterning are presented. In addition to that, the method for gain medium deposition method is also discussed. The technical procedures of the stated fabrication techniques are summarized together with their benefits and comparisons to the traditional fabrication techniques.

1. Introduction

Today, laser become one of the most significant inventions that has been used in all area such as medicine, electronics, computer hardware, technology, communication, and even in the scientific research area. Laser can be defined as a device that produces and amplifies a narrow beam light in a certain defined electromagnetic spectrum region. Laser has its own distinctive properties compared to normal light source like mercury and tungsten lamp as the light waves of laser can travel in longer distance with a very small divergence. To make a laser works, there are three main elements required as shown in Figure 1, namely, gain medium, feedback resonator and excitation source [1]. The gain medium of the laser basically can be a liquid, a solid, or a gas that use the energy level of atoms or molecules to increase the light wave power during its propagation [1]. The gain medium is placed in a resonator and the light repeatedly goes through the gain medium as it bounces back and forth in between the mirrors. The output coupler mirror reflects most of the light, but transmits small portions as the output laser beam [2]. For organic laser, feedback structure is used instead of mirror to make it compatible for the periodic structure of organic gain medium [2,3]. The organic compounds construct a new type of semiconductor electronic materials to be utilized in various applications such as organic light-emitting diodes (OLEDs), organic photovoltaic (PV), thin film transistors (TFTs) organic spin-valve devices as well as optical switches.

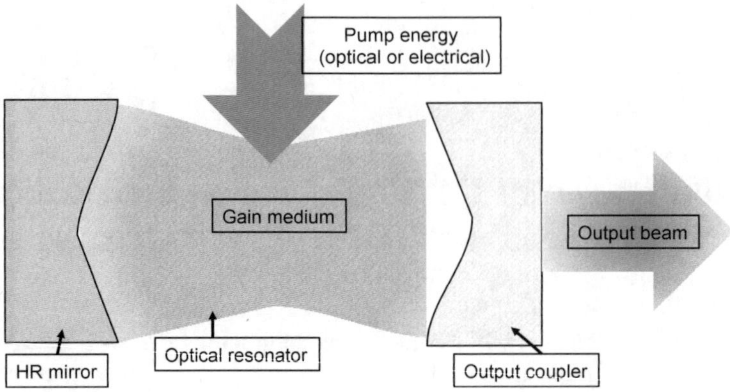

Figure 1. A simple laser building block.

The latest developments of organic laser are focusing on the usage of organic materials in photonic device due to its attractive processing capabilities, broad absorption and emission spectra, and high spectral coverage [4,5,6]. In contrary to the inorganic material, less harmful organic material for laser offers the potential of large emission coverage or highly tunable, convenient and inexpensive structure cost [7,8,9]. These advantages could help the organic laser to easily adjust their optical characteristic. “Organic” term in laser is normally used as there is organic gain medium utilized in the structure. In 1992, Moses et al introduced the first organic laser where they demonstrated the laser utilized conjugated polymer as the gain medium [10]. However, to date more examples were demonstrated where both gain medium and the optical feedback structure are both “organic” [11] There are few optical feedback structures that have been demonstrated for organic lasers for example planar cavities [12,13], distributed feedback and photonic crystal structure [14,15], whispering gallery mode resonators [16,17], plasmonic structure [18], organic crystal based cavities and a few more [1,19].

It is observed that organic distributed feedback (DFB) based laser source gained particular interests in the state-of-art organic laser technology [14,15] due to its excellent characteristics such as well-defined directional output [20,21,22], the emitter can be tuned to cover the whole visible spectral range [23], low operating threshold [24], high reflectivity as well as simple fabrication steps. DFB lasers consists of two main features, the feedback mechanism and gain medium are integrated and distributed throughout the structure [25,26]. The optical feedback and the output coupling for this type of laser are achieved via DFB resonator. As shown in Figure 2, this DFB resonator is consists of grating-patterned substrate and the organic gain material is then coated on top of it. This corrugated structure replaced the diffraction grating where the light traveling in one direction will be diffracted in a new direction [1,2]. The maximum coupling between forward and backward propagation along the waveguide produced the light feedback of the laser.

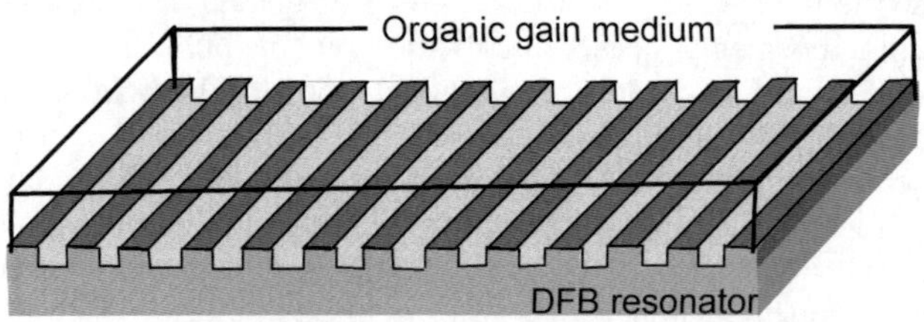

Figure 2 Distributed feedback structure with corrugated substrate and gain medium deposited on top of the substrate.

IOP Conf. Series: Journal of Physics: Conf. Series **914** (2017) 012032 doi:10.1088/1742-6596/914/1/012032

In term of fabrication, resonator DFB is conventionally formed evaporating using electron beam lithography (EBL). But, even though EBL is well known to be one of the most precise, and versatile methods for nano-patterning, it is a time consuming method, requires high vacuum, as well as advanced control electronics, which makes cost-inefficient to be utilized [27,28,29]. On the other hand, organic gain medium is normally being processed via inexpensive solution process approach for example spin-coating method. However, using this method, the gain material will cover the whole substrate [30-31]. Regardless of the targeted application, the fabrication technology of resonator and gain medium is the paramount important to fully exploiting the potential associated with the organic materials. Thus, suitable solution process approach to produce conjugated polymer is required. To the author's best knowledge, it is somewhat difficult to find the literatures that summarized all the fabrication techniques available for DFB type of organic laser. Thus, in this paper, the fabrication methods for organic laser using DFB corrugated structure will be presented.

2. Fabrication techniques for DFB corrugated structure

Miniaturized organic laser light sources shown a promising performance when integrated in all-polymer lab on chip and sensing application [29,32-34]. Low-cost fabrication of DFB corrugation nanostructures can be achieved via the fabrication technique that we discuss here. For DFB patterning, we will discuss about four different techniques namely Laser Direct Writing Techniques, Ultrafast Photo Excitation Dynamic Laser, Laser Interference Lithography and Nanoimprint Lithography. We will continue with the deposition method of organic gain in the next section which will cover two different approaches, which are thermal deposition and solution process method.

2.1 Laser Direct Writing Techniques

The first technique is Laser Direct Writing (LDW). This technique is a computer controlled two-dimensional (2D) and three-dimensional (3D) pattern beams by transparent material, causing the two and more photons being absorbed and polymerized locally [35]. The focused UV laser beam pulses pass through multi-photon polymerization (MPP). The beam of ultrafast laser is tightly focused inside the volume of a ribbon support and interacts with an ink coating containing a suspension of the material to be deposited. This technique is depicted as a method using a pulsing laser to create synthetic nano-diamond films and patterns from graphite, with potential applications from biosensors to computer chips [36]. Figure 3 illustrates the process of Laser Direct Writing (LDW).

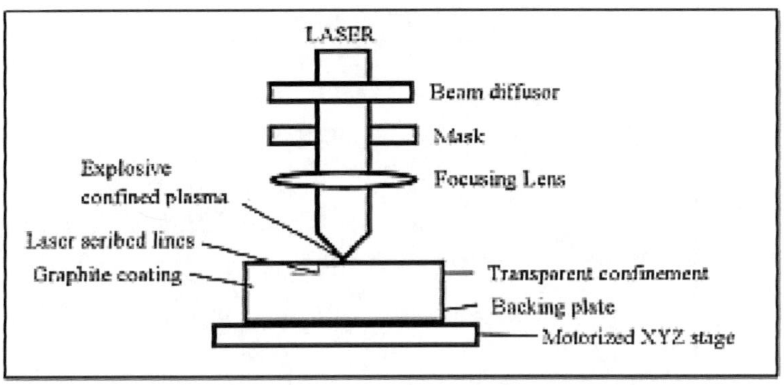

Figure 3. Laser Direct Writing (LDW) Process. Adapted from [36].

LDW utilizes laser induced forward transfer to output patterns of complex shapes. It is a

IOP Conf. Series: Journal of Physics: Conf. Series **914** (2017) 012032 doi:10.1088/1742-6596/914/1/012032

non-lithographic digital microfabrication of microelectronics, which is faster, economical and highly versatile than the convention fabrication method of lithography [37]. It also allows deposition under normal room temperature, high temperature and pressure chamber. It combines the ability of remove ablation, modification and addition controlled using computer aids to process complex and delicate materials in the processes of metal cutting and welding, using the computing technology as a latest method. This method opens a path to new technology such as Scaffold 3D printing [38].

2.2 Ultrafast Photo Excitation Dynamic Laser

Ultrafast photo excitation dynamics of the radical laser is photo excited concentrated laser beam, which were investigated by femtosecond laser spectroscopy. Femtosecond pump probe technique is a powerful tool for the investigation of ultrafast processes in condensed matter [39]. Ultrafast dynamics and the laser action of organic semiconductors is one of the most recent fabrication methods, which involves the laser action in pi-conjugated polymer films, solutions and microcavities [40,41]. In 2D and 3D lasers, ultrafast femtosecond laser is applying the technique of pump probe spectroscopy. The laser pulse is divided into two portions, a stronger beam (pump) for exciting the sample, generating a non-equilibrium state, and a weaker beam (probe) for monitoring the pump-induced changes in the optical reflectivity or transmission. The lasers pulses yield information about the relaxation of electronic states from the time delay between arrivals of pump and probe pulses [42].

2.3 Laser Interference Lithography

Laser Interference Lithography (LIL) or known as holographic lithography become an attractive technique especially for microscale and nanoscale manufacturing. This technology has been used to fabricate photonic crystals slabs (PCS) [43], templates of high-density and periodic structure for Micro-electromechanical Systems (MEMS) [44]. This technique uses mask-less exposure in which uses the interference of two or more electromagnetic waves of the same length with different angle laser beams to form an interference fringe pattern on substrate of photoresist layer. There are critical process parameters that affect the LIL process, which are beam power, exposure dosage, intensity distribution, and angle between beams [45]. The general principle of two-beam LIL is illustrated in Figure 4. In this set-up, the organic semiconductor laser is optically pumped with an operation of 266 nm wavelength by diode pumped solid-state (DPSS). The beam splitter produced two interference beams and directed both beams toward the sample of photoresist layer using rotatable concave mirrors. On the photoresist coating layer, a standing wave is formed at angle + θ and – θ thus resulting an interference pattern with a period of $\lambda/2\sin\theta$, where λ is the wavelength of operation light [46].

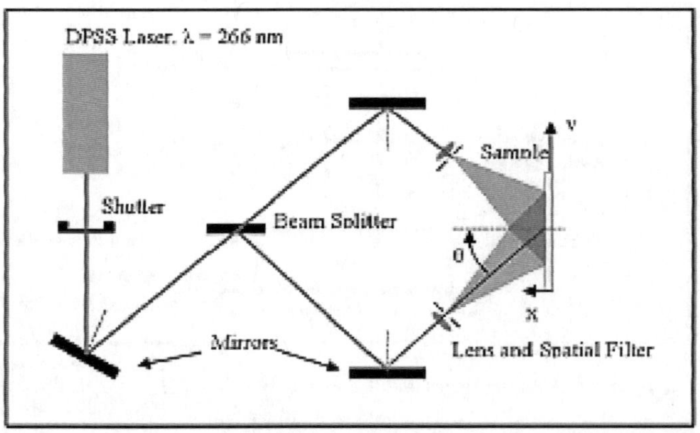

Figure 4. Two-laser beams interference [46].

LIL technology is mostly used for mass production as it offers low cost compared to the other beam technologies, high fabrication resolution compared to other optical technologies, and high processing rate. In the same time, it is able to fabricate large area of pattern which up to hundreds of mm in diameter. LIL also offers high-efficiency for parallel writing technique because it able to fabricate periodic structure by recording the interference pattern on the target material in one step [44]. Furthermore, the faster periodic pattern generation of NIL on the larger area offers the creation of large-size of 2D PCs without losing performance aspect.

2.4 Nanoimprint Lithography

Nanoimprint Lithography (NIL) is another non-conventional lithographic of nanofabrication technique that has been used for mass production and for high-throughput polymer nanostructure patterning at great precision with low costs [44]. Nowadays, there are several techniques of NIL exist such as thermal NIL (hot embossing), UV-assisted NIL, roll-to-roll nanoimprinting [46], and soft lithography [47]. In this technique, NIL creates a replication of patterned mold from a master by mechanical deformation of the polymer layer or the resist material on substrates. The master can be used for multiple times to replicate the pattern into a desired gain medium. One of the inventions by B. Guilhabert et al is on fabrication and characterization of mechanically flexible 4x4 array organic semiconductor laser (OSL) by using soft lithography from a 1-D silica master granting reproduction. This process used blue-emitting tris (trifluorene) truxene oligomers as gain medium. The fabrication process created a grating-on-mesa array structures on the suitable polymer by masking certain area of a 1-D SiO2 master grating with a patterned photoresist (PR) [47]. In addition, NIL can be used in a wide range of applications since it can imprint a structure of functional device especially in electronics, data storage, photonic, and biotechnology [48].

Roller-Nanoimprinting technique is beneficial to fabricate a surface nanorelief large-format nanostructure mass production with higher production rate and lower cost. NIL currently used in production of semi-industrial of photonic and microelectronics device since it can reach ~1 m/min of fabrication speed, which is 104-105 times faster compared to a traditional electron beam (e-beam) lithography. In addition, this technique offers improvement on light absorption, low molecular weight polymers and better control on chain alignment orientation in conjugated crystalline. The device efficiency also increased by 500% when the power conversion efficiencies is set to the optimum value [49].

3. Fabrication techniques for Gain medium deposition

Thus far, the thin film deposition method is the main tool for development of integrated organic lasers that intend at producing a wave-guiding layer of the gain medium on top of low refractive index substrate. This technique is commonly used in fabricating DFB organic laser as this type of laser usually required thin film of less than 200 nm [50]. Among thin film technology available to date are thermal sublimation [51], plasma enhanced chemical vapour deposition (CVD) method [52], pulsed laser deposition (PLD) [53,54], electrospray deposition [55,56], atmospheric-pressure ion deposition [57] matrix-assisted pulsed laser evaporation [58,59] as well as solution process method [60-62]. Among all these technology, the most prevailing one is the solution process, as it is well known to be inexpensive and facile method.

3.1 Solution Process methods

Figure 5 depicts three different low cost solution process approaches namely (a) spin-coating, (b) dip-coating as well as (c) doctor blading for gain material deposition [63]. The thickness of the thin film could be controlled by the solvent concentration, spin speed (for spin coating), spin duration as well as gap height for doctor-blade process. However, the process to control the layer thickness and the film

IOP Conf. Series: Journal of Physics: Conf. Series **914** (2017) 012032 doi:10.1088/1742-6596/914/1/012032

uniformity is more challenging especially to control the film thickness at the nanometer level as it is a trial and error process where film characterization is required through profilometer or interfermetric. In addition to that, via spin-coating or doctor blading, the solvent on top of the corrugated structure may erode the nano-patterned polymer substrate.

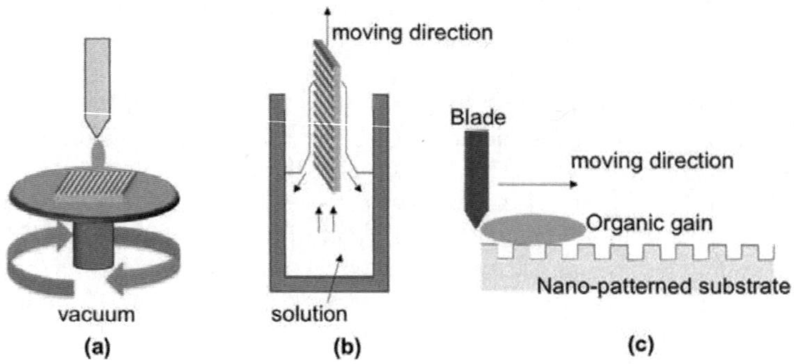

Figure 5. (a) Spin coating, (b) dip coating and (c) doctor blading method.

As an alternative to this, the lateral control can be achieved using printing techniques such as the use of ink-jet printing.

3.2 Ink-jet printing method

To date, ink jet printing has widely been used in various organic electronic device fabrication [64-66]. This printing method is also been applied in organic laser fabrication in order to achieve an accurate controlled gain medium deposition on top of substrate [67]. It also allows the deposition of few different organic materials to be deposited on to one substrate [68].

In ink jet printing process, a mixture of high boiling and low boiling solvents are utilized for dissolving the polymer to optimize film uniformity. In contrast to solution based deposition technique, ink jet printing allows the deposition of thin films with arbitrary lateral shapes and small lateral dimensions [69] as depicted in Figure 6. Thus, a precise organic gain medium deposited on defined grating trenches could be achieved.

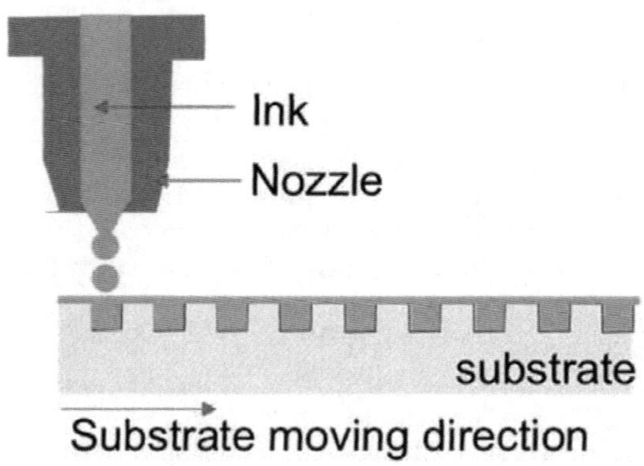

Figure 6 Ink-jet printing technique

4. Conclusion and Outlook

To date, various organic gain medium for laser has been designed and synthesized to be integrated with the resonators architectures which demonstrate tremendous improvement in spectral properties, lifetime, low operating threshold as well as the improve gains produced. The sizes of laser components are shrunk into smaller scale and the recent fabrication equipment are also smaller, compact, modifiable, and faster in response time. The cyber world has upgraded the organic laser to have higher dimensions with the vitality of computing modeling. The organic laser is an essential optical application in the emerging organic semiconductor field due to its three main reasons, which are smaller in material size, lower in producing and maintenance cost, and increasing demands of high energy organic laser in the field of communication engineering, medical treatment engineering and optical engineering.

In order to fabricate highly efficient, smaller as well as compact organic laser, appropriate fabrication technique must be properly considered. The detailed descriptions on the recent fabrication techniques of the organic laser development are summarized in this paper together with their procedural techniques and advantages over the conventional methods. Although intense research has been made in order to improve the laser fabrication, there are still more challenges to come. More attention should be paid in choosing the fabrication technology for resonator and gain medium fabrication regardless of the targeted applications.

References

[1] Chénais S and Forget S 2012 Recent Advances in Solid-State Organic Laser *Polym. Int.* **61** 390-406

[2] Samuel, I.D. and Turnbull, G.A., 2004. Polymer lasers: recent advances. *Materials today*, *7* 28-35

[3] Huang, W., Chen, L. and Xuan, L., 2014. Efficient laser emission from organic semiconductor activated holographic polymer dispersed liquid crystal transmission gratings. *RSC Advances* *4* (73) 38606-38613.

[4] Foucher C, Guilhabert B, Herrnsd orf J, Laurand N and Dawson M D 2014 Diode-Pumped, Mechanically-Flexible Organic Lasers Fully Encapsulated With Ultra-Thin Glass Membranes *2014 IEEE Photonics Conf.* **3** 536–37

[5] Bettotti, P., 2014. Hybrid materials for integrated photonics. *Advances in Optics* **2014** 1-24

[6] Dong, H., Zhang, C. and Zhao, Y.S., 2017. Host–guest composite organic microlasers. *Journal of Materials Chemistry C.* 1-3

[7] Clark J and Lanzani G 2010 Organic photonics for communications *Nat. Publ. Gr.* **4** 438–446

[8] Bogue R 2011 Developments in organic lasers and their role in sensing *Sens. Rev.* **31** 13–17

[9] Vannahme C, Klinkhammer S, Christiansen M, Kolew A, Kristensen A, Lemmer U and Mappes T 2010 All-polymer organic semiconductor laser chips: parallel fabrication and encapsulation *Opt. Express* **18** 24881–87

[10] Moses, D. (1992). High quantum efficiency luminescence from a conducting polymer in solution—a novel polymer laser-dye. Applied Physics Letters, 60(26), 3215–3216.

[11] Palatnik, A., Bitton, O., Aviv, H. and Tischler, Y.R., 2016. Influence of gain material concentration on an organic DFB laser. *Optical Materials Express*, *6*(9), pp.2715-2724.

[12] Schulzgen, A., Spiegelberg, C., Morrell, M.M., Mendes, S.B., Nabor, M.F., Mash, E.A., Allemand, P.M., Kippelen, B. and Peyghambarian, N., 1998, May. A vertical cavity surface emitting polymer laser. In *Lasers and Electro-Optics, 1998. CLEO 98. Technical Digest. Summaries of papers presented at the Conference on* IEEE 6-7 IEEE

[13] Dodabalapur, A., Berggren, M., Slusher, R.E. and Bao, Z., 1997, November. Design of materials and cavities for organic lasers. In *Lasers and Electro-Optics Society Annual Meeting, 1997. LEOS'97 10th Annual Meeting. Conference Proceedings., IEEE* **2** 171

[14] Baumann, K., Stöferle, T., Moll, N., Raino, G., Mahrt, R.F., Wahlbrink, T., Bolten, J. and Scherf, U., 2010. Design and optical characterization of photonic crystal lasers with organic gain material. *Journal Of Optics*, **12** (6) 065003.
[15] Klein, S., Barsella, A., Stortz, V., Taupier, G., Fort, A. and Dorkenoo, K.D., 2005, June. Photopolymerization techniques for the design of DFB organic lasers. In *Lasers and Electro-Optics Europe, 2005. CLEO/Europe. 2005 Conference on* p.261 IEEE
[16] Zhang, C., Zou, C.L., Yan, Y., Wei, C., Cui, J.M., Sun, F.W., Yao, J. and Zhao, Y.S., 2013. Self Assembled Organic Crystalline Microrings as Active Whispering Gallery Mode Optical Resonators. *Advanced Optical Materials*, **1** (5) 357-361
[17] Melnikau, D., Savateeva, D., Chuvilin, A., Hillenbrand, R. and Rakovich, Y.P., 2011. Whispering gallery mode resonators with J-aggregates. *Optics express*, **19** (22) 22280-22291
[18] Wu, H.Y., Liu, L., Lu, M. and Cunningham, B.T., 2016. Lasing emission from plasmonic nanodome arrays. *Advanced Optical Materials.*
[19] Chakaroun, M., Coens, A., Fabre, N., Gourdon, F., Solard, J., Fischer, A., Boudrioua, A. and Lee, C.C., 2011. Optimal design of a microcavity organic laser device under electrical pumping. *Optics express*, **19**(2) 493-505
[20] Wu, C., Khanal, S., Reno, J.L. and Kumar, S., 2016. Terahertz plasmonic laser radiating in an ultra-narrow beam. *Optica*, **3**(7) 734-740
[21] S.L. Lee, I.F. Jang, C.Y.Wang, C.T. Pein, T.T Shih, 2000 Monolitically integrated multi-wavelength sampled grating DBR lasers for dense WDM applications. *IEE J. Seleced Topics Quant. Electron.* **6** 197
[22] R. Kaiser, F.Fidorra, H.Heidrich, P.Albrecht, W.Rehbein, S.Malchow, H.Schroeter-Janssen, D.Franke, G.Sztefka, 1994, *6th Inter. Conf. on InP and Related Materials* P.474 CA
[23] Klinkhammer, S., Liu, X., Huska, K., Shen, Y., Vanderheiden, S., Valouch, S., Vannahme, C., Bräse, S., Mappes, T. and Lemmer, U., 2012. Continuously tunable solution-processed organic semiconductor DFB lasers pumped by laser diode. *Optics express*, **20**(6)6357-6364
[24] Tsiminis, G., Wang, Y., Shaw, P.E., Kanibolotsky, A.L., Perepichka, I.F., Dawson, M.D., Skabara, P.J., Turnbull, G.A. and Samuel, I.D., 2009. Low-threshold organic laser based on an oligofluorene truxene with low optical losses. *Applied Physics Letters*, **94**(24)165
[25] Shank, C.V., Bjorkholm, J.E. and Kogelnik, H., 1971. TUNABLE DISTRIBUTED FEEDBACK DYE LASER. *Applied Physics Letters*, **18**(9) 395-396
[26] Aoyagi, Y., Aoyagi, T., Toyoda, K. and Namba, S., 1975. Tunable distributed feedback dye laser. *Applied Physics Letters*, **27**(12) 687-688
[27] Li, L., Cao, B. and Chen, X. 2014 Demonstration of a low-cost cascade tunable semiconductor DFB laser In *Optical Communications and Networks (ICOCN), 2014 13th International Conference* 1-4
[28] Tsuji, Y., Yanagisawa, M., Yoshinaga, H., Inoue, N. and Nomaguchi, T., 2011. Highly uniform fabrication of diffraction gratings for distributed feedback laser diodes by nanoimprint lithography *Japanese Journal of Applied Physics*, ***50*** (6S) 06GK06
[29] Liu, X., Prinz, S., Besser, H., Pfleging, W., Wissmann, M., Vannahme, C., Guttmann, M., Mappes, T., Koeber, S., Koos, C. and Lemmer, U. 2014. Organic semiconductor distributed feedback laser pixels for lab-on-a-chip applications fabricated by laser-assisted replication. *Faraday discussions*, *174,* 153-164.
[30] Mhibik, O., Chénais, S., Forget, S., Defranoux, C. and Sanaur, S., 2016. Inkjet-printed vertically emitting solid-state organic lasers. *Journal of Applied Physics*, **119** (17) 173101.
[31] Wan, W., Huang, W., Pu, D., Qiao, W., Ye, Y., Wei, G., Fang, Z., Zhou, X. and Chen, L., 2015. High performance organic distributed Bragg reflector lasers fabricated by dot matrix holography. *Optics express*, **23**(25) 31926-31935
[32] Llobera, A., Juvert, J., González-Fernández, A., Ibarlucea, B., Carregal-Romero, E., Büttgenbach, S. and Fernández-Sánchez, C., 2015. Biofunctionalized all-polymer photonic lab on a chip with integrated solid-state light emitter. *Light: Science & Applications*, **4**(4), e271.

[33] Williams, G., Backhouse, C. and Aziz, H., 2014. Integration of organic light emitting diodes and organic photodetectors for lab-on-a-chip bio-detection systems. *Electronics*, **3**(1) 43-75.
[34] Mappes, T., Lenhert, S., Kassel, O., Vannahme, C., Schelb, M. and Mohr, J., 2008, July. An all polymer optofluidic chip with integrated waveguides for biophotonics. In *IEEE/LEOS Summer Topical Meetings, 2008 Digest of the* (pp. 215-216). IEEE.
[35] Häfner M, Pruss C and Osten W 2011 Laser direct writing recent developments for the making of diffractive optics *Opt. & Photonik* **4** 40–43
[36] Nian Q, Wang Y, Yang Y, Li J, Zhang M Y, Shao J, Tang L and Cheng G J 2013 Direct laser writing of nanodiamond films from graphite under ambient conditions *Scientific Report* **4** 6612
[37] Selimis A, Mironov V and Farsari M 2015 Direct laser writing: Principles and materials for scaffold 3D printing *Microelectronic Engineering* **132** 83-89
[38] Rekstyte, S., Balciunas, E., Baltriukiene, D., Rutkunas, V., Bukelskiene, V., Gadonas, R. and Malinauskas, M., 2013, May. Direct laser fabrication of composite material 3D microstructured scaffolds. In *Lasers and Electro-Optics Europe (CLEO EUROPE/IQEC), 2013 Conference on and International Quantum Electronics Conference* (pp. 1-1). IEEE.
[39] Horn W, Kroesen S and Denz C 2014 Two-photon fabrication of organic solid-state distributed feedback lasers in rhodamine 6G doped SU-8 *Appl. Phys. B Lasers Opt.* **117** 311–315
[40] Vardeny Z V 2009, *Ultrafast Dynamics and Laser Action of Organic Semiconductors* (Boca Raton: CRC Press)
[41] Agranovich V M and Leo K **177** 177–178
[42] Shah, J., 2013. *Ultrafast spectroscopy of semiconductors and semiconductor nanostructures* (Vol. 115). Springer Science & Business Media.
[43] Abrams N, Mallouk T E, Divliansky I and Mayer T S 2006 Fabrication of TiO 2 -Organic Hybrid Dot Arrays Using Nanosecond Laser Interference Lithography *the American Ceramic Society* **3510** 3507–10
[44] Liu Q, Duan X and Peng C 2014 *Novel Optical Technologies for Nanofabrication* (New York: Springer)
[45] Seo J, Park J H, Ma Z, Choi J and Ju B 2014 Nanopatterning by laser interference lithography: applications to optical devices *Journal of Nanoscience and Nanotechnology* ***14*** (2) 1521-32
[46] Liu X 2015 *Organic Semiconductor Lasers and Tailored Nanostructures for Raman Spectroscopy* KIT Scientific Publisher.
[47] Guilhabert B, Laurand N, Herrnsdord J, Chen Y, Kanibolotsky A L, Orofino C, Skabara P J and Dawson M D 2012 Mechanically flexible organic semiconductor laser array *IEEE Photonics Journal* ***4*(3)** 684-690
[48] Wang Y 2014 *Low Threshold Organic Semiconductor Lasers: Hybrid Optoelectronics and Applications as Explosive Sensors* (New York: Springer) Chapter 2 9-29
[49] Comoretto D 2015 *Organic and Hybrid Photonic Crystals* (New York: Springer) 303–320
[50] Andrews, J.H., Crescimanno, M., Singer, K.D. and Baer, E., 2014. Melt processed polymer multilayer distributed feedback lasers: Progress and prospects. *Journal of Polymer Science Part B: Polymer Physics*, **52**(3) 251-271
[51] Berggren, M., Dodabalapur, A., Slusher, R.E. and Bao, Z., 1997. Light amplification in organic thin films using cascade energy transfer. *Nature*, **389** (6650) 466-469
[52] Kozlov, V.G., Parthasarathy, G., Burrows, P.E., Forrest, S.R., You, Y. and Thompson, M.E., 1998. Optically pumped blue organic semiconductor lasers. *Applied physics letters* **72** (2) 144-146
[53] Meier, M., Mekis, A., Dodabalapur, A., Timko, A., Slusher, R.E., Joannopoulos, J.D. and Nalamasu, O., 1999. Laser action from two-dimensional distributed feedback in photonic crystals. *Applied Physics Letters* **74**(1) 7-9
[54] Gill, D.S., Anderson, A.A., Eason, R.W., Warburton, T.J. and Shepherd, D.P., 1996. Laser operation of an Nd: Gd3Ga5O12 thin film optical waveguide fabricated by pulsed laser deposition. *Applied physics letters*, **69**(1) 10-12

[55] Ju, J., Yamagata, Y. and Higuchi, T., 2009. Thin Film Fabrication Method for Organic Light Emitting Diodes Using Electrospray Deposition. *Advanced Materials*, **21**(43) 4343-4347

[56] Chrisey, D.B., Pique, A., McGill, R.A., Horwitz, J.S., Ringeisen, B.R., Bubb, D.M. and Wu, P.K., 2003. Laser deposition of polymer and biomaterial films. *Chemical Reviews*, **103**(2) 553-576

[57] Saf, R., Goriup, M., Steindl, T., Hamedinger, T.E., Sandholzer, D. and Hayn, G., 2004. Thin organic films by atmospheric-pressure ion deposition. *Nature materials*, **3**(5) 323-329

[58] Wu, P.K., Ringeisen, B.R., Callahan, J., Brooks, M., Bubb, D.M., Wu, H.D., Piqué, A., Spargo, B., McGill, R.A. and Chrisey, D.B., 2001. The deposition, structure, pattern deposition, and activity of biomaterial thin-films by matrix-assisted pulsed-laser evaporation (MAPLE) and MAPLE direct write. *Thin Solid Films*, **398** 607-614

[59] Piqué, A., Wu, P., Ringeisen, B.R., Bubb, D.M., Melinger, J.S., McGill, R.A. and Chrisey, D.B., 2002. Processing of functional polymers and organic thin films by the matrix-assisted pulsed laser evaporation (MAPLE) technique. *Applied Surface Science*, **186** 1 408-415

[60] Navarro-Fuster, V., Calzado, E.M., Boj, P.G., Quintana, J.A., Villalvilla, J.M., Díaz-García, M.A., Trabadelo, V., Juarros, A., Retolaza, A. and Merino, S., 2010. Highly photostable organic distributed feedback laser emitting at 573 nm. *Applied Physics Letters*, **97**1(7) 171104

[61] Turnbull, G.A., Andrew, P., Jory, M.J., Barnes, W.L. and Samuel, I.D.W., 2001. Relationship between photonic band structure and emission characteristics of a polymer distributed feedback laser. *Physical Review B*, **64** (12) 125122.

[62] Heliotis, G., Xia, R., Turnbull, G.A., Andrew, P., Barnes, W.L., Samuel, I.D.W. and Bradley, D.D., 2004. Emission Characteristics and Performance Comparison of Polyfluorene Lasers with One and Two Dimensional Distributed Feedback. *Advanced Functional Materials*, **14**(1) 91-97

[63] Diao, Y., Shaw, L., Bao, Z., & Mannsfeld, S. C. (2014). Morphology control strategies for solution-processed organic semiconductor thin films. *Energy & Environmental Science*, **7** (7), 2145-2159.

[64] Calvert, P., 2001. Inkjet printing for materials and devices. *Chemistry of materials*, **13** (10) 3299-3305.

[65] Rogers, J. and Katz, H., 1999. Printable organic and polymeric semiconducting materials and devices. *Journal of Materials Chemistry* **9** (9) 1895-1904

[66] Rigas, G.P., Payne, M.M., Anthony, J.E., Horton, P.N., Castro, F.A. and Shkunov, M., 2016. Spray printing of organic semiconducting single crystals. *Nature Communications*, *7*.

[67] X. Liu, S. Klinkhammer, K. Sudau, N. Mechau, C. Vannahme, J. Kaschke, T. Mappes, M. Wegener, U. Lemmer 2012 Ink-Jet-Printed Organic Semiconductor Distributed Feedback Laser *Appl. Phys. Express* **5** **(7)**, 072101

[68] Geffroy, B., Le Roy, P. and Prat, C. 2006. Organic light emitting diode (OLED) technology: materials, devices and display technologies. *Polymer International* **55**(6) 572-582

[69] Tekin, E., Holder, E., Kozodaev, D., & Schubert, U. S. 2007. Controlled Pattern Formation of Poly [2 methoxy 5 (2′ ethylhexyloxyl)–1, 4 phenylenevinylene](MEH–PPV) by Ink Jet Printing. *Advanced Functional Materials*, **17** 2, 277-284.

Modelling Dynamic Behaviour and Spall Failure of Aluminium Alloy AA7010

N Ma'at[1], M K Mohd Nor[1,*], A E Ismail, K A Kamarudin, S Jamian, M N Ibrahim, M K Awang

[1]Crashworthiness and Collision Research Group, Mechanical Failure Prevention and Reliability Research Center, Engineering Mechanics Department, Faculty of Mechanical and Manufacturing Engineering, Universiti Tun Hussein Onn Malaysia

Abstract. A finite strain constitutive model to predict the dynamic deformation behaviour of Aluminium Alloy 7010 including shockwaves and spall failure is developed in this work. The important feature of this newly hyperelastic-plastic constitutive formulation is a new Mandel stress tensor formulated using new generalized orthotropic pressure. This tensor is combined with a shock equation of state (EOS) and Grady spall failure. The Hill's yield criterion is adopted to characterize plastic orthotropy by means of the evolving structural tensors that is defined in the isoclinic configuration. This material model was developed and integration into elastic and plastic parts. The elastic anisotropy is taken into account through the newly stress tensor decomposition of a generalized orthotropic pressure. Plastic anisotropy is considered through yield surface and an isotropic hardening defined in a unique alignment of deviatoric plane within the stress space. To test its ability to describe shockwave propagation and spall failure, the new material model was implemented into the LLNL-DYNA3D code of UTHM's. The capability of this newly constitutive model were compared against published experimental data of Plate Impact Test at 234m/s, 450m/s and 895m/s impact velocities. A good agreement is obtained between experimental and simulation in each test.

1. Introduction

Orthotropic material is a subset of anisotropic material, where their properties is change when measure from different direction. The familiar example of orthotropic materials is sheet of aluminium alloy formed by squeezing thick sections of metal between heavy rollers. Aluminium Alloy 7010 in temper conditions T7651 has been extensively used as structure components in aero industry due to its high strength and high resistance to stress corrosion cracking and good fracture toughness. The ability to predict finite strain deformation and failure in such materials therefore, becoming more and more important and has attracted attention designer and the user of metal structures for many years [1]. This is a great challenge in understanding the behaviour of these materials from quasi-static to high strain rate regimes as adopted in broad engineering applications [2]. Moreover, there are numerous mechanics of materials issues that have yet to be solved, related to orthotropic elastic and plastic behaviour. Based on this motivation, this work is conducted to develop a constitutive formulation to predict the behaviour of commercial aluminium alloys undergoing large deformation including spall failure.

2. New constitutive formulation

The new hyperelastic-plastic constitutive formulation for orthotropic materials in the finite strain regime is formulated based on the multiplicative decomposition of the deformation gradient **F**.

$$\mathbf{F} = \mathbf{F}_e \cdot \mathbf{F}_p \tag{1}$$

where $\mathbf{F}_e$ and $\mathbf{F}_p$ is the deformation gradient representing the elastic and plastic part of deformation [15]. This concept distinguishes the proposed constitutive model from hypoelastic-plastic material models (when elastic strains are small compared to the plastic strains). The intermediate (generally non-Euclidean) configuration corresponds to elastically unloaded material, known as the elastic reference configuration which can be physically obtained by elastic unloading of material (unstressed condition).

The deformation gradients $\mathbf{F}_e$ and $\mathbf{F}_p$ are not uniquely defined because the intermediate configuration is not unique, arbitrary local material rotations can be superposed to intermediate configuration preserving it unstressed configuration, therefore making the intermediate configuration non-unique. This issue has been discussed extensively in the [3], [4], [5] and [6] to eliminate the non-uniqueness of this configuration. Using the multiplicative decomposition Equation (1), the associated Green-Lagrange strain tensor $\mathbf{E}$ can be defined as

$$\mathbf{E} = \frac{1}{2}\left(\mathbf{F}^{\mathrm{T}} \cdot \mathbf{F} - \mathbf{I}\right) = \frac{1}{2}(\mathbf{C} - \mathbf{I}) \tag{2}$$

where $\mathbf{C}$ refers to the right Cauchy-Green strain tensor. This formulation is uniquely determined except for arbitrary rigid body rotations that are superposed in the intermediate configuration while preserving the unstressed condition. As discovered in [12], the experimental work shows a strong correlation between elastic and plastic material symmetries. Therefore, to resolve the non-uniqueness of the intermediate configuration of this formulation, the isoclinic configuration is adopted in this constitutive formulation.

This unique configuration defines the principal directions of material elastic and plastic orthotropy coincide and are not influenced by inelastic deformation. Symbol $(\widehat{\ })$ is used in this work upon each of kinematic and kinetic variables formulated in the isoclinic configuration [7].

The structural tensors $\mathbf{M}_{ii}|i = 1,2,3$ [8] are introduced in this work to construct the orthotropy symmetric group ϑ of the new constitutive model. Introduction of the structural tensors $\widehat{\mathbf{M}}_{ii}\big|i = 1,2,3$ in the isoclinic configuration assumes the principal directions of material elastic orthotropy (material symmetry) aligned with the unit (director) vectors $\widehat{\mathbf{n}}_1$, $\widehat{\mathbf{n}}_2$ and $\widehat{\mathbf{n}}_3$. As depicted in Figure 1, the structural tensors can be pushed forward from an initial configuration $\mathbf{\Omega}_o$ to elastically unloaded configuration $\overline{\mathbf{\Omega}}_p$ as $\mathbf{M}_{ii} = \mathbf{F}_p \mathbf{M}_{ii} \mathbf{F}_p^{-1}$.

A triad of unit vectors that represent the material symmetries $\widehat{\mathbf{n}}_1$, $\widehat{\mathbf{n}}_2$ and $\widehat{\mathbf{n}}_3$ is presented by two orthogonal axes with arrows in Figure 1. It can be seen that the plastic deformation gradient $\widehat{\mathbf{F}}_p$ and $\mathbf{F}_p$ are defined with respect to $\widehat{\mathbf{\Omega}}_i$ and $\overline{\mathbf{\Omega}}_p$ respectively. The elastic part $\mathbf{F}_e$ of the deformation gradient $\mathbf{F}$ contains elastic stretching and rotation. To avoid any rotation to the principal axes of material orthotropy, the plastic rotation $\mathbf{R}_p$ is assigned to $\mathbf{F}_p$. Therefore the principal axes remains fixed or unaltered by plastic deformation in the chosen isoclinic configuration $\widehat{\mathbf{\Omega}}_i$.

The rotation and distortion during elastoplastic deformation are contained in the elastic part of $\mathbf{F}$ as a result of isoclinic configuration [9]. It is further emphasized in [9] that the elastic deformation can be compatible with plastic deformation when plastic or elastic deformation is homogeneous. The plastic deformation is assumed incompressible:

$$\det\left(\widehat{\mathbf{F}}_p\right) = 1 \tag{3}$$

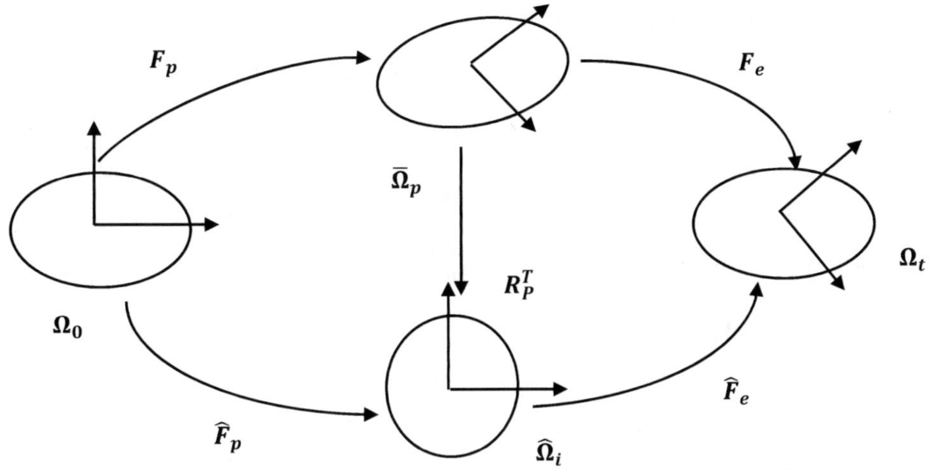

Figure 1 Definition of the chosen isoclinic configuration $\widehat{\boldsymbol{\Omega}}_i$.

2.1 New Generalised pressure

A new generalised pressure $\tilde{P}$ for orthotropic materials defined in [10] is adopted in the proposed formulation. See for instance [14], [15] and [1], it is proven that this newly orthotropic pressure is capable to produce a good agreement with respect to published experimental data. This generalised pressure is defined as

$$\tilde{P} = -\sqrt{\frac{(c_{11}+c_{12}+c_{13})^2+(c_{12}+c_{22}+c_{23})^2+(c_{13}+c_{23}+c_{33})^2}{3}}\,\varepsilon_v = -3K_\psi\varepsilon_v \tag{4}$$

where C_{ij} represent elastic stiffness tensor of orthotropic materials Therefore, the new stress tensor decomposition can be defined as

$$\tilde{S}_{ij} = \sigma_{ij} - \tilde{P}_{ij} \tag{5}$$

The parameter K_ψ reduces to the conventional bulk modulus in the limit of material isotropy. ψ_{ij} which is used defines the direction of the new volumetric axis in stress space then can be defined as

$$\psi_{(ii)} = \frac{c_{i1}+c_{i2}+c_{i3}}{\sqrt{\frac{(c_{11}+c_{12}+c_{13})^2+(c_{12}+c_{22}+c_{23})^2+(c_{13}+c_{23}+c_{33})^2}{3}}} \tag{6}$$

Using the above formulation, any considered orthotropic materials will uniquely define their own deviatoric plane (yield surface) within the principal stress space. It should be noted that ψ_{ij} becomes δ_{ij} when dealing with isotropic materials.

2.2 New Mandel stress tensor

Mandel stress tensor $\hat{\Sigma}$ generally can be defined as

$$\boldsymbol{\Sigma} = \mathbf{C}\cdot\boldsymbol{S} \tag{7}$$

The Mandel stress tensor in respect to isoclinic configuration can be defined as follow

$$\bar{\boldsymbol{\Sigma}} = \mathbf{F}_e^T \cdot \boldsymbol{\tau} \cdot \mathbf{F}_e^{-T} = \det(\mathbf{F}) \cdot \mathbf{F}_e^T \cdot \boldsymbol{\sigma} \cdot \mathbf{F}_e^{-T} \tag{8}$$

The new Mandel stress tensor defined in a deviatoric plane can be defined as

$$\hat{\boldsymbol{\Sigma}}' = \underbrace{\det(\mathbf{F}) \cdot \hat{\mathbf{F}}_e^T \cdot \boldsymbol{\sigma} \cdot \hat{\mathbf{F}}_e^{-T}}_{\hat{\boldsymbol{\Sigma}}} - \underbrace{\det(\mathbf{F}) \cdot \hat{\mathbf{F}}_e^T \cdot \frac{\boldsymbol{\sigma\psi}}{\boldsymbol{\psi\psi}} \cdot \boldsymbol{\psi} \cdot \hat{\mathbf{F}}_e^{-T}}_{\hat{\boldsymbol{\Sigma}}_p} = \det(\mathbf{F}) \cdot \hat{\mathbf{F}}_e^T \cdot \mathbf{S} \cdot \hat{\mathbf{F}}_e^{-T} \tag{9}$$

where $\hat{\boldsymbol{\Sigma}}_p$ defines the volumetric part (pressure) of the proposed Mandel stress tensor.

2.3 *Coupling with Equation of State*

An appropriate constitutive formulation intended to describe the anisotropic material response under shock loading should consider strength effect and the equation of state (EOS). Based on this motivation, the proposed formulation in this work is combined with the Mie-Gruneisen EOS.

$$P_{EOS} = \frac{\rho_0 c^2 \mu \left[1 + \left(1 - \frac{\Gamma}{2}\right)\mu - \frac{\Gamma}{2}\mu^2\right]}{\left[1 - (S_1 - 1)\mu - S_2 \frac{\mu^2}{\mu+1} - S_3 \frac{\mu^3}{(\mu+1)^2}\right]^2} + (1 + \mu) \cdot \Gamma \cdot E \qquad \text{when } \mu > 0 \tag{10}$$

$$P_{EOS} = \rho_0 c^2 \mu + (1 + \mu) \cdot \Gamma \cdot E \qquad \text{when } \mu < 0 \tag{11}$$

where

E is the internal energy per initial specific volume
S_1, S_2, S_3 are the coefficients of the slope of the $U - u_p$ curve
γ_0 is the Gruneisen gamma for the un-deformed material,
a is the first order volume correction to γ_0,
$c, S_1, S_2, S_3, \gamma_0, a, \rho_0$ represent the material properties supplied by the user to characterize this EOS.

2.4 *Elastic Free Energy Function*

The elastic orthotropy is defined using of the Helmholtz free energy as a function of evolving structural tensors.

$$\hat{\varphi}_e = \hat{\varphi}_e\left(\hat{\mathbf{E}}_e, \hat{\mathbf{M}}_{11}, \hat{\mathbf{M}}_{22}\right) = \hat{\varphi}_e\left(\mathbf{Q}\hat{\mathbf{E}}_e\mathbf{Q}^T, \mathbf{Q}\hat{\mathbf{M}}_{11}\mathbf{Q}^T, \mathbf{Q}\hat{\mathbf{M}}_{22}\mathbf{Q}^T\right) \tag{12}$$

where $\mathbf{Q}$ is orthogonal rotation tensor.

2.5 *Orthotropic Yield Criterion*

The yield function used to model the dependence on plastic anisotropy in this formulation is defined using the structural tensors $\hat{\mathbf{M}}_{ii} | i = 1,2,3$ of the material symmetry group ϑ. The Hill's yield function is adopted to define the plastic anisotropy of the proposed constitutive model.

$$\hat{f} = \sqrt{\hat{\boldsymbol{\Sigma}}' : \hat{\mathbb{h}} : \hat{\boldsymbol{\Sigma}}'} - \hat{f}(\alpha) = 0 \tag{13}$$

where $\hat{\mathbb{h}}$ refers to a fourth-order tensor defined in the chosen isoclinic configuration $\hat{\boldsymbol{\Omega}}_i$. The incremental of flow stress during plastic deformation is defined by $\hat{f}(\alpha)$ that is controlled by isotropic hardening.

2.6 Evolution Equations

The evolution of the proposed constitutive model is defined using the second law of thermodynamics framework. The formulation can be expressed using the Clausius-Plank inequality as

$$\widehat{\mathcal{D}} = \dot{\lambda}\left(\sqrt{\widehat{\boldsymbol{\Sigma}}' : \mathbb{h} : \widehat{\boldsymbol{\Sigma}}'} - \alpha : \frac{\partial f}{\partial \alpha}\right) \geq 0 \tag{14}$$

where α and $\dot{\lambda}$ define isotropic hardening and plastic rate parameter respectively. The plastic multiplier $\dot{\lambda}$ is calculated from the consistency (loading-unloading) condition.

$$\dot{\lambda} = \frac{\hat{f}_{\widehat{\boldsymbol{\Sigma}}'} : \mathbf{C}_{el}^{\widehat{\boldsymbol{\Sigma}}_D} : \widehat{\mathbf{D}}}{-\hat{f}_\alpha \cdot \sqrt{\frac{2}{3}}\widehat{\mathbf{K}}_f + \hat{f}_{\widehat{\boldsymbol{\Sigma}}'} : \mathbf{C}_{el}^{\widehat{\boldsymbol{\Sigma}}_D} : \mathrm{sym}\hat{\mathbf{r}}} \tag{15}$$

The relationship between stress and strain increments is concluded by the elastoplastic tagent modulus $\mathbf{C}_{\widehat{\boldsymbol{\Sigma}}'}$

$$\mathbf{C}_{\widehat{\boldsymbol{\Sigma}}'} = \mathbf{C}_{el}^{\widehat{\boldsymbol{\Sigma}}'} - \frac{\left(\mathbf{C}_{el}^{\widehat{\boldsymbol{\Sigma}}_D} : \mathrm{sym}\hat{\mathbf{r}}\right) \otimes \left(\hat{f}_{\widehat{\boldsymbol{\Sigma}}'} : \mathbf{C}_{el}^{\widehat{\boldsymbol{\Sigma}}_D}\right)}{-\hat{f}_\alpha \cdot \sqrt{\frac{2}{3}}\widehat{\mathbf{K}}_f + \hat{f}_{\widehat{\boldsymbol{\Sigma}}'} : \mathbf{C}_{el}^{\widehat{\boldsymbol{\Sigma}}_D} : \mathrm{sym}\hat{\mathbf{r}}} \tag{16}$$

2.7 Grady Failure Model

Grady spall model known as an energy-based failure model is adopted in the proposed formulation [11]. This model assumes the material spall when the strain energy reaches a certain level. Two mechanisms are investigated in this Grady model, which are brittle and ductile fracture.

$$p_{s\,(Ductile)} = \sqrt{2\rho c_0^2 \sigma_Y \varepsilon_{fail}} \quad \text{Ductile Failure} \tag{17}$$

$$p_{s\,(Brittle)} = \sqrt[3]{3\rho c_0 K_c^2 \dot{\varepsilon}} \quad \text{Brittle Failure} \tag{18}$$

where:

ρ	= density
c_0	= bulk sound speed
σ_Y	= yield stress
ε_{fail}	= Critical strain failure
K_c	= Fracture toughness
$\dot{\varepsilon}$	= Rate of volumetric dilatation

For a certain strain rate there is a transition point between ductile and brittle spall. This critical strain rate, ε_{crit} can be calculated from:

$$\dot{\varepsilon}_{crit} = \sqrt{\frac{8B_0^2(\sigma_Y \varepsilon_{fail})^3}{9\rho K_c^4}} \tag{19}$$

where:

ρ	= density
σ_Y	= yield stress

IOP Conf. Series: Journal of Physics: Conf. Series **914** (2017) 012033 doi:10.1088/1742-6596/914/1/012033

ε_{fail} = Critical strain failure
K_c = Fracture toughness
B = Isotropic/kinematic hardening

3. New Material Model Implementation

The proposed constitutive formulation was implemented into DYNA3D finite element code of UTHM's version and name Material Type 92 referring to the works published in [16] and [17]. It should be noted that this is the first ever algorithm implementation of a newly constitutive formulation performed in the code. As discussed previously, a new generalized pressure is adopted in the proposed formulation. Therefore, few modifications are required to uniquely characterize the DYNA3D of UTHM's version with the new orthotropic pressure. A new definition of EOS was introduced to the code by decoupling the volumetric parts from the full stress tensor. Equations 17 until 19 are directly adopted for spall model implementation into the code.

4. Numerical Simulation

At this stage, the capability to predict the deformation behaviour under high pressure and shockwave is assessed against the experimental data of Plate Impact test published in [12]. Thus, a set of experiments plate impact test of Aluminium Alloy 7010 was used. The finite element FE model used in this analysis to simulate the Plate Impact test is discussed in this section. Figure 2 shows the configuration used to simulate each Plate Impact test analysis in this work.

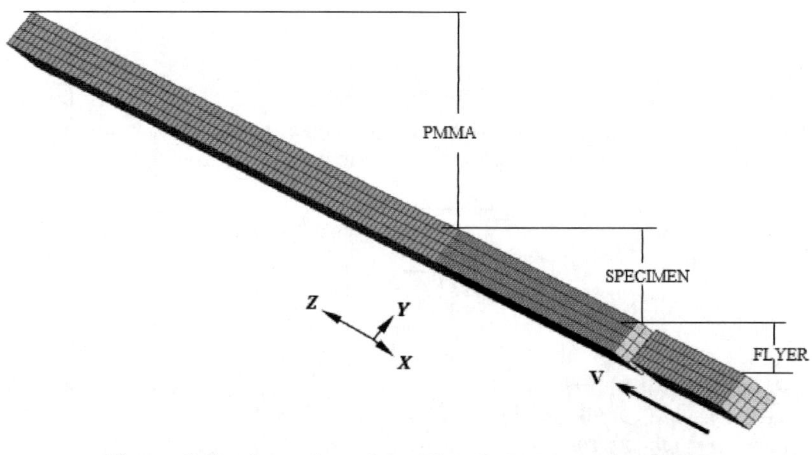

Figure 2 Configuration of the Plate Impact test simulation

It can be seen that the model is divided into three parts of rectangular bars with 4x4 solid elements for its cross section. The PMMA block, specimen and flyer is modelled with 100 solid elements (12mm in length), 75 solid elements (10mm in length) and 25 solid elements (2.5mm in length) respectively. As can be observed, the model orientation is parallel to the impact axis, (Z axis). The mesh applied of this simulation was set to allow a 1D wave propagation along the length of the bars during the impact. From this figure, it is noticed that symmetrical planes were adopted on all sides of the bars. A non-reflecting boundary condition was used to define the back of this block (PMMA), to ensure that no release wave was reflected from the back of the PMMA block into the test specimen. There was a contact interface defined in between the flyer (impactor) and the test specimen. A time history block was prescribed to the elements at the top of PMMA bar to ensure the stress time histories was recorded during impact.

5. Results and Discussion

The Z stress in the top elements of the PMMA bar was compared against the experimental data in short transverse and the longitudinal (rolling) directions of the specimen, performed at three different impact velocities: 234m/s, 450m/s and 895m/s. Table 1 shows the material properties of each parts including the Gruneisen EOS and the Grady failure model adopted in this analysis. The flyer is prescribed as Aluminium 6082-T6 [12].

The results obtained for each impact velocity in respect to both longitudinal and transverse directions are shown in Figures 3 until 8. Referring to the results in the Figures 3 until 8, it can be observed that the proposed Mat92 is capable of describing the elastic-plastic loading-unloading behaviours of the Al7010 since a close prediction is obtained in each test. The comparison between Mat 92 and the experimental data are summarised in Table 2.

Table 1 Material properties used in the Plate Impact test analysis

Parameters		Materials		
		Al7010	Al6082	PMMA
Young's modulus	E_a	70.6 GPa	–	–
	E_b	71.1 GPa	–	–
	E_c	70.6 GPa	–	–
Poisson's ratio	v_{ba}	0.342	–	–
	v_{ca}	0.342	–	–
	v_{cb}	0.342	–	–
Shear modulus	G_{bc}	26.31 GPa	26.8 GPa	2.3 GPa
	G_{ab}	26.48 GPa	26.8 GPa	2.3 GPa
	G_{ac}	26.48 GPa	26.8 GPa	2.3 GPa
Yield stress in a – direction	σ_y	564 MPa	250 MPa	70 MPa
Tangent plastic modulus in a – direction	H	0.13 GPa	130 MPa	300 MPa
Pressure cutoff	p_{cut}	–	2.5 GPa	–
Density	ρ	2.81 gcm^{-3}	2.7 gcm^{-3}	1.18 gcm^{-3}
Hill's parameters	R	1	–	–
	P	0.719	–	–
	Q_{bc}	1	–	–
	Q_{ba}	1	–	–
	Q_{ca}	1	–	–
Gruneisen parameters	C	5200 ms^{-1}	5240 ms^{-1}	2180 ms^{-1}
	s_1	1.36	1.4	2.088
	s_2	0.00	0.00	−1.124
	s_3	0.00	0.00	0.00
	Γ	2.2	1.97	0.85
	a	0.48	0.48	0.00
Grady parameters	ρ	2.81 gcm^{-3}	–	–
	c_0	0.52	–	–
	ε_{fail}	0.5	–	–
	K_c	0.0025	–	–
	B	$\rho/c_0{}^2$	–	–

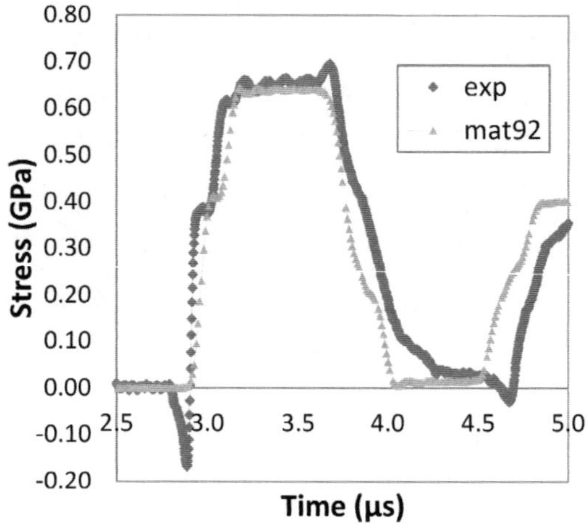

Figure 3 Longitudinal Stress at 234ms^{-1} impact in longitudinal direction

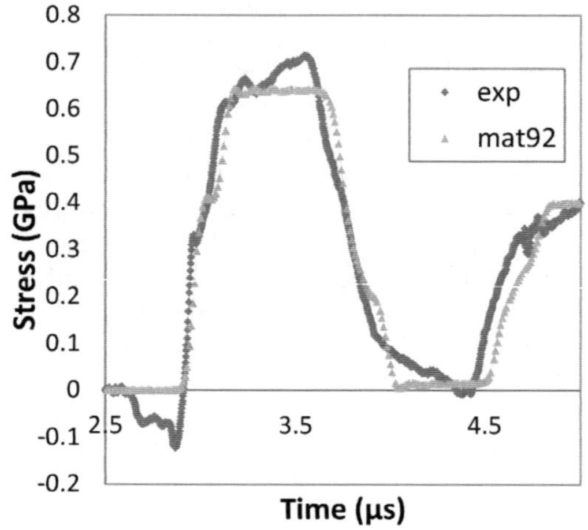

Figure 4 Longitudinal Stress at 234ms^{-1} impact in transverse direction

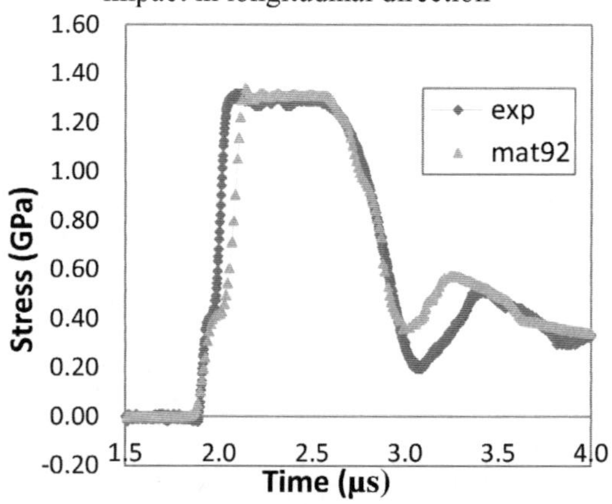

Figure 5 Longitudinal Stress at 450ms^{-1} impact in longitudinal direction

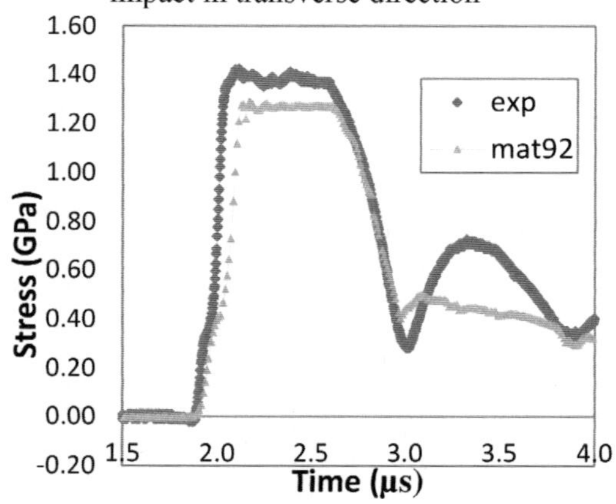

Figure 6 Longitudinal Stress at 450ms^{-1} impact in transverse direction

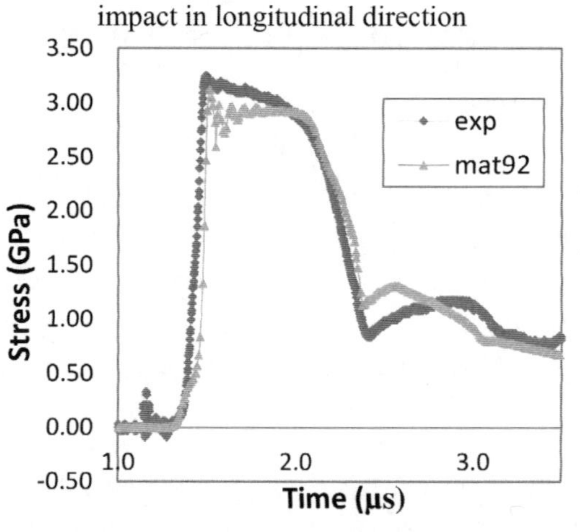

Figure 7 Longitudinal stress at 895ms^{-1} impact in longitudinal direction

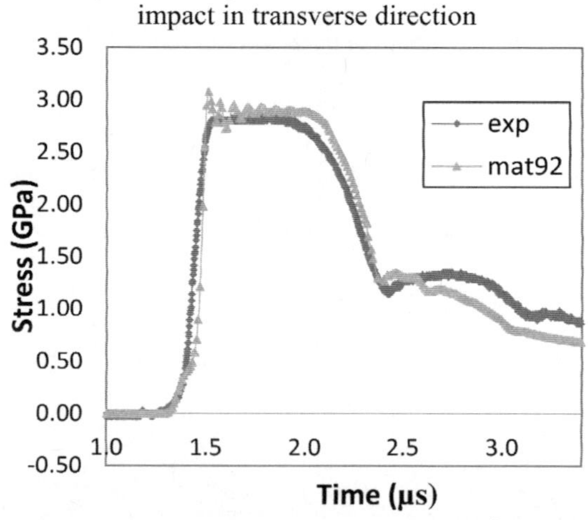

Figure 8 Longitudinal stress at 895ms^{-1} impact in transverse direction

Table 2 Mat92 vs. Plate Impact Test Data

Impact Velocity/ Direction	Analysis Criteria		
	HEL (GPa)	Hugoniot Stress Level (GPa)	Pulse (μs)
$234ms^{-1}$ (Longitudinal):			
Simulation	0.40	0.64	1.15
Experiment	0.39	0.65	1.65
$234ms^{-1}$ (Transverse):			
Simulation	0.40	0.64	1.18
Experiment	0.33	0.63	1.44
$450ms^{-1}$ (Longitudinal):			
Simulation	0.41	1.32	1.13
Experiment	0.43	1.31	1.16
$450ms^{-1}$ (Transverse):			
Simulation	0.38	1.28	1.08
Experiment	0.39	1.38	1.13
$895ms^{-1}$ (Longitudinal):			
Simulation	0.36	2.95	1.07
Experiment	0.21	3.25	1.10
$895ms^{-1}$ (Transverse):			
Simulation	0.32	2.89	1.12
Experiment	0.19	2.80	1.13

In general, the shape of generated pulse shows a good agreement compared to the experimental data. The Hugoniot Elastic Limit (HEL) is described by the initial slope of the longitudinal stress increment. An adequate level of anisotropy of the material is reflected by a different HEL value in each direction. In addition, the generated pulse and the Hugoniot stress levels are closely agreed with the experimental data to confirm the elastic-plastic formulation including the newly implemented orthotropic pressure of the proposed constitutive model is capable to describe the behaviour of orthotropic material undergoing finite strain deformation.

The simulated longitudinal stress at the interface between the target material and PMMA block (recorded in the elements at the back of the specimen), for plate impact at $234ms^{-1}$ is presented in Figure 3 and Figure 4. It can be clearly seen that the tensile wave failure or spall (demonstrated by the reloading of the longitudinal stress after the first loading-unloading pulse) is not generated with a lower impact velocity ($234ms^{-1}$) . However, a clear spall criterion can be observed when higher impact velocities ($450ms^{-1}$ and $895ms^{-1}$) are applied.

At $450ms^{-1}$ impact velocity, a clear break in slope in the start rising part of trace (HEL) can be seen in a trace as shown in Figure 5 and Figure 6. The value of the HEL shows a slight difference. Specifically, the value in the longitudinal directions are 0.41 GPa and 0.38GPa, respectively. The associated value obtained in the experiment are 0.39 GPa and 0.33 GPa, respectively. A very close agreement is observed in respect to Hugoniot stress levels. This confirmed the capability of the implemented orthotropic pressure of Mat92 to capture shockwaves in orthotropic materials.

In addition, a clear pull back signals (spall) is developed in both traces; measured 0.31 GPa in the longitudinal and 0.42 GPa in the transverse direction. The values obtained using the Grady's spall criterion are 0.21 GPa and 0.10 GPa in the longitudinal and transverse directions, respectively. It can be observed the values are smaller than the values in experiment. The constitutive model however, shows sensitivity to the direction of impact. This behaviour is depicted by the point where the spall starts. The similar behaviour is obtained experimentally.

Figure 7 and Figure 8 show the results for $895ms^{-1}$ impact velocity. It is clear that the shape of the pulse including the Hugoniot stresses show a good agreement with the experimental data in both impact directions. This is summarised in Table 2. As observed in 450ms^{-1} impact velocity, again the proposed constitutive model predict a smaller pullback signals in impact direction.

Referring to the results, the model as it stands is capable to simulate elastic-plastic, shockwave propagation and spall failure in orthotropic materials of aluminium alloy AA7010. The general pulse shape, the Hugoniot stress level and the EOS are predicted satisfactorily. A higher HEL is observed in the longitudinal direction compared to the short transverse. The Hugoniot stress levels show a very good prediction capability. The pulse shape including the width are in line with the experimental data. The adopted Grady's spall failure model is obviously capable to predict spall in the materials. However, it must be emphasised that the criterion is not capable to provide a very good accuracy to deal with a complex spall strength evolution developed at different impact velocity. Further works, therefore, are required in both experimental and constitutive modelling aspects.

6. Conclusion

The proposed constitutive model is one of integration between elasticity, plasticity, equation of state (EOS) and spall failure concepts. The formulation is combined with the new stress tensor decomposition defined specifically for orthotropic materials.

The proposed mathematical formulation is implemented into the DYNA3D of UTHM's version using the suitable stress update algorithm. All parameters defined for the new constitutive model are defined and implemented in detail into the DYNA3D finite element code by modification of several routines in the code.

The final part is a comparison against range of Plate Impact Test data at 234ms^{-1}, 450ms^{-1} and 895ms^{-1} impact velocities [12]. A good agreement is obtained in each test. At the end of this work, it can be concluded that the proposed formulation of the new constitutive model has been successfully implemented. The developed formulation is the key novelty of this work that represent shockwave propagation including spall failure of orthotropic materials in a unique alignment of the deviatoric plane within the principal stress space.

7. Acknowledgement

Authors wish to convey a sincere gratitude to Universiti Tun Hussein Onn Malaysia (UTHM), and Ministry of Higher Education Malaysia for providing the financial means during the preparation to complete this work under Fundamental Research Grant Scheme (FRGS), Vot 1547.

8. References

[1] Mohd Nor M K, 2016 Modelling Inelastic Behaviour of Orthotropic Metals in a Unique Alignment of Deviatoric Plane within The Stress Space, International Journal of Non-Linear Mechanics 87, pp. 43–57
[2] Chen Y, Asay J R., Dwivedi S K and Field D P, 2006 Spall Behaviour of Aluminum with Varying Microstructures, Journal Applied Physic 99, pp. 1–13
[3] Mandel J, 1972 Plasticité Classiqueet Viscoplastiécism Lecture Notes, Springer-Verlag, Wien,
[4] Asaro R J, 1983 Micromechanics of Crystals and Polycrystals. Advances in Applied Mechanics 23, 1–115
[5] Dafalias Y F and Rashid M M, 1989 The Effect of Plastic Spin On Anisotropic Material Behaviour, International Journal of Plasticity. Vol. 5, pp. 227-246
[6] Belytschko T, Guo Y, Liu W K and Xiao S P, 2000 A Unified Stability Analysis of Meshless Particle Methods, International Journal Numerical method in Engineering, Vol. 48, Issue 9
[7] Mohd Nor M K, 2016 Modeling of Constitutive Model to Predict The Deformation Behaviour of Commercial Aluminum Alloy AA7010 Subjected To High Velocity Impacts, ARPN Journal of Engineering and Applied Sciences, ISSN 1819-6608
[8] Boehler J P, 1977 On irreducible Representations for Isotropic Scalar Functions, ZAMM 57, pp. 323–327

[9] Schröder J, Gruttmann F and Löblein J, 2002 A Simple Orthotropic Finite Elasto–Plasticity Model Based on Generalized Stress–Strain Measures, Computational Mechanics 30, pp. 48,64
[10] Vignjevic R, Campbell J C, Bourne N K and Djordjevic N, 2008 Modeling Shock Waves in Orthotropic Elastic Materials, Journal of Applied Physics 104, pp. 044904
[11] Grady D, 1988 Incipient Spall, Crack Branching, and Fragmentation Statistics in the Spall Process, Journal de Physique Colloques, 49 (C3), pp.C3-175-C3-182
[12] De Vuyst T A, 2003 Hydrocode Modelling of Water Impact, Phd Thesis Cranfield University
[13] Man C, 1995 On The Correlation of Elastic and Plastic Anisotropy in Sheet Metals, Journal Elastic-Plastic 39 (2) pp. 165–173
[14] Mohd Nor M K, Vignjevic R and Campbell J, 2013 Plane-Stress Analysis of the New Stress Tensor Decomposition, Applied Mechanics and Material, Vol. 315, pp. 635-639
[15] Mohd Nor M K, Vignjevic R. and Campbell J, 2013 Modelling of Shockwave Propagation in Orthotropic Materials, Applied Mechanics and Materials Vol. 315, pp 557-561
[16] Mohd Nor M K, Ma'at N, Kamarudin K A and Ismail A E, 2016 Implementation of Finite Strain-Based Constitutive Formulation in LLLNL-DYNA3D to Predict Shockwave Propagation in Commercial Aluminum Alloys AA7010, International Engineering Research and Innovation Symposium (IRIS), Materials Science and Engineering 160, pp. 012023
[17] Mohd Nor M K and Ma'at N, 2016 Simplified Approach to Validate Constitutive Model Formulation of Orthotropic Materials Undergoing Finite Strain Deformation, Journal of Engineering Applied Science 11, pp. 2146-2154

Effect of triggering angles on the crushing mechanisms of hybrid woven kenaf/aluminum hollow cylinders

Al Emran Ismail*, Fazimah Mat Noor, Zaleha Mohamad, Mohd Amran Madlan, M. Zulafif Rahim, M. Rasidi Ibrahim, Sufizar Ahmad, Mohd Nasrull Abdol Rahman, Salihatun Md Salleh, Azmahani Sadikin, Shahruddin Mahzan, Nik Hisyamudin Muhd Nor

Faculty of Mechanical and Manufacturing Engineering, Universiti Tun Hussein Onn Malaysia, 86400 Johor, Malaysia.

Corresponding author: *emran@uthm.edu.my / al_emran@hotmail.com

Abstract. This paper presents the effect of triggering angles constructed on the top of hybrid woven kenaf/aluminium hollow cylinders on the energy absorption performances. The crushing performances of aluminium tubes can be found widely in open literature. However, lack number of work on the hybridizing the aluminium tubes with woven kenaf fibre is found. Woven kenaf mats are produced and bathed with polymeric resin before they are wrapped around the aluminium tubes twice. Different fibre orientations, $\pm\theta^o$ are used where $\theta = 0$, 15, 30 and 45. Once the hybrid composite hardened, one of their end are chamfered using different angles of 0°, 30°, 45° and 60°. The tubes are quasi-statically compressed in order to obtain their force-displacement responses and crashworthiness parameters are extracted and discussed with the relation of fibre orientations and chamfering angles. It is found that the chamfering angles are only affected the force-displacement curves during the first stage of elastic deformation whereas there is no obvious effect in the second stage. However, varying the fibre orientations are slightly increased the force-displacement curves especially when the fibre is orientated with 30°. Based on the fracture mechanism observations, most of composite experienced large fragmentation indicating that the composites absorbed the crushing energy ineffectively.

1. Introduction

Natural fibers are now a bright potential in reinforcing polymer composites due to their ecofriendly aspect and sustainability. These kind of fibers are considered to have tremendous number of applications since they have good mechanical and physical properties over synthetic fiber such as relative low weight, low cost, good mechanical performances and etc. [1]. Applications of natural fiber for automotive parts can be found in [2, 3]. One of the interest point to be considered is the utilization of natural fiber for the energy absorbing structures made of natural fibers.

Meredith et al. [4] developed conical shaped geometry fabricated using natural fiber before it is subjected impact loading. It is found that the cones exhibited high values of SEA: unwoven hemp 54.3 J/g, woven flax 48.5 J/g and woven jute 32.6 J/g. The SEA was influenced primarily by fiber volume fraction, V_f where a high V_f leads to high SEA. Significant variability in SEA resulted from the variation in fiber strength and V_f as a result of the vacuum assisted resin transfer molding manufacturing process. Laban and Mahdi [5] studied the performances of energy absorption of cotton fiber/epoxy composite square and rectangular tubes where the effect of cross-sectional aspect ratios are emphasized. The experiment also showed that the cross-sectional aspect ratio significantly affects the load carrying and energy absorption capability. The buckling failure mode has been identified as the primary failure mode for the rectangular tubes under the different loading conditions. Tubes with cross-sectional aspect ratio of 2.0 have the best load carrying capacity.

Ismail [6] presented the tensile strength of woven yarn kenaf reinforced polyester composites. Two important parameters are studied such as fiber orientations and number of layers. According to the results, it is shown that fiber orientations greatly affected the ultimate tensile strength but it is not for modulus of elasticity for both types of layers. It is estimated that the reductions of both ultimate tensile strength and Young's modulus are in the range of 27.7-30.9% and 2.4-3.7% respectively, if the inclined fibers are used with respect to the principal axis.

Lateral crushing performances of woven kenaf composites can also be found in [7]. As-received kenaf yarn is firstly submerged into a resin bath before it was wrapped around the cylindrical mould. Two important parameters were considered during the composite preparation stage which were fiber orientations (0^0, 5^0 and 10^0) and number of layers (1, 2 and 3 layers). Then, the hardened composite tubes were positioned horizontally prior to quasi-static compression. Lateral crushing resistances of the composite tubes are compared in term of specific energy absorptions, force ratios and deformation modes. Lastly, these crashworthiness parameters are discussed relating with fiber orientations and number of layers. Axial energy absorption capabilities of woven kenaf fiber reinforced composites are available in [8-11]. Works on fiber hybridizations with the metallic cylinders can be found in [12, 13]. For an example Ismail [12] hybridized aluminum tubes with kenaf yarn composites and compressed quasi-statically and obliquely. It is found that specific energy absorption performances increased when the numbers of layers are increased, fiber orientations are not played an important role in increasing the capabilities of specific energy absorptions, force ratios decreased when the numbers of layers are increased but it is significantly not affected the energy absorption performances and fiber orientations seemed not to affect the force ratios. However when fiber orientations increased, force ratios are slightly decreased.

. On the other hand, Ismail [13] also hybridized the steel tubes with kenaf yarn before it is compressed quasi-statically. The kenaf yarn fibres were firstly wetted with a polymeric resin and it was then wound around the steel tubes. The tube diameter was kept constant 50 mm in diameter and fiber orientations were varied. The steel/kenaf hybrid tubes are aligned vertically and compressed quasi-statically. The force-displacement curves were recorded automatically and crashworthiness parameters were determined and related with the hybridization parameters. It was found that the present of kenaf yarn fiber around the steel tubes capable to slightly enhance the crushing performances. During progressive collapses, fiber splaying mode was observed with large composite wall fragmentations.

This paper presents the effect of different triggering angles on the crushing performances of hybrid woven kenaf/aluminum tubes under axial compression. There are four chamfering angles are selected such as 0°, 30°, 45° and 60°. Aluminum tubes are wrapped with two layers of woven kenaf mats where they are firstly wetted with polyester resin where each layer is oriented using different angles. Once the composite hardened (after 24 hours of curing time), they are quasi-statically compressed in order to obtain their force-displacement responses. Energy absorption performances are calculated according to the area under the curves and the crushing behavior are evaluated and analyzed.

2. Methodology

An industrial grade aluminum hollow tube is used with a thickness of 1mm and 50mm external diameter. The length of tube is 100mm where it is long enough to prevent global buckling. As-received kenaf yarn (as in Figure 1(a)) is weaved into a plain woven mat before it is wrapped around the outer surface of aluminum tube. The weaving process is conducted using an in-house facility as in Figure 1(b) and 1(c). While Figure 1(d) revealed the completed woven kenaf mat. Woven kenaf mat is firstly immersed into a resin bath where a proper care is given in ensuring resin penetrates into the fibers uniformly. The wet fiber mat is wrapped around the tube twice and each layer oriented according to the specific orientation as tabulated in Table 1. Figure 1(e) shows the completed hybrid woven kenaf/aluminum hollow tube. This sample is positioned vertically between two flat rigid plates. The bottom plate is stationary while the upper plate is moved downward. The sample is compressed quasi-statically with a speed of 1.5 mm/min of cross-head displacement. The force versus displacement curves are recorded automatically where the area under the curves represent the capability of energy absorption, E and it can be represented as:

$$E = \int_0^L P ds \qquad (1)$$

where P is a force, ds is a small element on the curve and L is the crushed length. Instead of energy absorption performance, other important parameter such as force ratio, ψ is also determined as in Equation (2).

$$\psi = \frac{P_{mean}}{P_{peak}} \qquad (2)$$

where P_{mean} is the average force during the crushing process and P_{peak} is the maximum or peak force just after the elastic deformation. Final crushed samples are also observed in order to study the effect of variations of fiber orientations on the crushing performances.

(a) (b) (c)

(d) (e)

Figure 1. (a) As-received kenaf yarn, (b) Weaving machine, (c) Weaving process in progress and (d) Completed woven kenaf mat and (e) Hardened hybrid aluminum composite tube.

Table 1. Sample parameters and compression conditions.

No.	Orientations	Layers	Chamfered angles
1	[±0]	2	0°
2		2	30°
3		2	45°
4		2	60°
5	[±15]	2	0°
6		2	30°
7		2	45°
8		2	60°
9	[±30]	2	0°
10		2	30°
11		2	45°
12		2	60°
13	[±45]	2	0°
14		2	30°
15		2	45°
16		2	60°

3. Results and discussion

Figure 2 shows the force-displacement responses of hybrid composites when upper edges are chamfered with different angles. In general, chamfering the tubes capable to increase the peak forces. However, it is strongly depend on the fiber orientations. For the fiber orientation of 0° and 15°, it is seemed that chamfering the upper edge with specific angles do not played an important role in increasing the peak force when compared with other orientations such as 30^0 and 45°. This is probably due to the fact that 0° and 15° fiber orientations capable to resist the deformation only in the lateral direction. For higher orientations (30° and 45°), the fiber not only constraint the deformation laterally but obliquely and therefore increasing the peak forces. In the second or plateau stages, the effect of triggering angles on the force-displacement responses are insignificant where there is no obvious distinct. However, the responses of hybrid composite tubes are relatively higher that empty tubes due to the present of layer of composites. This is indicated that chamfering angles played an important role only in the first stage. Many research reports have found similar mechanisms regardless of material type [7-13].

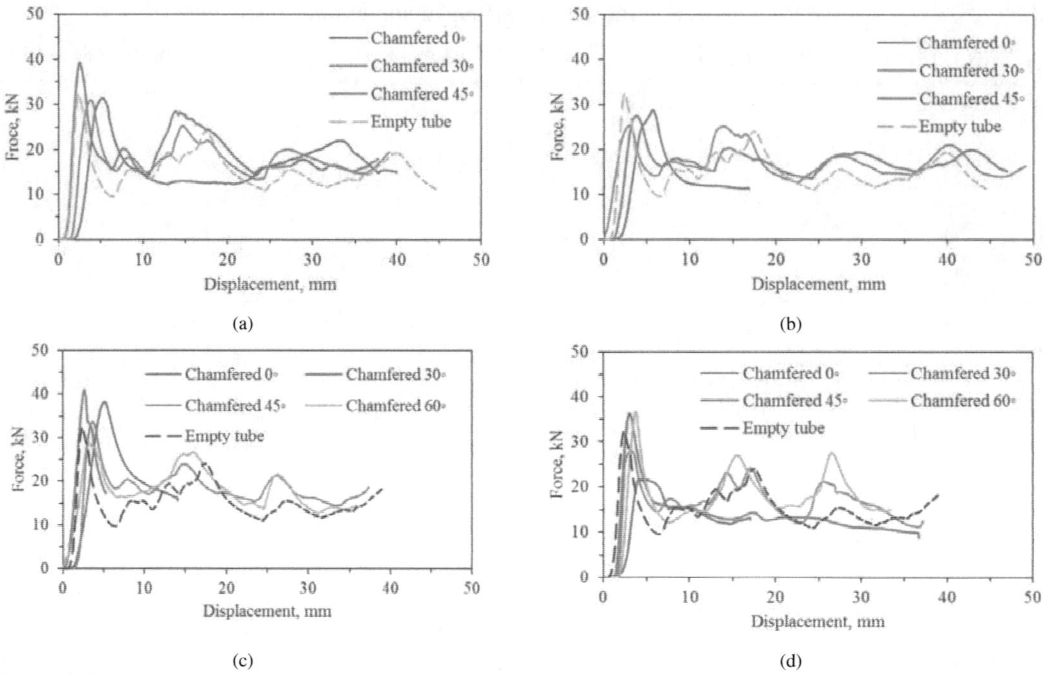

Figure 2. Effect of chamfering angles on the force-displacement curves for different fibre orientations, (a) 0°, (b) 15°, (c) 30° and (d) 45°.

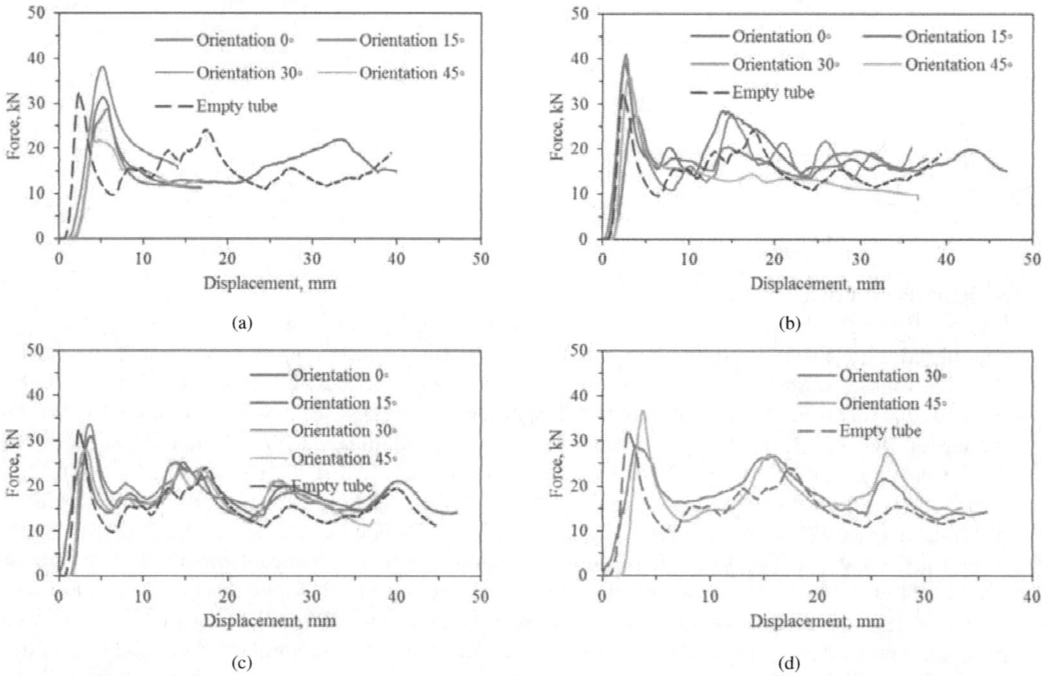

Figure 3. Effect of fibre orientations on the force-displacement curves for different chamfering angles, (a) 0°, (b) 30°, (c) 45° and (d) 60°.

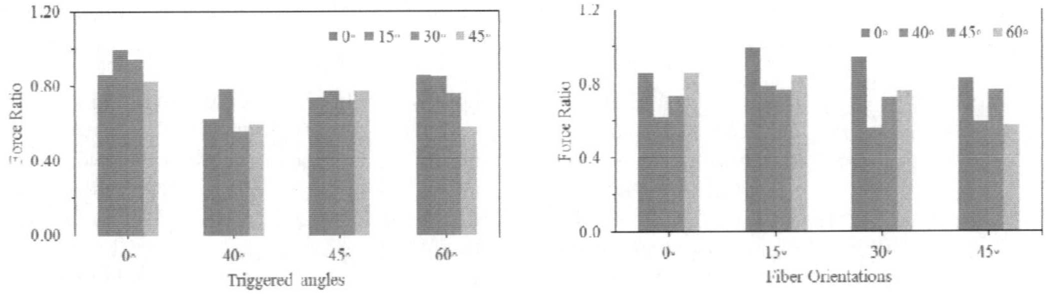

(a) (b)

Figure 4. (a) Effect of triggering angles on the force ratios and (b) Effect of fibre orientations on the force ratios.

(a) (b) (c)

(d) (e)

Figure 5. Hybrid tube deformation at different compression stages (a) initial condition, (b) 25 mm deformation, (c) 45 mm deformation, (d) 55mm deformation and (e) an aerial view of 55mm final deformation.

Figure 3 shows the effect of fiber orientations on the force-displacement curves when chamfering angles are varied. In general, it is seemed that increasing the chamfering angles slightly reduced the peak forces. Figures 3(a) and 3(b) reveal the force-displacement responses for the hybrid tubes chamfered with 0° and 30°, respectively. It is observed that hybridizing the aluminum tubes with woven kenaf mat capable to increase the peak forces compared with empty tubes, it is also observed that 30° fiber orientation produced the highest peak forces. This is probably if the fiber oriented at 30°, it is capable to resist the deformation in both directions laterally and axially. Therefore strengthening the hybrid tubes. When the chamfering angles are further increased, there is no significant effect on the peak forces are observed indicating that the present of composite layer wrapped around the outer surface of the tubes do not contributed to the enhancement of the peak forces as compared with the peak forces of empty tubes.

Figure 4 reveals the effect of chamfering angles and fiber orientations on the force ratios, respectively. In general, there are no strong relationship between chamfering angles and fiber orientations on the force ratios. However, Figure 4(a) shows that the chamfering seemed to slightly decrease the force ratios for all type of fiber orientations. This is due to the fact that when the upper edge is chamfered, it is acted as a triggering mechanism where the tube is easily crushed in the very initial stage as compared with untampered edge. Therefore, the force ratio is relatively lowered can be produced. While Figure 4(b) obviously observed that 45° fiber orientation offered the lowest force ratios probably due to no fiber oriented at 0° worked to restraint the tube deformation laterally.

Figure 5 shows the sequences of the tube deformations under quasi-static axial compression. Figure 5(a) shows the initial condition of the tube while Figures 5(b) and 5(c) show the crushing mechanisms when the tubes are crushed approximately at 10 and 20 mm, respectively. On the other hand, Figures 5(d) and 5(e) show the final crushed geometry when the tube is crushed in half of its middle height. It is revealed that large composite fragmentations are observed indicating that they are absorbed the crushing energy insufficiently.

4. Conclusion

Based on the experimental results, there are no strong relationship between the effect of chamfering and orientation angles on the force ratios. All chamfering angles offered better force ratios except for the case of 30°. Lower force ratio indicating that the hybrid tubes seemed to collapse catastrophically or they experienced a sudden drop. On the other hand, the effect of fiber orientations on the force ratios are almost insignificant. However, 15° fiber orientation showed slightly higher than others. In term of fracture mechanism for all cases, large composite fragmentations are observed revealing that the composite ineffectively absorbed the crushing energy.

Acknowledgement

Authors acknowledge Universiti Tun Hussein Onn Malaysia and Office for Research, Innovation, Commercialization and Consultancy Management (ORICC) for sponsoring this work.

References

[1] Layth Mohammed, Ansari MNM, Grace Pua, Mohammad Jawaid and Saiful Islam M 2015 A review on natural fiber reinforced polymer composite and its application *International Journal of Ploymer Science* **2015** 243947.

[2] Davoodi MM, Sapuan SM, Ahmad D, Aidy A, Khalina A and Jonoobi M 2011 Concept selection of car bumper beam with developed hybrid bio-composite material *Materials & Design* **32** 4857-4865.

[3] Pickering K 2008 Ed. Properties and Performance of Natural-Fibre Composites Woodhead Publishing, Cambridge UK.

[4] James Meredith, Richard Ebsworth, Stuart R. Coles, Benjamin M. Wood, Kerry Kirwan 2012 Natural fibre composite energy absorption structures *Composites Science and Technology* **72** 211-217.

[5] Othman Laban and Elsadig Mahdi 2016 Energy Absorption Capability of Cotton Fiber/Epoxy Composite Square and Rectangular Tubes *Journal of Natural Fibers* **13** 726-736.

[6] Ismail AE and Che Abdul Aziz MA 2015 Tensile strength of woven yarn kenaf fiber reinforced polyester composites *Journal of Mechanical Engineering and Sciences* **9** 1695-1704.

[7] Roslan MN, Ismail AE, Hashim MY, Zainulabidin MH, Khalid SNA 2014 Modelling analysis on mechanical damage of kenaf reinforced composite plates under oblique impact loading *Applied Mechancis and Materials* **465-466** 1324-1328.

[8] Ismail AE and Sahrom MF 2015 Lateral crushing energy absorption of cylindrical kenaf fiber reinforced composites *International Journal of Applied Engineering Research* **10** 19277-

19288.

[9] Ismail AE, Zainulabidin MH, Roslan MN, Mohd Tobi AL and Muhd Nor NH 2014 Effect of velocity on the impact resistance of woven jute fiber reinforced composites *Applied Mechanics and Materials* **465-466** 1277-1281.

[10] Ismail AE and Hassan MA 2014 Low velocity impact on woven kenaf fiber reinforced composites *Applied Mechanics and Materials* **629** 503-506.

[11] Ismail AE 2016 Kenaf fiber reinforced composites tubes as energy absorbing structures *International Journal of Mechanical and Mechatronics Engineering* **16** 49-53.

[12] Ismail AE and Mohd Tobi AL 2016 Axial energy absorption of woven kenaf fiber reinforced composites *ARPN Journal of Engineering and Applied Sciences* **11** 8668-8672.

[13] Eyvazian A 2016 Mathematical model for thin-walled corrugated tube under axial compression *MATEC Web of Conferences* **40** 02010.

[14] Eyvazian A, K Habibi M, Hamouda, AM and Hedayati R 2014 Axial crushing behavior and energy absorption efficiency of corrugated tubes *Materials & Design* **54** 1028-1038.

[15] Ismail AE 2016 Oblique crushing performances of kenaf/aluminum composites tubes *International Journal of Applied Engineering Research* **11** 5997-6002.

[16] Ismail AE 2015 Axial crushing energy absorption capability of steel/kenaf fibre hybrid cylindrical tubes *International Journal of Engineering and Technology* **7** 1098-1104.

IOP Conf. Series: Journal of Physics: Conf. Series **914** (2017) 012035 doi:10.1088/1742-6596/914/1/012035

Structural characterization and mechanical properties of polypropylene reinforced natural fibers

M A A Karim[1, a], I Zaman[1, 2, b], S A M Rozlan[1], N I C Berhanuddin[3], B Manshoor[3], M S Mustapha[3], A Khalid[4], S W Chan[5]

[1]Noise and Vibration Analysis Research Group, Faculty of Mechanical and Manufacturing Engineering, Universiti Tun Hussein Onn Malaysia, Johor, Malaysia.
[2]Mechanical Failure Prevention and Reliability Research Center, Universiti Tun Hussein Onn Malaysia, 86400 Batu Pahat, Johor, Malaysia.
[3]Faculty of Mechanical and Manufacturing Engineering, Universiti Tun Hussein Onn Malaysia, 86400 Batu Pahat, Johor, Malaysia.
[4]Faculty of Engineering Technology, Universiti Tun Hussein Onn Malaysia, 86400 Batu Pahat, Johor, Malaysia
[5]Faculty of Technology Management and Business, Universiti Tun Hussein Onn Malaysia, 86400 Batu Pahat, Johor, Malaysia

Corresponding author: aizat2412@yahoo.com [a], izzuddin@uthm.edu.my [b]

Abstract. Recently the development of natural fiber composite instead of synthetics fiber has lead to eco-friendly product manufacturing to meet various applications in the field of automotive, construction and manufacturing. The use of natural fibers offer an alternative to the reinforcing fibers because of their good mechanical properties, low density, renewability, and biodegradability. In this present research, the effects of maleic anhydride polypropylene (MAPP) on the mechanical properties and material characterization behaviour of kenaf fiber and coir fiber reinforced polypropylene were investigated. Different fractions of composites with 10wt%, 20wt% and 30wt% fiber content were prepared by using brabender mixer at 190 °C. The 3wt% MAPP was added during the mixing. The composites were subsequently molded with injection molding to prepare the test specimens. The mechanical properties of the samples were investigated according to ISO 527 to determine the tensile strength and modulus. These results were also confirmed by the SEM machine observations of fracture surface of composites and FTIR analysis of the chemical structure. As the results, the presence of MAPP helps increasing the mechanical properties of both fibers and 30wt% kenaf fiber with 3wt% MAPP gives the best result compare to others.

1. Introduction

Over the last few decades, there is a growing trend in using natural fibers as reinforcement for polymeric composites due to their attractive properties such as lightweight, renewability, low density, high specific strength, non-abrasively, combustibility, non toxicity, low cost and biodegradability [1]. Natural fibers such as abaca, jute, kenaf, coir, ijuk have all been proved to be good reinforcement in thermoset and thermoplastics matrices [2-5]. These days, the use of natural fibers reinforced composites is well known in various applications such as automotive components, building materials, and the aerospace industry due to the economical and environmental concerns over traditional

IOP Conf. Series: Journal of Physics: Conf. Series **914** (2017) 012035 doi:10.1088/1742-6596/914/1/012035

inorganic reinforcements and fillers [6,7]. Figure 1 presents the different components made of lignocellulosic reinforced polymer composite in a Mercedes Benz sedan [8].

Nowadays, the global plastic production and consumption is rising progressively. The increased demand shows that the usage is important and also an integral part of our life [9]. A vigorous effort to explore more environmentally friendly, sustainable materials to change the current glass fiber and carbon fiber composites are required, hence able to reduce the reliance on petroleum based fuel and products [10]. Presently, due to high stiffness and strength properties of synthetic fibers like glass, carbon and aramid, thus it is being used extensively in polymer-based composites. Nevertheless, these fibers have serious downsides in terms of their biodegradability, initial processing costs, recyclability, energy consumption, machine abrasion, health hazards, and others [11]. Therefore, the interest has recently moved to the fabrication and properties of natural fiber reinforced materials.

Figure 1. Automobile components made of natural fiber reinforced composites [8].

Despite of the advantage of natural fibers, the use of natural fibers as reinforcement agents, particularly in hydrophobic polymeric matrices, also have several drawbacks such as poor wettability, incompatibility with some polymeric matrices and high moisture absorption by the fibers because of the hydrophilic nature of natural fibers [12]. Recently, there are several chemical treatment methods have been studied to improve the interfacial bonding such as acetylation, mercerization, isocyanate treatment, acrylation, silent treatment and others. Among these methods, mercerization or alkaline treatment may be considered to be the most economical technique [13]. The surface of fibers becomes rough and enhances mechanical interlocking with resin after the chemical treatments which lead to decrease in moisture absorption as well as improving the wettability of fibers by matrix. This results in imparting better mechanical properties to the composite [14].

The compatibility between the fibers and matrix also can be enhanced by modifying the matrix with a compatibilizers that adheres well to both the fibers and matrix. One of the most suitable compatibilizers available for use in natural fiber reinforced polypropylene composites is Maleic anhydride polypropylene (MAPP). The presence of MAPP in natural fibers reinforced PP composites acts as a bridge between the non-polar polypropylene matrix and the polar fibres by chemically bonding with the cellulose fibers through the MA groups, and bonding to the matrix by means of polymer chain entanglement as shown in Figures 2 [15,16].

In this investigation, a study was performed to determine the effect of Maleic Anhydride Polypropylene (MAPP) on the mechanical properties and material characterization behavior of kenaf fiber/coir fiber composites. Composites were developed from polypropylene, kenaf fiber (KF) and coir fiber (CF) (which was treated with NaOH), and MAPP by using injection molding process.

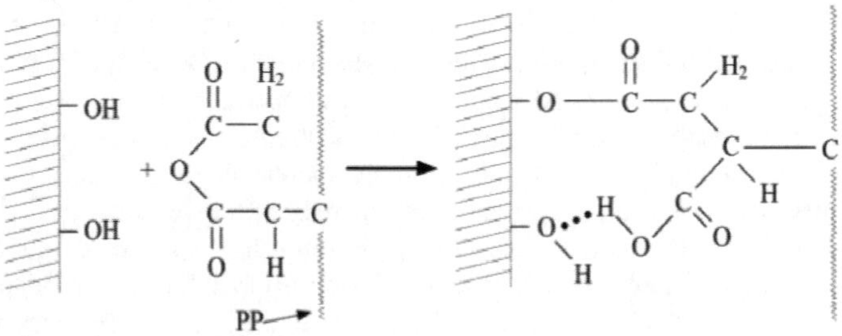

Figure 2. Formation of maleated polypropylene with natural fiber [16].

2. Experimental

2.1 Material

Thermoplastic polypropylene (PP) was obtained from Lotte Chemical Titan (M) Sdn Bhd (Malaysia) in the form of copolymer pellets as the matrix material. For the reinforcement, kenaf fiber (KF) was obtained from Institut Penyelidikan dan Kemajuan Pertanian Malaysia (MARDI) in the form of coarse fibers from their core and the coconut coir fiber (CF) was received from Nature At Work Sdn Bhd. The maleic anhydride polypropylene, MAPP (Orevac CA100) as a compatibilizer was supplied from Syarikat Saintifik Bersatu (M) Sdn Bhd.

2.2 Fiber Preparation

The natural fiber of KF and CF were immersed in 5 % of NaOH water solution at room temperature for 24 h before washing it with running water until water used for washing the fiber no longer gave any alkalinity reaction as shown in Figures 3a [17]. The washed fibers were then dried in an open air for one day before drying it finally in an oven for another 24 h at 80°C as shown in Figures 3b [18,19]. The treated fibers were then milled by rotor mill machine to get fine sizes random fibers by using ring sieves 1 mm, 0.5 mm and 0.2 mm at 12 rpm as shown in Figures 3c. The milled fibers were sieved for 30 min using a sieve with 200 μm mesh size as shown in Figures 3d [20].

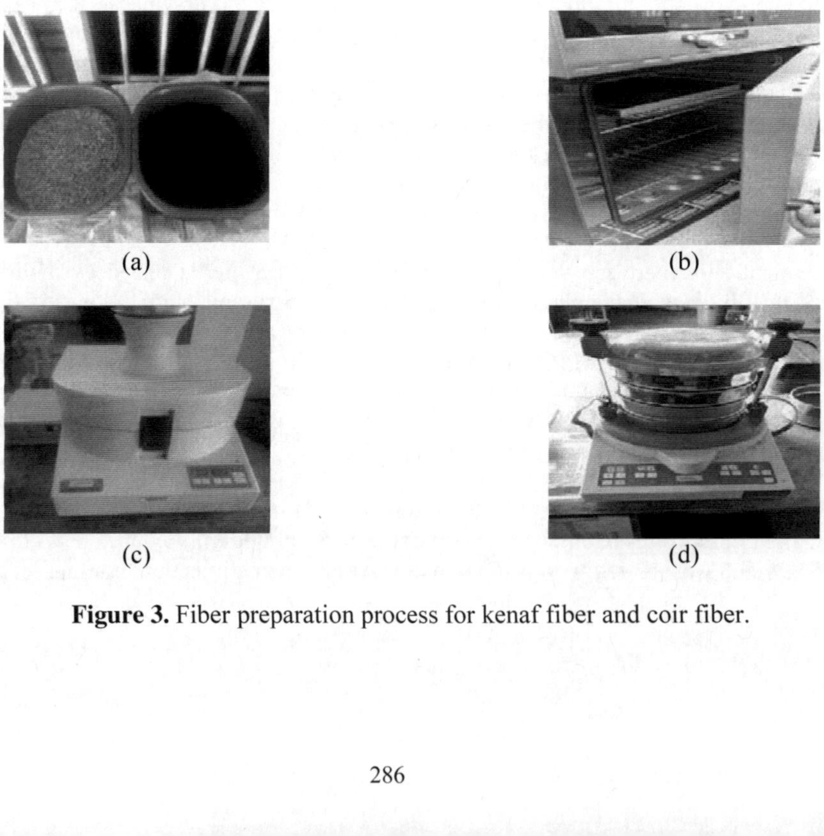

Figure 3. Fiber preparation process for kenaf fiber and coir fiber.

2.3 Composite Fabrication

Mixing of fibers and polymer were prepared with few sets of different composition of 0 %, 10 %, 20 %, and 30 % of the fibers in the composites. The MAPP (3wt%) was added during mixing as a compatibilizer as shown as in Table 1. The fibers and polypropylene were mixed by using twin screw Brabender Plastograph machine at 190 °C melt temperature setting and at a speed of 20 rpm. Then, the compound was crushed by using granulator machine to make a form of pellet shape. The test samples were prepared by using horizontal screw type injection molding (Nissei Plastic Industrial Co. Ltd, model NP7-1F).

Table 1. Composition of kenaf fibers and coir fibers

Sample	Composite	Polypropylene (wt%)	Fibers (wt%)	MAPP (wt%)
Pure	1	100	0	0
Unmodified	2	90	10	0
	3	80	20	0
	4	70	30	0
Modified	5	87	10	3
	6	77	20	3
	7	67	30	3

2.4 Characterization of composites

2.4.1 *Scanning electron microscope*

The morphology of the KF and CF reinforced PP composites with and without MAPP were investigated by using scanning electron microscope SEM (Hitachi, Japan) at 10 kV. The fractured surfaces of tensile test samples were studied with SEM after being sputter coated with gold.

2.4.2 *Fourier Transform Infrared Spectroscopy*

The FTIR spectra of composites with and without MAPP were recorded in the range of 4000 to 600 cm^{-1} at a resolution of 4 cm^{-1} with 32 scans using an FTIR machine Model Perkin-Elmer Spectrum 100 FT-IR Spectrometer (Perkin-Elmer, Norwalk, CT, USA).

2.4.3 *Mechanical Properties*

Tensile test was conducted according to ISO 527 at a test speed 5 mm/min by using Shimadzu Universal Testing Machine (Model AG-1 10kN, Japan) for both natural fiber composites with and without compatibilizer [21]. An average of five tests was reported.

3. Result and Discussion

3.1 Morphology Analysis

The morphology of composites were studied the effect of the MAPP on the interface between coir fibers and kenaf fibers reinforced polypropylene with different composition were investigated by referring the fracture surface of the tensile specimens. For unmodified composition, it shows there were poor fiber matrix adhesion which is justified by the presence of gaps between both of the fibers and the matrix and fiber pull-outs as shown in Figures 4. Meanwhile, the presence of the MAPP compatibilizer gives a better improvement of the wettability of both fibers surface by the polymer. The gaps were less distinctly for both composites which is indicating good adhesion and better wettability between the fibers and matrix as shown in Figures 5. The good dispersion during blending avoids the unoccupied area; hence it was reflected by giving the good mechanical properties compared to composites without compatibilizer [9].

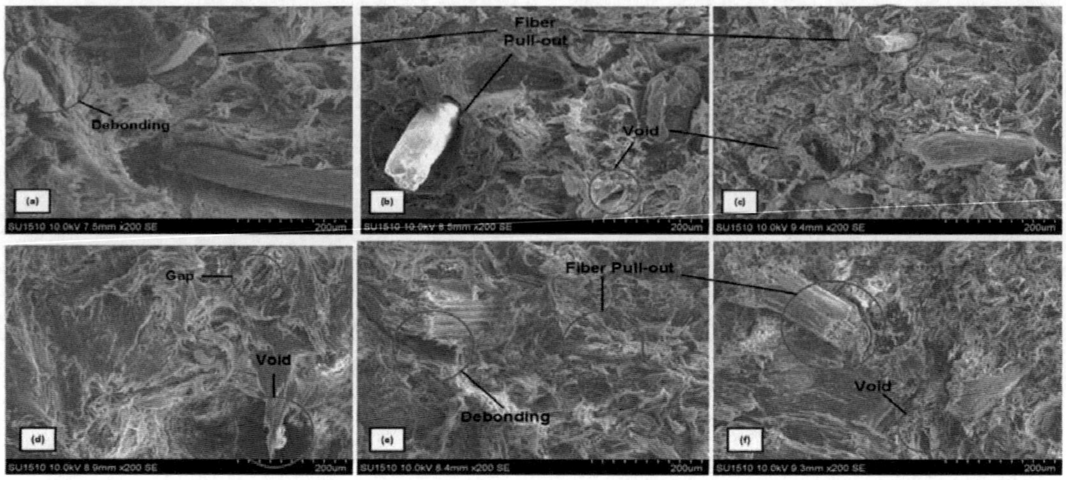

Figure 4. Image of fractured surface of composites with (a) unmodified 10% coir fiber (b) unmodified 20% coir fiber (c) unmodified 30% coir fiber (d) unmodified 10% kenaf fiber (e) unmodified 20% kenaf fiber, and (f) unmodified 30% kenaf fiber.

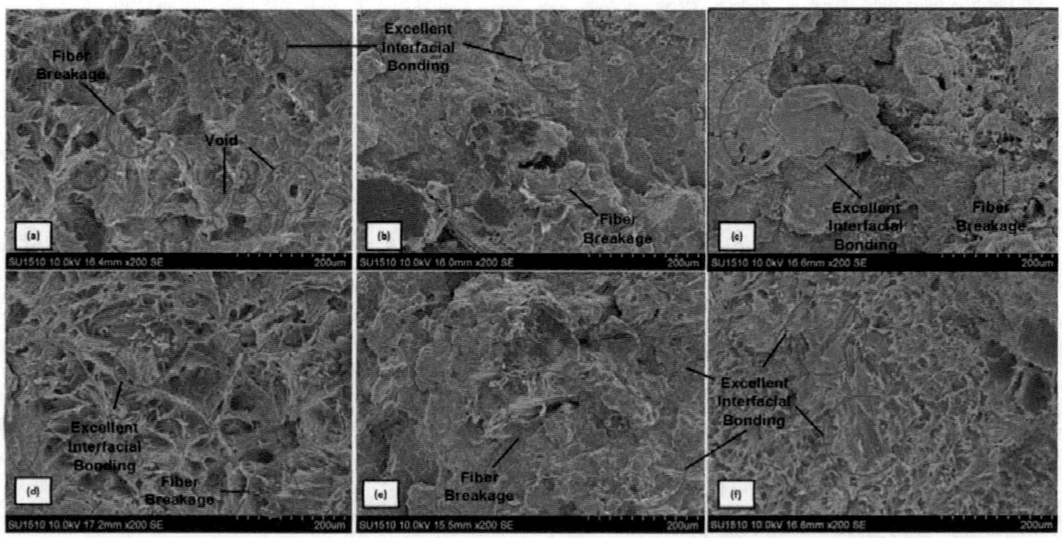

Figure 5. Image of fractured surface of composites with (a) modified 10% coir fiber (b) modified 20% coir fiber (c) modified 30% coir fiber (d) modified 10% kenaf fiber (e) modified 20% kenaf fiber, and (f) modified 30% kenaf fiber.

3.2 FTIR Analysis

The FTIR spectra of unmodified and modified CF and KF were shown in Figure 6 and Figure 7. In Figure 6, the spectrum of the CF fibers are dominated by the peak at 2918. The peak at around 2900 cm^{-1} is due to the C-H asymmetric and symmetric stretching from aliphatic saturated compounds which is corresponding to the aliphatic moieties in cellulose and hemicellulose [22]. On the other hand, the peaks in the region 1455 cm^{-1} are due to C-H groups from various lignin. The intensity of the peak at 1376 cm^{-1} is sharply reduced after the presence of MAPP which is indicating C-H groups in methyl and phenolic alcohol and results from bending vibration in the molecule. Apart from that, the peaks found at 1261 cm^{-1}, 1100 cm^{-1} and 807 cm^{-1} for unmodified CF seems disappear at peak for modified CF. Based on the FTIR spectra, a new peak was found at 1729 cm^{-1} for modified CF. This peak shows the evidence of the presence of ester linkage (C–O) between CF and MAPP.

Figure 7 shows the FTIR spectra of unmodified and modified KF. The peaks at 2917 cm^{-1} is assigned to C-H stretching vibrations. Also, the absorption band of 1455 cm^{-1} and 1376 cm^{-1} were represented to –CH2 and –CH3 bending vibration. A new peak also have been found at 1737 cm^{-1} for modified KF. This result expressed that the MA functional groups of MAPP can be reacted with the hydroxyl groups of the kenaf fiber to produce covalent bonding and esterification reaction [23].

It can be summarized that FTIR only observed the possibility of interaction, functional group change on rigidity imposed by composite formation. Moreover, the treatments with alkali and the addition of MAPP had removed most of the lignin and hemicelluloses components from the fiber surface [24]. Furthermore, the H-bonding shift and differences in the intensities of the absorptions be the main factor that contribute to the mechanical properties of the natural fiber composites.

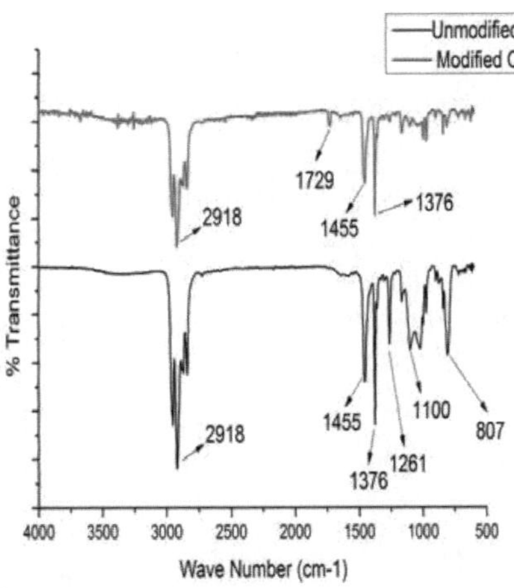

Figure 6. FTIR spectrums of unmodified CF and modified CF

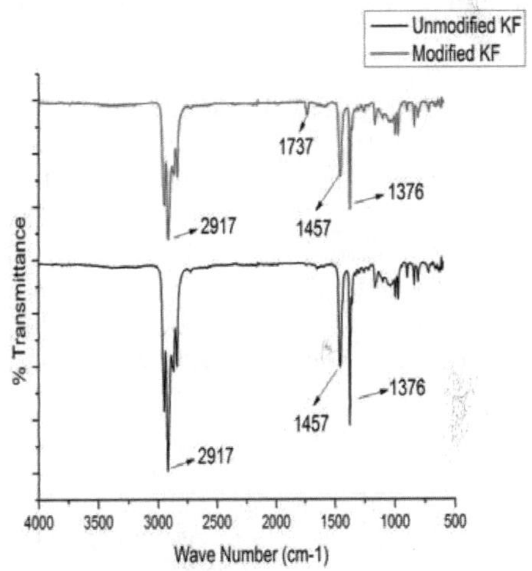

Figure 7. FTIR spectrums of unmodified KF and modified KF

3.3 Mechanical Properties Analysis

The mechanical properties of natural fiber reinforced polymer composites are one of major importance for all applications. Figure 8 shows the effect of MAPP on the tensile strength and tensile modulus of PP/CF composites. Obviously, the tensile strength of the CF composites modified with MAPP increased as the fiber content increased compared to unmodified CF composites, in which were found decrease with the increased fiber content. The incorporations of MAPP was enhanced the tensile strength of CF composites, which may be due to a good compatibility at interfacial regions between natural fiber and matrix as shown in Figure 8a [25].

The tensile modulus of PP/CF composites were illustrated in Figure 8b. It were observed that tensile modulus of PP/CF modified with MAPP for 30% fiber content found to be 12 % and 94 % better compared to unmodified CF at 30% fiber content and neat PP. The results indicate that the MAPP was able to enhance the stiffness of the composites. It may be due to the anhydride groups can react with the hydroxyl groups of CF and form ester bonds, which improves the interfacial adhesion between the matrix and fillers [26].

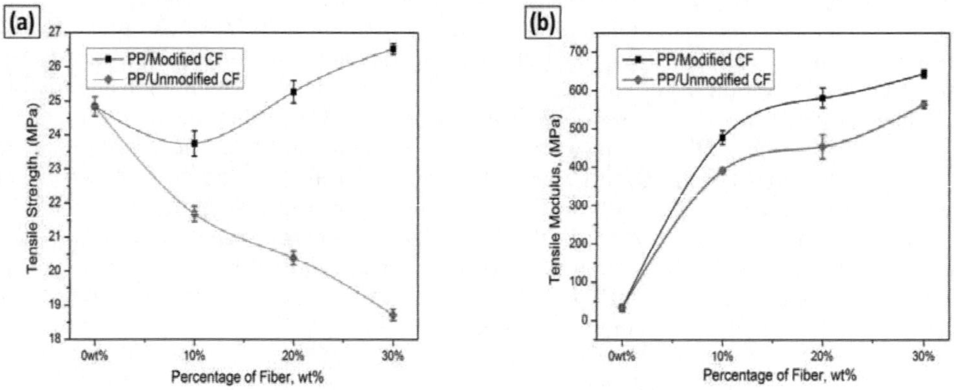

Figure 8. (a) Tensile strength of PP/Unmodified CF and PP/Modified CF (b) Tensile modulus of PP/Unmodified CF and PP/Modified CF.

Figure 9a shows the influence of MAPP on the tensile strength of PP/KF composites. The tensile strength of PP/KF composites modified with MAPP increased with the increased of fiber fraction compared to unmodified PP/KF. The improvement in tensile strength may be attributed to the hydrophilic anhydride groups of MAPP, which was promoting the dispersion and surface wetting of KF in the matrix. The tensile modulus of PP/KF with modified MAPP also gave the good improvement compare to the unmodified PP/KF as shown in Figure 9b. For the 30% fiber content, tensile modulus of modified PP/KF composites increased by 14.2 % and 96 % compared to unmodified PP/KF and pure PP. This is a common behaviour which is natural lignocellulosic fillers have been found as having elastic modulus higher than PP [27].

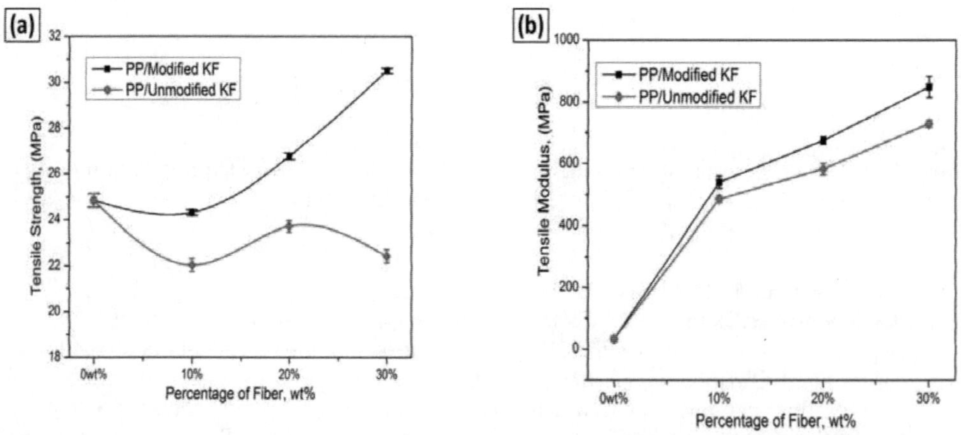

Figure 9. a) Tensile strength of PP/Unmodified KF and PP/Modified KF (b) Tensile modulus of PP/Unmodified KF and PP/Modified KF.

As the result, these compatibilizers act on both fiber and PP matrix to achieve superior fiber matrix adhesion. It formed carbon-carbon covalent bond with the matrix and reinforced with hydroxyl groups of the fiber, hence providing effective interlocking. Compatibilizer also diminished fibers fractures and enhanced the tensile strength. The reaction of anhydride of compatibilizer with the hydroxyl groups of fibers also lessen the compatibility problems between the fiber and matrix [28].

4. Conclusion

In the present study, the effect of different fiber fraction of PP/CF and PP/KF composites, as well as the effect of the addition of MAPP as a comptibilizer on the microstructure and mechanical properties were investigated. Based on the finding results, it can be concluded that the presence of the MAPP as a comptibilizer showed good compatibility for KF and CF between the matrices. The 30 % KF reinforced PP with 3 % of MAPP was found to be good results compare to unmodified PP/KF, unmodified PP/CF and modified PP/CF. Furthermore, the SEM and FTIR results were clearly revealed that the interfacial adhesion of the KF and CF and PP matrix was significantly improved due the addition of MAPP.

Acknowledgements

The author thank to Centre for Graduate Studies, and Office for Research, Innovation, Commercialization and Consultancy Management, Universiti Tun Hussein Onn Malaysia and Ministry of Higher Education for the financial support under FRGS Grant Vot 1546.

References

[1] Zaman I, Mohd Tobi A L, Manshoor B, Khalid A, and Mohd Amin N A 2016 *ARPN Journal of Engineering and Applied Science* **11** 2308-2313.
[2] Sambu M, Yahya M N, Latif H A, Zaman I, and Ghazali M I 2016 *International Review of Mechanical Engineering* **10** 284-288.
[3] Madlan M A, Ghazali M I, Zaman I, Kasron M Z and Ying T C 2017 *IOP Conference Series: Materials Science and Engineering* **166** 012025.
[4] Latif H A, Yahya M N, Zaman I, Sambu M, Ghazali M I, and Hatta M N M 2016 *ARPN Journal of Engineering and Applied Sciences* **11** 6508-6513.
[5] Sambu M, Yahya M N, Latif H A, Zaman I, and Ghazali M I 2016 *International Review of Mechanical Engineering* **10** 284-288.
[6] Osita O, Ignatius O, and Henry U 2016 *International Journal of Engineering and Technologies* **7** 68-77.
[7] Berhanuddin N I C, Rozlan S A M, Zaman I, Mustapa M S, Abdullah M E, and Bahrudin I A 2017 *Materials Science Forum* **889** 14-18.
[8] Monteiro S N, Satyanarayana K G, Ferreira A S, Nascimento D C O, Lopes F P D, Silva I L A, Bevitori A B, Inácio W P, Bravo Neto J and Portela T G 2010 *SciELO Brasil* **15** 488-505.
[9] Arsad A, Suradi N L, Rahmat A R, and Danlami J M 2013 *International Journal of Plastics Technology* **17** 149-162.
[10] Zampaloni M, Pourboghrat F, Yankovich, S A, Rodgers B N, Moore J, Drzal L T, Mohanty A K and Misra M 2007 *Composite Part A: Applied Science and Manufacturing* **38** 1569-1580
[11] Kabir M M, Wang H, Lau K T, and Cardona F 2012 *Composites Part B: Engineering* **43** 2883-2892.
[12] Karthikeyan A, Balamurugan K, and Kalpana A 2014 *Iranian Journal of Science and Technology. Transactions of Mechanical Engineering* **38** 157-166
[13] Gu H 2009 *Materials & Design* **30** 3931-3934.
[14] Zahari W Z W, Badri R N R L, Ardyananta H, Kurniawan D, and Nor F M 2015 *Procedia Manufacturing* **2** 573-578.
[15] Beckermann G W, and Pickering K L 2008 *Composites Part A: Applied Science and Manufacturing* **39** 979-988.
[16] Pickering K L, Efendy M A, and Le T M 2016 *Composites Part A: Applied Science and Manufacturing* **83** 98-112.
[17] Zaman I, Manshoor B, and Khalid A 2015 *Applied Mechanics and Materials* **786** 174 -178.
[18] Zaman I, Ismail A E, and Awang M K 2009 *Journal of Science and Technology* **1** 55-71.
[19] Latif H A, Yahya M N, Zaman I, Sambu M, Ghazali M I, and Hatta M N M 2016 *ARPN Journal of Engineering and Applied Sciences* **11** 7670-7676.

IOP Conf. Series: Journal of Physics: Conf. Series **914** (2017) 012035 doi:10.1088/1742-6596/914/1/012035

[20] El Sheikeil Y A, Salit M S, Abdan K, and Zainudin E S 2011 *BioResources* **6** 4662-4672.

[21] Brown R 2002 *Handbook of Polymer Testing* (United Kingdom: Rapra Technology).

[22] Bledzki A K, Mamun A A, and Volk J 2010 *Composites Science and Technology* **70** 840-846.

[23] Kim H S, Lee B H, Choi S W, Kim S, and Kim H J 2007 *Composites Part A: Applied Science and Manufacturing* **38** 1473-1482.

[24] Punyamurthy R, Sampathkumar D, Ranganagowda R P G, Bennehalli B, Badyankal P V, and Venkateshappa S C 2014 *International Journal of Chemistry* **35** 1699-1706.

[25] Saad M J 2010 *Journal of Science and Technology* **2** 33-44.

[26] Gao H, Xie Y, Ou R, and Wang Q 2012 *Composites Part A: Applied Science and Manufacturing* **43** 150-157.

[27] Rosa S M L, Santos E F, Ferreira C A, and Nachtigall S M B 2009 *Materials Research* **12** 333-338.

[28] John M J, and Anandjiwala R D 2008 *Polymer composites* **29** 187-207.

Enhancement of mechanical properties of epoxy/graphene nanocomposite

N I C Berhanuddin[1], I Zaman[2, a], S A M Rozlan[3], M A A Karim[3], B Manshoor[3], A Khalid[4], S W Chan[5], Q Meng[6]

[1]Structural Integrity and Monitoring Research Group, Faculty of Mechanical and Manufacturing Engineering, Universiti Tun Hussein Onn Malaysia, Johor, Malaysia.
[2]Mechanical Failure Prevention and Reliability Research Centre, Universiti Tun Hussein Onn Malaysia, 86400 Batu Pahat, Johor, Malaysia.
[3]Faculty of Mechanical and Manufacturing Engineering, Universiti Tun Hussein Onn Malaysia, 86400 Batu Pahat, Johor, Malaysia.
[4]Faculty of Engineering Technology, Universiti Tun Hussein Onn Malaysia, 86400 Batu Pahat, Johor, Malaysia.
[5]Faculty of Technology Management and Business, Universiti Tun Hussein Onn Malaysia, 86400 Batu Pahat, Johor, Malaysia.
[6]School of Engineering, University of South Australiam Mawson Lakes Campus, South Australia, 5095 Australia
Email: izzuddin@uthm.edu.my [a]

Abstract. Graphene is a novel class of nanofillers possessing outstanding characteristics including most compatible with most polymers, high absolute strength, high aspect ratio and cost effectiveness. In this study, graphene was used to reinforce epoxy as a matrix, to enhance its mechanical properties. Two types of epoxy composite were developed which are epoxy /graphene nanocomposite and epoxy/modified graphene nanocomposite. The fabrication of graphene was going through thermal expansion and sonication process. Chemical modification was only done for modified graphene where 4,4'-Methylene diphenyl diisocyanate (MDI) is used. The mechanical properties of both nanocomposite, such as Young's modulus and maximum stress were investigated. Three weight percentage were used for this study which are 0.5 wt%, 1.0 wt% and 1.5 wt%. At 0.5 wt%, modified and unmodified shows the highest value compared to neat epoxy, where the value were 8 GPa, 6 GPa and 0.675 GPa, respectively. For maximum stress, neat epoxy showed the best result compared to both nanocomposite due to the changes of material properties when adding the filler into the matrix. Therefore, both nanocomposite increase the mechanical properties of the epoxy, however modification surface of graphene gives better improvement.

1. Introduction

Epoxy polymer have outstanding properties including good thermal and mechanical properties, compatible with most fibre, chemical and wear resistance and low cost [1]. Due to this properties, it makes them more attractive to be used in wide application such as adhesives, coatings, structural material or as the matrix of the composite [2,3]. As a thermoset material, epoxy has a high degree of crosslinking which make them very useful in term of high rigidity and strength. However, because of the crosslinking structure, it also make epoxy brittle and vulnerable to cracks, and has limitation on its application. Due to this limitation of epoxy, many researchers are interested to improve the epoxy toughness. The effective method to enhance this problems is by adding fillers into the polymer.

Graphene is a single layer of graphite in the form of a hexagonal lattice of carbon atoms bonded in the sp^2 configuration [4]. It is also known as the thinnest material in the universe which allows many new and exciting applications potential. Historically, graphene becomes well-known materials when Nobel prized winners, Prof. A.K. Geim and Prof. K. S. Novoselov fabricates the graphene materials by using a sticky tape to peel off the graphite layer by layer [5]. Graphene has excellent properties such as; higher electrical and thermal conductivities than copper materials and superior mechanical properties which are 1 TPa in Young's modulus [6], intrinsic strength values is 139 GPa [7] and 130 MPa in ultimate strength [8]. Graphene are preferred as a nanofiller compared to other other materials because of its layered structural fillers which has high surface area which is up to 2630 m^2/g [9]. According to Park and his colleagues, graphene shows a property enhancement at a small loading of graphene contents due to its high aspect ratio. In some case the graphene content less than 0.5 wt% already give the best enhancement of the composite [10].

Polymer/graphene based nanocomposite have attracted most academia and industry in the recent years due to its remarkable enhancement in the polymer properties at low filler fractions [11]. Nanocomposite is leading to a reduction in the filler size of composite from microscale to nanoscale. The effects of reduction increased the interphase and reduced surface interparticle distance. In this past few decades, polymer contains layered-structure molecule, for example, layer graphene which demonstrated a capability of mechanical toughness improvement due to its good properties [12,13].

In this study, various amounts of expanded graphite are used which is 0.5 wt%, 1.0 wt% and 1.5 wt%. The modification interface of graphene was developed by using chemical modification which is to study the effects and differences of unmodified and modified interface of graphene towards the mechanical properties of nanocomposite.

2. Experiments

2.1 Materials

Graphite carbons was purchased by Asbury Carbons, Asbury NJ. Epoxy resin, DER 331 which is a reaction product of epichlorohydrin and bisphenol A, and the hardener, Jointmine 925-3 were supplied by Epochemie International Pte Ltd, Singapore. The 4,4'-Methylene diphenyl diisocyanate (MDI) was purchased by Sigma-Aldrich.

2.2 Synthesis of graphene materials

2.2.1 Thermal expansion. Silica crucible was preheated in the laboratory furnace at 1000°C. 0.5 gram of raw graphite was poured into the silica crucible by using spatula. After 60 seconds, the silica crucible was taken out and place on a ceramic plate for 2 minutes. Then, after the expanded graphite was cooled down, it was transferred into a container. Figure 1 shows the raw graphite material and expanded graphite. The operator must be extra careful when conducting this process and should wear personal protective equipment such as goggle, heat resistant glove and closed shoes.

(a)

(b)

Figure 1. (a) Raw graphite (b) expanded graphite.

2.2.2 Sonication process. Expanded graphite was put in acetone solution inside a metal container for 30 minutes. The metal container was covered and this process was done in the ultrasonic bath at the frequency, 50-60 Hz. The ratio of expanded graphite to acetone is 1:150. The acetone is used as a solvent to make sure the expanded graphite were able to split into graphene platelets. This sonication process was done in order to make sure that the expanded graphite dispersed well. Figure 2 shows the sonication bath used in this experiment.

Figure 2. Sonication bath.

2.2.3 Fabrication of epoxy - graphene nanocomposite

The desired amount of epoxy was added into the mixture and mix manually for 5 minutes followed by sonication process for 1 hour. Then, the mixture was transferred into a beaker and stirred by using mechanical stirrer at 300 rpm for 2 and a half hours where the temperature of the hot plate was 120°C. Then the mixture was degassed inside the vacuum oven for 15 minutes to eliminate air bubbles. The presence of air bubbles will affect the mechanical properties of the composite. Hardener was added into the mixture and mixed manually for 5 minutes. Then, the mixture was poured into the mold. Make sure that the mold already been waxed. The mold was cured inside the drying oven at the preheated temperature of 80°C for 3 hours and followed by 120°C for 12 hours. The curing time as shown in Figure 3.

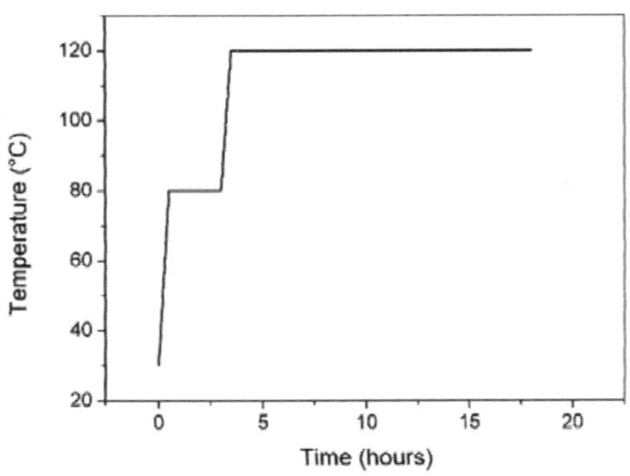

Figure 3. Curing time for the samples.

2.3 Synthesis of epoxy - modification graphene nanocomposite

The expanded graphite which already undergoing the sonication process was transferred into a round bottom flask equipped with a condenser as shown in figure 4. The weight ratio of MDI towards expanded graphite was 0.5:1. The MDI was poured into the mixture, and followed by mixing at ~80°C for 6 hours by using a magnetic stirrer. For the next process, the mixture was transferred into a metal container and epoxy was added, followed by sonication process for 1 hour. Then, the mixture was poured into a beaker and was stirred at 300 rpm for 2 hours and 30 minutes at the temperature of 120°C. Later, the mixture was degassed for 15 minutes and hardener was added. The mixture was poured into the mold and followed by curing process inside the drying oven, similar to step in Section 2.3.

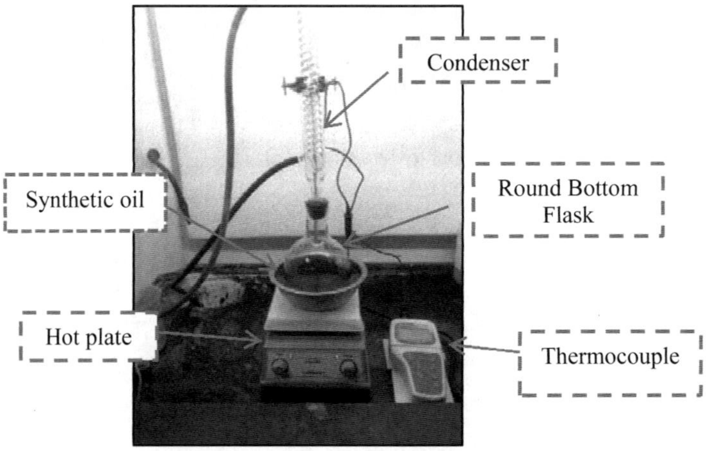

Figure 4. Modification process.

2.4 Characterisation

2.4.1 Modification of filler. Infrared spectra were measured with a Fourier transform infrared spectroscopy (FTIR), within the region of 4000 cm^{-1} – 600 cm^{-1} at 4 cm^{-1}, using a minimum of 32 scans collected at 2 cm^{-1} resolutions.

2.4.2 Mechanical property. Tensile test was performed on dumbbell samples at a strain rate of 0.5 mm/min according to ASTM D-638. This testing was conducted by using Gotech Testing Machine which the maximum load is 100 kN at room temperature. An extensometer was used to get the accurate displacement data for the Young's modulus measurement which was calculated from the range 0.01 mm to 0.17 mm. The gauge length of the extensometer is 25 mm.

3 Result and discussion

3.1 Fourier Transform Infrared Spectroscopy (FTIR)

Figure 5 illustrate FTIR spectra of raw graphite, expanded graphite and modified graphene. Raw graphite spectra show a weak absorption at 2116 cm^{-1}, relating to the presence of C=C. For the expanded graphite, it shows the highest absorbance which is 3723 cm^{-1}, corresponding to the presence of O-H group. It also shows the presence of weak C=C at 2116 cm^{-1} and strong wave number at 870 cm^{-1} caused by the bending vibration of C-H. After expanded graphite was modified by 4,4'-Methylene diphenyl diisocyanate (MDI), the spectra show the lowered absorbance at a strong band 3305 cm^{-1} lies under O-H group. Modified graphene also showed a medium absorbance at 2889 cm^{-1}

caused by stretching vibration of C-H. Then the absorbance increase at strong band at 2323 cm^{-1} of stretching vibration represent the O=C=O. The last spectra band for modified graphene is 1553 cm^{-1}, strong stretching vibration of Nitro compound, N-O.

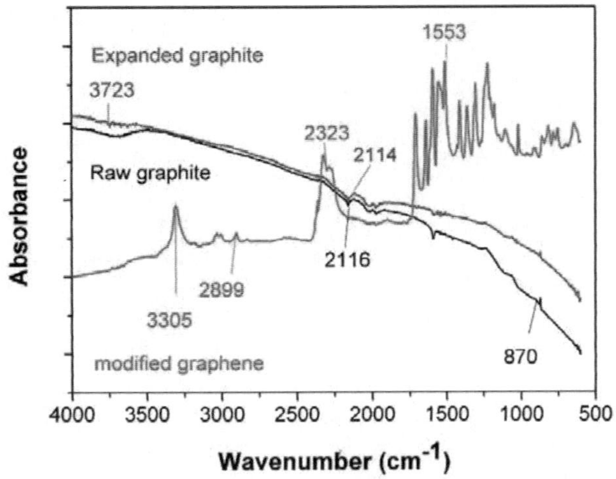

Figure 5. FTIR spectra of raw graphite, expanded graphite and modified graphene.

3.2 Mechanical Properties

Table 1 shows the result of Young's modulus and tensile strength for tensile test of the neat epoxy and nanocomposite with and without modification for three proportions of graphene contents while Figure 6 shows its comparison in graph form. The Young's modulus of 0.5 wt% of modified graphene indicates the best enhancement compared to the neat epoxy and other weight percentage of modified and unmodified epoxy/graphene nanocomposite. The value of 0.5 wt% of modified graphene increased the Young's modulus of neat epoxy from 0.675 to 8 GPa, with the increment of 91.56%. For 0.5 wt% of modified towards unmodified graphene, the differences was 25%; with the values are 8 GPa and 6 GPa, respectively. Nevertheless, the elasticity begins to degrade when adding more graphene contents. Thus, 0.5 wt% was the optimum value for both modified and unmodified nanocomposite. The result shows that, at 0.5 wt% of modified graphene provide a significant toughening effect caused by its high interface strength.

Both nanocomposite shows reduction of tensile strength compared to the neat epoxy. By adding ductile material of fillers, will change the composite from brittle to more ductile properties [13-15]. The result gained also proves the theory, while there are increment in the young's modulus. In comparison of all weight percentage used, 0.5 wt% of modified graphene obtained the best value. However, the value is lesser than neat epoxy, where the neat epoxy gives the tensile strength of 39.82 MPa while for 0.5 wt% of modification is 23.01 MPa. This is due to the changes of mechanical properties when adding filler to the neat epoxy. From both results, it shows that the interface modification of graphene gives better enhancement in term of its tensile properties of the nanocomposite.

IOP Conf. Series: Journal of Physics: Conf. Series **914** (2017) 012036 doi:10.1088/1742-6596/914/1/012036

Table 1. Result of Young's modulus and tensile strength

Graphene Content (wt%)	Young's Modulus (GPa)		Tensile Strength (MPa)	
	Unmodified	Modified	Unmodified	Modified
0	0.675		39.82	
0.5	6.0	8.0	16.96	23.01
1.0	3.2	5.5	12.28	18.26
1.5	3.0	5.4	8.75	13.39

(a)
(b)

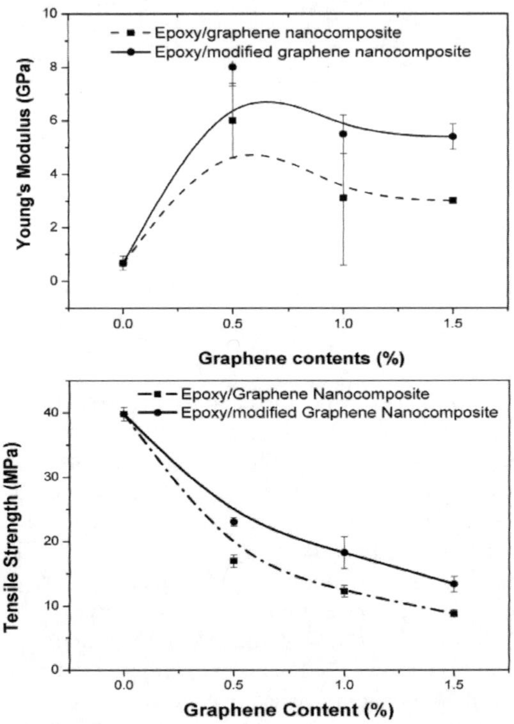

Figure 6. Value of (a) Young's modulus and (b) tensile strength of epoxy/graphene and epoxy/modified graphene

4 Conclusion

In this study, two types of epoxy nanocomposite was developed by using graphene materials in order to study the effects of modification of graphene interface towards its mechanical properties. Graphene was prepared from raw graphite and facing through three main process which are thermal expansion, sonication process and chemical modification. The chemical modification is done through grafting the 4,4'-Methylene diphenyl diisocyanate (MDI) into expanded graphite to produce modified graphene. At 0.5 wt% of modified graphene created 91.6% increment of Young's modulus at tensile properties compared to the neat epoxy. The differences value of maximum stress of modified nanocomposite and neat epoxy were 42.21%. It shows a decrement in the value when added filler because of the changes of neat epoxy mechanical properties which become more ductile. However, 0.5 wt% of modification graphene gives better result compared to unmodify with the increment of 25% and 26.29% for Young's modulus and maximum stress, respectively. Based on the experimental results, adding graphene will increase the elasticity of the neat epoxy where the modified one gives better result. Therefore, modification interface of graphene creates a strong interface between neat epoxy and graphene, where can restrain micro crack occurred in the epoxy/modified graphene nanocomposite.

Acknowledgements
Special thanks to Centre for Graduate Studies, and Research, Innovation, Commercialization and Consultancy Management, University Tun Hussein Onn Malaysia for the financial support under. Research Support Grant Scheme (RSGS), vot U105.

References

[1] Ma, J., La, L.T.B., Zaman, I., Meng, Q., Luong, L., Ogilvie, D., & Kuan, H.-C. 2011 *Fabrication, structure and properties of epoxy/metal Nanocomposites.* Macromol. Mater. Eng. **296**, 465–474.

[2] Ma, J., Meng, Q., Zaman, I., Zhu, S., Michelmore, A., Kawashima, N., Wang, C.H., and Kuan, H.C., *Development of polymer composites using modified, high-structural integrity graphene platelets.* Composites Science and Technology, 2014. **91**: p. 82-90.

[3] Meng, Q., Zaman, I., Hannam, J.R., Kapota, S., Luong, L., Youssf, O., and Ma, J., *Improvement of adhesive toughness measurement.* Polym. Testing, 2011. **30**: p. 243–250.

[4] Mukhopadhyay, P., & Gupta, R. K. 2012 *Graphite, Graphene, and their polymer nanocomposites*: CRC Press

[5] Geim, A. K., & Novoselov, K. S. 2007 *The rise of graphene.* Nature materials, **6** 183-191.

[6] Huang, X., Qi, X., Boey, F., & Zhang, H. 2012 *Graphene-based composites.* Chemical Society Reviews **41** 666-686

[7] Berhanuddin, C., Idani, N., Rozlan, M., Azmirul, S., Zaman, I., Mustapa, M. S., Abdullah, M. E., & Bahrudin, I. A. 2017 *Effect of Thermal Expansion and Sonication on Mechanical Properties and Adhesive Toughness Measurement of Polymer/Graphene Composite.* Materials Science Forum **889** 14-18.

[8] Saleem, H., Edathil, A., Ncube, T., Pokhrel, J., Khoori, S., Abraham, A., & Mittal, V. 2016 *Mechanical and Thermal Properties of Thermoset–Graphene Nanocomposites.* Macromolecular Materials and Engineering **301** 231-259

[9] Yasmin, A., Luo, J.-J., & Daniel, I. M. 2006 *Processing of expanded graphite reinforced polymer nanocomposites.* Composites Science and Technology **66** 1182-1189

[10] Park, Y. T., Qian, Y., Chan, C., Suh, T., Nejhad, M. G., Macosko, C. W., & Stein, A. 2015 *Epoxy toughening with low graphene loading.* Advanced Functional Materials **25** 575-585

[11] Kuilla, T., Bhadra, S., Yao, D., Kim, N. H., Bose, S., & Lee, J. H. 2010. *Recent advances in graphene based polymer composites.* Progress in polymer science **35** 1350-1375

[12] I. Zaman, B. Manshoor, A. Khalid, & S. Araby 2014 *From Clay to Graphene for Polymer Nanocomposites—a Survey.* Journal of Polymer Research **21** 429

[13] Araby, S., Meng, Q., Zhang, L., Zaman, I., Majewski, P., & Ma, J. 2015 *Elastomeric composites based on carbon nanomaterials.* Nanotechnology **26** 112001.

[14] Ye, L., Mai, Y., & Su, Z. 2004 *Composite Technologies for 2020.* Woodhead, Cambridge, UK, **57**

[15] Zaman, I., Manshoor, B., Khalid, A., Meng, Q., & Araby, S. 2014 Interface modification of clay and graphene platelets reinforced epoxy nanocomposites: a comparative study. J. Mater. Sci. **49** 5856-5865.

Optimization of Sound Absorbers Number and Placement in an Enclosed Room by Finite Element Simulation

S F Lau[1], *M H Zainulabidin*[1,3], M N Yahya[1], I Zaman[1], N A Azmir[1], M A Madlan[1], M Ismon[1], M Z Kasron[1] and A E Ismail[2]

[1] Noise and Vibration Analysis Research Group (NOVIA), Faculty of Mechanical & Manufacturing Engineering, Universiti Tun Hussein Onn Malaysia, Parit Raja, Batu Pahat, Johor, Malaysia
[2] Mechanical Failure Prevention and Reliability Research Center (MPROVE), Faculty of Mechanical & Manufacturing Engineering, Universiti Tun Hussein Onn Malaysia, Parit Raja, Batu Pahat, Johor, Malaysia

Correponding author: *hafeez@uthm.edu.my

Abstract. Giving a room proper acoustic treatment is both art and science. Acoustic design brings comfort in the built environment and reduces noise level by using sound absorbers. There is a need to give a room acoustic treatment by installing absorbers in order to decrease the reverberant sound. However, they are usually high in price which cost much for installation and there is no system to locate the optimum number and placement of sound absorbers. It would be a waste if the room is overly treated with absorbers or cause improper treatment if the room is treated with insufficient absorbers. This study aims to determine the amount of sound absorbers needed and optimum location of sound absorbers placement in order to reduce the overall sound pressure level in specified room by using ANSYS APDL software. The size of sound absorbers needed is found to be $11\ m^2$ by using Sabine equation and different unit sets of absorbers are applied on walls, each with the same total areas to investigate the best configurations. All three sets (single absorber, 11 absorbers and 44 absorbers) has successfully treating the room by reducing the overall sound pressure level. The greatest reduction in overall sound pressure level is that of 44 absorbers evenly distributed around the walls, which has reduced as much as 24.2 dB and the least effective configuration is single absorber whereby it has reduced the overall sound pressure level by 18.4 dB.

1. Introduction

Sound is influential to human being; we use sound to communicate to each other, to learn effectively, to receive information precisely, experiencing through hearing makes life significant. Unfortunately, excessive or unwanted sound may become a noise which creates unpleasant environment. Noise is defined as disagreeable or undesired sound where the differentiation of sound and noise is highly subjective as perceived by one [1]. Excessive unwanted sound contributes to noise pollution where it may become a serious matter which tends to disrupt the natural rhythm of life. Noise is believed to cause annoyance and disturb activities and communication which contributing to stress and illness [2].

Noise pollution can be controlled through systematic approach by the application of engineering principles. The methods of control include control at the source, control at the transmission paths and the control at the receiver [3]. Noise control meant to reduce sound emissions which aim to bring human comfort, as for environmental considerations or legal compliance. Sound absorbers are used to control the noise within a particular space where they reduce the sound level by converting it into heat when sound waves are impinged on them. Most of the commercially available sound absorbing materials are made from synthetic fibers such as minerals and polymers [4]. Even though synthetic sound absorbing materials have excellent acoustical performance, but they are quite expensive to manufacture and to install.

The sound in a room is a combination of direct sound and indirect sound, so what we hear are the direct sound from the speaker and also the indirect reflected sound that bounce off the medium it encountered. Reflected sound may results in standing waves as it interferes with incident waves. Sound reflection will also causes reverberation where the sound is prolonged and the details of the listened contents will be masked. These phenomena cause sound distortion and greatly affects the acoustic quality.

Excessive sound reflection contributes to poor room acoustic and causes intolerable reverberation. Peters et al. [5] recommended that for small lecture room which consists of less than 50 people, the reverberation time should be kept less than 0.8 second for a good learning environment.

Giving a room proper acoustic treatment is both art and science. Different rooms may require different treatment as over-treated or wrong types of material can have adverse effects. Acoustic design brings comfort in the built environment and also reduces noise level by using the sound absorbers. The pressure distribution in a room is closely related to the placement of absorbents. Mounting sound absorbing materials flat and evenly distributed on the walls as shown in Figure 1 can help to reduce the effect of standing waves effectively [6]. Besides, there is a little more absorption and more diffuse sound environment by placing the absorbers not only on ceiling, but also on the walls [7]. Finite element method (FEM) had been used to investigate the proper locations of absorbers in different polygonal small rooms at low frequencies and the results showed that the absorptive materials are best located at the pressure antinodes around the sidewalls for low frequency modes below 100 Hz [8].

This study is aimed to determine the effect of installing sound absorbers in a small room on the overall sound pressure level. Consequently, this study is aim to determine the amount of sound absorbers needed and its optimum placement by using ANSYS APDL.

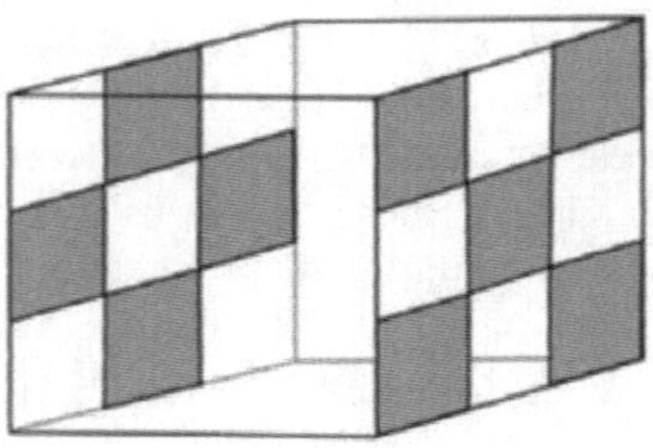

Figure 1. Placement of sound absorbers in staggered position for maximum effectiveness

2. Methodology

The model of rectangular room with dimensions of $3m$ high × $4m$ wide × $5m$ long was built in ANSYS as illustrated in Figure 2 with a total volume of $60m^3$. All the surfaces were made from unpainted concrete which have absorption coefficients of 0.01.

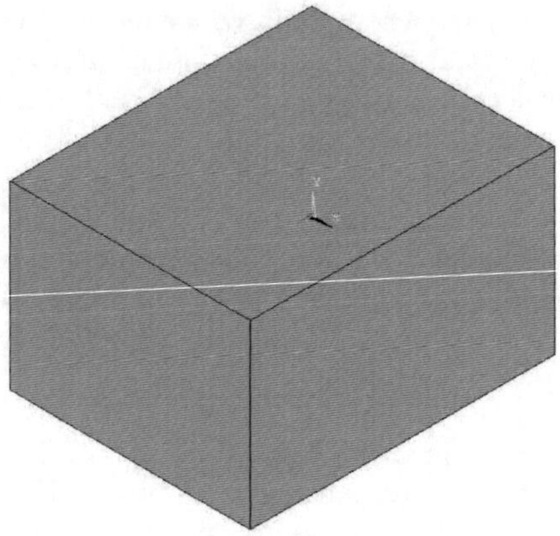

Figure 2. Room model

Since the recommended maximum reverberation times for small room consists of less than 50 person would requires RT_{60} less than 0.8, where it is suitable for speech, so the room was modified by introducing the correct amount absorptive materials, through utilization of Eqn. (1).

$$RT_{60} = 0.161 \times \frac{V}{A} \quad (1)$$

where RT_{60} is the time required for reflections of a direct sound to decay by 60 dB, V is the room volume in m^3 and A is the total absorption in Sabins [9].

Throughout this study, the absorptive materials were assumed to have an absorption coefficient of $\alpha = 1$. Hence, the additional absorption A needed for the room is 11.135 Sabins. This is corresponding to an area of $11.135\ m^2$ of absorptive materials. For simplification, the total area of absorptive materials used was $11\ m^2$.

Absorptive materials of total area $11\ m^2$ were applied to the walls, either on single wall, two opposing walls, at the corners or evenly distributed on every walls. The absorber patches of $11\ m^2$, $1\ m^2$ and $0.25\ m^2$ were used to treat the room to see the effect of number of absorbers to the pressure distribution of the room. Different absorbers patches used were of the same total area. Table 1 shows the number of absorptive patches used during the room treatment.

Table 1. Number of absorptive materials used

Unit area (m^2)	Number of absorptive materials	Total area (m^2)
11	1	11
1	11	11
0.25	44	11

Type of element used to model the fluid medium was FLUID30 and the element size used 0.1 m and hence this made a total of 60000 finite elements. ANSYS recommends that at least 12 elements per wavelength should be used for FLUID30 in order to obtain the acceptable accurate results [10]. For uniformity purpose, the materials properties were evaluated at atmospheric pressure of $1\ atm$ and 20°C where the density of air was equal to $1.2041\ kg/m^3$ and speed of sound in the air was $343.24\ m/s$. The finite element model was analyzed by harmonic analysis at octave band middle frequency of 16, 32, 63, 125, 250, 500, 1000 and 2000 Hz.

A 0.2 Pa pressure loading source was located at the front of the room. Six receivers were used to measure the pressure response inside the room, with each of them located at equidistant from both side walls as shown in Figure 3.

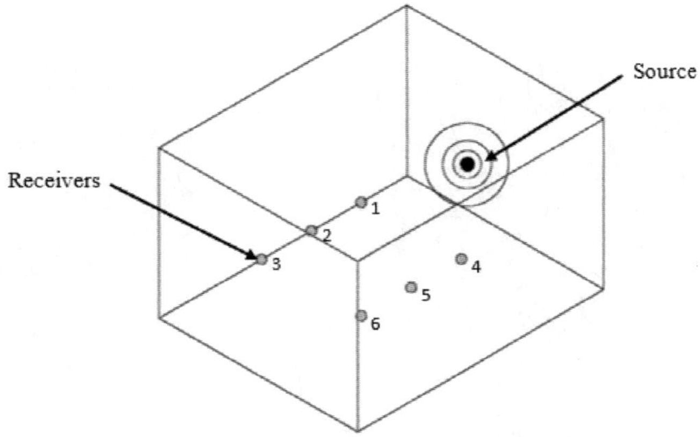

Figure 3. Position of sound source and receivers

The absorbers installation of different configurations caused a reduction of sound pressure level (SPL) in the room. The total SPL for each of the frequency for each configurations was obtained by combining the individual SPL values from the six receivers, using Eqn. (2).

$$Total\ SPL = 10log\left(10^{\frac{SPL_1}{10}} + 10^{\frac{SPL_2}{10}} + 10^{\frac{SPL_3}{10}} + \cdots + 10^{\frac{SPL_6}{10}}\right) \quad (2)$$

The overall SPL over the frequency range for each configurations was obtained by combining the individual total SPL values of the eight frequencies obtained by Eqn. (2), by using Eqn. (3).

$$Overall\ SPL = 10log\left(10^{\frac{Total\ SPL_{16}}{10}} + 10^{\frac{Total\ SPL_{32}}{10}} + 10^{\frac{Total\ SPL_{63}}{10}} + \cdots + 10^{\frac{Total\ SPL_{2000}}{10}}\right) \quad (3)$$

The configurations of the absorbers are as follows:

a) Single absorber on one side wall
b) 11 absorbers installed identically on two side walls
c) 11 absorbers installed oppositely on two side walls
d) 11 absorbers evenly distributed on walls
e) 11 absorbers evenly distributed around corner walls
f) 44 absorbers installed identically on two side walls
g) 44 absorbers installed oppositely on two side walls
h) 44 absorbers evenly distributed on walls
i) 44 absorbers evenly distributed around corner walls

Figures 4(a)-(i) show the absorbers arrangement for each configuration.

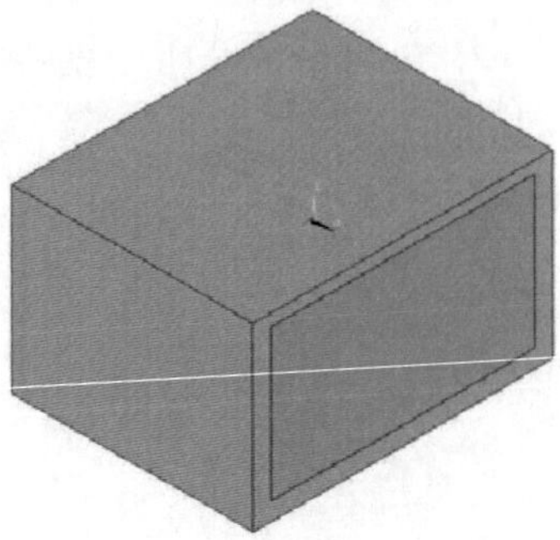

(a)

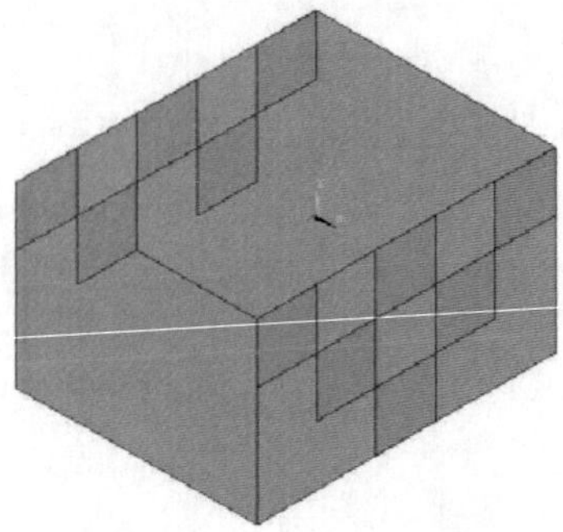

(b)

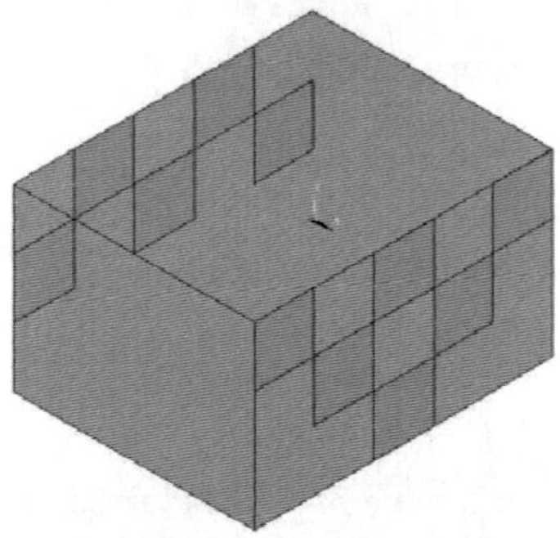

(c)

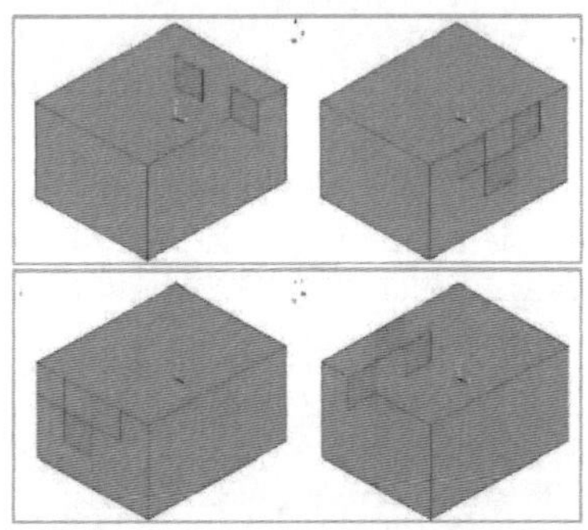

(d)

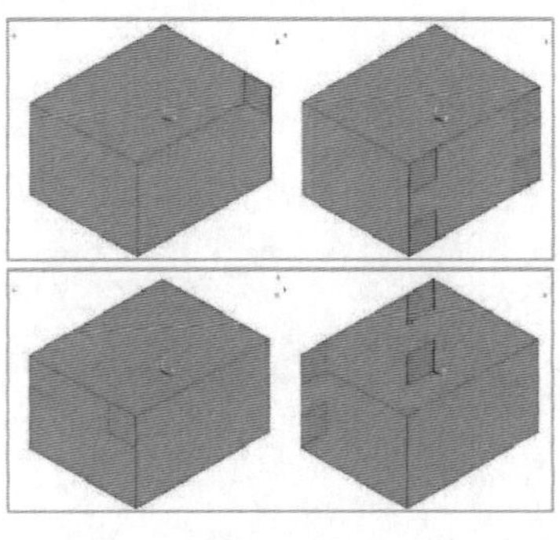

(e)

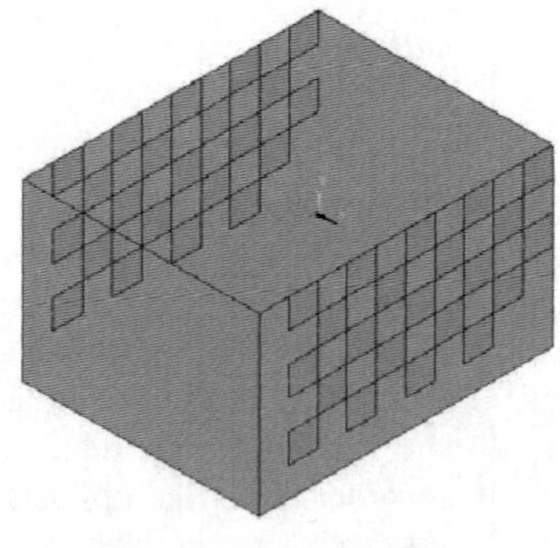

(f)

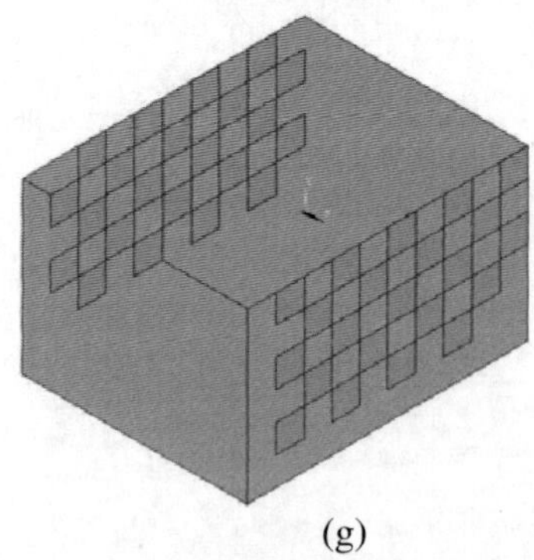

(g)

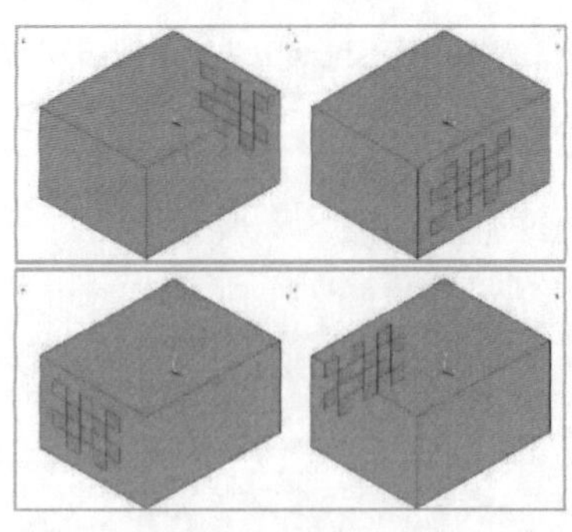

(h)

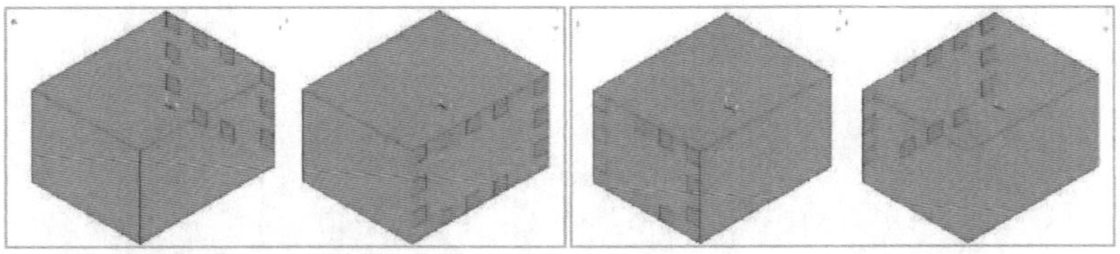

(i)

Figure 4. Sound absorbers arrangement

3. Results and Discussion

The results in the form of continuous contour plot within the room for bare room and every absorber configurations as shown in Figures 5(a) to 5(j) are viewed in the general postprocessor (POST1) using command PLNSOL. From these plots, the distribution of the sound pressure level in the room can be clearly pictured based on the color indicators shown.

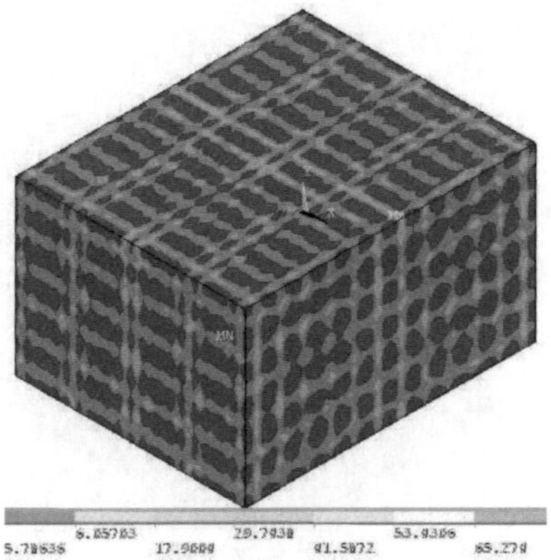

(a)

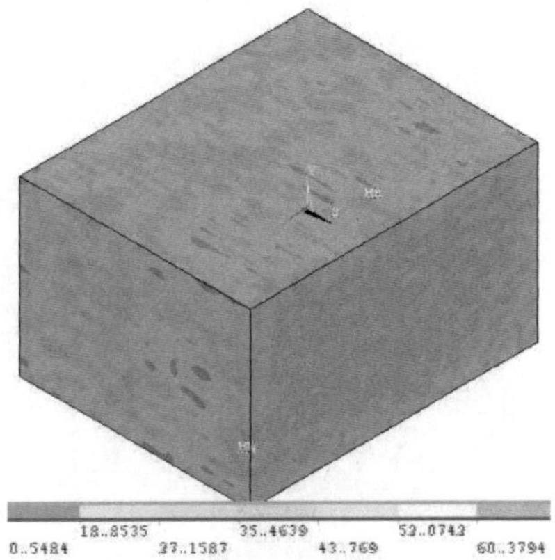

(b)

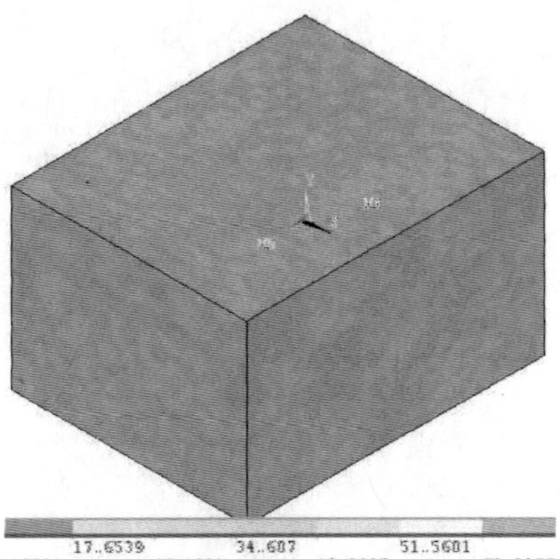

(c)

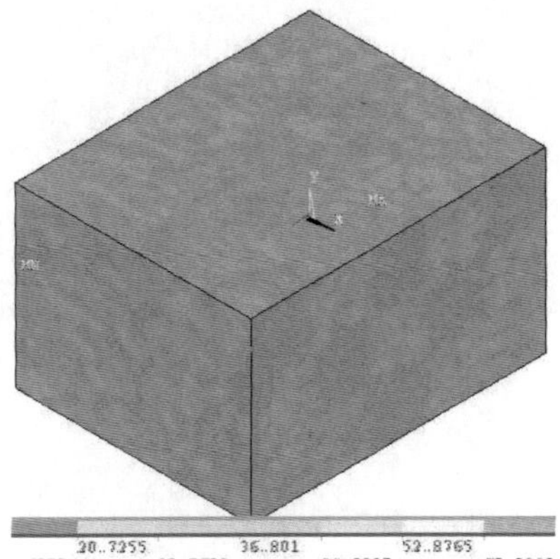

(d)

IOP Conf. Series: Journal of Physics: Conf. Series **914** (2017) 012037 doi:10.1088/1742-6596/914/1/012037

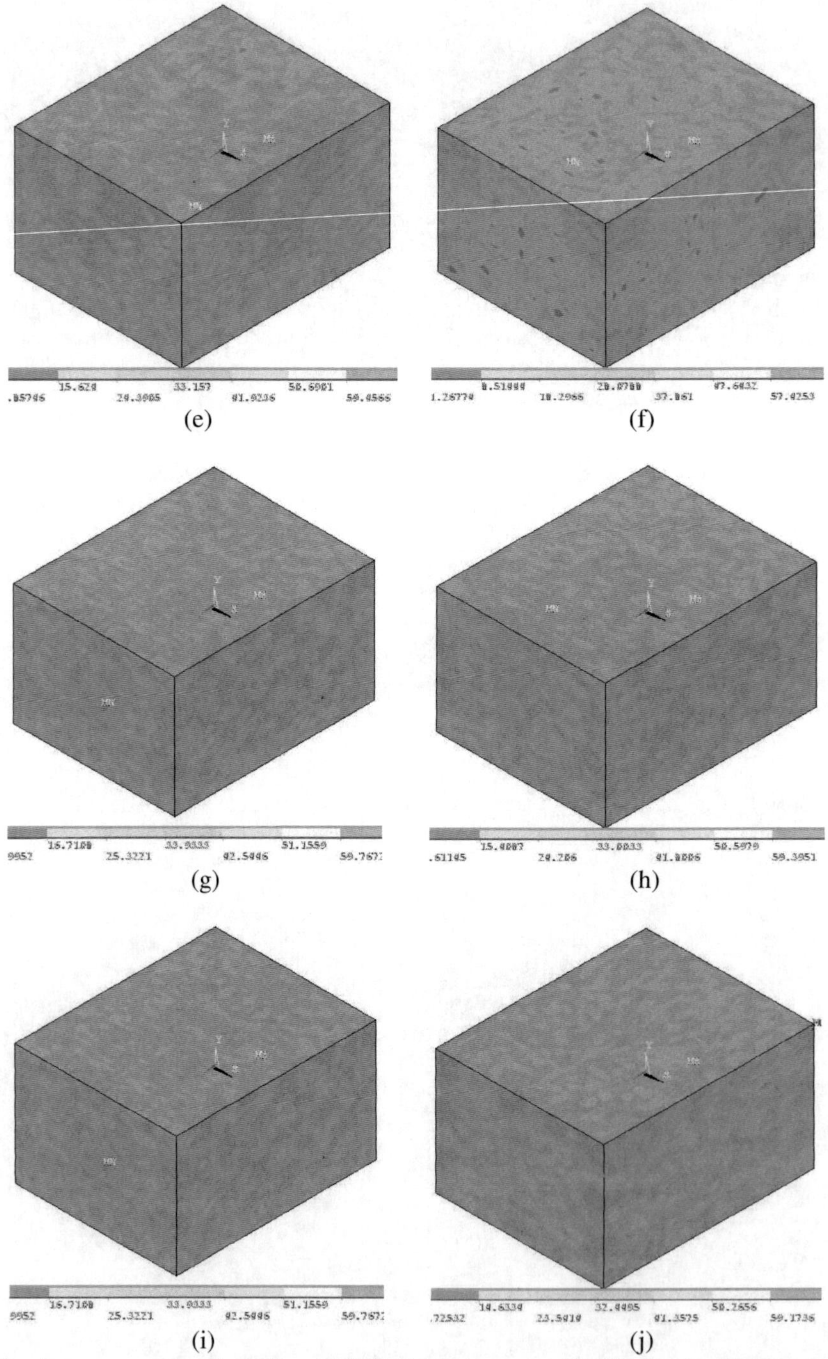

Figure 5. Continuous contour plots of room SPL: (a) Bare room; (b) Single absorber on one side wall; (c) 11 absorbers installed identically on two side walls; (d) 11 absorbers installed oppositely on two side walls; (e) 11 absorbers evenly distributed on walls; (f) 11 absorbers evenly distributed around corner walls; (g) 44 absorbers installed identically on two side walls; (h) 44 absorbers installed oppositely on two side walls; (i) 44 absorbers evenly distributed on walls; (j) 44 absorbers evenly distributed around corner walls

Table 2 illustrates data relating to overall sound pressure level (SPL) in the room. There are a total of 9 different configurations for the absorber placement: 1 for the 11 m^2 unit area, 4 for the 1 m^2 and another 4 for the absorber of 0.25 m^2 unit area. SPL for bare room is included for comparison.

Table 2. Overall sound pressure level (SPL) for every absorbers configurations

Configuration	Overall SPL (dB)			
	Bare Room	**1 Absorber**	**11 Absorbers**	**44 Absorbers**
Bare Wall	92.5	-	-	-
One Side Wall	-	74.1	-	-
Both Side Walls - Identical	-	-	70.7	69.9
Both Side Walls - Opposition	-	-	71.0	70.3
Evenly Distributed	-	-	69.4	68.3
Evenly Distributed Around Corner Walls	-	-	71.9	69.6

Referring to Table 2, it can clearly be noticed that the SPL in room has been reduced significantly after absorber has been installed regardless of any configurations, whereby the initial overall SPL for bare room is as high as 92.5 dB. For the ease of comparison, the SPL data is presented in column chart as depicted in Figure 6.

Figure 6 shows the overall SPL of the room for single unit absorber on a wall compared to those of multiple absorbers installed on multiple walls. Different trends can be seen for various absorber configurations, while 44-absorber configuration showed the greatest reduction in overall SPL especially the one that evenly distributed in the room.

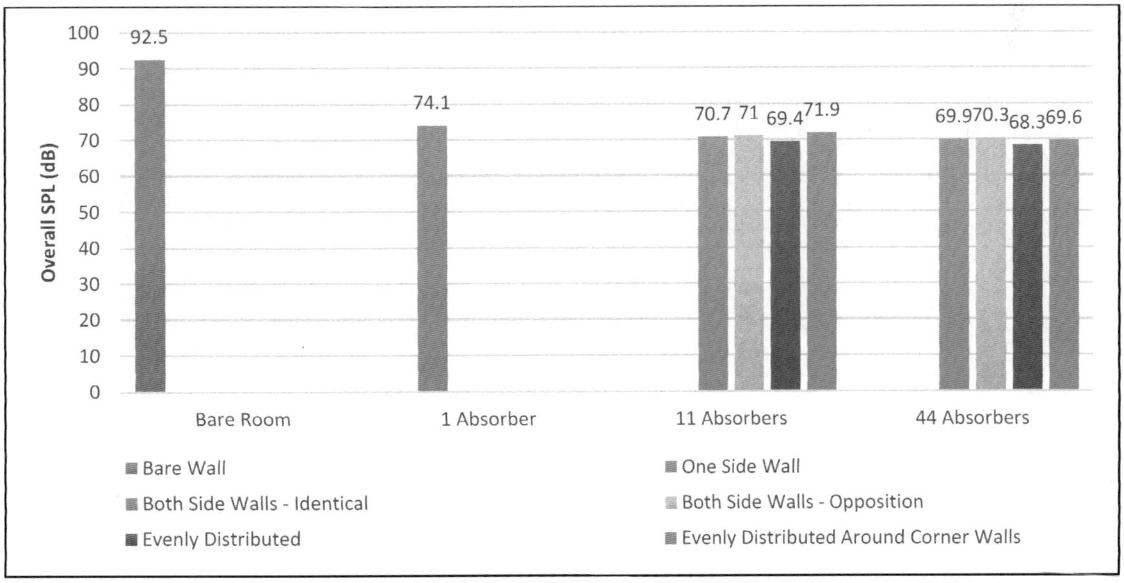

Figure 6. Overall SPL for different absorber configuration

By looking at the trend, single unit absorber is the least effective in reducing the overall SPL. It is only capable to bring down the overall SPL by 18.4 dB, from 92.5 dB to 74.1 dB. This may due to the reflected sound waves that bounce off the two opposing parallel walls because the absorber is only installed on a single wall. There is superposition of sound waves emitted from a source interfering constructively with its reflected waves which is in opposite direction, forming the standing waves which contributes to higher overall SPL than the other configurations.

The overall SPL is further reduced when the 11 m^2 absorber is divided into smaller pieces: 11 absorbers and 44 absorbers. By using 11 absorbers of 1 m^2 unit area, the overall SPL has went down to values ranging from 69.4 dB to 71.9 dB for different configurations, as shown in Figure 7 above. The best configuration for 11 absorbers is by distributing them evenly around the walls as this configuration has resulted in a decrement in overall SPL by 23.1 dB, which is from 92.5 dB to 69.4 dB. It is followed by installing absorbers identically in both side walls, oppositely in both side walls and evenly distributed around the corner walls.

For the case of 44 absorbers, the absorbers placement when they are evenly distributed around the walls contributes to the lowest overall SPL, 68.3 dB, where it has decreased as much as 24.2 dB as compared to overall SPL of bare room. It is then followed by placing the absorbers evenly around the corner walls, identically in both side walls and oppositely in both side walls.

To sum up, all three categories (single absorber, 11 absorbers and 44 absorbers) has successfully treating the room by reducing the overall SPL. The greatest reduction in overall SPL would be by using 44 absorbers evenly distributed around the walls and the least effective would be by using one whole absorber. Table 3 summarizes the results comparison for the overall SPL received by the receivers and their percentage of reduction.

Table 3. Comparison results for overall SPL received by the receivers

No. of Absorbers (patches)	& Configurations	Overall SPL (dB) Before	After	Percentage of Reduction (%)
1	One Side Wall	92.5	74.1	19.9
11	Both Side Walls - Identical		70.7	23.5
	Both Side Walls - Opposition		71.0	23.2
	Evenly Distributed		69.4	25.0
	Evenly Distributed Around Corner Walls		71.9	22.2
44	Both Side Walls - Identical		69.9	24.5
	Both Side Walls - Opposition		70.3	24.0
	Evenly Distributed		68.3	26.2
	Evenly Distributed Around Corner Walls		69.6	24.8

4. Conclusion

The rectangular room model with dimensions of $3m$ high × $4m$ wide × $5m$ long with the predefined conditions as mentioned above has been modified by introducing the correct amount of absorptive materials. By taking the recommended maximum RT_{60} for small room consists of less than 50 person into account, the number of sound absorbers needed to treat the room is found to be11 m^2. This number indicates the total coverage areas that must be introduced into the room in order to reduce RT_{60} to less than 0.8 seconds, where it is suitable for speech. After the room simulation was carried out using ANSYS APDL for each and every absorber configurations, the data shown that the optimum location of sound absorbers placement is by placing them on every wall with even distribution. The absorbers that were evenly distributed in the room contribute to the biggest reduction in overall sound pressure level (SPL) that received by all the receivers. This is true for both of the 11-absorber and 44-absorber cases. For 11 absorbers that is evenly distributed around the walls, the overall SPL in the room has reduced to 69.4 dB, where there is a total reduction of 23.1 dB equivalent to 25 % of decrement in overall SPL as compared to bare room. On the other hand, 44 absorbers evenly distributed around the walls has successfully reduced the overall SPL by 24.2 dB or 26.2 %, from 92.5 dB for bare room to 68.3 dB for this configuration. To conclude on this study, the total amount of sound absorbers needed to treat the room is $11\ m^2$ and the optimum location or best configuration would be placing the absorbers evenly in staggered position on every wall by using 44 absorbers. More absorber division means different area on the walls can be covered and the problem with standing waves can be prevented as well because every walls are now covered by some absorbers to lessen the superposition of reflected sound waves which promotes higher overall sound pressure level (SPL).

5. References

[1] Hansen C H, and Sehrndt C H 2001 Fundamentals of acoustics *Occupational Exposure to Noise: Evaluation, Prevention and Control. World Health Organization*
[2] Stansfeld S A and Matheson M P 2003 Noise pollution: non-auditory effects on health. *British Medical Bulletin* 68(1) 243-257
[3] Abdul Rahman I, Lee K S, Ng K K, Omar A, Leong S, Mohd Noor J, Ong C H, Mohd Mokhtar H A and Abdul Rahman M 2005 Guidelines for Control of Occupational Noise Putrajaya: Department of Occupational Safety and Health, Malaysia
[4] Arenas J P and Crocker M J 2010 Recent Trends in Porous Sound-Absorbing Materials *Sound & Vibration* 44(7) 12–18
[5] Peters R J, Smith B J and Hollins M 2011 Acoustics and Noise Control, 3rd Edition. Harlow, Essex: Pearson Education Limited
[6] Acoustic Treatment 101: Getting Your Room to Sound Great. (n.d.). Retrieved on April 16, 2016, from E-home Recording Studio: http://ehomerecordingstudio.com/acoustic-treatment-101/
[7] Campbell C, Nilsson E and Svensson C 2015 The Same Reverberation Time in Two Identical Rooms Does Not Necessarily Mean The Same Levels of Speech Clarity and Sound Levels When We Look at Impact of Different Ceiling and Wall Absorbers *Energy Procedia* 78 1635–1640
[8] Shen X and Shen Y 2004 Optimization of The Locations of Absorptive Material in Four Typical Small Rooms With Finite Element Method *Architectural and Building Acoustics* 2411–2414
[9] Talbot-Smith M 2002 Sound Engineering Explained Woburn: Focal Press
[10] Howard C Q and Cazzolato B S 2014 Acoustic Analyses Using MATLAB and ANSYS Boca Raton FL: Taylor & Francis Group

Acknowledgement

The work described in this paper was fully supported by Fundamental Research Grant Scheme (FRGS), Phase 1/2015, Vot. 1549 from the Ministry of Higher Education of Malaysia.

IOP Conf. Series: Journal of Physics: Conf. Series **914** (2017) 012038 doi:10.1088/1742-6596/914/1/012038

Modelling of Rigid Walled Enclosure Couple to a Flexible Wall using Matlab and Ansys APDL

I Zaman[1,2,a], S A M Rozlan[2], N A Azmir[2], M Ismon[2], M A Madlan[2], M N Yahya[2], M H Zainulabidin[2], M S M Sani[3], M F M Noh[4]

[1]Mechanical Failure Prevention and Reliability Research Centre, Universiti Tun Hussein Onn Malaysia, 86400 Batu Pahat, Johor, Malaysia.
[2]Noise and Vibration Analysis Research Group, Faculty of Mechanical and Manufacturing Engineering, Universiti Tun Hussein Onn Malaysia, Johor, Malaysia.
[3]Automotive Engineering Centre, Universiti Malaysia Pahang, 26600 Pekan, Pahang, Malaysia.
[4]Kolej Kemahiran Tinggi MARA Kuantan, KM8 Jalan Gambang, 25150 Kuantan Pahang, Malaysia

Corresponding author: [a]izzuddin@uthm.edu.my

Abstract. Generally, solutions to improve the noise problems in enclosure are to redesign or modifying the system such as increasing the thickness of the wall panels, enhancing the elasticity of the structure, and increase the damping mechanism of the wall structure. In this paper, the application of vibroacoustic modelling of enclosure coupled to a flexible wall was presented. The sound pressure characteristics of rigid walled enclosure, such as natural frequency and mode shape were determined using two approaches which are finite element simulation of Ansys® and mathematical model. The mathematical equations derived in Matlab® such as rigid walled enclosure and rigid walled enclosure coupled to flexible wall were used to validate finite element analysis (FEA). The result indicates that the theory and FEA display in a good agreement. Thus, proved that the FE model was accurate and can be applied in further research such as sound pressure and noise attenuation in enclosure.

1. Introduction

These days, noise and vibration has turned into a vital issue for modern society to find a better life satisfaction. This in turn makes vibration and acoustic characteristics as the critical criteria to be examined and considered in many engineering design issues. For instance, in the car and aviation industries, the level of vibration and noise has become an important asset in manufacturing because of its impact on passenger comfort [1].

Theoretically, noise in real applications mostly involves structural vibration generating sound waves. This combination is known as vibroacoustic. Vibroacoustic study shows that when an elastic structure is in contact with the liquid, the structural vibration and acoustic pressure in the fluid are influenced by the mutual vibroacoustic coupling interaction [2]. Despite the fact that these issues have been known from time to time [2-4], there is an absence of studies in the field of vibroacoustic especially when it comes to modeling analysis.

Lately, vibroacoustic is observed to be a major contributor to health disorders. These disorders are basically due to the vibration and noise occurring simultaneously. The effect of vibration can cause sleeping deprivation, exhaustion and headache soon after or during exposure [5]. The symptoms are the same as many people experience after a long boat or car trip. After consistent exposure every day for several years, whole body vibration can affect the entire body and cause several health disorders. For example, sitting in the ship or sea vehicles can cause motion sickness when vibration exposure happens in the frequency range of 0.1-0.6 Hz [5]. Studies of bus drivers have discovered that exposure to whole body vibration may contribute to muscle and circulatory problems, breathing and back pain. The joined effects of body posture, postural fatigue, dietary habits and whole-body vibration are the possible reasons for these problems.

Research shows that whole body vibration can increase oxygen uptake, heart rate, respiratory rate, which in turn, can prompt changes in blood and urine [6]. For some other cases, the effect of noise affects the hearing organs (cochlea) in the inner ear. That is why noise-induced hearing loss is indeed sensory neural type of hearing loss. The workers in noisy environments who are also exposed to vibration (for example, a person operating a jack hammer) may experience hearing loss greater than those exposed to a similar level of noise but not to vibration [7].

Therefore, this paper expects to preseent the problem solving of vibroacoustic system of rigid walled enclosure coupled to a flexible wall by using analytical approach plotted in Matlab and finite element analysis of Ansys. Similir to our previous analytical modelling study [8-10], the result of this review will give a supportive reference to future analysts who adapt an approach to minimize the sound pressure level of rigid walled enclosure attached to a flexible wall.

2. Modelling

This section describes the parametric study of rigid walled enclosure with a flexible wall and determination of a mathematical model which permits the estimation of sound pressure in enclosure attached to a flexible wall.

2.1 Parametric Study

Figure 1 shows the schematic diagram of rigid walled enclosure with attached a flexible wall. Force is applied on the flexible wall at the coordinate (1, 0.1, 0.1) m, while sound pressure, P is measured at the coordinate (0, 0.5, 0) m. The parameters of enclosure, air and flexible wall used in this analysis are shown in Tables 1.

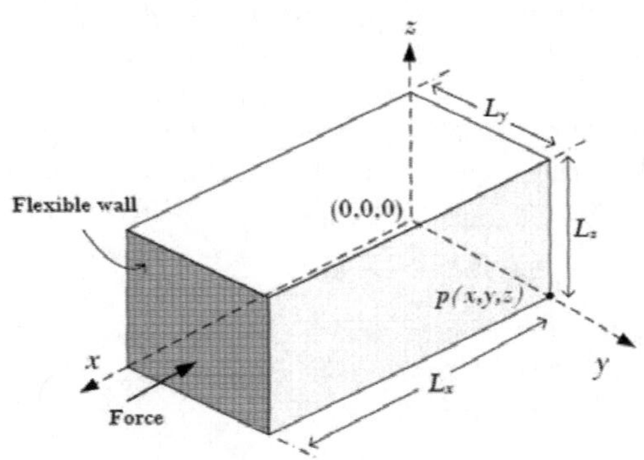

Figure1. Schematic diagram of rigid walled enclosure with a flexible wall

Table 1. Parameters applied in the analysis of an enclosure with a flexible wall

Medium	Parameters	Description	Value	Units
Enclosure	L_x	Length	1.0	m
	L_y	Width	0.5	m
	L_z	Height	0.3	m
Air	ρ_a	Density	1.21	kg/m^3
	c	Speed of sound	344	m/s
Flexible wall	B	Width	0.2	m
	h	Thickness	1.0×10^{-2}	m
	E	Young's modulus	2.1×10^{11}	GPa
	v	Poisson ratio	0.3	-
	ρ	Density	7.85×10^3	kg/m^3
	ζ	Damping ratio	1	%

2.2 Mathematical Derivation

The acoustic pressure, p at x inside the enclosure and the structural vibration velocity, u at y are given by [11,12]:

$$p(x) = \sum_n a_n(\omega)\psi_n(x) \tag{1}$$

$$u(y) = \sum_m b_m(\omega)\psi_m(y) \tag{2}$$

where $\Psi_n(x)$ and $a_n(\omega)$ are the amplitude of uncoupled acoustic mode shape function and acoustic pressure and, respectively. Similarly, $\Psi_m(y)$ and $b_m(\omega)$ are the respective complex amplitude of uncoupled vibration mode shape function and vibration velocity

The amplitude of the nth acoustic mode is shown as in Eqn. (3) [11,12].

$$a_n(\omega) = \frac{j\omega\rho_a c^2}{\Lambda_n(\omega_n^2 - \omega^2 + 2j\xi_n\omega_n\omega)} \times \left[\int_v q_{vol}\psi_n dV - \int_s u_n\psi_n dS\right] \tag{3}$$

where $\Lambda_n = \int_v \psi_n^2(x)dV$, ω_n and ξ_n are the natural frequency and damping ratio of the nth acoustical mode Ψ_n, respectively, ρ_a is the fluid density, c is the speed of sound in fluid, q_{vol} is the volume velocity of sound source and u_n is the vibration velocity of structure.

In this case, the damping ratio of enclosure is neglected and the volume speed of sound source is not considered for this study. Thus, Eqn. (3) can be further improved as given by:

$$a_n(\omega) = \frac{j\omega\rho_a c^2 \int_s u_n\psi_n dS}{\Lambda_n(\omega_n^2 - \omega^2)} \tag{4}$$

Meanwhile, the amplitude of the mth structural mode is given by [12,13]:

$$b_m(\omega) = \frac{j\omega}{M_m(\omega_m^2 - \omega^2 + 2j\zeta_m\omega_m\omega)} \times \left[\int_s p_m\psi_m dS - \int_s F_m\psi_m dS\right] \tag{5}$$

where $M_m = \int_s \psi_n^2 dS$, ω_m and ζ_m are the natural frequency and damping ratio of the mth simply-supported wall mode Ψ_m, respectively, p_m is the pressure on the surface of the flexible wall and F_m is the external force imposed on the flexible wall.

By employing mode summation method, the amplitude of the nth acoustic mode can be written as [11]:

$$a_n(\omega) = \frac{j\omega\rho_a c^2}{\Lambda_n(\omega_n^2 - \omega^2)}\left[-\sum_{m=1}^{\infty} b_m(\omega)C_{mn}\right] \tag{6}$$

Likewise, by employing mode summation method, the complex amplitude of the mth structural mode is simplified as [13]:

$$b_m(\omega) = \frac{j\omega}{M_m(\omega_m^2 - \omega^2 + 2j\zeta_m\omega_m\omega)} \times \left[\sum_{n=0}^{\infty} a_n(\omega)C_{mn} - \sum_{m=1}^{\infty} F_m\psi_m\right] \tag{7}$$

where C_{mn} is defined in Eqn. (8) as the modal coupling between the the structural mode and acoustic mode [13]:

$$C_{mn} = \int_s \psi_n \psi_m dS \tag{8}$$

The analytical solution for the modal coupling between acoustic mode and the structural mode has been derived by [13], and it is given as follows:

$$C_{mn} = 2(-1)^{n_3} S\sqrt{\varepsilon_{n_1}\varepsilon_{n_2}\varepsilon_{n_3}} \left\{\frac{m_1}{n_1^2 - m_1^2}\right\}\left\{\frac{m_2}{n_2^2 - m_2^2}\right\} \times \left\{\frac{-1^{n_1+m_1} - 1}{\pi}\right\}\left\{\frac{-1^{n_2+m_2} - 1}{\pi}\right\} \tag{9}$$

where n_1,n_2 and n_3 defines cavity modal integers, m_1 and m_3 are the structural modal integers and ε_{n1}, ε_{n2} and ε_{n3} are the normalization factors. By solving equation above, the sound pressure inside the enclosure with attached flexible wall can be determined.

3. Result and Discussion

In this study, the analytical equation shown before was plotted using Matlab® and later was validated with numerical analysis of Ansys®. Figure 2 displays the results of sound pressure level in an enclosure with attached a flexible wall when the wall is imposed by a point force, at different discrete frequencies varied from 0 to 500 Hz with 1 Hz step size. There are 400 structural modes and 1000 acoustic modes were applied in the mode superposition method in Matlab® in order to get an accurate outcomes. Once more, for this analysis, it is discovered that Matlab® and Ansys® results accomplishes a better agreement when the finite element model of the enclosure was discretised into an adequate number of elements and substeps. The frequency modes are identified in the frequency range of 0 to 500 Hz which represented by the total number of the peaks [14].

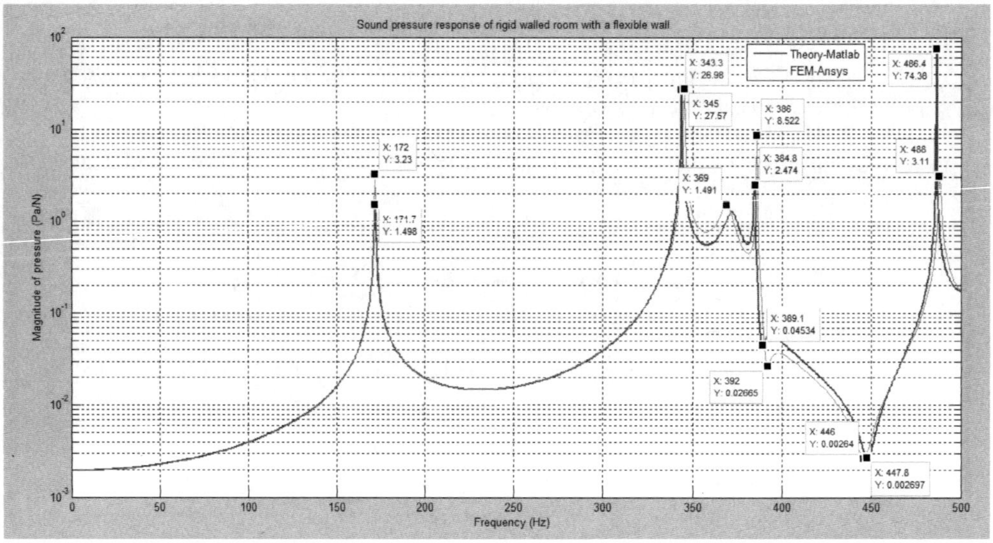

Figure 2. Magnitude of sound pressure response in a rigid walled cavity with flexible wall

Table 2 indicates the mode frequency values of rigid walled enclosure with a flexible wall obtained from theoretical model and numerical Ansys®. The outcome shows that the percentage error between theoretical model and numerical Ansys® is too small which less than 10%, thus can be negligible. Meanwhile, Figure 3 shows the characteristics mode shape of rigid walled enclosure with a flexible panel that obtained from the modal analysis of Ansys®.

Table 2. Errors of rigid walled enclosure with a flexible wall between theory and FEM

Mode n^{th}	Ansys®	Matlab®	Error (%)
1	172	171.7	0.17
2	345	343.3	0.49
3	369	372	0.81
4	386	384.8	0.31
5	392	389.1	0.73
6	446	447.8	0.40
7	488	486.4	0.33

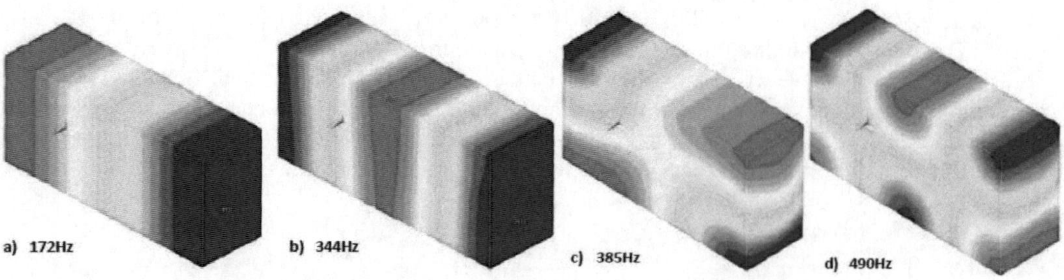

Figure 3. Mode shape of enclosure with a flexible wall

4. Conclusion

This paper attempted to predict the sound pressure of a rigid walled enclosure with a flexible wall by deriving the mathematical model. Matlab® was used for plotting the analytical analysis while Ansys® for numerical analysis. The result shows that the error difference between both methods are below 1% which can be neglected. Thus, this prove the theoretical as well as numerical model prediction can be used for the other research related.

Acknowledgements

The authors wish to acknowledge Ministry of Higher Education Malaysia and Research, Innovation, Commercialization and Consultancy Management, Universiti Tun Hussein Onn Malaysia for the partial support under Fundamental Research Grant Scheme (FRGS) vote 1546.

References

[1] Jang H-K and Griffin M 2000 *Effect of phase, frequency, magnitude and posture on discomfort associated with differential vertical vibration at the seat and feet.* Journal of Sound and Vibration **229** 273-286
[2] Zhou Q and Wang D 2015 *Vibro-acoustic coupling dynamics of a finite cylindrical shell under a rotor-bearing-foundation system 's nonlinear vibration excitation.* Journal of Sound and Vibration **347** 150-168
[3] Li Y, Wang X, Huang R and Qiu Z 2015 *Active vibration and noise control of vibro-acoustic system by using PID controller.* Journal of Sound and Vibration **348** 57-70
[4] Conlon S C, Fahnline J B and Semperlotti F 2015 *Numerical analysis of the vibroacoustic properties of plates with embedded grids of acoustic black holes.* The Journal of the Acoustical Society of America **137** 447-457
[5] Burdzik R and Konieczny Ł, Application of vibroacoustic methods for monitoring and control of comfort and safety of passenger cars. in Proceedings of the Solid State Phenomena, 2014, p. 20-25.
[6] Nassiri P, Ebrahimi H, Monazzam M, Rahimi A and Shalkouhi P J 2014 *Passenger noise and whole-body vibration exposure—a comparative field study of commercial buses.* Journal of Low Frequency Noise, Vibration and Active Control **33** 207-220
[7] Stansfeld S, Brown B and Haines M 2000 *Noise and health in the urban environment.* Reviews on environmental health **15** 43-82
[8] Zaman I, Rozlan S A M, Yusoff A, Madlan M A and Chan S W 2017 *Theoretical modelling of sound radiation from plate.* IOP Conference Series: Materials Science and Engineering **166** 012023
[9] Rozlan S A M, Zaman I, Manshoor B, Khalid A, Chan S W and Sani M S M 2016 *Theoretical modelling of a beam with attached spring-mass-damper system* MATEC Web of Conferences **90** 01030
[10] Salleh M M and Zaman I 2016 *Finite element modelling of fixed-fixed end plate attached with vibration absorber.* ARPN Journal of Engineering and Applied Sciences **11** 2336-2339
[11] Fahy F J and Gardonio P 2012 *Sound and structural vibration: radiation, transmission and response* (UK: Academic press)
[12] Hansen C, Snyder S, Qiu X, Brooks L and Moreau D 2012 *Active control of noise and vibration* (CRC Press)
[13] Fahy F J 2000 *Foundations of engineering acoustics* (Academic press)
[14] Zaman I, Salleh M M, Ismon M, Manshoor B, Khalid A, Sani M S M and Araby S 2014 *Study of passive vibration absorbers attached on beam structure.* Applied Mechanics and Materials **660** 511-515

IOP Conf. Series: Journal of Physics: Conf. Series **914** (2017) 012039 doi:10.1088/1742-6596/914/1/012039

Effect of Rotating Mold Speed on Microstructure of Al LM6 Hollow Cylinder Fabricated Using Centrifugal Method

Maximus Kohnizio Mahli,[1] Saifulnizan Jamian,[2] Al Emran Ismail,[2] Nik Hisyamudin Muhd Nor,[1] Mohd Khir Mohd Nor,[2] Kamarul Azhar Kamarudin

[1]Faculty of Mechanical Engineering & Manufacturing, Universiti Tun Hussein Onn Malaysia, 86400 Batu Pahat, Malaysia

[2]Crashworthiness and Collisions Research Group, Faculty of Mechanical and Manufacturing Engineering, Universiti Tun Hussein Onn Malaysia, 86400 Batu Pahat, Malaysia

Abstract. Al LM6 hollow cylinder is fabricated using horizontal centrifugal casting which produce a very fine grain on the outer surface of the structure. In this study, the effect of motor speed and pouring temperature on the microstructure of Al LM6 hollow cylinder is determined. The speed of the motor used during casting are 1300rpm, 1500rpm and 1700rpm and the pouring temperature are 690°C, 710°C and 725°C. The Al LM6 hollow cylinder is produced by pouring the molten Al LM6 into a cylindrical casting mold which is connected with a shaft and it is rotated by motor until it is solidified. Then, the cross-section is observed using OM and SEM/EDS. From the microstructure observation, the distributions of Si are more concentrated at the inner parts and the size of Si is bigger at the inner parts. The result shows that the Si particles at the inner part which is fabricated at the highest motor speed (1700rpm) have the most Si particles compared with the Si particles that are casted with other motor speeds.

1 Introduction

Aluminium is the most abundant metal on the earth and comprising one-twelfth of the earth crust. Aluminium has the weight of one-third of the steel weight and therefore it can be made thicker and stronger while the weight is reduced in all applications. The versatility of aluminium makes it to be used widely as it has a favourable strength property. Aluminium does not rust and it can be combined with other metal to form an aluminium alloys because there is a transparent layer of aluminium oxide present on its surface when it exposed to the atmosphere [1].

Al LM6 is an aluminium alloy that normally used in the industry such as aerospace, automotive, construction and many more because it has high endurance, high impact strength, light-weight and exhibits outstanding resistance to corrosion under both ordinary atmospheric and marine conditions. This aluminium alloys are rather difficult to machine because of the hardness is increased with the Silicon (Si) content in the alloy but the ductility is reduced. With Si content in the Al LM6, it improves its fluidity which is good to produce any product according to the desired shape. This aluminium alloy can be cast into thinner mold and more intricate parts than any of the other types of casting alloys [2].

Centrifugal castings process normally being used to produce pipes, cylinder liners, piston rings, rollers, pulleys and others. Centrifugal casting process has a high reliability because it reduces the shrinkage and gas porosity in the structure [3]. One of the methods of producing functionally graded materials (FGMs) is by using centrifugal casting. The presence of FGMs in the alloy is due to the difference in density of different phases such as Al and Si particles [4]. Table 1 shows the comparison of density for Al and Si.

Table 1. Density of Al and Si [2]

Material	Al	Si
Density (g/cm^3)	2.70	2.34

Radhika and Raghu [5] investigate the microstructure at the outer of the FGM composites, Al/B_4C, Al/SiC, Al/Al_2O_3 and Al/TiB_2, with reinforcement which are produced by using centrifugal casting. The ingot of Al−12Si−Cu alloy is cut into small pieces and then melted in the furnace in an inert gas atmosphere to reduce the defect in the cast. The reinforcement is preheated at 300°C and then added to the molten aluminium alloy. The mixing is stirred at 200rpm to produce vortex. Then, the mixing is poured into die that is preheated at 350°C of the centrifugal casting setup and it is rotated with a motor speed of 1300 rpm. The process continues until it solidified and the cast part is removed. The hollow cylinder produced has the dimension of 150mm outer diameter and length 150mm with 18 mm of thickness. The result of microstructure study shows that the reinforcement present at the outer surface and the hardness test using Vicker's hardness test shows that the outer region have higher hardness than in the middle and inner region except for B_4C that shows variation of hardness in different region. The abrasive wear test shows that the FGM with reinforcement TiB_2 have high wear resistance while FGM with reinforcement B_4C have less wear resistance. Therefore, the FGM composite have better mechanical and wear properties by using centrifugal casting.

Jamian et al [6] has investigates the distribution of functionally graded natural fiber/epoxy (FGNF/epoxy) hollow cylinder that are fabricated using centrifugal casting method. Banana trunk is used as its natural fiber. The difference of density between natural fiber and the epoxy causes natural fiber particle to move at the outer radius of the hollow cylinder during casting. An epoxy cylinder without fiber will be used as the reference. For centrifugal casting process, three different rotating speed will be used which are 1145rpm, 1187rpm and 1240rpm. Based on the testing result, the natural fiber is graded from inner to outer surface of the FGNF/epoxy cylinders that is produced by using centrifugal casting method. The rotation of the motor and the natural fiber composition affected the graded distribution in the FGNF/epoxy cylinders. It is also found that the outer surface has a higher value of hardness, density and strength along the cylinder and reduced gradually towards the inner surface. This is because the composition of natural fiber at the outer surface is higher than at the inner surface. In conclusion, centrifugal casting methods are effective in producing a better mechanical strength.

Jamian et al [7] has investigates the distribution of Si particles of Al LM6 hollow cylinder which is produced by using centrifugal casting method. In the process, the Al LM6 ingot will be melted before pouring it into the rotating mold. The speed of rotation of the mold is 1700rpm (100G) and the pouring temperatures of the Al LM6 are 690°C, 710°C and 725°C. Based on the microstructure result using OM, it shows that Si particle content are the same at all parts in the sample. However, the particle size at the inner surface is bigger than in the middle and at the outer surface of the sample because of the difference in density between Si and Al and the expansion of Si particles during the cooling process of the sample. The size of Si particles is different for all the three samples casted with different melt temperature. It is found that the melt temperature, the density and the cooling rate affect the size and distribution of Si particles.

Based on the previous research [5-7], it shows that centrifugal casting process is the most efficient process to produce a better mechanical and wear properties. Centrifugal casting also helps in eliminate the gas porosity present during solidification process which will give the material better mechanical properties. Besides that, pouring temperature also affect the properties of the material. In this study, the

IOP Conf. Series: Journal of Physics: Conf. Series **914** (2017) 012039 doi:10.1088/1742-6596/914/1/012039

distribution of Si particles inside the Al LM6 hollow cylinder fabricated using centrifugal casting is observed. The effect of mold rotating speed on the Si particles distribution is investigated.

2 Methodology

In this research, nine Al LM6 hollow cylinders are fabricated using centrifugal casting method. The main casting parameters considered are pouring temperature and mold rotating speed. The casting condition for the each casting is shown in Table 2. The volume of the hollow cylinder fabricated is 5.5×10^{-5} m^3. The process starts by cutting the Al LM6 ingot into small pieces with the mass 160g. Then, it is put inside the crucibles and melted using induction furnace. After that, the molten Al LM6 is poured inside the rotating mold with the desired motor speed. The process is left for 2 hours to let the molten Al LM6 solidified. Next, the solidified Al LM6 is removed from the mold and left for 5 hours to let it cool down. The sample is cut to be grinded, polished and etched before observed by using OM and SEM/EDS to study its microstructure. The etching method used is wet chemical etching and the solution mixture used is NaOH 5% + H_2O 95%.

Table 2. Parameter of centrifugal casting of Al LM6 Hollow Cylinder

Specimen	Parameter	
	Mold Rotating Speed (rpm)	Pouring Temperature (°C)
1	1300 (57G)	690
2		710
3		725
4	1500 (75G)	690
5		710
6		725
7	1700 (100G)	690
8		710
9		725

3 Results and Discussion

Figure 1 shows the Al LM6 hollow cylinder that is fabricated using centrifugal casting method. As can be observed in Figure 1(a), there are scars on the outer surface due to the vibration of the centrifugal casting machine during solidification process. Figure 1(b), there are layers of Al present because of the high rotational speed in the mold. The inner surface of the sample is defected due to the hot tear during shrinkage process as shown in Figure 1(c).

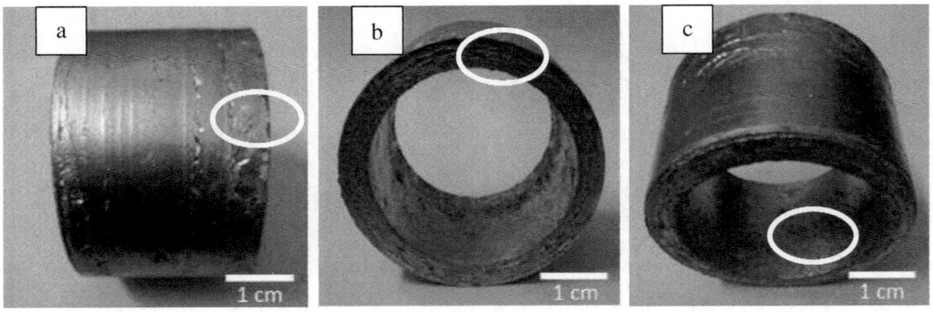

Figure 1: The physical surface from three surface views (a) outer (b) front (c) inner.

Microstructure study is carried out to observe the distribution of Si and Al inside the sample with different parameter used. The distributions are observed at the outer, middle and inner parts of the sample. Figure 2 shows the microstructure of the as-received Al LM6. The Si particles are in the needle shape with the average length of 50μm and distributed homogenously. Meanwhile, Figures 3 to 11 show the microstructure of the sample fabricated under different casting condition at 1300rpm motor speed. Figures 7 to 9 show the sample fabricated at 1500rpm motor speed and Figures 10 to 12 show the sample fabricated at 1700rpm speed.

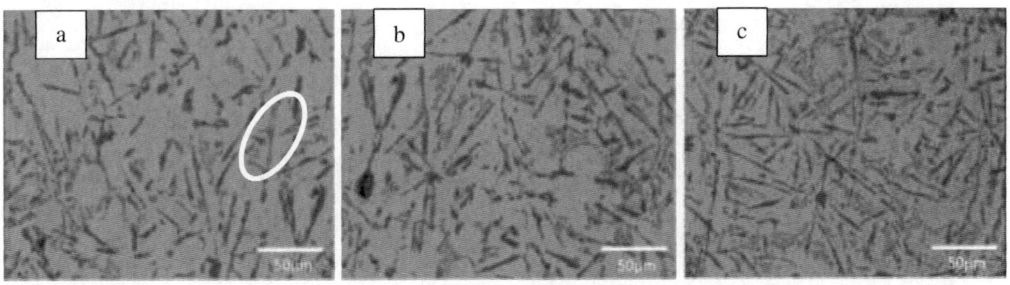

Figure 2: Microstructure of Al-LM6 before casting (a) outer; (b) middle; (c) inner

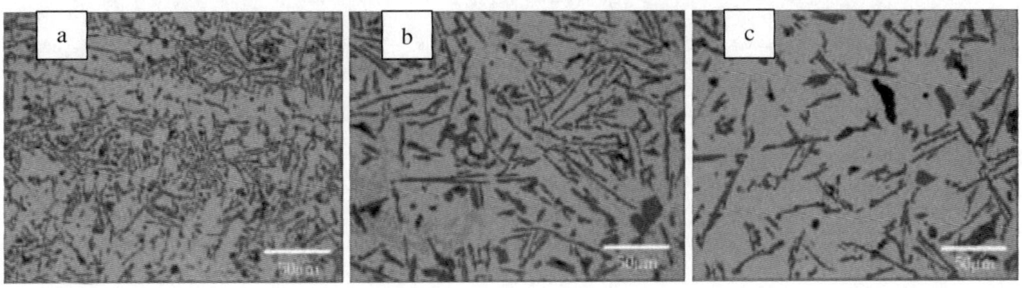

Figure 3: Microstructures of Al-LM6 cylinder with motor speed 1300rpm and melting temperature of 690 °C. (a) outer; (b) middle; (c) inner

IOP Conf. Series: Journal of Physics: Conf. Series **914** (2017) 012039 doi:10.1088/1742-6596/914/1/012039

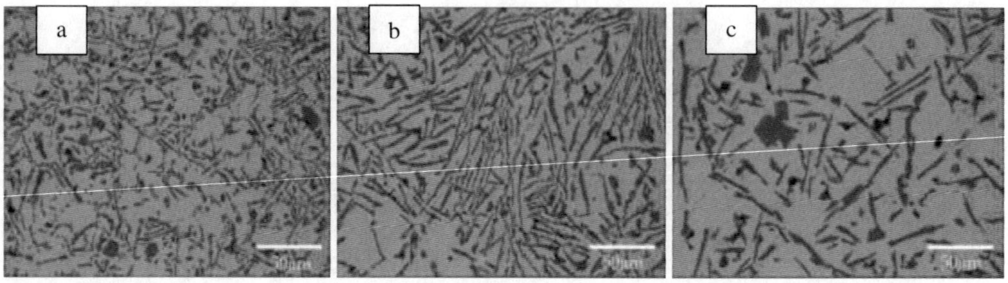

Figure 4: Microstructures of Al-LM6 cylinder with motor speed 1300rpm and melting temperature of 710 °C. (a) outer; (b) middle; (c) inner

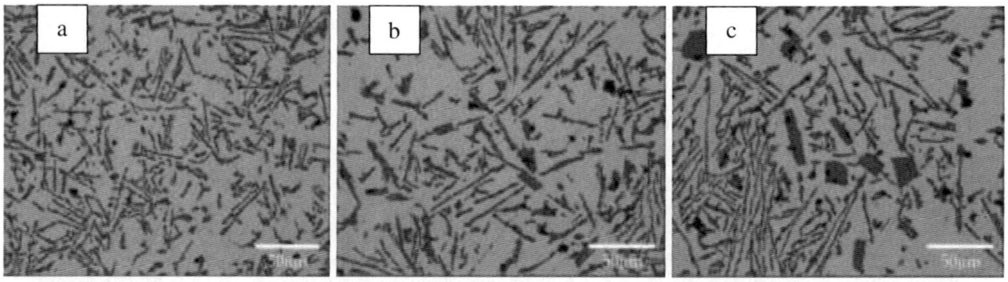

Figure 5: Microstructures of Al-LM6 cylinder with motor speed 1300rpm and melting temperature of 725 °C. (a) outer; (b) middle; (c) inner

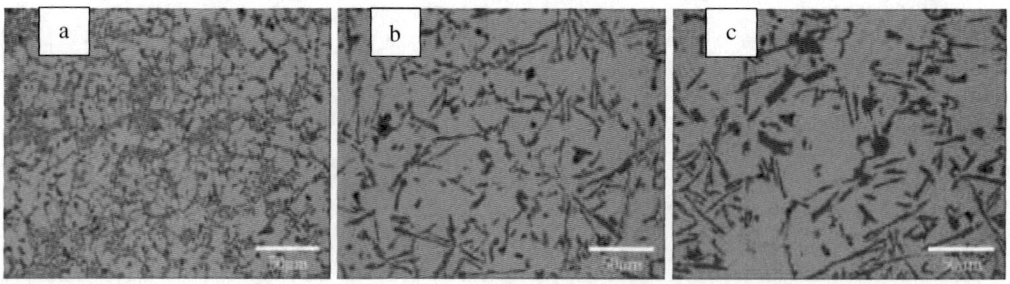

Figure 6: Microstructures of Al-LM6 cylinder with motor speed 1500rpm and melting temperature of 690 °C.(a) outer; (b) middle; (c) inner

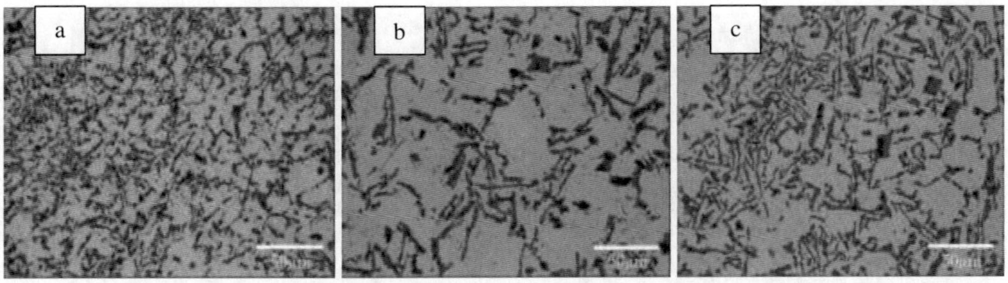

Figure 7: Microstructures of Al-LM6 cylinder with motor speed 1500rpm and melting temperature of 710 °C.(a) outer; (b) middle; (c) inner

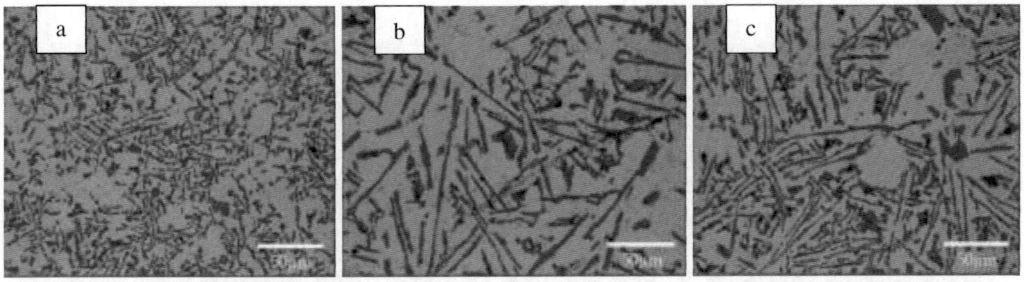

Figure 8: Microstructures of Al-LM6 cylinder with motor speed 1500rpm and melting temperature of 725 °C.(a) outer; (b) middle; (c) inner

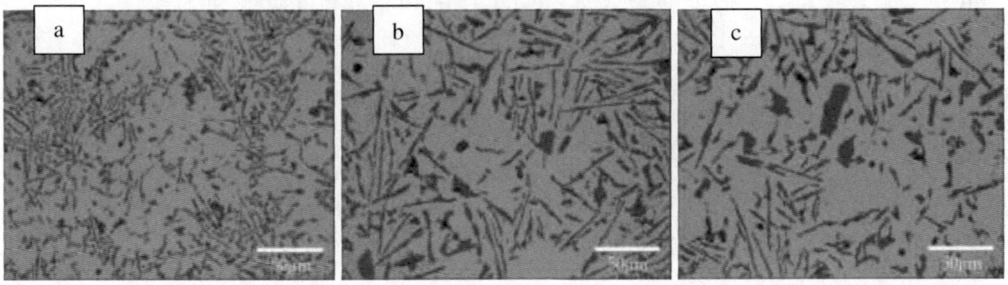

Figure 9: Microstructures of Al-LM6 cylinder with motor speed 1700rpm and melting temperature of 690 °C. (a) outer; (b) middle; (c) inner

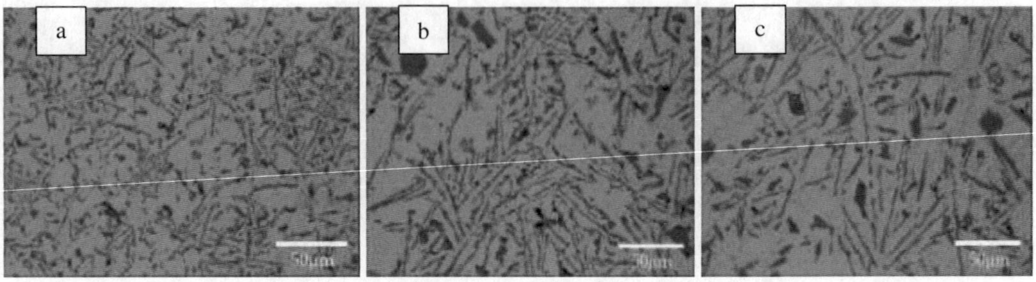

Figure 10: Microstructures of Al-LM6 cylinder with motor speed 1700rpm and melting temperature of 710 °C. (a) outer; (b) middle; (c) inner

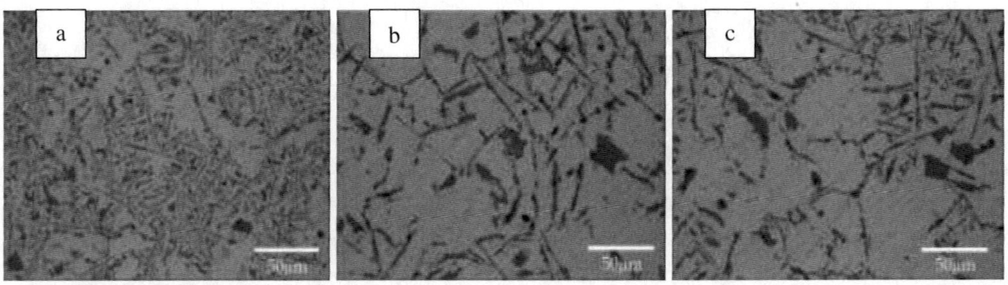

Figure 11: Microstructures of Al-LM6 cylinder with motor speed 1700rpm and melting temperature of 725 °C. (a) outer; (b) middle; (c) inner

It can be seen in the Figures 3 to 11 that the Si particles are more concentrated at the inner surface while the Al particles are more concentrated on the outer surface. This is because Si has lower density compared to Al which makes the Si to move inward. Besides that, the size of Si in the inner are bigger than the outer surface due to the difference in cooling rate where the cooling rate at the inner surface is much slower than the outer which gives it more time to combine and form a larger size of Si. The Si particles are free to move and distributed unevenly at the sample due to its properties of high viscosity and melting temperature [8]. Moreover, the porosity present at the inner surface is because the gas bubbles are trapped during casting process because it has low density [9]. Comparing the distribution of Si particles at the inner surface for the sample fabricated at 1300 rpm (Figure 3(a)), 1500 rpm (Figure 6(a)) and 1700 rpm (Figure 9(a)) with 690°C pouring temperature, clearly observed that the concentration of Si particles increase as the mold rotating speed increase.

The percentage of Si accumulation can be determined by using Energy Dispersive Spectroscopy (EDS). Figures 12 – 14 show the image taken from EDS for sample 8 and Table 3 shows the weight and atomic percentage for both Al and Si content in the sample. EDS is used to analyse the composition of Si and Al in the sample. From the EDS result, the weight percentage of Si is the highest at the inner part which is 20.47% and it decreases at the middle which is 16.56%. At the outer parts, the weight percentage of SI is the lowest which is 16.51%. The atomic percentage for Si in the inner part is 19.83%, at the middle part is 16.01% and at the outer part is 15.96%. Based on the microstructure observation, the size of Si is bigger at the inner part than the size of Si at the outer part. Therefore, the data shows that the Si particles are more concentrated at the inner part of the sample. Other elements are also presented during the EDS analysis however the value are less than 1% which is too small compared with the Si and Al content [10].

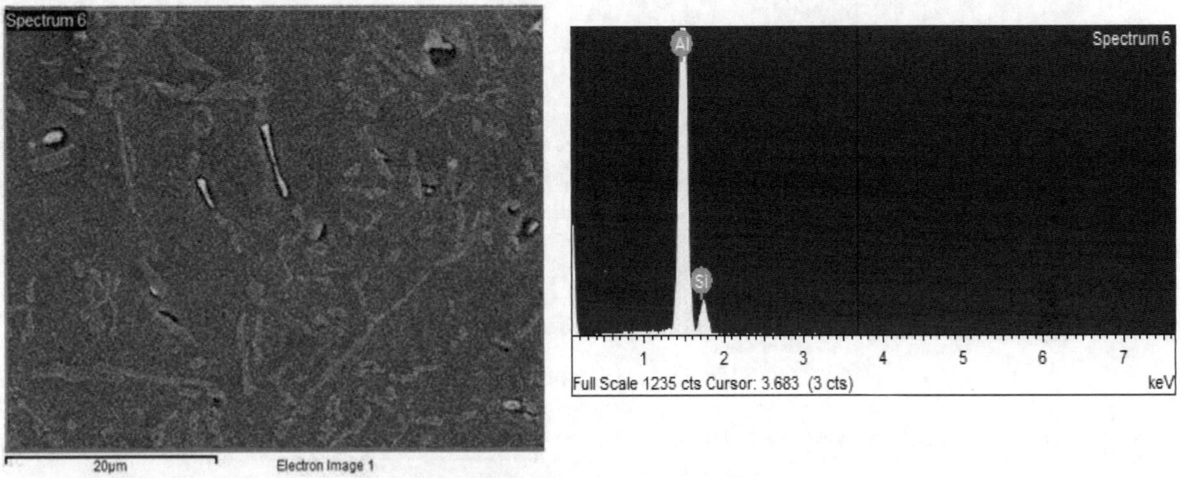

Figure 12: EDS of outer part of sample 8

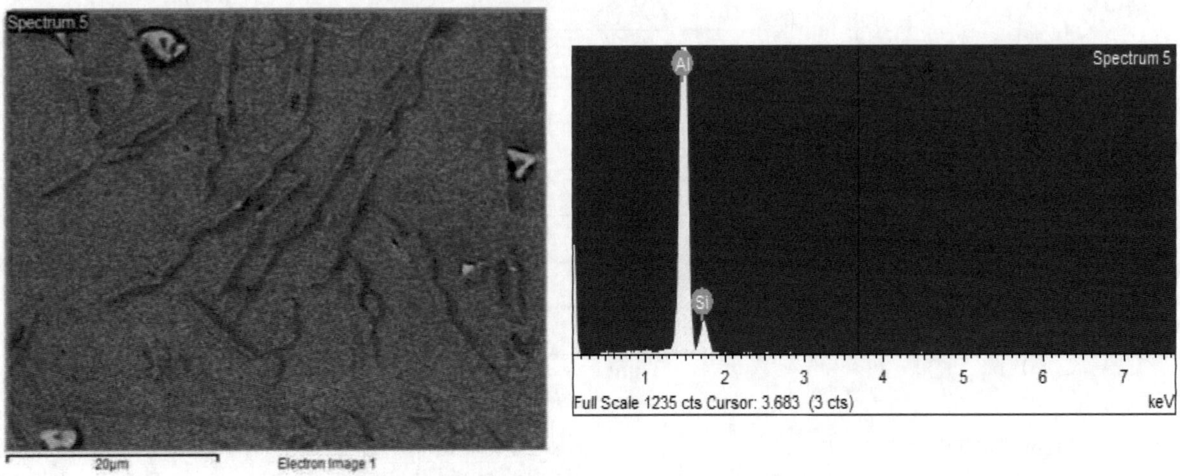

Figure 13: EDS of middle part of sample 8

IOP Conf. Series: Journal of Physics: Conf. Series **914** (2017) 012039 doi:10.1088/1742-6596/914/1/012039

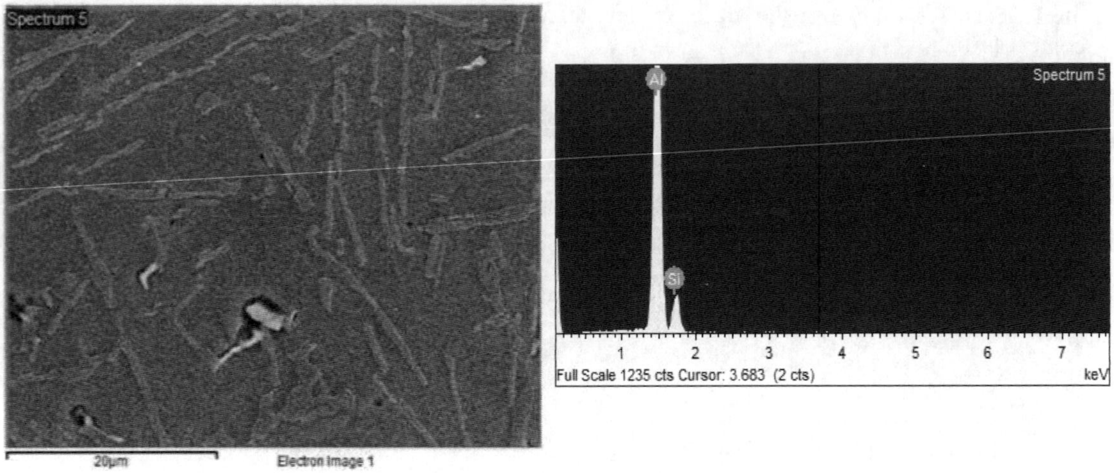

Figure 14: EDS of inner part of sample 8

Table 3. Weight and atomic percentage value for sample 8 at three different parts.

Part	Element	Weight %	Atomic %
Outer	Al K	83.49	84.04
	Si K	16.51	15.96
	Totals	100.00	100.00
Middle	Al K	83.44	83.99
	Si K	16.56	16.01
	Totals	100.00	100.00
Inner	Al K	79.53	80.17
	Si K	20.47	19.83
	Totals	100.00	100.00

4 Conclusion

In conclusion, this study have achieve its objectives to study the effect of pouring temperature and speed of motor rotation to produce an Al LM6 hollow cylinder by using horizontal centrifugal casting method. From the microstructure study, it shows that the Si particles are distributed at the whole part of the sample and the Si particles are more concentrated at the inner part of the sample while Al are more concentrated at the outer part of the sample. This is because Al has higher density compared to Si which makes the Al to move outward. Besides that, the particles of Si at the inner part are bigger than the outer part for all of the samples because of the difference in cooling rate in which the inner part will have a slow cooling rate that gives the Si more time to combine to form a larger particles.

Acknowledgement
This research is funded by Universiti Tun Hussein Onn Malaysia (UTHM) under Graduate Research Grant (GPPS) with Vot no. U801.

References

[1] L. S. Maurice, *Behavior and Design of Aluminium Structures*, McGraw-Hill, The University of Michigan, USA (1993).

[2] R. Brown, Foseco non-Ferrous foundryman's handbook. In Butterworth- Heinemann (Ed), Aluminium casting alloys, Oxford, 8 – 40 (1999).

[3] Madhusudhan, S. Narendranaath, G. C. Mohankumar & P. G. Mukunda, effect of mould wall thickness on rate of solidification of centrifugal casting, International Journal of Engineering Science and Technology, **2(11),** 6092-6096 (2010).

[4] G. Chirita, D. Soares, F.S. Silva, Advantages of the centrifugal casting technique for the production of structural components with Al–Si alloys, Materials and Design, **29**, 20–27 (2008).

[5] N. Radhika & R. Raghu, Development of functionally graded aluminium composites using centrifugal casting and influence of reinforcements on mechanical and wear properties, Transactions of Nonferrous Metals Society of China, **26 (4)**, 905-916 (2016).

[6] S. Jamian, S. N. Ayob, R. Z. Abidin & N.H. Muhd. Nur, Fabrication of functionally graded natural fibre/epoxy cylinder using centrifugal casting method, ARPN Journal of Engineering and Applied Sciences, **11 (4)**, 2327-2331 (2016).

[7] S. Jamian, S. Yutiman & H. Hehsan, Al LM6 Hollow Cylinder Fabricated Using Centrifugal Casting, ARPN Journal of Engineering and Applied Sciences, **11 (18)**, 11221-11225 (2016).

[8] F. Bonollo, A. Moret, S. Gallo, & Mus C., Cylinder liners in aluminium matrix composite by centrifugal casting, La metallurgia italiana, **6**, pp 49-55 (2004).

[9] D.Rajan, R.M. Pillai & B.C. Pai, Characterization of centrifugal cast functionally graded aluminum-silicon carbide metal matrix composites", Materials Characterization, **61**, pp 923–928 (2010).

[10] B. Rabindra, D. Chatterjee & G. Sutradhar, Effect of Reinforcement Particles on the Fluidity and Solidification Behavior of the Stir Cast Aluminum Alloy Metal Matrix Composites, American Journal of Materials Science, **2(3)**, pp 53-61 (2012).

IOP Conf. Series: Journal of Physics: Conf. Series **914** (2017) 012040 doi:10.1088/1742-6596/914/1/012040

Technology Model of Aquaculture Production System

K.W. Hor, S.M Salleh[1,2], Abdullah, Mohd Ezree[3], Zaman, I. [1,2], M.H. Hatta[4], S. Ahmad[2], A.E Ismail[1,2], W.A.W Mahmud

[1]Mechanical Failure Prevention and Reliability (MPROVE), [2]Faculty of Mechanical and Manufacturing Engineering, [3]Faculty of Civil and Environment Engineering, [4]Faculty of Diploma Studies, Universiti Tun Hussein Onn Malaysia, Parit Raja, Batu Pahat, Johor, Malaysia

Corresponding author: saliha@uthm.edu.my

Abstract. The high market demand has led to the rapid growth in fish farming. The young generation are inexperienced in determining the estimated results of fish farming and the preparation of fish pond during the period of fish farming. These need a complete guide as their reference which includes the knowledge of fish farming. The main objective of this project is to develop a practical design of real pond appropriate with aquaculture technology and fish farming production. There are three parts of study in this project which include fish farming cage, growth of fish and water quality of fish farming pond. Few of experiments were carried out involved the collection data in terms of growth of fish and parameters of water quality.

1. Introduction

For the last thirty years, supply and demand for fish has been increased tremendously. As human population become larger, the consumption per capita of global demand for fish increased. This demand has met by a rapid growth in production and increased global trade. Aquaculture becomes the fastest growing component nowadays. Aquaculture expanded about 12 times of the world food fish production. Aquaculture supplied the world with 59.9 million tonnes of fish in 2010 and increased production of 63.6 million tonnes for 2011 indicate increased production [1]. As expected in the future, aquaculture production will overcome the supply of food. Factors affecting fisheries production are the preparation, management of fish ponds and fish breeding types. Therefore, monitoring of pool quality parameters and water quality parameters is required. Fish pond management which includes fish food size, water quality and fish growth. Comprehensive preparation before planting fish will help to get a good return.

This study aims to provide information on the preparation and management of real fish ponds. It will also estimates results in fish harvest over a certain period of time. Aquaculture production systems emphasize the methods of water quality and growth of fish in real cage. Practically caged fish farming is very easy and complex and it can be done if the farmers able to access native or artificial lakes. The usage of cage system is appropriate when school of fish share a good water quality and environment. However, most modern farmers from present generation have face problems in understanding the information of water quality, fish pond size and food distribution that had been conducted previously. All previous information can be obtained through experienced farmers and rarely documented. It will only slow the pace of fish research development which potentially attracting the new generation involved. This study provides guidance to farmers regarding pH, salinity, temperature, ammonia concentration and oxygen in water. The rate of growth and harvesting of fish can also be estimated from harvest time. This can help increase aquaculture production and make the aquaculture industry as one of vibrant industry in Malaysia.

2. Model Description

An aquaculture cage system was built with low cost material of PVC piping structure. It has a portable features with cage size of 6m x 6m x 1m. The size of 4" catfish had been released into the pond with approximately about 1500 fish. The parameter of water quality in fish pond need to be monitored to ensure no potential contamination occured due to feeding or from other factors. Thus, it is important to test the every parameters of the water like pH, salinity, temperature, concentration of oxygen and concentration of unionized ammonia. Consequently, these parameters of the water quality can provide health precaution for the fish. The final step is measure the growth of fish. The initial weight of fish, initial length of fish, initial number of fish and date of fish farming start were measured and recorded. Mortality of fish, length and weight were measured every week. The amount of fish food was controlled by automatic feeding machine. The machine operated with timer and feed the fish two times per day.

3. Methodology

Several steps used to develop of aquaculture cage system, control feeding system, investigate the parameter of water quality and growth rate of catfish.

3.1 Fish Feeding Machine

The microcontroller, PLC uses timer to control operating time at one presetting duration. Figure 1 (a) shows an automatic feeding machine with a maximum capacity of 6 kg. Other components and devices used in automatic fish feeding systems include toggle switches, emergency stop buttons, relays, warning lights, adjustable logical controllers (PLCs) and power supply units. However, the machine can be set to 3kg, 4kg, 5kg and 6kg for fish food distribution. Experiments were conducted to determine the quantity or weight of the fish pallets that would fall into the transmission plate when the motor was switched on. The total weight of the minimum discharged food is 3kg, but the weight of the pelleted food needed for catfish is about 500g. The electronic timer works to control the timing of the motor to remove the amount of food it deserves. The periods of 10s, 15s, and 20s were used. Each time period is repeated six times to get accurate results.

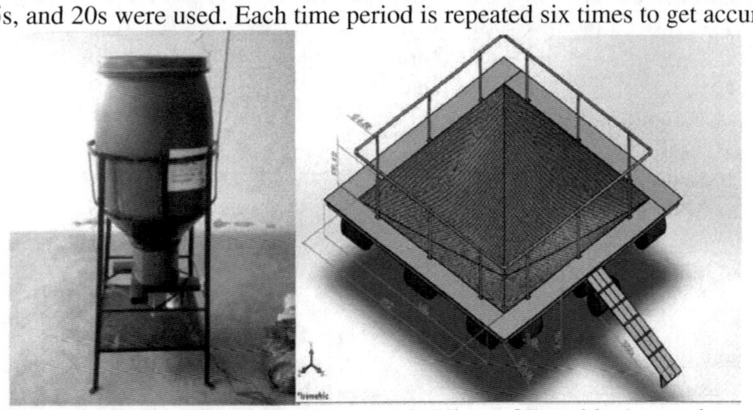

Figure 1: a) Automatic Feeding Machine b) Isometric View of Portable Aquaculture Cage System

3.2 Cage System

The pH and DO water levels were measured to ensure the appropriate water quality. Once an aquaculture cage is built, the feeding system and pump will be installed. Fish cage design is done using the sketch design and Solidwork as shown in Figure 1 (b). This cage system is left for at least two to four days to ensure there is no leakage, sufficient water flow, and no flaws exists.

3.3 Water Quality Parameters

The measurement of water quality include pH, dissolved oxygen, ammonia-hydrogen, salinity and temperature of the fish ponds as described below; and the instruments used as depicted in Figure 2.

i) pH : accurate result obtained by fully immerged the device in the middle water of cage
ii) Temperature: the thermometer time was allowed to come to equilibrium and read immediately.
iii) Salinity: to measure salinity, the protective cap was taken off and the meter immersed into the sample and swirled slowly
iv) Dissolved Oxygen (DO): The tip of probe was immersed into the water to make sure the velocity of the water coming into contact with the probe at least 0.2 - 0.3m/s.
v) Concentration of Unionized Ammonia: measurement of concentration of unionized ammonia in similar way to measure water salinity. The meter was immersed into the water and swirled slowly.

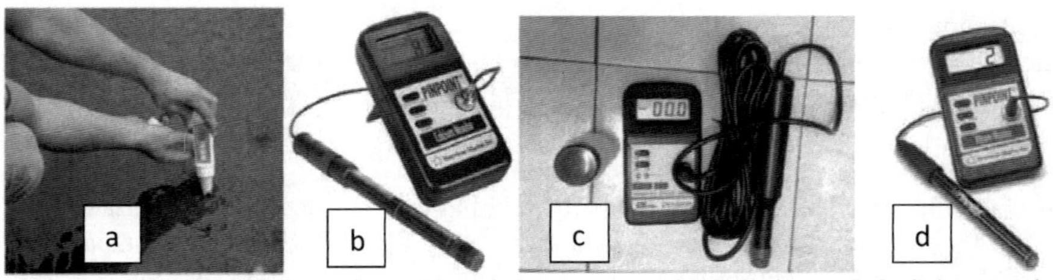

Figure 2: Instrumentation of a) pH and Temperature b) Salinity c) DO d)Ammonia

4. Results and Discussions

4.1 Experiment on weight of pelleted feed distribution

The results for this experiment were repeated six times each when motor operates for 10, 15 and 20 s. The average reading have been evaluated. From experiments conducted, it was found that the average weight of fish pallet released from motor for time period of 10s, 15s, and 20s are 331.67g, 504.17g and 660.83g, respectively. In the case of fish food distribution for 500g, the motor required about 15s to complete the operation. The cage system and fish feeder machine as shown in Figure 3. The time taken approximately to distribute food at different weight is shown in Table 1.

Table 1: Approximate time to distribute different weight of fish food.

Weight of Fish Food (g)	Approximate Time (s)
330	10
360	11
400	12
430	13
460	14
500	15
530	16

Figure 3 : On-site cage system

4.2 Growth rate of fish

In fish growth period for 11 weeks, the weight of the Patin fish has been recorded. The average weight for three fish was taken and recorded. Figure 4 shows detailed data of changes to the weight of the Patin fish. It also shows that the weight of the Patin fish increases in the line of curvature. The average length of three fish was taken and recorded. The measurement data were recorded and plotted as in Figure 4.The average length of the initial length is 6.3 cm whereas the final length is 22.7 cm. The average length of fish increased by 16.4 cm. This shows the change in length of fish increased by 1.5 cm a day. Figure 4 shows that the length of fish increased 1.5 cm per day. The result of length changes in eleven weeks shows that the fish grow is linearly increased.

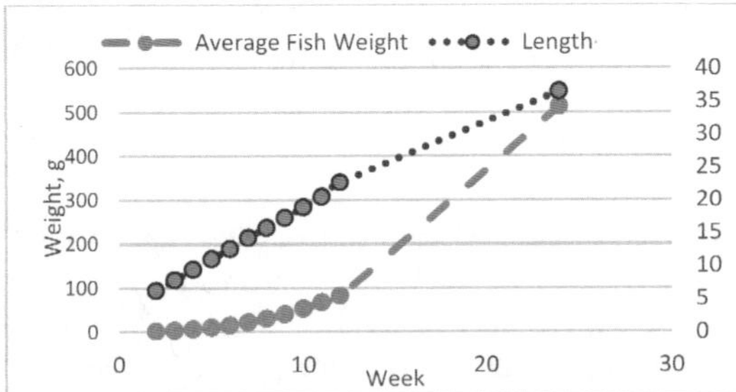

Figure 4 : Weight and length changes of Patin fish

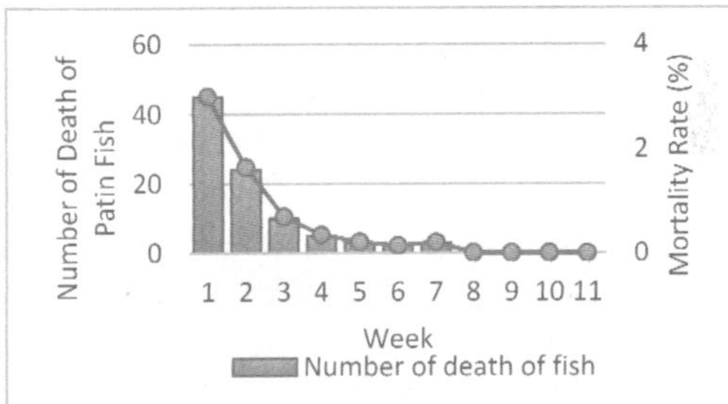

Figure 5: Mortality of Patin fish

For mortality, the number of death of Patin fish was recorded week by week. The mortality rate is calculated with the formula $M\% = 100\%/N * n_i$ is number of dead fish for day i, N is the number of fish for day i-1 and M is the extinction of fish every day.From Figure 5, it shows that the number of death of fish in week 1st was decreases compared with week 7th. In addition, there was no death of fish found in the continuous week until week 11th . The fish death is most probably due to phenomenon of "New Tank Syndrome". This is common name given when fish become ill or die in a new environment. Most of the fish cannot adapt in new environment with different temperature or concentration of dissolved oxygen. Besides, there were some predator like fish and bird near the cage which may cause the death of fish.

The feeding frequency was 2 times per day to ensure optimum growth performance of Patin fish. The time set of feeding was about 9.30am to 10.30am and 4.30pm to 5.30pm. However, 500 grams of pallete have been distributed inially and will changed week by week according to the number of fish remaining and the appetite of fish.

4.3 Parameters of water quality

The results for water quality include pH level, temperature, concentration of oxygen and concentration of unionized ammonia and water salinity. The pH level of water can provide information on potential contamination. Consequently, testing the pH of the water can be an important public health precaution for the fish. The result of pH of the water was from 6.89 to 7.10 during the period of the experiment. The average pH value for each week is 6.98. The growth rate of fish was the best at pH 7.5 – 8.0. For best Patin fish grow at pH 5.0 – 7.0, however the ideal pH of water is pH 6. The value of pH should be maintained in the range at 6.5 – 7.0 for the perfect cycle.

Meanwhile the measurement of water temperature is almost same which is in range from 29.8 °C to 30.3 °C. Therefore, the average temperature is 29.9 °C per week. The result shows that the temperature of the pond is suitable for the Patin fish as the optimum water temperature for the life of Patin fish between

28 °C – 32 °C. Patin fish prefer deep with low temperature fluctuations. Life was disrupted if the water temperature drops to 14 °C or up to 35 °C [6]. Figure 6 shows the measurement of pH level and its temperature during the period of fish farming.

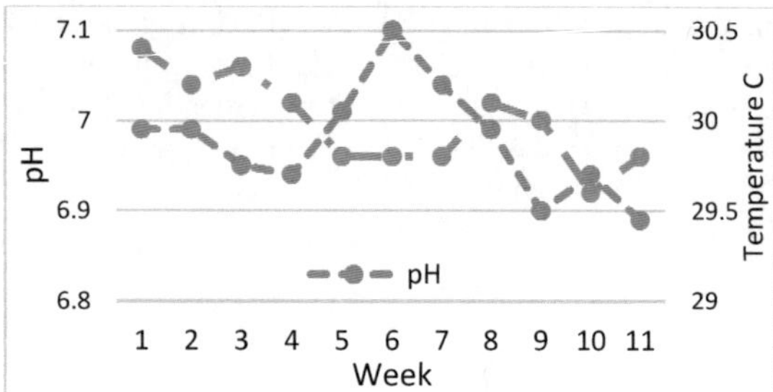

Figure 6: Measurement of temperature

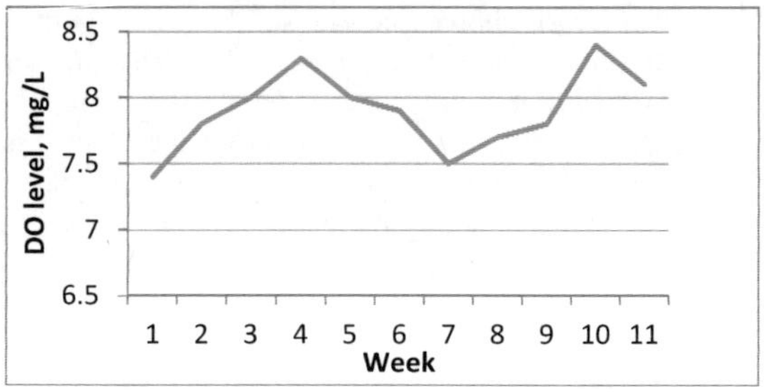

Figure 7: Concentration of dissolved oxygen

The concentration of dissolved oxygen obtained from the experiment is from 7.4 mg/L to 8.4 mg/L. The concentration of dissolved oxygen is at average of 7.9 mg/L per week. The most suitable range of concentration of dissolved oxygen for Patin fish is from 4 mg/L to 15 mg/L [7]. Figure 7 shows that the measurement of concentration of dissolved oxygen in the experiment during the period of fish farming. It was also found that salinity and unionized ammonia were measured at 0.1ppm and no value or zero reading respectively for every week. This showed that the condition of water is good and suitable for the Patin fish farming because the optimum water salinity for Patin fish is less than 4 ppt [7].

5. Conclusions

The project successfully develop a practical cage and determine the water quality with use of feeder machine to feed the fish. From the project, it had concluded that the cage size of 6m x 6m x 1m is directly proportional to the number of fish and size of fish. For water quality, the pH is between 5.0 – 7.0, the water salinity should be less than 4.0 ppt, the ideal temperature for Patin fish life is from 28 °C to 32 °C, the concentration of dissolved oxygen is from 7.4 mg/L to 8.4 mg/L and the concentration of unionized ammonia is less than 0.02 mg/L. These parameters will help to reduce the mortality of fish and help to increase the growth rate of the fish. The operational time of feeding machine is concluded with the weight of the fish pellets distribution.

6. Reference

[1] Nations, F. a. (2012). The State of World Fisheries and Aquaculture.
[2] Chua, T. E. (1979). *Site selection, structural design, construction, management, and production of floating cage culture systems in Malaysia.* Proceeding of the SEAFDEC/IDRC , 65-80.
[3] Nelson, J. S. (2006). *Fishes of the World.* John Wiley & Sons, Inc. , 120-122
[4] Nehemia, A. &. (2012). Length-Weight relationship and condition factor of fish grown in marine and fresh water ponds. *Agriculture and Biology Journal of North America* , 117-124.
[5] Wang, Y. &. (2006). Model and Software Development for Predicting Fish Growth in Trout Raceways. 130-135.
[6] Djarijah, N. (2001). Budidaya Jamur Tiram Putih.
[7] Rottmann, R. W. (1994). *Management of Water Quality for Fish.* University of Florida

7. Acknowledgement
We gratefully acknowledge the financial support from contract grant of University Tun Hussein Onn Malaysia Vot No. A053, U397 and Knowledge Transfer Program grant Vot No.1081.

Improvement of automatic fish feeder machine design

How Chui Wei, S.M Salleh[1], Abdullah, Mohd Ezree[3], Zaman, I. [1], M.H Hatta[4], Md Zain, B.A.[1], S. Mahzan[1], M.N.A. Rahman[2], W.A.W Mahmud,

[1]Mechanical Failure Prevention and Reliability (MPROVE), [2]Faculty of Mechanical and Manufacturing Engineering, [3]Faculty of Civil and Environment Engineering, [4]Faculty of Diploma Studies, Universiti Tun Hussein Onn Malaysia, Parit Raja, Batu Pahat, Johor, Malaysia

Corresponding author: saliha@uthm.edu.my

Abstract. Nation Plan of action for management of fishing is target to achieve an efficient, equitable and transparent management of fishing capacity in marine capture fisheries by 2018. However, several factors influence the fishery production and efficiency of marine system such as automatic fish feeder machine could be taken in consideration. Two latest fish feeder machines have been chosen as the reference for this study. Based on the observation, it has found that the both machine was made with heavy structure, low water and temperature resistance materials. This research's objective is to develop the automatic feeder machine to increase the efficiency of fish feeding. The experiment has conducted to testing the new design of machine. The new machine with maximum storage of 5 kg and functioning with two DC motors. This machine able to distribute 500 grams of pellets within 90 seconds and longest distance of 4.7 meter. The higher speed could reduce time needed and increase the distance as well. The minimum speed range for both motor is 110 and 120 with same full speed range of 255.

1. Introduction

Sufficient proteins containing all essential amino acids, essential fats, vitamins and minerals which are very important to gain a healthy balance diet. Seafood such as fish can be considered a source of these nutrition. The State of World Fisheries and Aquaculture 2016 reported that a significant growth in fisheries and aquaculture production in the past fifty years, has enhanced world's capacity to consume diversified and nutritious food [1]. In Malaysia, aquaculture industry contributed 268,500 tan (about 16 percent) to the national seafood supply, valued at about RM1.3 billion. The Department of Fisheries Malaysia has established the National Plan of Action for the Management of Fishing Capacity in Malaysia (NPOA Fishing Capacity) which was published in 2008. The long term objective of the Nation Plan is to achieve an efficient, equitable and transparent management of fishing capacity in marine capture fisheries by 2018 [2]. However, there are several factors influence the fishery production and efficiency of marine system such as the tool and machine use in fisheries management. According to the World Fish Centre, the global fish production was increased to 160 million tan in order to satisfy the high demand and the demand has growth to 238 million tan. Fish farming is one of the marine sectors to overcome the deficiency of supply. First automatic fish feeder machine was designed under intellectual pattern by Anthony Christopher Halford in 15 April 1999. Then, it had been improved by Chang et al. in 2004 by installing with programmable logic controller

(PLC) [3]. Others than that, some of the fish feeder machine was designed with solar energy such as Koisan automatic solar powered fish feeder and FIAP solar feeder.

2. Background Study

Generally, there are two types of fish which are saltwater fish and freshwater fish. Others than fishing at the sea side, fish farming is very popular in Malaysia likes fish farming with cage system, pond system, integrated recycling systems and integrated recycling systems. Fish farming usually spending higher cost, time consuming, higher human capital needed and the limited fish production as the sizes of cage system. Machinery could be taken into consideration to improve the efficiency and reduce the requirement of human capital. The machine that innovated is purposely used for fish farming especially for fish farming with cage system. There were various type of fish feeder machine had been developed.

2.1 Functionality, advantage and disadvantages of current machines

Two latest machine has been chosen as the benchmark and reference for this research. Both of the machines as shown in Figure 1(b) and (c) had been located at the real marine agriculture site to testing the functionality and suitability of machine. The functionality, advantage and disadvantages of current machines are describes as follow;

2.1.1 Machine A

Figure 1(a) and (b) show the machine fish feeder machine with solar energy and the processes of functioning the solar energy fish feeder machine respectively. Initially, the solar panel with 30 watts is used to collect the solar energy from sun and monitor to recharge battery through solar manager. Battery works as the storage of solar energy. After that, the timer for open the gate and fish feeder schedule is setting in the system. Once the timing for fish feeding has reach, the system will automatically functioning the motor and blower to drop out the fish food.

2.1.2 Machine B

Figure 1(c) shows the automatic fish feeder machine with Programmable Logic Controller (PLC). This machine was setting to 24 hour timer with direct electricity supply. Initially, the switch plug need to plug on to supplied the electricity to the control panel of the machine. The machine was tested with various quantity of fish pellets with minimum 3kg to the maximum 6kg was chosen. The determination of quantity need to be set based on the requirement. The motors stopped operation after specific time period which setting in the system depends on the quantity of pellets. There were 90 s need for 3kg fish pellets, 120 s for 4kg, 150 s for 5kg and 180 s for 6kg.

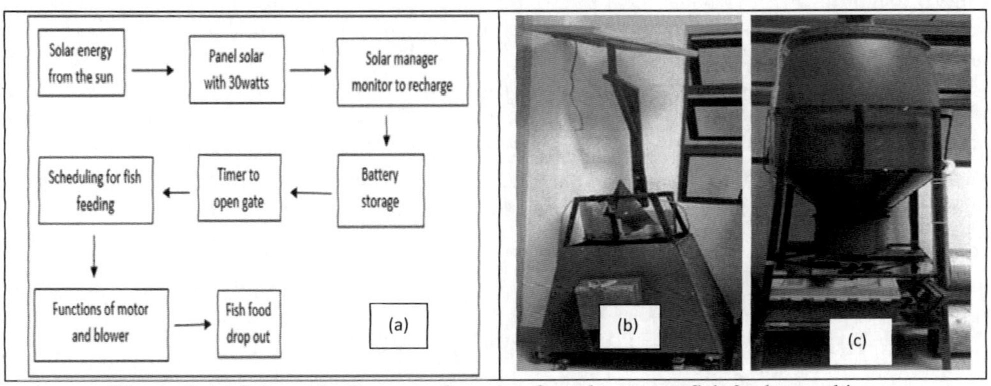

Figure 1. (a) Schematic diagram of process for solar energy fish feeder machine (b) Machine A with solar panel (c) Machine B with PIC controller

IOP Conf. Series: Journal of Physics: Conf. Series **914** (2017) 012041 doi:10.1088/1742-6596/914/1/012041

2.1.3 Advantage and Disadvantages

In machine A, the storage container was broken and expansion of fish food blocked the distribution due to hot weather temperature. The heavy solar panel of Machine A was lack of flexibility as it require to adjust and fix with specific angle in order it able to receive the solar energy. Machine A was installed with wheel set to increase the portability, but wheel setting was not suitable for outdoor marine system. The materials used for making the machine was lack of water and temperature resistance that lead to rusting and broken of fish food storage container. The machine body was very heavy and hard to mobilize the machine. Besides that, the fish food was expand due to hot weather and blocking the distribution pipe which only able to have one-point distribution.

Machine B was lack of portability with heavy machine body and height of machine was inconvenient for farmer to refill the fish pellets to the storage as well. For this, there is a need for improvement of machine design and materials of automatic fish feeder machine to have light weight, high water and temperature resistance and low cost. Based on the machine, the machine with direct electricity supply to the control panel make the machine inconvenient to use for outdoor fish farming activities. Besides that, the materials used also lack of water resistance with heavy components. But, the machine has better food distribution that multi-point and long distance location. The quantity of fish pellets was flexible and controllable as well. As overall observation, the solar energy and wireless battery supply make the machine convenient and suitable for outdoor activities of fish farming rather than indoor fish farming in tanks. However, there are much more issues needs to take in consideration such as temperature resistance when running outdoor fish farming activities. Overall, the existing machines has located and tested at the marine agriculture site. Based on the observation to the existing fish feeder machine, it has found that the machine A and B was made in heavy structure, low water and temperature resistance materials and non-uniform distribution of fish food. The listed advantages and disadvantages has shown in Table 1.

Table 1. Advantages and disadvantages of Machine A and B

Machine A		Machine B	
Advantages: • Far location food distribution • Controllable the portion of food	Disadvantages: • Heavy body component • Low water resistance • Wiring battery supply	Advantages: • Solar recharge • Wireless	Disadvantages : • Fish food blocking at the distributor • Low temperature resistance • One-point distribution • Low water resistance • Heavy body component

3. Methodology

3.1 Design Procedure

Design procedure of fish feeder machine include several steps to be carried out as shown in Figure 2. Brainstorming requires five main characteristics on the new machine, such as lightweight, good water and temperature resistant, weight resistant and low cost as well. After brainstorming, the new design of machine has sketching out especially the interior design of machine. Materials selection and cost analysis is related with each another. The chosen materials must be aligned with the desire characteristics of machine. Cost analysis has been study on every single materials before fabrication. One the fabrication is done, the characteristic and some functionality of machine is tested. The data and results is recorded and analyzed as well. Finally, the last step is recommendation and conclusion of the research.

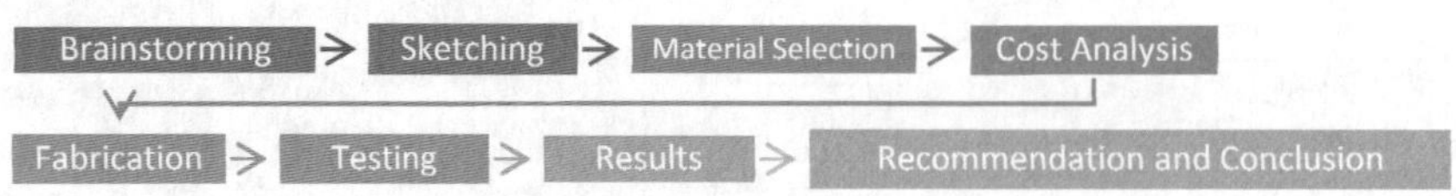

Figure 2. Design Procedure

IOP Conf. Series: Journal of Physics: Conf. Series **914** (2017) 012041 doi:10.1088/1742-6596/914/1/012041

3.2 Functionality experiment procedure
Four experiments have been conducted to test the ability of new automatic machine. These include storage, food drop time, speed range and distance of food dispersed.

3.2.1 Storage - The minimum speed rate for DC motor at upper part is 110 with 255 full speed. The motor will not function when the speed is setting under 110. At the same time, the gravity of machine also take in consideration when it fill with full tank of pellets.

3.2.2 Time needed for food drop -This experiment will started with lowest speed range. Then, recording the time taken until the food has drop all completely of the quantity pellets. The experiment is repeated with different speed rate.

3.2.3 Ability for food dispense - The time setting for both motor is based on the ratio 1:1.5 which 1 for motor 1 and 1.5 for motor 2. This experiment is testing with full storage tank of pellets. After setting all these, the motor was switched on and it will stop based on the time period that has been set. Then, measure the weight of pellets that has spread out from the PVC sockets. The experiment is repeated with different time period.

3.2.4 Distance of food dispersed - This experiment will started with lowest speed range of lower part motor. The motor will stop based on the time has set. Lastly, researcher need to choose three distance point of pellets spread out (most far, median and nearest) and measure the distance. This experiment is repeated with different speed range of lower part of motor.

4. Results and Discussion

4.1 Design improvement
Based on the Table 2, there are the solutions to eliminate the limitation of current machine. The PVC is chosen as it is lightweight, good in water and temperature resistance. Besides, PVC is low cost materials.

Table 2. Recommendation for machine improvement and materials

Limitation	Improvement	Potential Solution
Low temperature resistance	Using high temperature resistance materials	PVC pipe
Low water resistance	Used high water resistance materials	PVC pipe
heavy body component	Used light materials	PVC pipe
Fish food expand due to hot weather and blocking the distributor	Using high temperature resistance materials	PVC pipe
One-point distribution	Design more distribution with strong momentum motor to spread food	Four distribution with high speed motor
Lack of flexibility	Modify to be flexible	Use PVC pipe
High fabrication cost	Used low cost materials	Used rubber or PVC

The new machine made by PVC pipes and PVC socket which assistant to easily wash. The interior distribution divider also make with PE rubber which is good in water resistance. Besides that, new machine also design with four distributions rather than one-point distribution. The sketching of machine as shown in Figure 3. As it can be seen that the funnel is located inside the container and composite panel is used to modify the funnel in order to fit properly with the container. The machine has two parts, one for food dispenser and the other part for distributer. The rotating blade plate will turn and pushing the fish food out from the PVC socket.

IOP Conf. Series: Journal of Physics: Conf. Series **914** (2017) 012041 doi:10.1088/1742-6596/914/1/012041

The funnel is located at the upper part of machine with the food drop which locates a dispenser. A DC metal gear motor is installed at the side of the rubber container. The motor installed out of rubber container for maintenance purposes. The dispenser is place together with the funnel and PVC which under the funnel. After that, the largest PVC pipe will connected together with rubber container. For the lower part of machine, the distribution divider was placed at the bottom part of PVC socket to ensure the food distribute into four points. Part A is a storage space for fish pellet and part B is the food space before fish food has drop out from the food gate as shown in Figure 4. The funnel has made specially to lead down the food to the food gate. The container is connected with the PVC pipe and dispenser at the lower part. The DC metal gear motor is installed at the side of rubber container. The dispenser will rotate and drop down the food. Meanwhile, the part C is the distributions space that food pellet will spread out. The part D is the PVC pipe which play a role as the stand of the machine and locates the DC geared motor with U support clip.

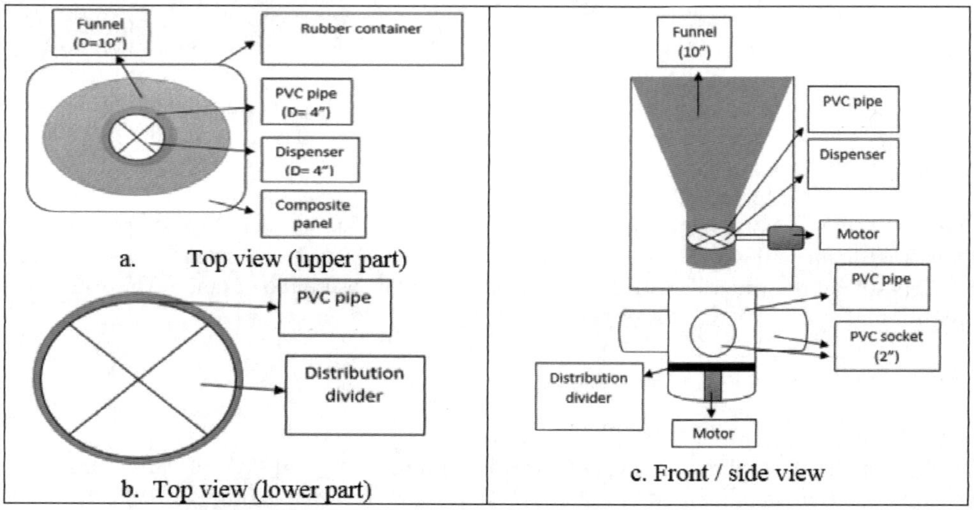

Figure 3. The sketch of current machine

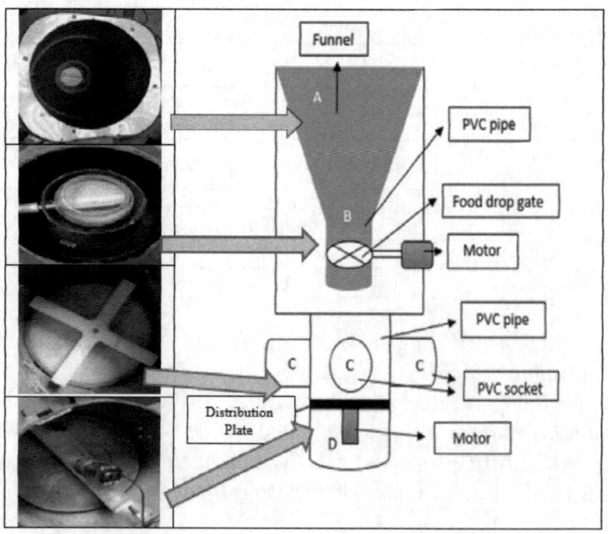

Figure 4. Current machine design

4.2 Experiment Result

4.2.1 Storage

Based on the experiment, the rubber container able to fill 5 kilograms of fish pellets. This quantity is the maximum amount for the container. The DC motor 1 and gravity of machine has been tested with the full tank quantity. The DC motor which using for upper part is DC metal gear motor with 15 RPM and 15 kg/cm torque. This motor able to functions when the storage is full of tank. Besides that, the RPM of motor is quite low but it has perfect match with time and speed of DC motor of lower part. The heavy storage of pellets or full tank of pellets not effect on the functionality of the machine.

4.2.2 Time needed for food drop

The experiment has been done to study the time needed with various speed rate of DC geared motor. The quantity amount of 600 grams of fish pellets has chosen to use for the experiment. Based on the results from Figure 5, 54.7s was required to drop out all 600g pellets with minimum speed rate. This motor was unable to function if it setting under 110 of rpm. As the speed increase, the time need is shorter. The shortest time is 25.14s with the speed 250 rpm. However, the full speed of motor is 255was used. The differences shows the time reduce is inconsistent.

4.2.3 Ability for food dispense

The experiment was proceed to test the ability of DC motor to distribute food at one time. Based on the results in Figure 6, there was 583g of fish pellets could spread out in 90s of DC motor and 60s of DC geared motor. According to the previous study, the Patin fish feeding schedule was twice a day and 500 grams for once time of feeding. For this, this machine could be very effective as it able to distribute more than 500 grams of pellets. DC geared was able to drop out 500 grams of food within one minute with lowest speed rate. For this, the motor is setting at 110 rpm for upper part and 255 rpm for lower part.

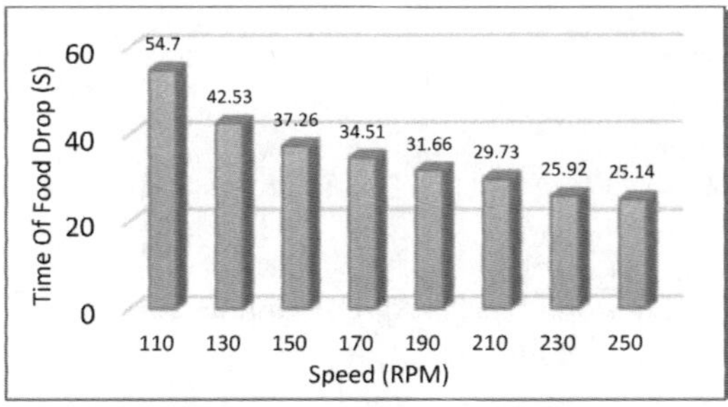

Figure 5. Time needed for food drop

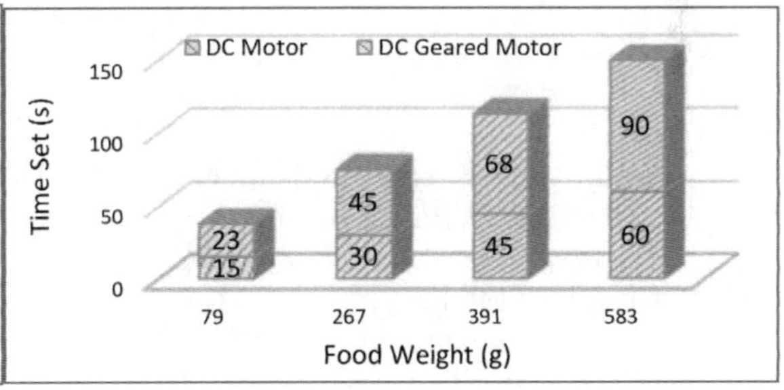

Figure 6. Time set for DC motors

4.2.4 Distance of food dispense

Another experiment was run to study the distance of food distribution. For this, the speed rate of upper part motor and quantity of food is fix. The time was setting around 5 seconds for upper motor and 8 seconds for lower motor. From Figure 7, the shortest distance is 0.12 meter, 1.09 m and 2.16 m with 150 rpm. The speed rate of 200 could distribute the pellets to 0.20 m, 2.26 m and 3.47 m. lastly, the speed rate 250 is able to distribute the pellets to distance 0.18 m, 2.28 m and 4.7 m. The full speed of motor at lower part is 255, the distance for distribution could be higher than 4.7 m when it function with full speed.

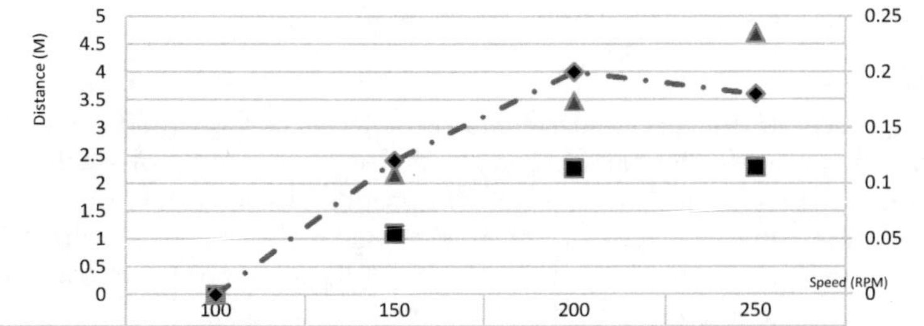

	100	150	200	250
Median	0	1.09	2.26	2.28
Longest	0	2.16	3.47	4.7
Nearest	0	0.12	0.2	0.18

Figure 7. Distance of food dispense

5. Conclusion

This research was conducted to study and improve the current automatic fish feeder machine with better materials and low cost. The materials selected must be good in water and temperature resistance. Besides that, the machine should design with smaller physical body to make it portable for outdoor fish farming. However, this machine storage is smaller than the previous machine as to make it portable. Based on the results and new design of this machine, most part of this machine is fabricate by using PVC rubber and PVC pipe. PVC rubber is very good in temperature and water resistance with low cost as well. There were two DC motors is used to functioning the machine. The functionality of food drop from food gate is faster than the food distribution at the lower PVC part. Matching the time period to match both motors is very important. Besides that, the maximum quantity for lower part PVC pipe is 500 grams. The food should not drop more than 500 grams as the quantity will effect to the functionality of food distribution. For complete cycle of distributions, the speed range of DC motor at upper part is 110rpm and 255rpm for motor at lower part. The machine could distribute 500 grams of fish pellets within 90s.

References

[1] To, C., Security, F., & All, N. F. O. R. (2016). *THE STATE OF WORLD FISHERIES AND AQUACULTURE.*
[2] *National Plan of Action For Management of Fishing*. (2015.).
[3] Chang, C. M., Fang, W., Jao, R. C., Shyu, C. Z., & Liao, I. C. (2005). Development of an intelligent feeding controller for indoor intensive culturing of eel, *32*, 343–353. http://doi.org/10.1016/j.aquaeng.2004.07.004
[4] Yeoh, S. J., Taip, F. S., Endan, J., Talib, R. A., & Mazlina, M. K. S. (2010). Development of Automatic Feeding Machine for Aquaculture Industry, *18*(July 2008), 105–110.
[5] Weiss, P. E. G. (1993). United States Patent 1191.
[6] Feeder, F. F., & Documents, F. P. (1983). United States Patent 1191.
[7] Moore, E. M. (1970). itlI ’

Acknowledgement

We gratefully acknowledge the financial support from contract grant of University Tun Hussein Onn Malaysia Vot No. A053, U397 and Knowledge Transfer Program grant Vot No.1081

Modelling and Control of Robotic Leg as Assistive Device

Yee Jingye[1,a], Badrul Aisham bin Md Zain[1,b]

[1]Faculty of Mechanical and Manufacturing Engineering, Universiti Tun Hussein Onn Malaysia, 86400 Parit Raja, Johor, Malaysia.

Email: yeejingye@gmail.com[a], aisham@uthm.edu.my[b]

Abstract. The ageing population (people older than 60 years old) is expected to constitute 21.8% of global population by year 2050. When human ages, bodily function including locomotors will deteriorate. Besides, there are hundreds of thousands of victims who suffer from multiple health conditions worldwide that leads to gait impairment. A promising solution will be the lower limb powered-exoskeleton. This study is to be a start-up platform to design a lower limb powered-exoskeleton for a normal Malaysian male, by designing and simulating the dynamic model of a 2-link robotic leg to observe its behaviour under different input conditions with and without a PID controller. Simulink in MATLAB software is used as the dynamic modelling and simulation software for this study. It is observed that the 2-links robotic leg behaved differently under different input conditions, and perform the best when it is constrained and controlled by PID controller. Simulink model is formed as a foundation for the upcoming researches and can be modified and utilised by the future researchers.

1. Introduction

Due to development of human civilisation, improvement in the area of nutrition, medical advancement, health care, sanitation, education and economic well-being people are able to live longer than ever[1], resulting in the increasing number of ageing population with the ageing symptoms on them which could reduce their mobility. On top of that, various health conditions such as multiple sclerosis, cerebral palsy, stroke or injury of spinal cord could also lead to acute or chronic impairment in human's locomotors. The wheelchairs or wheeled type robots would face difficulty climbing an obstacle with the height more than half of the diameter of the wheel, yet a leg-type robot would be able to do so easily[2]. Hence, a promising solution for this problem will be a robotic assistive device for the lower limbs. The main function of a lower-extremity exoskeleton in locomotion assistance is to provide external torque at the respective positions of human joints to replace the deficient motor function of the patients[3].

This study is to provide a start-up platform for the basic understanding towards the dynamic modelling of a lower limb powered-exoskeleton. By developing the dynamic model of the proposed lower limb powered-exoskeleton, the coordinates of the different critical points of the link could be determined so that they could be managed and controlled correctly. This study is to provide the elderlies, patients, and victims who are suffering from the gait impairment with a new hope of standing up again, which in turn will benefit the overall population of the earth.

A group of researchers from Sun Yat-sen University came out with a mathematical model for their 3-links exoskeleton by using Euler-Lagrangian approach. They incorporated the kinematic model, together with friction model and motor model into their dynamic model. The kinematic model is about the exoskeleton and ideal joint torque, the friction model describes the compensation for the friction in the servo-system, and the motor model presents the conversion of electrical energy to kinetic energy[4].

The existing powered exoskeletons employs different control strategies. Berkeley Lower Extremity Exoskeleton (BLEEX) employs a mixture of position control for stance-leg and positive feedback based sensitivity amplification controller for the swing leg. Lower Extremity Powered Exoskeleton (LOPES) has 3 different modes of choice, namely "patient-in-charge" mode, "robot-in-charge" mode, and "therapist-in-charge" mode. For Active Leg Exoskeleton (ALEX), 2 force-torque sensors are mounted at the shank segment and thigh segment respectively[3]. The force-field controller employed receives the feedback of force and torque to assist-as-needed. MoonWalker uses 3 pressure sensors under each foot. As for KIT-EXO-1, it utilises the interaction forces acting in between the user's body and the exoskeleton as the input for the controller, which is used to predict the upcoming motion. The forces are collected using sensors attached at different part of the exoskeleton, especially at the knee joints[5].

Proportional-integral-derivative (PID) controllers are said to be the most adopted controllers in industrial settings due to its relative ease of use and the satisfactory performance in different processes. A PID controller consists of the 3 different control actions, the proportional part, the integral part, and the derivative part. PID controller has to be tuned to provide the best optimised performance for a mechanical control system either by traditional rule-based tuning methods or automated tuning.

2. Mathematical model, Simulink model, and parameters

2.1 Mathematical Model

2.1.1 *Robotic leg.* Lagrange equation is used to derive the equations of the model and the equation of kinetic and potential energy are substituted in. Torque of Motor 1, T_1 and Torque of Motor 2, T_2 will be

$$T_1 = A\ddot{\theta}_1 + B\ddot{\theta}_2 - D\dot{\theta}_2^2 - 2D\dot{\theta}_1\dot{\theta}_2 + G_1 \quad (1)$$
$$T_2 = C\ddot{\theta}_2 - B\ddot{\theta}_1 + D\dot{\theta}_1^2 + G_2 \quad (2)$$

where:

$$A = m_1 l_{c1}^2 + I_1 + m_2\left(l_1^2 + l_{c2}^2 + 2l_1 l_2 \cos\theta_2\right) + I_2 \quad (3)$$
$$B = m_2\left(l_{c2}^2 + l_1 l_{c2}\cos\theta_2\right) + I_2 \quad (4)$$
$$C = m_2 l_{c2}^2 + I_2 \quad (5)$$
$$D = m_2 l_1 l_{c2}\sin\theta_2 \quad (6)$$
$$G_1 = m_1 g l_{c1}\sin\theta_1 + m_2 g\left[l_1\sin\theta_1 + l_{c2}\sin(\theta_1 + \theta_2)\right] \quad (7)$$
$$G_2 = m_2 g l_{c2}\sin(\theta_1 + \theta_2) \quad (8)$$

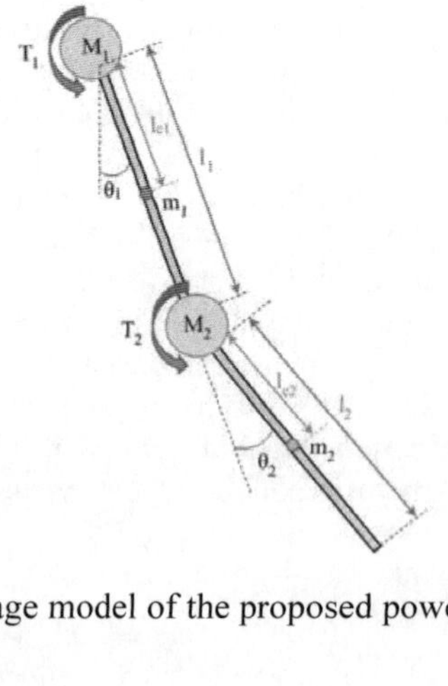

Figure-1: 2-linkage model of the proposed powered-exoskeleton

IOP Conf. Series: Journal of Physics: Conf. Series **914** (2017) 012042 doi:10.1088/1742-6596/914/1/012042

2.1.2 *Robotic Leg Dimension.* The average height of Malaysian male is 165 cm[6]. From a height estimation formulae list[7], the normal length of a femur, tibia, and fibula can be estimated by reverse calculation. Thus, for the height of 165cm, the length of femur is 44.01 cm, while the length for tibia and fibula is 35.76 cm and 35.85 cm respectively. The dimension are adapted as followed:

Table-1: Dimension of robotic leg

Link	Length [cm]	Diameter [cm]
1 (Femur)	44	2.2
2 (Tibia+Fibula)	36	1.8

2.1.3 DC motor. From the study of previous researchers[4], the DC motor can be modelled accordingly. The output torque of DC motor is proportional to the current. The voltage of the DC motor could be obtained from differentiating the output torque with respect to time and incorporating Kirchhoff law. Combining the output torque of DC motor with the voltage of the DC motor, the equation of DC motor will be:

$$T_m = \frac{K_m V_m - \frac{d(T_m)}{dt} L_m - K_m K_v \omega}{R} \tag{9}$$

where T_m = *DC motor output torque*
V_m = *DC motor voltage*
K_v = *Back-EMF constant*
R = *DC motor armature's resistance*
K_m = *DC motor torque sensitivity*
L_m = *DC motor armature's inductance*
ω = *DC motor output angular velocity*

2.2 Simulink Model of Subsystems

The robotic leg, DC motor, and the system are modelled in Simulink according to equation (1) to (11). The overall Simulink model of the robotic leg is shown in Figure 2.

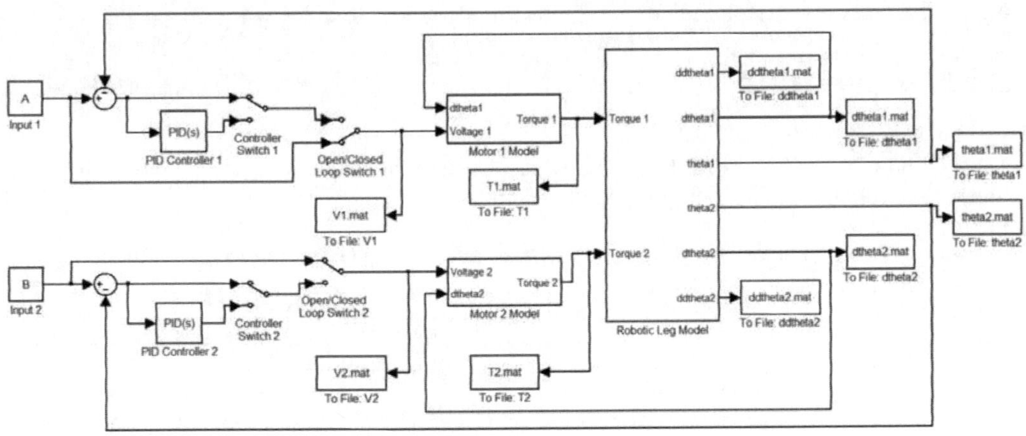

Figure-2: Full Simulink model of the robotic leg

2.3 Parameters of simulation

The parameters for the simulation condition is shown in Table 2.

Table-2: Parameters of the variables for simulation

Variable	Value	Variable	Value
Mass 1, m_1	0.46832 [kg]	Torque sensitivity of motor, K_m	4.32 [N m/A]
Mass 2, m_2	0.2565 [kg]	Armature's inductance of motor, L_m	0.0025 [H]
Length of link 1, l_1	0.44 [m]	Back-EMF constant, K_v	0.287 [v/rad s^{-1}]
Length of link 2, l_2	0.36 [m]	Armature's resistance of motor, R	4 [Ω]
Gravitational acceleration, g	9.81 [m/s^2]	Torque sensitivity of motor, K_m	4.32 [N m/A]

3. Results

3.1 Simulation Conditions

Table-3 and Table-4 shows the conditions for the simulation of the cases.

Table-3: Cases of applying constraint to one of the links

Case	Input A		Input B	
A	Step input at t = 1s, Open-loop	No controller	Unconstrained	
B	Step input at t = 1s, Closed-loop	No controller	Unconstrained	
C	Step input at t = 1s, Closed-loop	PID controller	Unconstrained	
D	Step input at t = 1s, Open-loop	No controller	Fixed to 0° using PID controller	
E	Step input at t = 1s, Closed-loop	No controller	Fixed to 0° using PID controller	
F	Step input at t = 1s, Closed-loop	PID controller	Fixed to 0° using PID controller	
G	Unconstrained		Step input at t = 1s, Open-loop	No controller
H	Unconstrained		Step input at t = 1s, Closed-loop	No controller
I	Unconstrained		Step input at t = 1s, Closed-loop	PID controller
J	Fixed to 0° using PID controller		Step input at t = 1s, Open-loop	No controller
K	Fixed to 0° using PID controller		Step input at t = 1s, Closed-loop	No controller
L	Fixed to 0° using PID controller		Step input at t = 1s, Closed-loop	PID controller

Table-4: Cases of applying constraint to both links

Case	Input A	Input B
M	Ramp input Slope = 1, Closed-loop, PID-controlled	Ramp input Slope = -1, Closed-loop, PID-controlled
N	Sine wave input Amplitude = $\frac{\pi}{3}$, Frequency = 1 rad/s, Closed-loop, PID-controlled	Sine wave input Amplitude = -$\frac{\pi}{2}$, Frequency = 1 rad/s, Closed-loop, PID-controlled

3.2 Simulation Results

Table-5 shows the overview of the simulated results from the cases of applying constraint to one of the links of the robotic legs with the supplied input and controlled input.

Table-5: Overview of results for Cases of applying constraint to one of the links

Case	Link	Final position	Steady-state error
A	1	0.5633 rad	No comparison is available for an open-loop system
B	1	0.3588 rad	Yes
C	1	$\frac{\pi}{3}$ rad	No
D	1	0.4555 rad	No comparison is available for an open-loop system
E	1	0.3134 rad	Yes
F	1	$\frac{\pi}{3}$ rad	No
G	2	Rotate clockwise continuously.	No comparison is available for an open-loop system
H	2	-1.2118 rad	Yes
I	2	-$\frac{\pi}{2}$ rad	No
J	2	Rotate clockwise continuously.	No comparison is available for an open-loop system
K	2	-1.1826 rad (67.76°)	Yes
L	2	-$\frac{\pi}{2}$ rad	No

In open-loop system, the links are uncontrolled and differ with different inputs. If the voltage supplied is large enough to overcome the gravitational acceleration, the link will keeps rotating. For a closed-loop system, the link will eventually come to an equilibrium position but with steady-state error. In a closed-loop system and being controlled by PID controller, the link will eventually stop at the desired angle and no steady-state error is observed. Though there are some overshoot for the cases, but with proper tuning and refining of the PID controller, the problem can be reduced and minimised. Table-6 shows the simulated results from the cases of applying constraint to both links of the robotic legs.

It is observed that different types input will lead to different movement pattern of the robotic leg. When the desired angular displacement is having a ramp input, the link will rotate at a constant angular velocity defined by the magnitude of the slope of the ramp. As for sine wave input, the link will rotate to a peak angular displacement defined by the amplitude of the sine wave, and the frequency of the oscillation is according to the frequency of the sine wave input.

Table-6: Overview of results for Cases of applying constraint to both links

Case	Link	Kinematics motion
M	1	Rotate with constant magnitude of angular velocity as specified by the inputs.
	2	
N	1	Rotate to peak of angular displacement according to the specified amplitude of sine wave inputs, with constant frequency as specified by the sine wave inputs
	2	

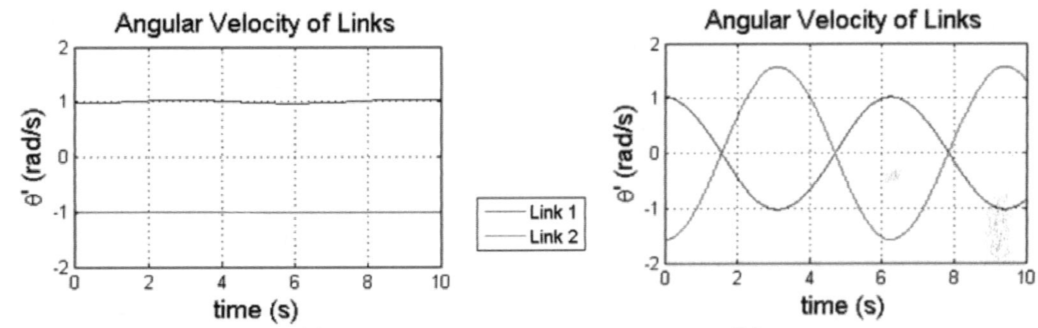

(a) (b)

Figure-3: Angular velocity of (a)Case M and (b)Case N

4. Conclusion

This study is a start up for future development of more functionable and workable powered-exoskeleton in Malaysia. Lagrangian equation is used to form the mathematical model of the 2-links robotic leg and the mathematical equation is successfully converted and transformed into the robotic leg model in the Simulink component in MATLAB software.

The simulation results show that with the application of the PID controller, the robotic leg is able to move as dictated by the desired parameters by controlling the voltage supplied to the DC motors of the robotic leg. It is desired that this platform can spearhead a future development on this topic, which can be very much beneficial to the generations to come, and to provide the underprivileged to undergo the normal human life just like the rest.

Acknowledgement

We gratefully acknowledge the financial support of the Office for Research, Innovation, Commercialization and Consultancy Management (ORICC) of UTHM under Geran Penyelidikan Pascasiswazah (GPPS) of Grant Vot No. U804.

IOP Conf. Series: Journal of Physics: Conf. Series **914** (2017) 012042 doi:10.1088/1742-6596/914/1/012042

References

[1] UNFPA and HelpAge International, "Ageing in the Twenty-First Century:A Celebration and AChallenge," 2012.

[2] B. Kwak and J. Bae, "Design and analysis of a rotational leg-type miniature robot with an actuated middle joint and a tail (RoMiRAMT)," *IEEE Int. Conf. Intell. Robot. Syst.*, vol. 2015–Decem, pp. 2148–2153, 2015.

[3] B. Chen, H. Ma, L.-Y. Qin, F. Gao, K.-M. Chan, S.-W. Law, L. Qin, and W.-H. Liao, "Recent developments and challenges of lower extremity exoskeletons," *J. Orthop. Transl.*, vol. 5, no. October, pp. 1–12, 2015.

[4] J. Wu, J. Gao, R. Song, R. Li, Y. Li, and L. Jiang, "The design and control of a 3DOF lower limb rehabilitation robot," *Mechatronics*, vol. 33, pp. 13–22, 2016.

[5] J. Beil, G. Perner, and T. Asfour, "Design and Control of the Lower Limb Exoskeleton KIT-EXO-1," in *International Conference on Rehabilitation Robotics (ICORR)*, 2015, pp. 119--124.

[6] S. Tang, "Singaporeans Are Tallest In ASEAN," *Business Insider Singapore*, 2014. [Online]. Available: http://www.businessinsider.sg/snapshot-singaporeans-are-tallest-in-asean/#4AXPFiL6cwhJlv6g.97. [Accessed: 29-Mar-2016].

[7] A. J. Bertino and P. Bertino, *Forensic Science: Fundamentals and Investigations*, 2nd ed. Boston: Cengage Learning, 2015.

IOP Conf. Series: Journal of Physics: Conf. Series **914** (2017) 012043 doi:10.1088/1742-6596/914/1/012043

Stress Intensity Factors of Slanted Cracks in Bi-Material Plates

Al Emran Ismail*, Kamarul Azhar Kamarudin, Nik Hisyamudin Muhd Nor

Faculty of Mechanical & Manufacturing Engineering, Universiti Tun Hussein Onn Malaysia, Batu Pahat, 86400 Johor, Malaysia.

*Corresponding email: emran@uthm.edu.my / al_emran@hotmail.com

Abstract. In this study, the stress intensity factors (SIF) of slanted cracks in bi-material plates subjected to mode I loading is numerically solved. Based on the literature survey, tremendous amount of research works are available studying the normal cracks in both similar and dissimilar plates. However, lack of SIF behavior for slanted cracks especially when it is embedded in bi-material plates. The slanted cracks are then modelled numerically using ANSYS finite element program. Two plates of different in mechanical properties are firmly bonded obliquely and then slanted edge cracks are introduced at the lower inclined edge. Iso-parametric singular element is used to model the crack tip and the SIF is determined which is based on the domain integral method. Three mechanical mismatched and four slanted angles are used to model the cracks. According to the present results, the effects of mechanical mismatch on the SIF for normal cracks are not significant. However, it is played an important role when slanted angles are introduced. It is suggested that higher SIF can be obtained when the cracks are inclined comparing with the normal cracks. Consequently, accelerating the crack growth at the interface between two distinct materials.

1. Introduction

The evaluation of stress intensity factor (SIF) to determine the structural reliability and durability is an important aspect in modern engineering world. There are many methods used to characterize the components or materials containing cracks such as J-integral, COD, CTOD, SIF and others [1-5]. For linear elastic materials, SIF is widely and successfully used. Generally, in order to determine the SIF, there are two groups of methods used such displacement extrapolation and the energy release rate methods. J-integral is also used and then it is converted to SIF. Another type of works can also be found in analyzing the cracks using SIFs [6-8].

For the crack positioned between two different materials, the displacement extrapolation method is difficult to use. Then, energy based method is frequently used. Matsumto et al. [9] used an interaction energy release rate based on the boundary element method (BEM) to compute the SIF for the interfacial cracks. They found that the proposed approach is applied to several cracks in dissimilar media and the results are compared with those obtained by the conventional approach based on the extrapolation method. The analytical displacement and stress solutions for an interface crack between two infinite dissimilar media subjected to uniform stresses at infinity are used to give the auxiliary field, in which the values of the stress intensity factors are known. It is demonstrated that the present method can give accurate results for the stress intensity factors of various bi-material interface cracks under coarse mesh discretization.

Ikeda et al. [10] developed the mixed mode fracture criterion in order to characterize the crack between dissimilar materials. Various types of crack geometries are experimentally tested. They found that the developed criterion is successfully characterized by the SIF determined at the interface cracks. Noda and Xu [11] investigated the controlling parameter affecting the SIF for a planar interfacial crack. The problem is formulated as a system of singular integral equations on the basis of the body force method. In the numerical analysis, unknown body force densities are approximated by the products of the fundamental density functions and power series, where the fundamental density functions are chosen to express singular behavior along the crack front of the interface crack exactly. The numerical results show that this numerical technique is successful, and the boundary conditions are satisfied precisely.

Noda [12] presented the SIF formulas for three-dimensional crack in homogenous and bonded dissimilar materials. The generalized stress intensity factors at the interface are expressed as a function of Dundurs parameters α and β. Proposed formulas are usefully evaluating defects with any aspect ratio under any combinations of the materials. Lan et al. [13] numerically studied the effect of material combinations and relative crack size to the SIFs at the crack tip of a bi-material bonded strip. The variations of stress intensity factors of a bi-material bonded strip are discussed with varying the relative crack size and material combinations. This investigation may contribute to a better understanding of the initiation and propagation of the interfacial cracks.

Therefore, this work presents the stress intensity factors when the cracks are inclined obliquely with the presence of mechanical mismatches. The preliminary works have been conducted [16, 17] and analyzed where the cracks are aligned eccentrically and on the multiple crack interactions in dissimilar materials. There are four slanted angles are used including the normal cracks and three types of mechanical mismatches are applied. The goal is to examine the SIFs response when these two important parameters are varied.

2. Calculating stress intensity factors

The evaluation of J-integral around the crack tip is based on the domain integral method initially introduced by Shih et al. [14]. This integral formulation uses area integration for 2D and volume integration for 3D problems which offer much better accuracy than contour integral and it is also much easier to implement numerically. Eq. (1) represents the 2D domain J-integral taking accounts the absence of thermal strain, path dependent plastic strain, body forces occur within the integration area [2].

$$J = \int A \left[\sigma_{ij} \frac{\partial u_j}{\partial x_1} - w\delta_{1i} \right] \frac{\partial q}{\partial x_i} dA \tag{1}$$

where qij is the stress tensor, uj is the displacement vector, w is the strain energy density, δij is the Kronecker delta, xi is the coordinate axis and q is referred to as the crack extension vector. The direction of q is similar with x-axis of the local coordinate specified at the crack tip and it is normally chosen as zero at nodes along the contour, Γ. It is also a unit vector for all nodes inside Γ except the mid-side nodes and known as virtual crack extension nodes. For 3D problems, the principal is similar to the 2D problems. However, domain integral representation of the J-integral becomes volume integration [15].

Two approaches for calculating stress intensity factor (SIF) are available in ANSYS software, interaction integral and displacement extrapolation methods. The first method is used because much easier to implement numerically and it is also offers better accuracy and fewer mesh requirement. This method is similar to the domain integral method for J-integral evaluation describe previously. The discussion on the domain integral methods can be found elsewhere [15]. The interaction integral is defined as in Eq. (2).

$$I=-\int_V q_{ij}\left[\sigma_{k,l}\varepsilon_{k,l}^{aux}\delta_{i,j}-\sigma_{k,j}^{aux}u_{k,i}-\sigma_{k,j}u_{k,i}^{aux}\right]dv \Big/ \int_s \delta q_n ds \tag{2}$$

where σ_{ij}, ε_{ij} and u_i are the stress, strain and displacement and $\sigma_{k,j}^{aux}$, $\varepsilon_{k,l}^{aux}$ and $u_{k,i}^{aux}$ are the stress, strain and displacement of the auxiliary field and q_i is the crack extension vector.

$$\sigma_{ij}^{aux}=\frac{K_I^{aux}}{\sqrt{2\pi r}}f_{ij}^{I}(\theta)+\frac{K_{II}^{aux}}{\sqrt{2\pi r}}f_{ij}^{II}(\theta)+\frac{K_{III}^{aux}}{\sqrt{2\pi r}}f_{ij}^{III}(\theta) \tag{3}$$

$$u_{ij}^{aux}=\frac{K_I^{aux}}{2\mu}\sqrt{\frac{r}{2\pi}}g_j^{I}(\theta,v)+\frac{K_{II}^{aux}}{2\mu}\sqrt{\frac{r}{2\pi}}g_j^{II}(\theta,v)+\frac{2K_{III}^{aux}}{\mu}\sqrt{\frac{r}{2\pi}}g_j^{III}(\theta,v) \tag{4}$$

$$\varepsilon_{i,j}^{aux}=\frac{1}{2}\left(u_{i,j}^{aux}+u_{j,i}^{aux}\right) \tag{5}$$

An expression for the energy release rate in terms of mixed-mode SIFs is defined as for plane strain condition as in Eqs. (6)-(8).

$$J=\frac{\left(K_I^2+K_{II}^2\right)\left(1-\upsilon^2\right)}{E}+\frac{K_{III}^2\left(1+\upsilon\right)}{E} \tag{6}$$

$$J=\frac{\left[\left(K_I+K_I^{aux}\right)^2+\left(K_{II}+K_{II}^{aux}\right)^2\right]\left(1-\upsilon^2\right)}{E}+\frac{\left(K_{III}^2+K_{III}^{aux}\right)\left(1+\upsilon\right)}{E} \tag{7}$$

$$J=J+J^{aux}+I \tag{8}$$

The interaction integral can be associated with the SIFs as Eq. (9)

$$I=\frac{2(1-\upsilon^2)}{E}\left(K_I K_I^{aux}+K_{II}K_{II}^{aux}\right)+\frac{1}{\mu}K_{III}K_{III}^{aux} \tag{9}$$

By setting $K_{II}^{aux}=1$ and $K_I^{aux}=K_{III}^{aux}=0$,

$$K_I=\frac{E}{2(1-\upsilon^2)}I \tag{10}$$

By setting $K_{II}^{aux}=1$ and $K_I^{aux}=K_{III}^{aux}=0$ and $K_{III}^{aux}=1$ and $K_I^{aux}=K_{II}^{aux}=0$ leads to the relationship between modes II and III SIFs with I, respectively.

$$K_{II}=\frac{E}{2(1-\upsilon^2)}I \tag{11}$$

$$K_{III}=\mu I \tag{12}$$

where, K_i is a stress intensity factor with i is a loading mode, i = 1, 2 and 3. J-integral can be represented as J and I is a interaction domain integral. While, E and μ is a modulus of elasticity and modulus of rigidity, respectively.

3. **Methodology**

Two inclined ends of plates of different mechanical properties are perfectly bonded. At a lower end of connection, there is an edge crack with a length of a as shown in Figure 1(a). The edge point of the crack can be represented as a middle point of the total length of 2L. The crack angles θ are varied in the range of 00 to 400 and the crack length of also varied. The relative crack length, a/L is selected in the range of 0.1 to 0.7. In order to investigate the influence of mechanical mismatch on the stress intensity factors (SIF), three different types of relative mechanical mismatches are used such as 1.0, 0.5 and 5.0.

In this work, ANSYS finite element program is used to model and solve the crack problems. It is also assumed that the plate is fulfilled the plain strain condition. The constructions of the finite element model (FEM) started with the coordinate identifications of each point. Once all points are identified, all points are properly connected to form the plate with a crack. Then, the connected lines are converted into an area or domain. Two areas are used, at the upper and lower regions of the inclined crack. It is also used since this work dedicated to study the effect of mechanical mismatch on the SIF. Before the model is meshed, the crack tip is firstly identified. It is important since at this point, an iso-parametric singular element is used and on the other hand, the plate is modelled with irregular elements.

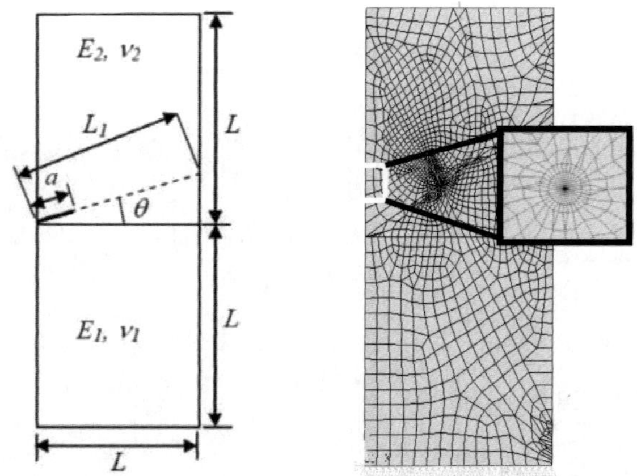

(a) (b)

Figure 1. (a) Schematic diagram of the plate with a crack and (b) FE model with an enlarged area of the crack tip.

The FEM is shown in Figure 1(b) with an enlarged area of the crack tip. The lower end of the plate is constrained in y-direction and in order to prevent the body rotation, at a bottom lower left end point is also constrained. The mode I loading is then applied at the upper end of the plate. In this paper, J-integral based on the domain integral method is used. Then, the calculated J-integral is converted into SIFs using the interaction integral formulation. This technique is used instead of the traditional technique which is the displacement extrapolation method due to ease of use. In order to generalize the values of SIFs, it is converted into normalized values or it is also known as the geometrical correction factor, F as in Eqs. (13) and (14):

$$F_I = \frac{K_I}{\sigma\sqrt{\pi a}} \tag{13}$$

$$F_{II} = \frac{K_{II}}{\sigma\sqrt{\pi a}}. \tag{14}$$

where, F is geometrical correction factor, K is a stress intensity factor, σ is a applied stress and a is a crack depth. Before further analysis is carried out, the present model must be validated with the existing model. Since it is hard to find the behavior of inclined crack in mechanical mismatch condition, then the crack in a similar mechanical properties condition is used. Figure 2 shows the comparison between the present works with the existing model. It is found that the present model is well agreed for all crack ranges.

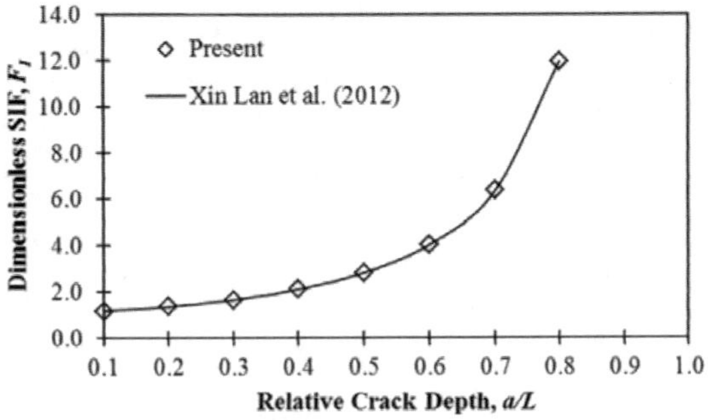

Figure 2. Model validations.

4. Results and discussion

4.1 Effect of Mechanical Mismatch on the SIFs

Two types of SIFs can be obtained simultaneously such as mode I and II SIFs. The values of SIFs are presented as normalized values F_I and F_{II}, respectively. Figure 3 shows the behavior of F_I when mechanical mismatches, α and inclined angles, θ are varied. For normal cracks, the values of F_I are almost flattened when the cracks are extended as in Fig. 3(a). It is also showed that the mechanical mismatches have no significant effect on the distribution of F_I. When slanted angles, θ are introduced, the effect of α are more pronounced as in Figs. 3(b)-3(d). In the region of relatively smaller cracks ($a/L < 0.3$), mechanical mismatch is also not contributed to increase the F_I where the SIFs are almost similar to each other's. When the slanted angles are increased, the distributions of F_I are diverged. It is seemed that the divergence of F_I is wider when slanted angle increased.

Fig. 4 shows the behavior of F_{II} for different slanted angles when mechanical mismatches are varied. The pattern of F_{II} is identical compared with F_I however they are in opposite directions. For normal cracks (θ = 00), the values of F_{II} are also flattened but it is slightly decreased when higher value of α and a/L is used as in Fig. 4(a). If slanted angles are introduced when modelled the cracks, the F_{II} is increased inversely. There is an interesting point occurs where if $\alpha \geq 5$, the values of F_{II} seemed not to alter for both slanted angles 30^0 and 40^0.

IOP Conf. Series: Journal of Physics: Conf. Series **914** (2017) 012043 doi:10.1088/1742-6596/914/1/012043

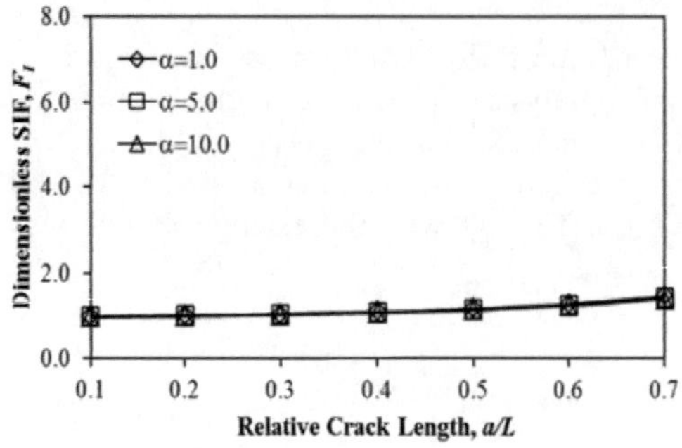

(a)

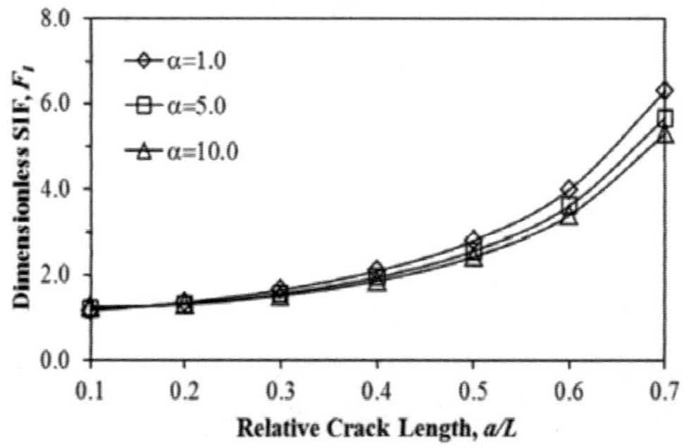

(b)

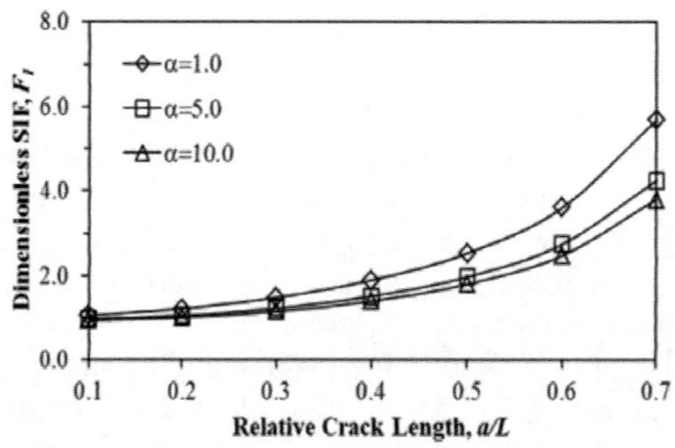

(c)

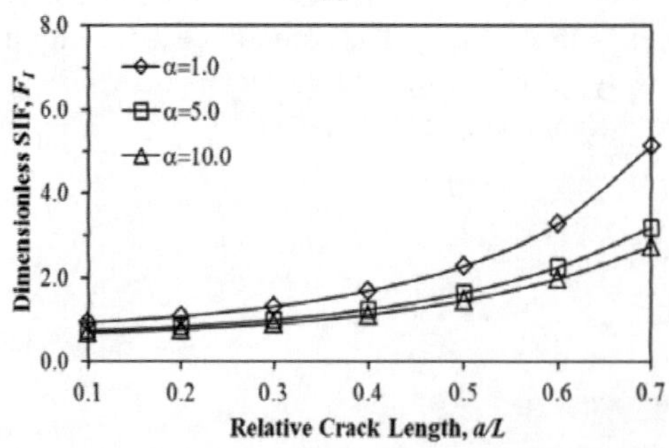

(d)

Figure 3. Effect of relative crack length on the dimensionless SIFs, F_I for different slanted angles, (a) 0^0, (b) 10^0, (c) 30^0 and (d) 40^0 when mechanical mismatched are varied.

IOP Conf. Series: Journal of Physics: Conf. Series **914** (2017) 012043 doi:10.1088/1742-6596/914/1/012043

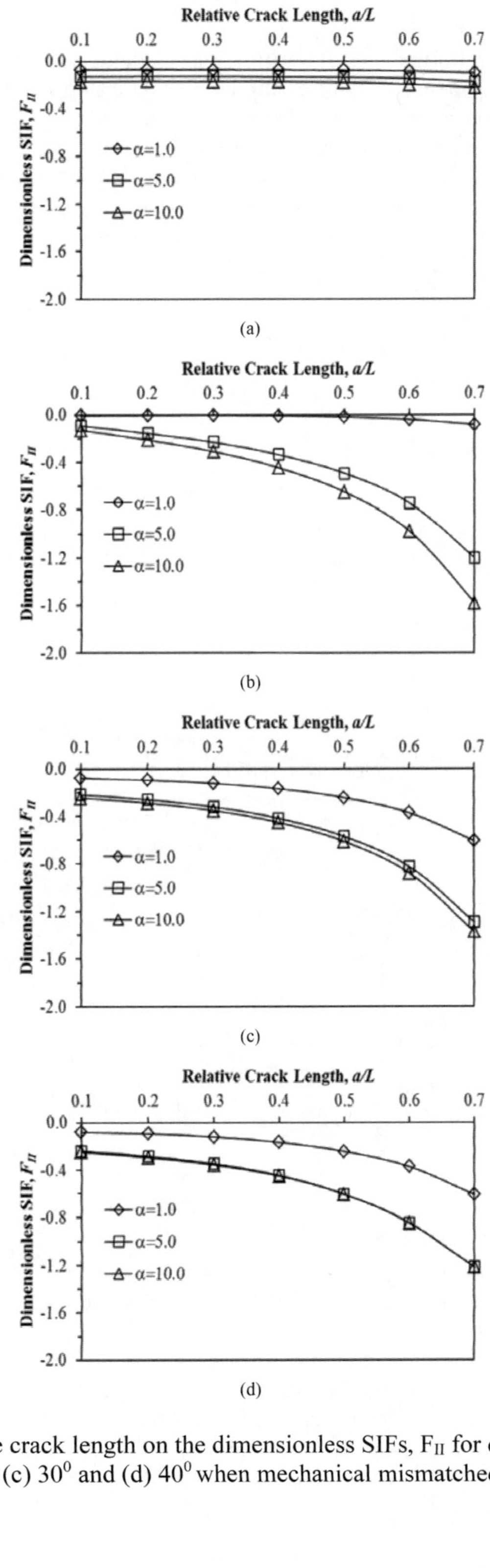

Figure 4. Effect of relative crack length on the dimensionless SIFs, F_{II} for different slanted angles, (a) 0^0, (b) 10^0, (c) 30^0 and (d) 40^0 when mechanical mismatched are varies.

IOP Conf. Series: Journal of Physics: Conf. Series **914** (2017) 012043 doi:10.1088/1742-6596/914/1/012043

4.2 Effect of Slanted angles on the SIFs

When mode I SIF, F_I is plotted against relative crack depth, a/L and the slanted angle is varied as shown in Fig. 5. The influence of F_I for similar mechanical properties is showed in Fig. 5(a) for various slanted angles. For normal cracks, the F_I is not affected when a/L is varied. If the slanted angle is introduced, the values of F_I are gradually increased. For 100 angle increment, F_I increased almost uniformly. This is probably due to the symmetrical effect of the cracks.

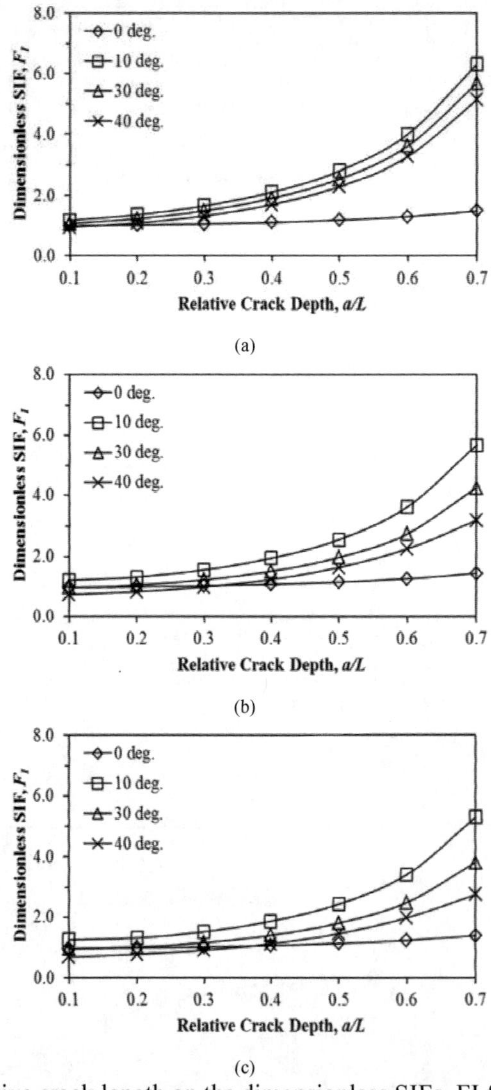

Figure 5. Effect of relative crack length on the dimensionless SIFs, FI for different mechanical mismatch, (a) $\alpha = 1$, (b) $\alpha = 5$ and (d) $\alpha = 10$ when slanted angles are varied.

Figs. 5(b) and 5(c) reveal the distributions of F_I for two mechanical mismatches, $\alpha = 5$ and 10. As the relative crack depth, a/L increased, the curves of F_I for different slanted angles diverged. It is also clearly showed that once the slanted angles are increased the F_I is slightly lowered and it is probably if $\theta > 40^0$, the SIFs obtained from slanted angles approached the SIFs of normal cracks.

4.3 Crack Mechanisms in Bi-Material Plates

Figs. 6 and 7 reveal the crack mechanisms of the 10^0 and 40^0 slanted angle cracks in bi-material plates. The mechanical properties of the bottom are higher than the upper ends. It is mean that the upper plate experienced relatively high mechanical deformation when compared with the lower end. According to Fig. 5, the F_I for 10^0 is higher than 40^0 regardless of their slanted angles. This behavior of cracks is due to the fact that for the almost flattened, the crack mount is capable to open wider than the higher value of slanted angle cracks. On the other hand, the stresses at the crack tip are nicely symmetrical. Comparing to the Fig. 7, the stresses at the crack tip is not symmetrical where the formation of mode II SIF, FII is significant. It is also revealed that when higher mechanical mismatch is used, the F_I is significantly reduced where F_I is approached the value of F_I for normal cracks.

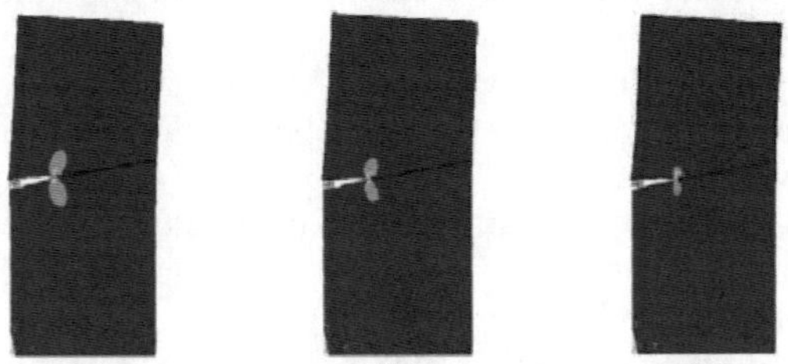

(a) (b) (c)

Figure 6 Crack mechanisms of the 10^0 slanted cracks in bi-material plates for different mechanical mismatches, α, (a) 1.0, (b) 5.0 and (c) 10.0.

(a) (b) (c)

Figure 7 Crack mechanisms of the 40^0 slanted cracks in bi-material plates for different mechanical mismatches, α, (a) 1.0, (b) 5.0 and (c) 10.0.

5. Conclusion

The following conclusions are drawn from this study:

i. For the normal crack, the distributions of F_I and F_{II} are not greatly dependent on the mechanical mismatched.
ii. When slanted angle cracks are introduced, higher SIFs can be obtained. However, when slanted angle is high, the SIFs approached the SIFs for normal cracks.
iii. The SIFs are reduced when higher slanted angles are used due to different crack mechanisms. Relatively, wider crack month occurred when lower slanted angle is used comparing with higher angle.

References

[1] J.C Newman Jr., M.A James, U. Zerbst. 2003. A review of the CTOA/CTOD fracture criterion. Engineering Fracture Mechanics. Engineering Fracture Mechanics 70:371-385.

[2] S. Courtin, C. Gardin, G. Benzine, H. Ben Hadj Hamouda. 2005. Advantages of the J-integral approach for calculating stress intensity factors when using the commercial finite element software ABAQUS. Engineering Fracture Mechanics 72:2174-2185.

IOP Conf. Series: Journal of Physics: Conf. Series **914** (2017) 012043 doi:10.1088/1742-6596/914/1/012043

[3] X.K Zhu, J.A Joyce. 2012. Review of fracture toughness (G, K, J, CTOD, CTOA) testing and standardization. Engineering Fracture Mechanics 85:2012:1-46.

[4] T. Tagawa, Y. Kayamori, M. Ohata, T. Handa, T. Kawabata, Y. Yamashita, K. Tsutsumi, H. Yoshinari, S. Aihara, Y. Hagihara. 2010. Comparison of CTOD standards: BS 7448-Part 1 and revised ASTM E1290. Engineering Fracture Mechanics 77:327-336.

[5] G.C Guinea, J. Planas, M. Elices. 2000. KI evaluation by the displacement extrapolation technique. Engineering Fracture Mechanics 66:243-255.

[6] A.E Ismail, A.K Ariffin, S. Abdullah, M.J Ghazali. 2011. Off-set crack propagation analysis under mixed mode loadings. Int. Journal of Automotive Technology 12(2):225-232.

[7] A.E Ismail, A.K Ariffin, S. Abdullah, MJ Ghazali, M. Abdulrazzaq, R. Daud. Stress intensity factors under combined bending and torsion moments. Journal of Zhejiang University Science A 13(1):1-8.

[8] A.E Ismail, A.K Ariffin, S. Abdullah, M.J Ghazali. 2014. Ungkapan kamiran-J retak permukaan pada bar silinder padu kenaan beban ragam I. Jurnal Teknologi (Sciences and Engineering) 68(1):7-17.

[9] T. Matsumo, M. Tanaka, R. Obara. 2000. Computation of stress intensity factors of interface cracks based on interaction energy release rates and BEM sensitivity analysis. Engineering Fracture Mechanics 65:683-702.

[10] T. Ikeda, N. Miyazaki, T. Soda. 1998. Mixed mode fracture criterion of interface crack between dissimilar materials. Engineering Fracture Mechanics 59:725-735.

[11] N.A Noda, C. Xu. 2007. Controlling parameter of the stress intensity factors for a planar interfacial crack in three-dimensional biomaterials. Int. Journal of Solids and Structures 45(3):1017-1031.

[12] N.A Noda. 2004. Stress intensity formulas for three-dimensional cracks in homogenous and bonded dissimilar materials. Engineering Fracture Mechanics 71(1):1-15.

[13] X. Lan, N.A Noda, Y. Zhang, K. Michinaka. 2012. Single and double edge interface crack solutions for arbitrary forms of material combination. Acta Mechanica Solida Sinica 25(4):404-416.

[14] C. F. Shih, B. Moran and T. Nakamura. 1986. Energy Release Rate along a Three-Dimensional Crack Front in a Thermally Stressed Body. Journal of Fracture 30(2):79-102.

[15] M.C Walters, G.H Pauline and R.H Dodds Jr. 2005. Interaction Integral Procedures for 3D Curved Cracks Including Surface Tractions, Engineering Fracture Mechanics 72:1635-1663.

[16] A.E Ismail. 2014. Stress intensity factor of eccentric cracks in bi-material plates. Applied Mechanics and Materials 663:98-102.

[17] A.E Ismail. 2014. Multiple crack interaction in bi-material plates under mode I tension loading. Applied Mechanics and Materials 629:57-61.

IOP Conf. Series: Journal of Physics: Conf. Series **914** (2017) 012044 doi:10.1088/1742-6596/914/1/012044

A Review on the Perforated Impact Energy Absorption of Kenaf Fibres Reinforced Composites

Al Emran Ismail*, S.N.A Khalid, Nik Hisyamudin Muhd Nor

Department of Engineering Mechanics, Faculty of Mechanical and Manufacturing Engineering, Universiti Tun Hussein Onn Malaysia, Batu Pahat, 86400 Johor, Malaysia.

Corresponding author: emran@uthm.edu.my / al_emran@hotmail.com

Abstract. This paper reviews the potential of mechanical energy absorption of natural fiber reinforced composites subjected to perforated impact. According to literature survey, several research works discussing on the impact performances on natural fiber reinforced composites are available. However, most of these composite fibers are randomly arranged. Due to high demand for sustainable materials, many researches give high attention to enhance the mechanical capability of natural fiber composites especially focused on the fiber architecture. Therefore, it is important to review the progress of impact energy absorption on woven fiber composite in order to identify the research opportunities in the future.

1. Introduction

Natural plant fibres can be economically and ecologically useful alternatives to reinforcement fibres in polymeric composites. Due to their low density and low cost in comparison to conventional fibres, natural fibre reinforced composites have great potential in engineering applications. A growing environmental awareness across the world has aroused interest in research and development of environmentally friendly and sustainable materials. Natural plant based fibres are used as reinforcements for composite materials and give various advantages as compared to conventional fibres.

In recent years, industries are attempting to reduce the dependence on petroleum based fuels and products due to the increasing of environmental consciousness. This leads to investigate environmentally friendly and sustainable materials to replace the synthetic fiber. The increasing of productions and consumption of plastics in every sector of our life and is lead to huge plastic wastes. Due to disposal problems, as well as strong regulations and criteria for cleaner and safer environment, have directed great part of the scientific research toward eco-composite materials. Among the different types of eco-composites, those which contain natural fibres (NF) and natural polymers have a key role to overcome a problem [1].

Few studies deal with structural based on natural reinforced composites. These studies are mainly oriented to the housing applications where structural panels and sandwich beams are manufactured out of natural fibres and used as roofs. Considering the high performance standard of composite materials in terms of durability, maintenance and cost effectiveness, the application of natural fibre reinforced composites as construction material holds enormous potential and is critical for achieving sustainability. Due to their low density and their cellular structure, natural fibre possess very good acoustic and thermal insulation properties and demonstrate many advantageous properties over glass or rockwool fibre, for instant in example handling and disposal material [2]. Recently, some of studies have been carried out to determine the effects of various variables on energy absorption capability of composites material under compressive loading [3, 4].

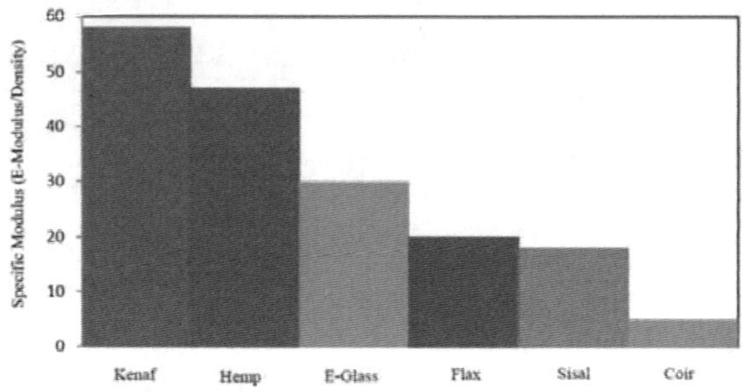

Figure 1. Comparison of specific modulus of kenaf fibre with several other fibres [5]

One of the main important aspects of the behaviour of natural plant fibre reinforced polymeric composites is their response to an impact load and the capacity of the composites to withstand it during their service life. Such damage may be caused by bumps or crashes and falling objects and debris. Some of the reported work [4, 5] has suggested that natural fibre composites are very sensitive to impact loading. The major drawback is its low impact strength as compared to glass fibre reinforced thermoplastic and thermosets composites. In the broader context, assessing the impact resistance of a composite material is always difficult since the damage manifests itself in different forms such as delamination at the interface, fibre breakage, matrix cracking and fibre pulls out. Due to their complexity, many of their characteristics still remain unresolved.

In this work, it is started with the review of energy absorption of natural fiber, included advantages of kenaf fibre. Then, it is discussed on the characteristic between woven and non-woven of natural fibre which is focusing the strength of various numbers of layers and orientations.

2. Energy Absorption Factor in Natural Fiber

2.1 Advantages of kenaf fiber

Natural fibre is widely used in composite application, and has many advantages which is low cost, low density, have high specific properties, non-abrasive and less harmful during handling. In the research of kenaf natural fibre is increasingly interested among researchers [5, 6]. In summary, a comparative study of mechanical properties of kenaf fibre reinforced composites and other natural fibre are collected together. The characteristic strength of natural fibre is important to determine material ability, especially under extreme and critical condition, which is directly connected with engineering performance.

Figure 1 illustrates that kenaf composite have higher modulus as compared to other fibres. In this research, kenaf fibres are selected based on their capability of modulus strength [7]. The mechanical properties of the kenaf fibre can be seen in Table 1. Flexural properties were not available from the manufacture hence values for a similar product from another manufacture are used in their research.

Table 1 Properties of Kenaf Fibre [7]

Property	Unit	Kenaf Fibre
Tensile strength	MPa	350-600
Elastic modulus	MPa	40,000
Elongation at break	%	2.5-3.5
Flexural strength	MPa	N/A
Flexural modulus	MPa	N/A
Density	Kg/m3	1500

2.2 Characteristic of woven and non-woven

Woven composite are widely used on glass fiber [8-10], butr glass fiber is very expensive and higher cost to manufacture. According Rizal et al., their research focus on the prediction of progressive failure in woven glass/epoxy composites. Most of the previous research, extensive used of plain weave on their fabric as shown in Figure 2. On the other hand, plain woven can absorb more energy, were perforated by higher force and their perforations were completed at higher deformation in comparison to other woven fabric during bursting strength and impact test.

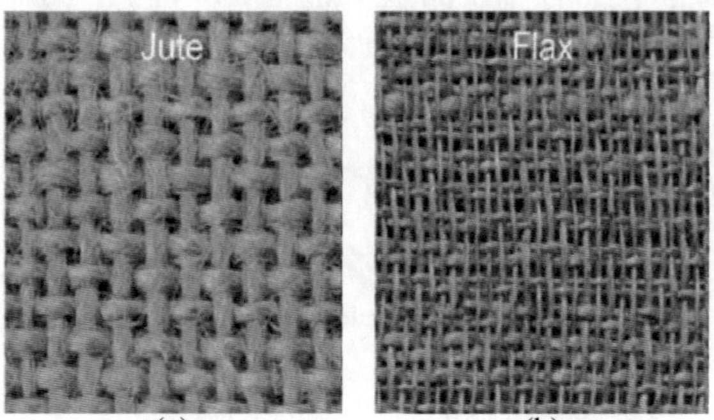

(a) (b)

Figure 2: Types plain woven of natural fibre (a) Jute (b) Flax [12]

IOP Conf. Series: Journal of Physics: Conf. Series **914** (2017) 012044 doi:10.1088/1742-6596/914/1/012044

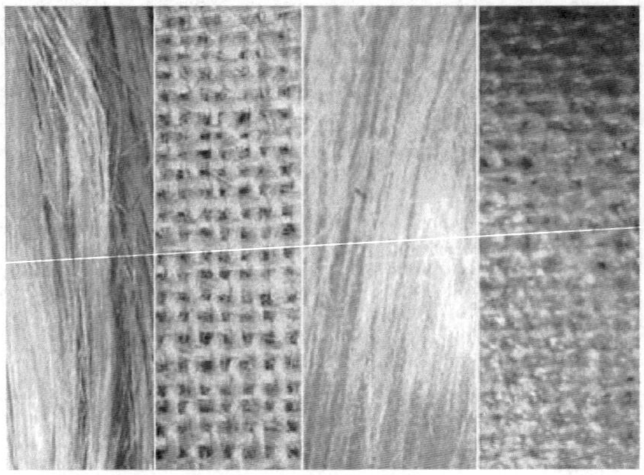

(a) (b) (c) (d)

Figure 3 Image of (a) unidirectional jute, (b) plain woven jute fabric, (c) non-woven jute fabric composite and (d) plain woven jute fabric composite [14]

According to some literature [13, 14], woven fibre is have good mechanical properties as compare to non-woven. Figure 3 shows the illustration image of unidirectional and woven of jute fabric. Table 2 shows the effect of NWJF (Non-Woven Jute Fabric) and WJF (Woven Jute Fabric) on impact strength of PLLA (Poly L-lactic Acid). The average impact strength of untreated WJF/PLLA composite in warp direction (16.4 kJ/m) is 85.9% higher than the impact strength of unreinforced PLLA. As a result, the strength of the WJF/PLLA fabric composite is greater than the strength in NWJF/PLLA. According to the result, it can be concluded that woven structure exhibited excellent mechanical behaviour under tensile, flexural, and impact loading as compare to non-woven composite. It also mentioned that, tensile, flexural and impact strength of WJF/PLLA composite is higher at warp direction as compared to weft direction.

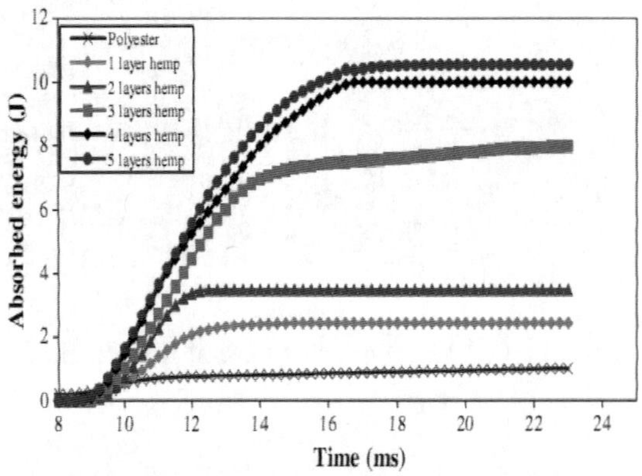

Figure 4 Absorb energy/time traces [16]

2.3 Characteristics of impact tests for different number of layers

According to previous researcher, [15] using different layers to identify capability of fracture toughness. In their research focuses on fabricated composite using long kenaf fiber with woven glass hybrid laminated. In this way, the laminated on layer are not be used in natural fiber only, but also both hybrid on natural and glass fiber. From this research, it shows that by using combination hybrid also can improve of their properties. Figure 4 shows the variation in energy absorbed with different layers [16]. From the Figure 4, there is an increase in the amount of energy absorption by specimen hemp reinforced compared without reinforcement. It can be seen that energy absorbed by 1 and 2 layers is less than 3 and 4 layers. As can be seen from the result, 1, 4 and 5 layers of hemp reinforced samples was absorb total energies of 10.558 J and 10.528 J, respectively, which is have a small significant effect. The trend shows that by increasing at layer up to its threshold value the energy absorbed also increase.

2.4 Influence of treatment and untreated

Many of previous research doing comparison properties of composite with treatment and untreated [17-19]. Some of them using distilled water, rain water, sea water [15] and sodium hydroxide (NaOH) [18, 19] for treatment. Many investigations have focused on the treatment of fibres to improve the bonding with resin matrix. As shown in Figure 5, it observed that by using 6% of NaOH to the kenaf fiber was contributed to the better performance of treated compare to untreated. The modulus flexural are increase with 67% for untreated and 74% for treated it can be explained with the fact that the waxy layers on the fibre surface are removed with the assistance of NaOH solution which in turn enhance the interfacial adhesion of the fibre with the matrix. This lead to the carrying loading by the fibre during loading condition; the load is transferred to the fibre and the fibre are be able to carry some of the load.

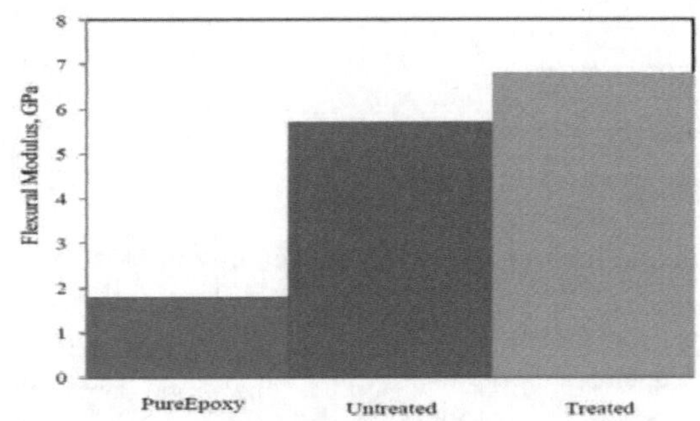

Figure 5 Flexural modulus of pure epoxy, untreated and treated kenaf fibre reinforced epoxy [19]

2.5 Influence of different orientation

According to the research by Sikarwar et al., it can be seen that the energy absorbing capacity are increase with increasing of laminates thickness [20]. In their research, the laminates (0/90) lay-up laminates is much higher than other of lay-up laminates. The (0/90/30/-45) absorb more energy than the (0/90/30/-60) orientation. Thus, it is understood that energy absorption is dependent both on failure strain well as on Young's Modulus. The result in their research can be seen as shown in Table 3.

2.6 Woven fragmentation with different densities

Research by Khalid et al., shows the woven fragmentation using difference densities and layers which is consist of Sample A,B,C and D [21]. Differences composite fragmentation pattern can be observed visually as shown in Table 4. Low speed impact causes fragmentation of the sample A(I) like a star and have fragmentation weft measuring over 20 mm where there is fragmentation fragment surrounding the point of impact on the composite panel. This occurs due to the occurrence of bending before failure on the composite panel. Significant difference can be seen with the fragmentation of the sample A(II) be round and have split weft measuring over 13 mm. Woven structure in high-density composite panels and balanced between warp yarns and weft yarns made of composite panel is fragile and will result in failure to make a full composite panel.

Table 2 Mechanical properties of jute fabrics/PLLA composite [14]

Lay-up sequence	Thickness	Lay-up sequence	Thickness
0/90	1.5	27.09	27.73
	3.0	36.75	38.25
	4.0	66.63	65.34
	5.0	92.43	108.37
0/90/30/-60	1.5	22.23	22.81
	3.0	39.78	37.50
	4.0	57.19	49.59
	5.0	70.38	60.48
0/90/45/-45	1.5	23.40	21.66
	3.0	33.84	37.50
	4.0	46.20	49.59
	5.0	67.33	58.59
30/-60/60/-30	1.5	18.90	19.44
	3.0	30.37	31.74
	4.0	42.93	39.78
	5.0	54.00	47.04

Table 3 Experimental and theoretically predicted energy absorption capacity of the glass/ epoxy laminates [20]

Sample	Tensile Strength (MPa)	Tensile Modulus (GPa)	Strain (%)	Flexural Strength (MPa)	Flexural Modulus (GPa)	Impact Strength (kJ/m^2)
1. Unreinforced PLLA.	40 ± 6.3	0.36 ± 0.001	3.4	42 ± 9.7	3.02 ± 0.8	8.82 ± 0.9
2. NWJF/PLLA composites.	55 ± 11.5	0.867 ± 0.02	6.0	67 ± 8.4	2.83 ± 1.1	13 ± 1.1
At warp direction:						
3. Untreated WJF/PLLA composites.	81 ± 13.5	1.12 ± 0.034	3.8	82 ± 12.0	4.3 ± 0.10	16.4 ± 1.8
4. Treated WJF/PLLA.	87 ± 8.5	1.42 ± 0.047	5.1	121 ± 13.4	5.3 ± 0.10	18.1 ± 2.3
At weft direction:						
5. Untreated WJF/PLLA composite.	71 ± 8.7	0.78 ± 0.063	4.1	81 ± 9.4	3.62 ± 0.08	14.3 ± 1.5
6. Treated WJF/PLLA	79.2 ± 9	0.91 ± 0.057	4.2	111 ± 8.1	4.72 ± 0.05	16.6 ± 1.8

IOP Conf. Series: Journal of Physics: Conf. Series **914** (2017) 012044 doi:10.1088/1742-6596/914/1/012044

Table 4 Visual fragmentation inspection [21]

Type of Sample	Single Layer Composites (I)	Double Layer Composites (II)	Type of Sample	Single Layer Composites (I)	Double Layer Composites (II)
Sample A	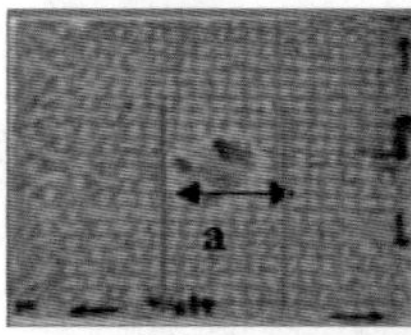 20 mm	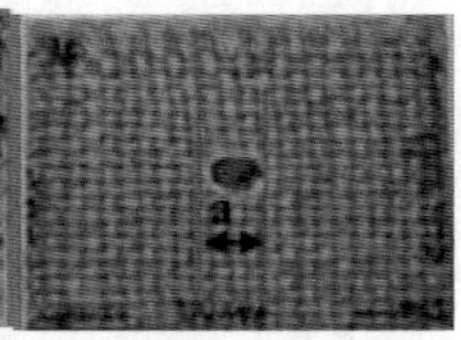 13 mm	Sample C	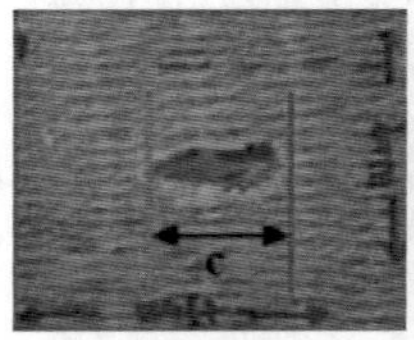 35 mm	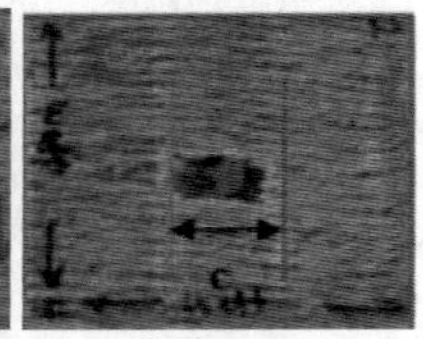 27 mm
Sample B	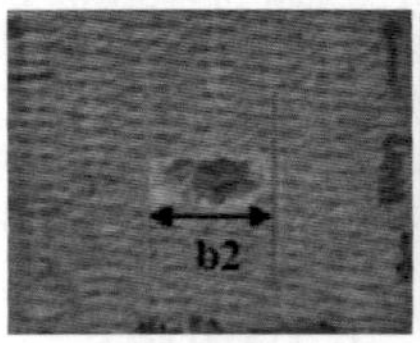 28 mm	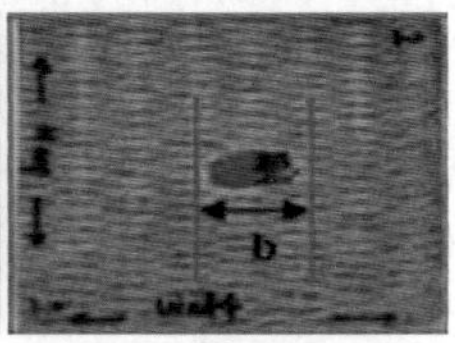 23 mm	Sample D	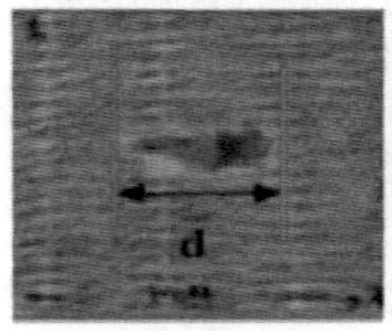 37 mm	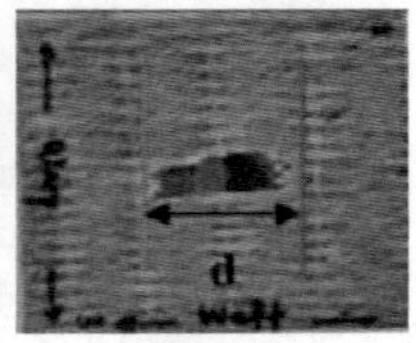 38 mm

Samples B(I), C(I) and D(I) show the effect of impact where the point of impact to be larger and elongated in the direction of the weft yarn has weft fragmentation during 28 mm, 35 mm and 37 mm. However, the sample B(II), C(II) and D(II) had the weft fragmentation during 23 mm, 27 mm and 38 mm. Damage is caused by decreases in tension warp yarns that support the weft yarns creating lacing effects more elongated in the direction weft yarns.

3. Research problem

In the latest years, there have been many researchers involved in this field of natural fibre reinforced plastics [22]. Most of them are based study on the mechanical properties of composites reinforced with short and non-woven fibres [18,23]. According to previous research [16], studied on low-velocity impact response of non-woven hemp fibre reinforced unsaturated polyester composites and only focus on low velocity impact of non-woven natural fibre using varying layers. Sikarwar et al., in their research focus on influence of fibre orientation and thickness on the response of glass/epoxy composites subjected to impact loading [20]. In experimental, glass/epoxy is laminated with orientation in order to determine the energy absorption capacity. Laminate with difference orientation [0/90], [0/90/30/-60], [0/90/45/-45], and [30/-60/60/-30] are used. Sas et al., also studied on orientation permeability of unidirectional fabric. In their study, the characterization of orientation is performed by using numerical analysis ANSYS software [24]. Although there has been research on orientation of material, but commonly uses a unidirectional.

4. Conclusion

Woven typed fiber orientation had not explored widely on kenaf natural fibre. Recently, most of the research focuses on tensile strength and flexural modulus of natural fibre [10, 25]. Energy absorption on woven natural fibre is better than non-woven were proved by previous researches on woven flax fibre has greater embodied energy than a glass fibre [26]. From the result, it can be summarized that woven on natural fiber are produced higher energy compared with non-woven. It is suggested, four layers are sufficient and comparable to the energy absorbed [14]. The effects of woven orientations are stronger than non-woven fibre. Dhakal et al., in their research, only focus on impact loading on fibre glass [16]. According to the author knowledge, lack of information on effect of orientation on impact loading by natural fibre from previous researcher and this field have opportunities in the future.

Acknowledgement
Authors acknowledged Universiti Tun Hussein Onn Malaysia (UTHM), MyBrain15 Scholarship, Office for Research, Innovation, Commercialization and Consultancy Management (ORICC), Kementerian Pendidikan Malaysia (KPM) for funding this project through Research Acculturation Grant Scheme (RAGS Vot R019)

References
[1] John Summerscales, Nilmini Dissanayake, Amandeep Virk, Wayne Hall. A review of bast fibres and their composites. Part 2 – Composites. Composites: Part A 41 (2010) 1336–1344.
[2] Igor M. De Rosa, Carlo Santulli, Fabrizio Sarasini. Acoustic emission for monitoring the mechanical behavior of natural fibre composites: A literature review. Composites: Part A 40 (2009) 1456–1469.
[3] S.H Masran, A.E Ismail, M.F Marian, A study of energy absorption performances of pultruded composites under quasi-static compressive loadings. Applied Mechanics and Materials (2014) 465:662-666.
[4] A.E Ismail, Effect of foam density and wall thickness interactions on the energy absorption performances. Applied Mechanics and Materials (2013) 315:665-669.
[5] H.M. Akil, M.F. Omar, A.A.M. Mazuki, S. Safiee, Z.A.M. Ishak, A. Abu Bakar. Kenaf fiber Reinforced Composites: A Review. Material and Design 32 (2011) 4107-4121.
[6] M.N Roslan, A.E Ismail, M.Y Hashim, M.H Zainulabidin, S.N.A Khalid, Modelling analysis on mechanical damage of kenaf reinforced composite plates under oblique impact loadings, Applied Mechanics and Materials 465 (2014) 1324-1328
[7 S. Rassmann, R.G Reid, R. Paskaramootrhy, Effect of processing condition on the mechanical and water absorption properties of resin transfer moulded kenaf fibre reinforced polyester composite laminated. Composites: Part A 41 (2010) 1612–1619
[8] M.A.A Ghani, Z. Salleh, Koay Mei Hyie, M.N Berhan, Y.M.D Taib, M.A.I Bakri. Mechanical properties of kenaf/fiberglass polyester hybrid composite. Procedia Engineering 41 (2012)1654-1659
[9] M.N Davoodi, S.M Sapuan, D. Ahmad, Aidy Ali, A. Khalina, Mehdi Jonoobi. Mechanical properties if hybrid kenaf/glass reinforced epoxy composite for passenger car bump beam. Materials and Composite 31 (2010) 4927-4932
[10] Hazizan Md. Akil, Ignor M. De Rosa, Carlo Santulli, Fabrizio Sarasini. Flexural behavior of pultruded jute/glass and kenaf/glass hybrid composite monitored using acoustic emission. Material Science and Engineering A 527 (2010) 2942-2950
[11] Rizal Zahari, Abdul Hannan Azmee, Faizal Mustapha, Mohd Sapuan Salit, Renuganth Varatharajoo, Azmine Shakrine. Prediction of progressive failure in woven glass/epoxy composite laminated panels. Jurnal Mekanikal (2008) No 25, 80-91
[12] James Meredith, Richard Ebsworth energy absorption structure. Natural fiber composite energy absorption structures. Composites Science and Technology 72 (2012) 211-217
[13] A.E Ismail, M.H Zainulabidin, M.N Roslan, A.L.M Tobi, N.H.M Nor. Effect of velocity on the impact resistance of woven jute fiber reinforced composites. Applied Mechanics and Materials (2014) 465:1277-1281.
[14] G.M. Arifuzzaman Khan, M. Terano, M.A. Gafur, M. Shamsul Alam. Studies On The Mechanical Properties Of Woven Jute Fabric Reinforced Poly(L -Lactic Acid) Composites. Engineering Science (2013).
[15] Z. Salleh, Y.M. Taib, Koay Mei Hyie, M. Mihat, M.N Berhan, M.A.A Ghani. Fracture behavior of pultruded jute/glass and kenaf/glass hybrid composites monitored using acoustic emission. Procedia Engineering 41 (2012)1667-1673

IOP Conf. Series: Journal of Physics: Conf. Series **914** (2017) 012044 doi:10.1088/1742-6596/914/1/012044

[16] H.N. Dhakal, Z.Y. Zhang, M.O.W. Richardson, O.A.Z. Errajhi. The Low Velocity Impact Response Of Non-Woven Hemp fibre Reinforced Unsaturated Polyester Composites, Composite Structures 81 (2007) 559–56.
[17] Bonnia N.N, Mahat M.M, Surip S.N, Anuar H., Hassan N.A, Ahmad S. Polyester/kenaf composite; effect of matrix modification. IEEE Engineering and Industrial Application (2012)
[18] Mohd Suhairi Meona, Muhamad Fauzi Othmana, Hazran Husaina, Muhammad Fairuz Remelia, Mohd Syahar Mohd Syawala. Improving tensile properties of kenaf fibers treated with sodium hydroxide. Procedia Engineering 41 (2012)1587-1592
[19] B.F Yousif, A. Shalwan, C.W Chin, K.C Ming. Flexural properties of treated and untread kenaf/epoxy composite. Material and Design 40 (2012) 378-385
[20] Rahul S. Sikarwar, R. Velmurugan, N.K Gupta. Influence of Fiber Orientation and Thickness on the Response. Composite: part B 60 (2014) 627-636.
[21] S.N.A.Khalid, M.R.Jamaluddin, M.N.Roslan, A.E.Ismail, M.H.Zainulabidin. Effects of Low-Velocity Impact on Different Density of Woven Kenaf Fiber Reinforced Composites. Journal of Applied Science and Agriculture, ISSN 1816-9112
[22] Q.N. Tran, C.A. Fuentes, C. Dupont-Gillain, A.W.Van Vuure, I. Verpoest. Understanding the Interfacial Compatibility and Adhesion of Natural Coir Fibre Thermoplastic Composites. Composite Science and technology 63 (2013) 23-30.
[23] A.K. Rana, A. Mandala, S. Bandyopadhyay, Short jute fiber reinforced polypropylene composites: effect of compatibiliser, impact modifier and fiber loading. Composite Science and technology 63 (2003)801-806.
[24] Hatice S. Sas, Eric B. Wurtzel, Pavel Simacek, Suresh G. Advani. Effect of Relative Ply Orientation on the through-thickness Permeability of Unidirectional Fabric. Composite Science and technology 96 (2014) 116-121.
[25] A.E Ismail, M.K Awang, M.H Sa'at, Tensile strength of natural fiber reinforced polyester composite, AIP Conf. Proc. 909 (2007) 174-179.
[26] Nilmini P. J., Dissanayake, J., Summerscales, S. M. Grove & M. M. Singh, Energy use in the production of flax fiber for the reinforcement of composites. J Nat Fiber 2009;6 (4):331-46.

IOP Conf. Series: Journal of Physics: Conf. Series **914** (2017) 012045 doi:10.1088/1742-6596/914/1/012045

Periodic Boundary Condition Technique on Carbon Fibre Composites

Kamarul Azhar Kamarudin*, **Al Emran Ismail**

Crashworthiness and Collisions Research Group, Faculty of Mechanical and Manufacturing Engineering
Universiti Tun Hussein Onn, Parit Raja, Batu Pahat, Johor.

*Corresponding author: kamarula@uthm.edu.my

Abstract. A three-dimensional finite element model of carbon fibre reinforced has been investigated numerically using periodic boundary condition method. This method was used to predict the elastic mechanical behaviour of a unit cell of such composites. Periodic boundary condition was used due to its capability to represent a single unit cell similar to the neighbouring unit cells with continuous physical elements. It is assumed that the paired nodes displaced continuously without separating or interrupting other nodes during the deformation step. From the study, it was revealed that the elastic modulus agreed well with the experimental results, indicating that the present model could be used effectively.

1. Introduction

Past decades have seen rapid development in micromechanics elastic modulus prediction such as the use of energy variation principles [1], energy balance approach[2] and mechanics approach [3]. Another technique known as periodic boundary condition (PBC) method has been studied extensively by Li [4], Li *et al.* [5], Xia *et al.* [6] and Tyrus *et al.* [7]. The early development of PBC used a 2D model to solve the problem. Li [4] developed a general packing approach that is able to accommodate unidirectional fibre of irregular cross sections. Later, Li and Wongsto [8] extended their research to 3D environment involving irregular geometries and local imperfection within composites. Xia *et al.* [6] presented unified periodic displacement boundary conditions in displacement-based FEM, which is applied into 2D multi-shaped models. Later in 2007, Tyrus *et al.* [7], demonstrated the study on 2D triangular and square shapes which focused on the effectiveness of computational speed.

In many publications, the analysis using periodic boundary condition is modelled to a single unit cell [8, 9], while few using multilayer composite [10, 11]. This paper attempts to reveal the elastic properties results from PBC in comparison with the experimental study, as well as the elastic prediction on unidirectional micromechanics of a single unit cell by using ABAQUS finite element software. In addition, the final results try to examine the application of periodic boundary condition method in a multi-layer unidirectional and bidirectional composite in three-dimensional model analysis.

2. Material and composite geometry

The unidirectional laminate of epoxy resin impregnated AS4 carbon fibre was used in the study and the properties are taken from King *et al.* [12] as shown in Table 1. The model is illustrate in Figure 1 and has dimension of 2x2x2 that represents length a, b and c, with fibre fraction measurement of 0.6. The model was design using Abaqus software and the element chosen for meshing purposes is C3D8R.

Table 1: Properties of composite [12]

Properties	Fibre AS4	Matrix
E_1(GPa)	235	4.8
E_2(GPa)	14	4.8
G_{12}(GPa)	28	1.8
ν_{12}	0.2	0.34
ν_{23}	0.25	0.34

2.1 Periodic Boundary Conditions for Repeating Unit Cells (RUCs)

Periodic methods are composed of fundamental building blocks called repeating unit cells (RUCs) which when assembled together side by side will form an infinite array. The displacement of each unit cell must be continuous, which means that the neighbouring RUC cannot be separated from each other after the deformation. In addition, traction distribution should also be similar at the opposite parallel boundaries of a RUC. This means that the element in forming macroscopic structure will produce the individual RUC, which can thus be assembled as a physically continuous body. To implement PBC technique, internal microstructure within the RUC is not important but only within the outer surface of the whole structure is considered.

To generalise the model into a cubic unit cell, subscripts 1 and 2(in Figure 1) indicate the position before the translational of *x, y, z* directions. Any parts of the given volume relatively have displacement of any point away from the existing cell tied with the number of cells $i,j,k = 0,\pm1, \pm2,\pm3\ldots$, which is related with:

$$(x_2, y_2, z_2) = (x_1 + ib, y_1 + jb, z_1 + kb) \quad (1)$$

and formed as:

$$\begin{aligned} u_2 - u_1 &= (ib)\bar{\varepsilon}_x + (jb)\bar{\gamma}_{xy} + (kb)\bar{\gamma}_{xz} \\ v_2 - v_1 &= (jb)\bar{\varepsilon}_y + (kb)\bar{\gamma}_{yz} \\ w_2 - w_1 &= (kb)\bar{\varepsilon}_z \end{aligned} \quad (2)$$

The equations established above are related with the displacement boundary condition that can be prescribed by macroscopic strains as loads, which can be treated as an independent degree of freedom to the system. With regards to the displacement of boundary conditions, positions *x*, *y* and *z* at origin *O* are constrained from displacements in all respective directions.

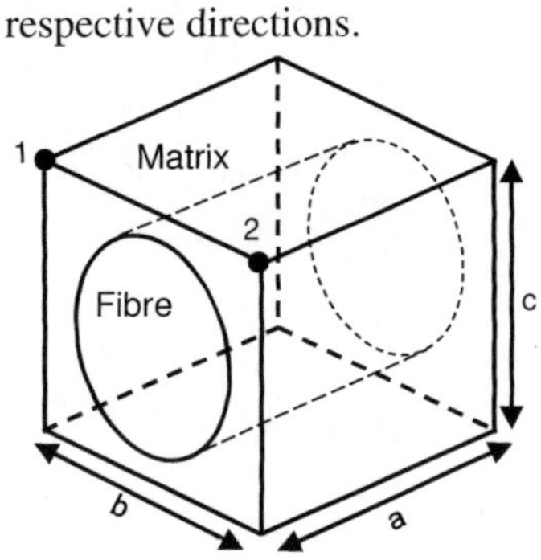

Figure 1: Illustrate of unit cell.

In addition, the *x*-axis is constrained by the rotation of the *z*-axis. By explaining mathematically at origin, *O*:

$$x, y, z = 0\,, \quad \frac{\partial u}{\partial z} = \frac{\partial v}{\partial z} = \frac{\partial w}{\partial y} = 0 \qquad (3)$$

Concentrated forces related forms are expressed as strains rather than displacements and are applied for the key degree of freedom. It is also equivalent to the key nodal forces. Macroscopic stresses that are related to the concentrated forces are applied on this key degree of freedom. These manners are similar when considering simple energy equivalent as the work done by concentrated forces over the macroscopic strain and is similar as the macroscopic stress over the macroscopic strain for the volume of unit cell.

For example, if a force in the x-direction is applied to the degree of freedom ε_x of the unit cell, while all other degrees of freedom are free from constrains, the work done by the force is taken as:

$$W = \frac{1}{2} F_x \varepsilon_x \qquad (4)$$

Meanwhile, the strain energy stored in the unit cell can be expressed in macroscopic stress and strains as:

$$E = \frac{1}{2}\int_V \sigma_x \varepsilon_x dV = \frac{1}{2} V \sigma_x \varepsilon_x \qquad (5)$$

where *V* is the volume of unit cell and work done W is equal to the strain energy. Equalizing both strain energy and work done on the unit cell gives simple stress relations obtained as:

$$\sigma_{x,} = \frac{F_x}{V}\,;\ \sigma_y = \frac{F_y}{V}\,;\ \sigma_z = \frac{F_z}{V} \qquad (6)$$

where *x, y,* and *z* are the directions of forces acting on the volume *V* of the unit cell. To make it simpler, it can be said that the magnitude of the concentrated load (F) is equal to the volume of the unit cell (*V*) times the stress it will generate macroscopically. In this study, the macroscopic stress is taken as 1 MPa. The effective material properties on the 'key nodes' for Young's modulus
parallel to the fibre directions E_x , E_y , E_z , G_{yz}, G_{xz} and G_{xy} and can be obtained as in Table 2:

Table 2: Elastic modulus

Direction	Apply	Displacement formed in	To determine
x	σ_x only	ε_x	$E_x = \frac{\sigma_x}{\varepsilon_x}$
y	σ_y only	ε_y	$E_y = \frac{\sigma_y}{\varepsilon_y}$
z	σ_z only	ε_z	$E_z = \frac{\sigma_z}{\varepsilon_z}$
yz	τ_{yz} only	γ_{yz}	$G_{yz} = \frac{\tau_{yz}}{\gamma_{yz}}$
zx	τ_{zx} only	γ_{zx}	$G_{zx} = \frac{\tau_{zx}}{\gamma_{zx}}$
xy	τ_{xy} only	γ_{xy}	$G_{xy} = \frac{\tau_{xy}}{\gamma_{xy}}$

IOP Conf. Series: Journal of Physics: Conf. Series **914** (2017) 012045 doi:10.1088/1742-6596/914/1/012045

3. Results and discussion

3.1 Unidirectional Composite using PBC approach

Figure 2 shows stress distribution of periodic boundary condition with regards to elastic constant. This scenario of boundary condition for PBC showed a continuous deformation along its boundaries. As mentioned earlier when the unit cell being assembled together side by side, it will form an infinite array of

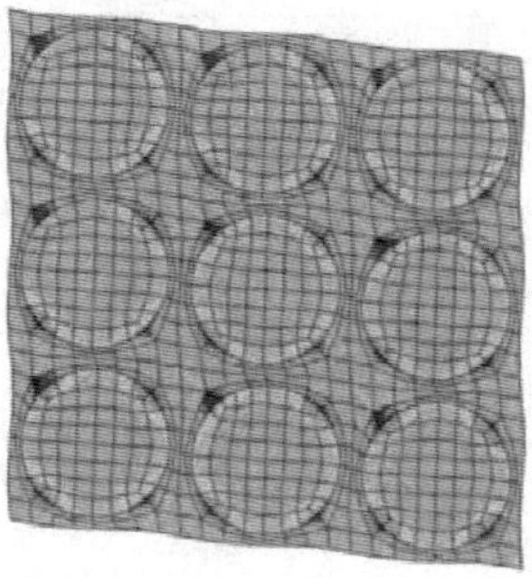

neighbouring unit cell as shown in
Figure **3**. The simulation model showed that the method were able to be used for different model geometry shapes other than single unit cell.

3.2 Bidirectional Composite using PBC approach

Applying the technique on another model, a unit cell of bidirectional fibrous composite was developed. Figure 4 shows a unit cell of fibre at cross angle of 0° and 90° respectively in its xz-plane. Using periodic method, the model was analysed and performed as illustrated in Figure 5. E_1 and E_3 produced a similar elastic value due to the symmetric fibre orientation while E_2 will approximately be 10% of the E_1 and E_3 values. The results of the 60% fibre volume bidirectional CFRP analysis are also shown in Table 3. There are numbers of similarities of the elastic properties which resulted from the PBC methods against experimental result by other researchers. The accuracy of the PBC method for elastic constants prediction against the existing results is tested.

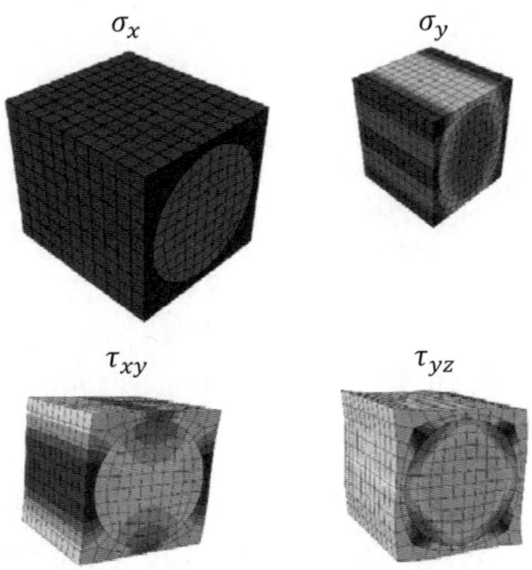

Figure 2: Stress distribution using periodic boundary condition method.

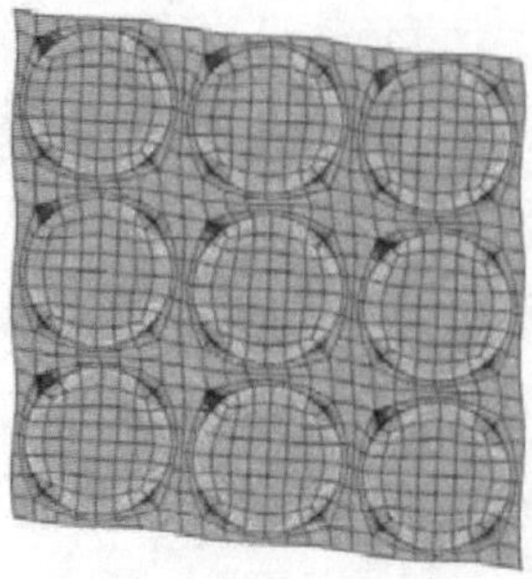

Figure 3: Shear stress images of multilayer unidirectional FRP arrangement.

Table 3: Comparison of elastic constants for Carbon Fibre Reinforced Plastic (CFRP)

Unidirectional Composite					Bidirectional Composite		
Elastic Constants	Sun and Vaidya [2]	Lee and Daniel [13]	Sun and Zhou [14]	Present	Elastic Constants	Kim *et al.* [15] V_f=60%	Present
E_x(GPa)	142.6	142	139	143.94	$E_1 = E_3$	65.9/61.8	76.14
E_y(GPa)	9.60	10.30	9.85	9.71	E_2	11.3	10.42
G_{xy}(GPa)	6.00	7.60	5.25	6.16	$G_{12} = G_{32}$	6.98/7.14	4.96
G_{yz}(GPa)	3.10	3.8	-	4.48	G_{13}	15.79	11.68
v_{xy}	0.25	-	0.3	0.25	v_{12}	0.46	0.35
v_{yz}	0.35	-	-	0.35	v_{13}	0.052	0.031

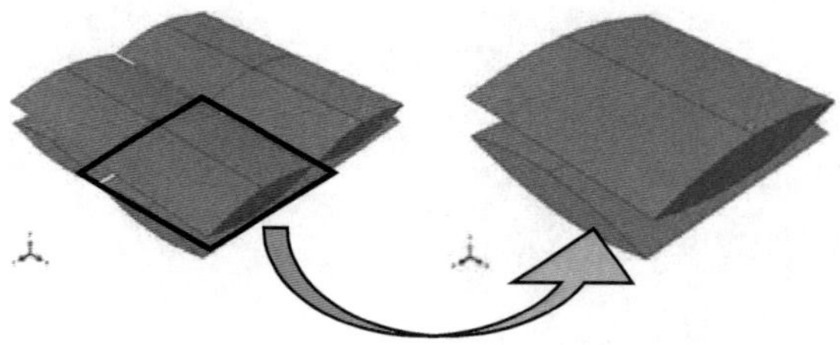

Figure 4: Bidirectional elliptical RUC fibre selection (left) full size (right) quarter size.

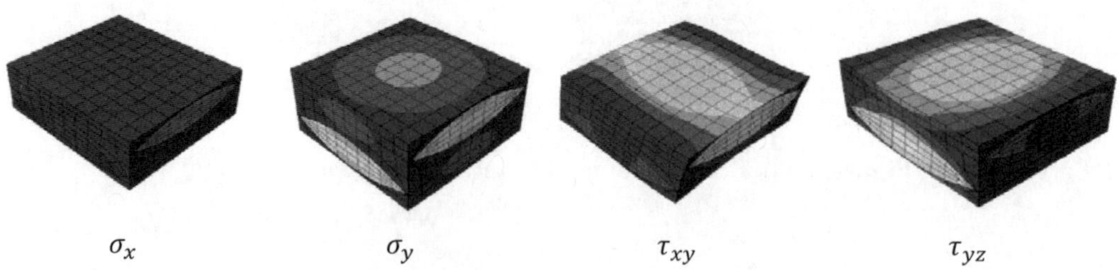

Figure 5: Stress distribution for bidirectional composite.

IOP Conf. Series: Journal of Physics: Conf. Series **914** (2017) 012045 doi:10.1088/1742-6596/914/1/012045

3.3 Weave Fibre Composite using PBC approach

One of the most common types of fabric that is used in the manufacturing of an aircraft parts is plain weave. Micromechanical FE analysis of plain weave composite can be found from many researchers [5], [16], [17]. Due to its material properties behaviour, applications for plain weave fabric are different compared to unidirectional fabric. Unidirectional has a higher strength and elastic modulus compared to plain weave by having flexibility to withstand force in different directions.

Figure 6: UD elliptical lenticular fibre

Plain weave is formed by interlacing two mutually perpendicular sets of yarns. The lengthwise threads are called warp and the crosswise threads is called fill or weft. The interlacing pattern of the warp and fill is known as the weave.

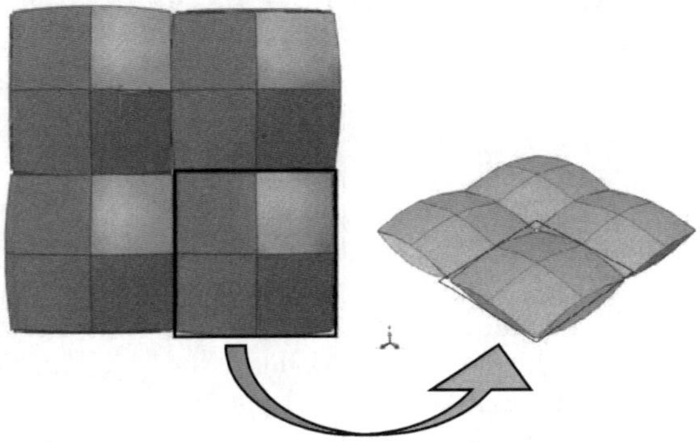

Figure 7: Symmetries model of unit cell based from plain weave.

Plain weave is formed by an orthogonal interlacing stands in a regular sequence of one over another. The mechanical behaviour of plain weave composites also depends on the fabric geometry, fibre volume fraction, weaving conditions and the type of weave and fibre interlacing.

In this model, the plain weave was made from the elliptical lenticular fibre geometry. The waviness of the lenticular throughout the model is assumed to be approximately 10 degree. Figure 6 shows the waviness degree which forms the lenticular for the single yarn. A local material orientation had been imposed to the model in the ABAQUS/Standard software.

The size of the unit cell of the plain weave as shown in Figure 7 shows one large model, can be reduced to a quarter shapes with symmetry geometry on every unit cell. Matrix dimension of 30(l) x 30(w) x 10.72(h) were used with the fibre and formed a 60% fibre volume. Table 4 shows the engineering constant results after running numerically using periodic boundary condition technique on the composite model. The plain weave composite having an orthotropic properties of $E_1 = E_3$, E_2, $G_{12} = G_{32}$and G_{13}. As expected, the properties of the plain weave composite are significantly positive correlation with Kim et al. [15]. Despite having slightly higher value, the result validates the use of periodic boundary condition method to predict elastic constant of a plain weave carbon fibre composite material. Figure 8 shows the stress distributions in three direction of plain weave carbon fibre unit cell.

Table 4: Plain weave micromechanical unit cell properties

Elastic Constants	Kim et al. [15]	Plain Weave Elliptical

		Composite V_f=60%
E_x(GPa)	65.9	68.8
E_z(GPa)	61.8	68.8
E_y(GPa)	11.3	10.2
G_{xy} (GPa)	6.98	5.2
G_{zy}(GPa)	7.14	5.38
G_{xz}(GPa)	15.79	17.45
$v_{xy} = v_{zy}$	0.46	0.35
v_{xz}	0.052	0.035

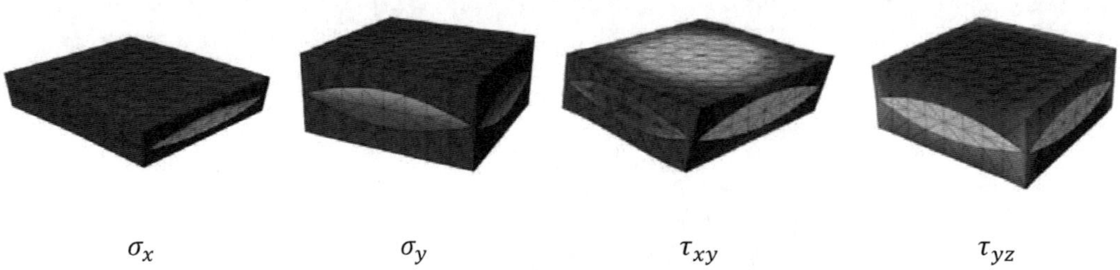

Figure 8: Stress distribution of plain weave fibre composites.

4. Conclusion

The focus of this paper is to compare elastic properties for reinforced composite fibre using periodic boundary condition method. The method was introduced on micromechanical unit cells with sets of boundary conditions applied. The translation of displacement along the boundary condition at the surrounding unit cell was symmetrical, including translations, reflections and rotations. Periodic boundary condition method shows a similar images on each unit cell which gives good agreement between the predicted results for all models compared to experimental from other studies.

Acknowledgement
The author would like to thank Universiti Tun Hussein Onn Malaysia and the Ministry of Higher Education Malaysia for their financial support.

References

[1] R. B. Hashin Z, "The Elastic Moduli of Fiber-Reinforced Materials.," *ASME, J. Appl. Mech.*, vol. 31, no. (2), pp. 223–232, 1964.
[2] C. T. Sun and R. S. Vaidya, "Prediction of composite properties from a representative volume element," *Compos. Sci. Technol.*, vol. 56, no. 2, pp. 171–179, Jan. 1996.
[3] M. B. RILEY and J. M. WHITNEY, "Elastic properties of fiber reinforced composite materials.," *AIAA J.*, vol. 4, no. 9, pp. 1537–1542, Sep. 1966.
[4] S. Li, "General unit cells for micromechanical analyses of unidirectional composites," *Compos. Part A Appl. Sci. Manuf.*, vol. 32, no. 6, pp. 815–826, Jun. 2001.
[5] S. Li, C. Zhou, H. Yu, and L. Li, "Formulation of a unit cell of a reduced size for plain weave textile composites," *Comput. Mater. Sci.*, vol. 50, no. 5, pp. 1770–1780, Mar. 2011.
[6] Z. Xia, C. Zhou, Q. Yong, and X. Wang, "On selection of repeated unit cell model and application of unified periodic boundary conditions in micro-mechanical analysis of composites," *Int. J. Solids Struct.*, vol. 43, no. 2, pp. 266–278, Jan. 2006.

[7] J. M. Tyrus, M. Gosz, and E. DeSantiago, "A local finite element implementation for imposing periodic boundary conditions on composite micromechanical models," *Int. J. Solids Struct.*, vol. 44, no. 9, pp. 2972–2989, May 2007.

[8] S. Li and A. Wongsto, "Unit cells for micromechanical analyses of particle-reinforced composites," *Mech. Mater.*, vol. 36, no. 7, pp. 543–572, Jul. 2004.

[9] N. Abolfathi, A. Naik, G. Karami, and C. Ulven, "A micromechanical characterization of angular bidirectional fibrous composites," *Comput. Mater. Sci.*, vol. 43, no. 4, pp. 1193–1206, Oct. 2008.

[10] A. Drago and M. Pindera, "Micro-macromechanical analysis of heterogeneous materials: Macroscopically homogeneous vs periodic microstructures," *Compos. Sci. Technol.*, vol. 67, no. 6, pp. 1243–1263, May 2007.

[11] N. Ohno, X. Wu, and T. Matsuda, "Homogenized properties of elastic–viscoplastic composites with periodic internal structures," *Int. J. Mech. Sci.*, vol. 42, no. 8, pp. 1519–1536, Aug. 2000.

[12] T. R. King, D. M. Blackketter, D. E. Walrath, and D. F. Adams, "Micromechanics Prediction of the Shear Strength of Carbon Fiber/Epoxy Matrix Composites: The Influence of the Matrix and Interface Strengths," *J. Compos. Mater.* , vol. 26 , no. 4 , pp. 558–573, Apr. 1992.

[13] Lee JW and Daniel IM, "Progressive transverse cracking of cross ply composite laminates," *J. Compos. Mater.*, vol. 24, pp. 1225–43, 1990.

[14] C. T. Sun and S. G. Zhou, "Failure of Quasi-Isotropic Composite Laminates with Free Edges," *J. Reinf. Plast. Compos.* , vol. 7 , no. 6 , pp. 515–557, Nov. 1988.

[15] B. C. Kim, D. C. Park, B. J. Kim, and D. G. Lee, "Through-thickness compressive strength of a carbon/epoxy composite laminate," *Compos. Struct.*, vol. 92, no. 2, pp. 480–487, Jan. 2010.

[16] I. Ivanov and A. Tabiei, "Three-dimensional computational micro-mechanical model for woven fabric composites," *Compos. Struct.*, vol. 54, no. 4, pp. 489–496, Dec. 2001.

[17] Y. C. Zhang and J. Harding, "A numerical micromechanics analysis of the mechanical properties of a plain weave composite," *Comput. Struct.*, vol. 36, no. 5, pp. 839–844, Jan. 1990.

Institute of Physics
Dirac House, Temple Back
Bristol BS1 6BE UK

ISSN: 1742-6588
ISBN 978-1-5108-5067-5

9 781510 850675

CURRAN ASSOCIATES INC.
proceedings.com

8th Annual International Conference on Science and Engineering (8th AIC 2018)

Connecting the World through Innovation and Sustainable Development

IOP Conference Series: Materials Science and Engineering
Volume 523

Aceh, Indonesia
12 - 14 September 2018

IOP
Institute of Physics